Recent Advances in Statistics
Papers in Honor of Herman Chernoff on His Sixtieth Birthday

Herman Chernoff

RECENT ADVANCES IN STATISTICS

PAPERS IN HONOR OF HERMAN CHERNOFF ON HIS SIXTIETH BIRTHDAY

EDITED BY

M. Haseeb Rizvi
Sysorex International, Inc.
Cupertino, California

Jagdish S. Rustagi
Department of Statistics
The Ohio State University
Columbus, Ohio

David Siegmund
Department of Statistics
Stanford University
Stanford, California

1983

ACADEMIC PRESS

A Subsidiary of Harcourt Brace Jovanovich, Publishers
New York London
Paris San Diego San Francisco São Paulo Sydney Tokyo Toronto

ACADEMIC PRESS RAPID MANUSCRIPT REPRODUCTION

ACADEMIC PRESS, INC.
111 Fifth Avenue, New York, New York 10003

United Kingdom Edition published by
ACADEMIC PRESS, INC. (LONDON) LTD.
24/28 Oval Road, London NW1 7DX

Library of Congress Cataloging in Publication Data

Main entry under title:

Recent advances in statistics.

1. Mathematical statistics--Addresses, essays, lectures. 2. Chernoff, Herman. 3. Mathematicians--United States--Biography. I. Chernoff, Herman. II. Rizvi, M. Haseeb. III. Rustagi, Jagdish S. IV. Siegmund, David, Date
QA276.16.R43 1983 519.5 83-5967
ISBN 0-12-589320-5

PRINTED IN THE UNITED STATES OF AMERICA

83 84 85 86 9 8 7 6 5 4 3 2 1

CONTENTS

IV. Statistical Graphics

V. Other Topics

CONTRIBUTORS

Numbers in parentheses indicate the pages on which the authors' contributions begin.

R. R. Bahadur (273), Department of Statistics, University of Chicago, Chicago, Illinois 60637

John Bather (19), Department of Mathematics, University of Sussex, Falmer, Brighton BN19QH, United Kingdom

P. K. Bhattacharya (143), Division of Statistics, University of California, Davis, California 95616

P. J. Bickel (511), Department of Statistics, University of California, Berkeley, California 94720

Morris H. DeGroot (175, 197), Department of Statistics, Carnegie-Mellon University, Pittsburgh, Pennsylvania 15213

Persi Diaconis (425), Department of Statistics, Stanford University, Stanford, California 94305

W. H. DuMouchel (529), Department of Mathematics, Massachusetts Institute of Technology, Cambridge, Massachusetts 02139

William F. Eddy (175), Department of Statistics, Carnegie-Mellon University, Pittsburgh, Pennsylvania 15213

Thomas S. Ferguson (287), Department of Mathematics, University of California, Los Angeles, California 90024

Jerome H. Friedman (425), Stanford Linear Accelerator Center, Stanford, California 94305

J. L. Gastwirth (143), Department of Statistics, George Washington University, Washington, D.C. 20052

Stephen B. Howell (557), Cancer Center, University of California at San Diego, La Jolla, California 92093

S. Iyengar (579), Department of Mathematics and Statistics, University of Pittsburgh, Pittsburgh, Pennsylvania 15260

Robert J. K. Jacob (449), Computer Science and Systems Branch, Naval Research Laboratory, Washington, D.C. 20375

Joseph B. Kadane (197), Department of Statistics, Carnegie-Mellon University, Pittsburgh, Pennsylvania 15213

Harold J. Kushner (211), Division of Applied Mathematics, Brown University, Providence, Rhode Island 02912

T. L. Lai (51), Department of Mathematical Statistics, Columbia University, New York, New York 10027

Lucien M. Le Cam (303), Department of Statistics, University of California, Berkeley, California 94720

E. L. Lehmann (593), Department of Statistics, University of California, Berkeley, California 94720

Clare Mahan (303), Public Health Program, Tufts University, Bedford, Massachusetts 01730

Richard A. Olshen (557), Department of Mathematics, University of California at San Diego, La Jolla, California 92093

Michael D. Perlman (339), Department of Statistics, University of Washington, Seattle, Washington 98195

John W. Pratt (601), Graduate School of Business Administration, Harvard University, Boston, Massachusetts 02163

John A. Rice (557), Department of Mathematics, University of California at San Diego, La Jolla, California 92093

Herbert Robbins (51), Department of Mathematical Statistics, Columbia University, New York, New York 10027

Pranab Kumar Sen (371), Department of Biostatistics, University of North Carolina, Chapel Hill, North Carolina 27514

David Siegmund (51), Department of Statistics, Stanford University, Stanford, California 94305

Gordon Simons (69), Department of Statistics, University of North Carolina, Chapel Hill, North Carolina 27514

Avinash Singh (303), Department of Mathematics and Statistics, Memorial University of Newfoundland, St. John's, Newfoundland, Canada

H. Solomon (579), Department of Statistics, Stanford University, Stanford, California 94305

Charles J. Stone (393), Department of Statistics, University of California, Berkeley, California 94720

Paul Switzer (95), Department of Statistics, Stanford University, Stanford, California 94305

W. R. Van Zwet (407), Department of Mathematics, University of Leiden, 2300 RA Leiden, The Netherlands

Howard Wainer (469), Educational Testing Service, Princeton, New Jersey 08541

Christine Waternaux (529), Biostatistics Department, Harvard School of Public Health, Boston, Massachusetts 02115

P. Whittle (237), Statistical Laboratory, Cambridge University, Cambridge, England, United Kingdom

Michael Woodroofe (69, 115), Department of Statistics, University of Michigan, Ann Arbor, Michigan 48104

S. Zacks (245), Department of Mathematical Sciences, State University of New York, Binghamton, New York 13901

PREFACE

This volume contains 27 papers contributed in honor of Herman Chernoff on the occasion of his sixtieth birthday. The contributors were selected because of their associations with Herman Chernoff or because of their activity in areas of statistics where Chernoff has made his major contributions. The papers are divided into five different categories: (i) sequential analysis including designs, (ii) optimization including control theory, (iii) nonparametrics including large sample theory, (iv) statistical graphics, and (v) other topics.

We are grateful to the following persons, who refereed the papers for the volume: R. J. Beran, P. K. Bhattacharya, D. Blough, P. Diaconis, K. Doksum, B. Efron, M. Hogan, B. Kleiner, C. N. Morris, J. Petkau, B. Pittell, D. Pregibon, C. R. Rao, W. J. Studden, L. Sucheston, J. Van Ryzin, M. T. Wasan, P. L. Zador, and J. V. Zidek.

We are also indebted to the secretarial staff of Sysorex International and the Departments of Statistics at The Ohio State University and Stanford University for taking care of large amounts of correspondence. In particular we appreciate the assistance of Penny Alexander, Chris Champagne, Judith Davis, Joan Hynes, Lillian Johnson, and Dolores Wills. The camera-ready copy was graciously prepared by Mona Swanger and Lauren Werling. We thank Judy Chernoff and W. H. DuMouchel, who provided some material for Herman Chernoff's biography. Finally we thank the staff of Academic Press for their cooperation.

HERMAN CHERNOFF:
AN APPRECIATION

Herman Chernoff was born in New York City on July 1, 1923. His parents, Max and Pauline Chernoff, were Jewish immigrants from Russia and they brought Herman up in the highest academic tradition. His early schooling was in New York City, and he graduated from Townsend Harris High School, a select preparatory school for the City College of New York. He received the B.S. degree from the City College of New York in 1943, majoring in Mathematics with a Minor in Physics. Before joining Brown University for graduate work in the Department of Applied Mathematics, he worked as a junior physicist with the United States Navy for a year and a half.

Herman Chernoff's interest in Statistics was stimulated by a paper of Neyman and Pearson that he read during his undergraduate years when he took some courses in Statistics. At Brown University he took courses in Applied Mathematics from many distinguished mathematicians, including L. Bers, William Feller, C. Loewner, Henry Mann, Paul Rosenbloom, and J. D. Tamarkin, and wrote a Master's thesis under the guidance of Professor S. Bergman on Complex Solutions of Partial Differential Equations, [1].

His studies at Brown University were interrupted by the United States Army, into which he was drafted in November 1945. With the help of a National Research Council Predoctoral fellowship, his application for discharge from the Army was approved in February 1946. He was attracted once more to Statistics through Abraham Wald's first paper on Decision Theory. He attended the summer session of 1946 at Raleigh, N.C., when the University of North

Carolina introduced its program in Statistics with lectures by R. A. Fisher, William Cochran, and J. Wolfowitz. After completing his nonthesis requirements for the Ph.D. at Brown, he went to Columbia University in January 1947 to work on his thesis in absentia. Under the supervision of Abraham Wald he wrote his thesis on Asymptotic Studentization in Testing of Hypothesis, [5].

Wald accepted Herman Chernoff as a student only after ascertaining that he had already published a paper based on his Master's thesis. He required Herman to take courses in Statistics at Columbia, where some of the modern big names in Statistics were teachers or students. Included among the former at Columbia University (besides Wald himself) were T. W. Anderson, R. C. Bose, J. L. Doob, E. Pitman, and J. Wolfowitz; among the latter were R. Bechhofer, J. Kiefer, I. Olkin, M. Sobel, C. Stein, and L. Weiss.

At Brown University Herman met a fellow graduate student in Mathematics, Judy Ullman, whom he married in 1947. They have now been happily married for more than 35 years and are the proud parents of two daughters, Ellen and Miriam. Judy herself has a Master's degree in Mathematics from Brown; her thesis was written under the guidance of Professor C. C. Lin, now at M.I.T. Due to a shortage of male students during the war years, Dean R. G. D. Richardson of Brown had recruited 15 women students to do graduate work in Mathematics; but despite his best efforts, Judy decided to relinquish the Ph.D. ambition in favor of family life.

After receiving his doctoral degree from Brown University, Herman Chernoff worked for a year and a half at the Cowles Commission for Research in Economics at the University of Chicago. Among the several statisticians at Cowles, he was greatly influenced by Herman Rubin, with whom he collaborated there, [12], and again later when they were colleagues at Stanford University, [26]. At the Cowles Commission he also came in contact with many economists, several of whom, such as K. J. Arrow,

M. Friedman, L. Klein, and T. C. Koopmans, were later to win Nobel prizes.

Before joining Stanford University as an Associate Professor in Statistics in 1952, Herman spent three years at the University of Illinois in the Department of Mathematics. At the invitation of Professors Kenneth J. Arrow and Albert Bowker he visited Stanford during 1951, and later on he decided to accept a regular position. He remained at Stanford for 22 years, becoming Professor in 1956. Several of his colleagues there have contributed to this volume: Persi Diaconis, Richard Olshen, David Siegmund, Herbert Solomon, and Paul Switzer.

In 1974, Herman moved to Massachusetts Institute of Technology, where he is currently Professor of Applied Mathematics and Co-Director of the Statistics Center. During his long academic career, he has held Visiting Professorships at various institutions, including Columbia University, the Center for Advanced Study in the Behavioral Sciences, the London School of Economics, the University of Rome, Tel-Aviv University, Hebrew University, and the Institute of Statistical Mathematics in Japan.

Professor Chernoff has been honored for his contributions by many professional societies and scientific organizations. He was President of the Institute of Mathematical Statistics and is an elected member of the American Academy of Arts and Sciences as well as of the National Academy of Sciences. He is a fellow of the American Statistical Association and of the Institute of Mathematical Statistics, and is an elected member of the International Statistical Institute. Recently he was awarded the Townsend Harris Medal by the Alumni Association of the City University of New York.

He holds memberships in many other professional societies, including Sigma Xi, Classification Society, Pattern Recognition Society, Bernoulli Society, and the American Mathematical Society.

He has served the statistical profession through his various editorial assignments and memberships in national committees. He has held memberships in Advisory Panels of the National Science Foundation, the Conference Board of the American Mathematical Sciences, and the National Research Council. Among the many invited addresses he has given to professional societies are the Wald Lectures for the Institute of Mathematical Statistics in 1968, lectures on Pattern Recognition and Sequential Analysis in the NSF Regional Research Conferences at New Mexico State University at Las Cruces, and the Fisher Memorial Lecture for the American Statistical Association.

Herman is known to be an inspiring teacher and has greatly influenced a large number of his doctoral students and postdoctoral research associates, who in turn have contributed vastly to the field of Statistics. Herman Chernoff's research spans a remarkably broad area in both theoretical and applied Statistics. What follows describes very briefly a few of his outstanding contributions. A complete listing of Chernoff's published works appears later.

Chernoff defined a measure of asymptotic efficiency of statistical tests and introduced the use of probabilities of large deviations into Statistics in [11]. Although Chernoff's measure of asymptotic efficiency has not been widely adopted, perhaps due to its computational complexity, the concept of large deviations has proved to be very important in probability and statistics. For example the theory of Bahadur efficiency in tests of hypotheses and the Donsker-Varadhan approach to a wide variety of difficult nonlinear probability problems, utilize extensively large deviations theory. Cramér had introduced an elegant theory of large deviations earlier. However, Chernoff's viewpoint and his perception of potential applications in Statistics have provided an important stimulus to research in large deviations theory.

Herman Chernoff and I. Richard Savage gave the first systematic large sample theory of an important class of non-parametric test statistics, by the elegant device of representing the statistics as an integral with respect to an empirical distribution function in [30]. They prove that linear rank statistics for two sample problems are asymptotically normal under quite general conditions and calculate the Pitman asymptotic efficiency of some important ones. The number of subsequent publications related to Chernoff-Savage theory runs in the hundreds.

The work on a measure of asymptotic efficiency and a paper on optimal designs for estimating parameters, [14], initiated Chernoff's lasting interest in optimal design. This led to a procedure which provides asymptotically efficient estimates for the sequential design of experiments, [32].

At every stage of a sequential experiment, one must decide whether to stop and announce a conclusion or to continue, in which case one must choose the next experiment to be performed. Although Chernoff shows that his proposals have desirable asymptotic properties, later numerical investigations indicate that alternatives, which are theoretically less appealing, can perform better for small to moderate sample sizes. Only within the last two years have R. Keener, a student of Chernoff, and G. Lorden given sufficiently penetrating asymptotic analyses to provide an expectation of obtaining procedures which are approximately optimal in both small and large samples. Since their approaches apply only to finite parameter spaces, the problem will undoubtedly provide additional challenges in the future.

Chernoff considered a problem of testing sequentially the sign of the mean of a normal distribution in [37]. He shows that by formulating the problem in continuous time one can use the techniques of partial differential equations to derive approximate solutions, whereas in the discrete time setting, a solution is accessible only by means of a substantial amount of machine computation. Chernoff has returned to the theme of this paper on numerous subsequent occasions to provide fairly explicit approximate solutions to difficult concrete problems in sequential statistics and stochastic control theory. A particularly interesting later development shows that even when one must ultimately resort to machine computation, the continuous time viewpoint allows one to determine the solutions to a number of related problems by pencil and paper modifications of a single numerical computation.

Among his many contributions to optimization in statistics, his paper on Gradient Methods of Maximization with J. B. Crockett, [21], proved to be a precursor of the variable metric gradient method of the Fletcher-Powell algorithm.

Chernoff's interest in foundations gave rise to his paper on Rational Selection of Decision Functions, [18], in which he provided a set of axioms for rational behavior. An earlier discussion of this work stimulated L. J. Savage to investigate the Bayesian Theory. Chernoff's book on Elementary Decision Theory, co-authored with L. E. Moses, [33], has been a pacesetter in the field and has been translated into three other languages.

Chernoff considered the problem of representing multi-dimensional data in a way that makes it easy to interpret, in [71]. His beautiful answer is to map the data onto a computer drawn cartoon of a human face, so one measurement gives the size of the eyes, another the curvature of the smile, and so on. Thus instead of a collection of (say) sextuples of numbers, one has a collection of faces to catalogue according to their similarities and differences. The Chernoff face is a convenient tool for the display and storage of multivariate data and can also serve as a valuable aid in exploratory discrimination and classification tasks. The human cartoon is perceived as a Gestalt and is memorable in its total effect. This work almost defies mathematical analysis. One is apt to ask: how stable is such a process under different assignments of features to the data? This has led to interesting experimental questions concerning changes in metric properties of the mapping under permutations of the coordinates which are mapped into given facial features. Chernoff and M. H. Rizvi, in [74], designed an experiment and addressed some of these questions in their empirical study.

It seems that one of the motivating factors of Herman Chernoff to move to Massachusettes Institute of Technology was to pursue important questions in pattern recognition. He still continues to contribute to questions of importance in control theory, image analysis and signal processing.

For these and other scientific accomplishments Herman Chernoff is justifiably held in high esteem. In addition, Herman's and Judy's warmth and humanity have earned them the love and admiration of their many friends, both within and outside the statistical profession.

PUBLICATIONS OF HERMAN CHERNOFF

[1] "Complex Solutions of Partial Differential Equations", (1946), *Amer. J. Math.*, 455-478.

[2] "A Note on the Inversion of Power Series", (1947), *Mathematical Tables and Other Aids to Computation*, 331-335.

[3] "The Effect of a Small Error in Observations on the Limited Information Estimates of Stochastic Equations", (1948), Cowles Commission Discussion Paper, Statistics No. 315, 1-8.

[4] "The Computation of Maximum Likelihood Estimates of Parameters of a System of Linear Stochastic Difference Equations in the Case of Correlated Disturbances", (1948), Cowles Commission Discussion Paper, Statistics No. 323, 1-9.

[5] "Asymptotic Studentization in Testing of Hypothesis", (1949), *Ann. Math. Statist.*, 268-278.

[6] "Computational Aspects of Certain Econometric Problems", (1951), *Proceedings of a Second Symposium on Large Scale Digital Calculating Machinery*. Harvard University Press, 348-350.

[7] "A Property of Some Type A Regions", (1951), *Ann. Math. Statist.*, 472-474.

[8] "An Extension of a Result of Liapounoff on the Range of a Vector Measure", (1951), *Proc. Amer. Math. Soc.*, 722-726.

[9] "Classification Procedures of the Form $\Sigma\phi(x_i) \leq k$ where $\phi(x)$ Assumes Only m Successive Integral Values", (1951), Technical Report No. 5, under Contract N6onr-25140 for the Office of Naval Research at Stanford University, 1-30.

[10] "A Generalization of the Neyman-Pearson Fundamental Lemma", (1952), (coauthor H. Scheffé). *Ann. Math. Statist.*, 213-225.

[11] "A Measure of Asymptotic Efficiency for Tests of a Hypothesis Based on the Sum of Observations", (1952), *Ann. Math. Statist.*, 493-507.

[12] "Asymptotic Properties of Limited Information Estimates Under Generalized Conditions", (coauthor H. Rubin), (1953). Studies in Econometric Method, Cowles Commission Monograph 14, Edited by Hood and Koopmans, Chapter VII, 200-212.

[13] "The Computation of Maximum Likelihood Estimates of Linear Structural Equations", (coauthor N. Divinsky), (1953). Studies in Econometric Method, Cowles Commission Monograph 14, Edited by Hood and Koopmans, Chapter X, 236-302.

[14] "Locally Optimal Designs for Estimating Parameters", (1953), *Ann. Math. Statist.*, 586-602.

[15] "Selection of a Distribution Function to Minimize an Expectation Subject to Side Conditions", (coauthor S. Reiter), (1954). Technical Report No. 23 under Contract N6onr-25140 for the Office of Naval Research at Stanford University.

[16] "On the Distribution of the Likelihood Ratio", (1954), *Ann. Math. Statist.*, 573-578.

[17] "The Use of Maximum Likelihood Estimates in χ^2 Tests for Goodness of Fit", (coauthor E. L. Lehmann), (1954). *Ann. Math. Statist.*, 579-586.

[18] "Rational Selection of Decision Functions", (1954), *Econometrica*, 442-443.

[19] "Use of Normal Probability Paper", (coauthor G. J. Lieberman), (1954), *J. Amer. Statist. Assoc.*, 778-785.

[20] "A Large Sample Bioassay Design with Random Doses and Uncertain Concentrations", (coauthor F. C. Andrews), (1954), *Biometrika*, 307-315.

[21] "Gradient Methods of Maximization", (coauthor J. B. Crockett), (1955), *Pac. J. Math.*, 33-50.

[22] "A Classification Problem", (1956). Technical Report No. 33 Under Contract N6onr-25140 for the Office of Naval Research at Stanford University, 1-19.

[23] "Large Sample Theory: Parametric Case", (1956), *Ann. Math. Statist.*, 1-22.

[24] "The Use of Generalized Probability Paper for Continuous Distributions", (coauthor G. J. Lieberman), (1956), *Ann. Math. Statist.*, *27*, 3, 806-818.

[25] "Sampling Inspection by Variables with no Calculations", (coauthor G. J. Lieberman), (1957), *Industrial Quality Control*, *XIII*, 7, 2-4.

[26] "The Estimation of the Location of a Discontinuity in Density", (coauthor H. Rubin), (1956). *Proceedings of the Third Berkeley Symposium on Math. Stat. and Prob., 1.* Contributions to the Theory of Statistics, Edited by J. Neyman, University of California Press, 19-37.

[27] "A Central Limit Theorem for Sums of Interchangeable Random Variables", (coauthor Henry Teicher), (1958), *Ann. Math. Statist., 29,* 1, 118-130.

[28] "Central Limit Theorems for Interchangeable Processes", (coauthors Blum, Rosenblatt, Teicher), (1958), *Canadian J. Math., 10,* 222-229.

[29] "The Distribution of Shadows", (coauthor J. F. Daly), (1957), *J. Math. Mech., 6,* 567-584.

[30] "Asymptotic Normality and Efficiency of Certain Non-Parametric Test Statistics", (coauthor I. R. Savage), (1958), *Ann. Math. Statist., 29,* 972-994.

[31] "An Antenna Propagation Problem", (1957), Technical Report No. 38, under Contract N6onr-25140 (NR-342-022) for the Office of Naval Research at Stanford University, 1-15.

[32] "Sequential Design of Experiments", (1959), *Ann. Math. Statist., 30,* 755-770.

[33] *Elementary Decision Theory,* (coauthor L. E. Moses), (1959), J. Wiley, New York.

[34] "A Compromise Between Bias and Variance in the Use of Non-Representative Samples", (1960), *Contributions to Probability and Statistics,* Edited by Olkin et al, Stanford University Press, Stanford, Chap. 14, 153-167.

[35] "Motivation for an Approach to the Sequential Design of Experiments", *Information and Decision Processes,* (1960), Edited by R. E. Machol, McGraw-Hill Book Co., New York, 15-25.

[36] "Sequential Experimentation", (1961), *Bull. of Int. Statist. Inst., 38,* 4, Tokyo, 3-9.

[37] "Sequential Tests for the Mean of a Normal Distribution", (1961), *Proc. of Fourth Berkeley Symposium on Math. Stat. and Prob.,* Edited by J. Neyman, University of California Press, *1,* 79-95.

[38] "The Scoring of Multiple Choice Questionnaires", (1962), *Ann. Math. Statist., 35,* 375-393.

[39] "Optimal Accelerated Life Designs for Estimation", (1962), *Technometrics, 4,* 381-408.

[40] "An Optimal Sequential Accelerated Life Test", (coauthors S. Bessler, A. W. Marshall), (1962), *Technometrics, 4,* 367-380.

[41] "Storage Policy for Incomplete Record Keeping", (coauthor G. Schwarz), (1971), Stanford, Technical Report No. 71, under Contract Nonr 225(52), 1-12.

[42] "Optimal Design of Experiments", (1963), *Proceedings of the Eight Conference on the Design of Experiments in Army Research Development and Testing,* 162-173.

[43] "Sequential Tests for the Mean of a Normal Distribution II (large t)", (1964), (coauthor John Breakwell), *Ann. Math. Statist., 35,* 162-173.

[44] "A New Approach to the Evaluation of Multiple Choice Questionnaires", (1962), *Bull. of Int. Statist. Inst., 39,* 289-294.

[45] "Preliminary Report on the Problem of Sequentially Testing Whether the Mean of a Normal Distribution is Postive or Negative", (1962), Technical Report under Contract Nonr 3439(00), 1-25.

[46] "Estimating the Current Mean of a Normal Distribution Which is Subjected to Changes in Time", (1964), (coauthor S. Zacks), *Ann. Math. Statist., 35,* 999-1018.

[47] "Estimation of the Mode", (1964), *Ann. Inst. of Statist. Math., 16,* 31-41.

[48] "Sequential Tests for the Mean of a Normal Distribution III (small t)", (1965), *Ann. Math. Statist., 36,* 28-54.

[49] "Sequential Tests for the Mean of a Normal Distribution IV (discrete case)", (1965), *Ann. Math. Statist., 36,* 55-68.

[50] "Limit Distributions of the Minimax of Independent Identically Distributed Random Variables", (coauthor Henry Teicher), (1964), Technical Report No. 50, Contract Nonr 225(72), 1-27.

[51] "Discussion of 'Asymptotically Optimal Tests for Multinomial Distributions' by W. Hoeffding", (1965), *Ann. Math. Statist., 36,* 405-407.

[52] "A Bayes Sequential Sampling Inspection Plan", (1965), (coauthor S. N. Ray), *Ann. Math. Statist., 36,* 1387-1407.

[53] "Sequential Decisions in the Control of a Spaceship", (1965), (coauthor J. Bather), Technical Report No. 6, under NSF Grant GP-3836, 1-34.

[54] "Sequential Decisions in the Control of a Spaceship (Asymptotic Results)", (1965), (coauthor J. Bather), Technical Report No. 8, under NSF Grant GP-3836, 1-20.

[55] "Sequential Decision in the Control of a Spaceship" (1967), (with John Bather), *Proc. Fifth Berkeley Symposium 3,* 181-207, Univ. of California Press.

[56] "Asymptotic Distribution of Linear Combinations of Functions of Order Statistics with Applications to Estimation", (1967), (coauthors J. L. Gastwirth and M. V. Johns), *Ann. Math. Statist., 38,* 52-72.

[57] "Sequential Decision in the Control of a Spaceship (Finite Fuel)", (1967), (with John Bather), *J. Appl. Prof., 4,* 584-604.

[58] "A Note on Risk and Maximal Regular Generalized Submartingales in Stopping Problems", (1967), *Ann. Math. Statist., 38,* 606-607.

[59] "Sequential Models for Clinical Trials", (1967), *Proc. Fifth Berkeley Symposium, 4,* 805-812, Univ. of California Press.

[60] "Bounds on the Efficiency of a Classification Procedure", (1967), Stanford University Technical Report No. 56 (Nonr 225(72) (NR-042-993), 1-19.

[61] "Decision Theory", (1968), *International Encyclopedia of the Social Sciences,* The Macmillan Co. and the Free Press.

[62] "Optimal Stochastic Control", (1968), Mathematics of the Decision Sciences, Part 2, Lectures in Applied Math., *12, Amer. Math. Soc.,* 149-172.

[63] "Optimal Stochastic Control", (1968), *Sankhyā, A 30,* 221-251.

[64] "Optimal Design Applied to Simulation", (1969), *Proceedings of the 37th Session of the ISI,* Book 2, *43,* 264-266.

[65] "Sequential Designs", (1969), *The Design of Computer Simulation Experiments,* edited by T. H. Taylor, Duke University Press, 99-120.

[66] "A Bound of the Classification Error for Discriminating Between Populations with Specified Means and Variances", (1971), *Studi di probabilita. statistics e ricerca operative in onore di Giuseppe Pompilj,* 205-211, Oderisi-Gubbio.

[67] "Metric Considerations in Cluster Analysis", *Proceedings of the Six Berkeley Symposium on Mathematical Statistics and Probability,* 621-629, Univ. of California Press, 1971.

[68] "The Selection of Effective Attributes for Deciding Between Hypotheses Using Linear Discriminant Functions", *Frontiers of Pattern Recognition,* 55-60, Academic Press, 1972.

[69] "The Efficient Estimation of a Parameter Measurable by Two Instruments of Unknown Precisions", *Optimizing Methods in Statistics,* Edited by J. S. Rustagi, 1-27, Academic Press, 1971.

[70] "A Note on Optimal Spacings for Systematic Statistics", (1971), Stanford University, Technical Report No. 70 (N00014-67-A-0112(NR 042-993)), 1-11.

[71] "The Use of Faces to Represent Points in k-Dimensional Space Graphically", (1973), *J. Amer. Statist. Assoc., 68,* 361-368.

[72] "Some Measures for Discriminating between Normal Multivariate Distributions with Unequal Covariance Matrices", *Multivariate Analysis* - II, Edited by P. R. Krishnaiah, 337-344, Academic Press, 1973.

[73] "Sequential Analysis and Optimal Design", *8th Regional Conference in Applied Mathematics,* 1-119, SIAM, 1972.

[74] "Effect on Classification Error of Random Permutations of Features in Representing Multivariate Data by Faces" (with M. H. Rizvi), *J. Amer. Statist. Assoc., 70,* 548-554, 1975.

[75] "Approaches in Sequential Design of Experiments, A Survey of Statistical Designs and Linear Models", *A Survey of Statistical Design and Linear Models,* Edited by J. N. Srivastava, 1973, 67-90, North Holland/American Elsevier, New York.

[76] "The Interaction Between Large Sample Theory and Optimal Design of Experiments", *On the History of Probability and Statistics,* Edited by D. B. Owen, Marcel-Dekker, New York, 1976.

[77] "An Optimal Stopping Problem for Sums of Dichotomous Random Variables", (with A. J. Patkau), (1976), *Ann. Prob., 4,* 875-889.

[78] "Some Applications of a Method of Identifying an Element of a Large Multidimensional Population", *Multivariate Analysis - IV,* Edited by P. R. Krishnaiah, 445-456, North-Holland, New York, 1977.

[79] "A Subset Selection Problem Employing a New Criterion", (with J. Yahav), *Statistical Theory and Related Topics, II,* Edited by S. Gupta, 93-119, Academic Press, New York, 1977.

[80] "Optimal Control of a Brownian Motion", (with A. J. Petkau), *SIAM J. App. Math., 34,* 1978, 717-731.

[81] "A Satellite Control Problem", (with A. J. Petkau), *Optimizing Methods in Statistics,* Edited by J. S. Rustagi , 1979, 89-124, Academic Press, New York.

[82] "Graphical Representations as a Discipline", *Graphical Representation of Multivariate Data,* Edited by P. Wang, 1979, 1-11, Academic Press, New York. Also in *Information Linkage Between Applied Mathematics and Industry,* Edited by P. Wang, 1979, 49-59, Academic Press, New York.

[83] "The Identification of an Element of a Large Population in the Presence of Noise", *Ann. Statist.,* 1980, *8,* 1179-1197.

[84] "Sequential Medical Trials Involving Paried Data", (with A. J. Petkau), *Biometrika,* 1981, *68,* 119-132.

[85] "A Note on an Inequality Involving the Normal Distribution", *Ann. Prob.,* 1981, *9,* 533-535.

[86] "Nonparametric Estimation of the Slope of a Truncated Regression", (with P. K. Bhattacharya and S. S. Yang), Office of Naval Research Report #18, July 7, 1980.

[87] "An Analysis of the Massachusetts Numbers Game", *Statistics and Related Topics,* Edited by M. Csörgö et al, 23-37, North-Holland, 1981.

[88] "How to Beat the Massachusetts Numbers Game. An Application of Some Basic Ideas in Probability and Statistics", *The Mathematical Intelligencer,* 1981, *3,* 166-175.

[89] "Chernoff Faces", *Encyclopedia of Statistical Sciences,* Edited by S. Kotz and N. L. Johnson, J. Wiley, New York, 1982, *1,* 436-438.

[90] "When it Seems Desirable to Ignore Data", Office of Naval Research Report No. 24, September, 1982, also in Festschrift for E. L. Lehman, 1982.

[91] Report of the Committee on Ballistic Acoustics, (N. Ramsey, Chairman), National Research Council, National Academy Press, Washington, D.C., 1-96, 1982.

I. SEQUENTIAL ANALYSIS INCLUDING DESIGNS

OPTIMAL STOPPING OF BROWNIAN MOTION: A COMPARISON TECHNIQUE

John Bather

School of Mathematical and Physical Sciences
University of Sussex
Brighton, England

I. INTRODUCTION AND SUMMARY

The aim here is to sketch a field first ploughed and sown some twenty years ago. It is a personal sketch in which I will try to indicate some, but by no means all, of the progress made in extending the field since then. My own interest began with Chernoff's original paper on sequential tests for the mean of a normal distribution (1961), when I noticed that a technique for comparing this problem with much simpler ones could provide useful bounds on the optimal sampling time (1962). The idea was something of a wild flower on the edge of the field at that time, but it has persisted in spite of the more powerful theoretical and computational machinery now in use and I will give two recent illustrations of it. The rustic image will serve its purpose if it suggests a soil that has proved fertile in some parts, but one that made the going rather heavy at times. This makes it difficult to record precisely when the main ideas first became clear. I prefer to recall Chernoff's quite insistence that the truth may lie a little deeper; this is a recurrent stimulus which I and many others will acknowledge with gratitude.

ISBN 0-12-589320-5

The next section gives a brief description of the general theory of optimal stopping for Brownian motion. It indicates why many sequential decision problems in discrete time lead to this type of stopping problem when they are formulated in continuous time. The solution of the limiting form of the problem can be constructed, in principle, by taking the union of certain open sets and this approach is closely related to the comparison technique. An explanation and the first illustration are given at the end of Section II.

Section III is concerned with the multi-armed bandit problem, as it has been called since it was first stated by Robbins (1952) in connection with clinical trials. The main result here is a theorem due to Gittins, which has been extended by him and others to deal with a large class of sequential allocation problems: see Gittins (1979). We consider several Markov processes, each of which is capable of producing a sequence of rewards, and the problem of choosing just one of them at each stage so as to maximize the total discounted expectation. Each separate process has an index, depending on its state, and the optimal allocation policy is then defined by choosing the process with highest index at every stage. There is a strong connection between multi-armed bandits and stopping problems because the index for a single Markov process must be determined by solving a family of stopping problems. In Section IV we shall investigate the Gittins index for a simple Markov process with normally distributed rewards, by applying the comparison technique.

The final section is about the buyer's problem: given several batches of similar goods and independent prior information on the unknown quality of each batch, what is the optimal policy for sequential sampling and when should we stop sampling and decide to buy one of the batches? Here, Gittins' result is not directly applicable, but it has been shown by Bergman (1981) that a policy based on indices is optimal if the number of batches is large and if we start from an assumption that each batch has the same prior distribution. The paper ends with another application of the comparison technique, due to Ferebee (1981), which leads to bounds on the index function for a version of the buyer's problem. Readers interested in further applications, in a different context, can refer to the joint papers with Chernoff (1966, 1967) and a recent study of similar stochastic control problems by Karatzas (1982).

II. STOPPING PROBLEMS FOR BROWNIAN MOTION

Consider a sequence of independent normal observations, each with unknown mean θ and known variance σ^2. It is a well-known fact that a normal prior distribution for θ leads, after sampling any number of observations, to a normal posterior distribution. Let us examine the effect of a typical observation ξ, when θ is represented by the distribution $N(u,v)$. It produces a transition $(u,v) \to (u',v')$, where

$$u' - u = \frac{v}{v + \sigma^2} (\xi - u), \quad v' = v - \frac{v^2}{v + \sigma^2} . \tag{2.1}$$

The reduction in v is deterministic: $v' = v - \delta v$, say, but the change in u is random and it is easily shown from (2.1) that $\delta u = u' - u$ has the distribution $N(0, \delta v)$. Thus, a sequence of observations can be represented by transitions in the (u,v) plane which are equivalent to those produced by a standard Brownian motion with time parameter $-v$. Of course, this stochastic process is only observable at certain irregular intervals of time, but is natural to study the continuous process for its own sake, since this enables us to use more powerful tools from the calculus.

It is perhaps surprising that the theory of optimal stopping for Brownian motion in m dimensions hardly depends on m. The next few paragraphs contain a summary of the general results established in my paper (1970), but many of the ideas there were first developed in Chernoff's investigation of the sequential testing problem. We shall return to the case m=1 for the applications, so readers who prefer the geometry of the plane may take m=1 throughout.

The mathematical representation of Brownian motion is usually called a Wiener process, defined as follows. Let $\{W(t),\ t\geq 0\}$ be an m-dimensional random process with mutually independent components. Each component $\{W_i(t)\}$ has independent normal increments and $W_i(0) = 0$, $E\{W_i(t)\} = 0$, $E\{(W_i(t_1) - W_i(t_2))^2\} = |t_1 - t_2|$, for $i = 1,2,\ldots,m$. Almost every trajectory is continuous. A stopping time T is a non-negative random variable which depends, without anticipation, on this process.

We introduce the following notation. Let $z = (u_1,u_2,\ldots,u_m,v)$, or simply $z = (u,v)$, denote a point of Euclidean (m+1)-space and let H be the open half-space where $v>0$. Suppose we are given a continuous function $r(z)$ which determines the reward for stopping at any point of H. Consider the process $\{Z(t)\}$ with initial point $z\varepsilon H$, defined by

$$Z(t) = (u + W(t),\ v - t), \quad (0\leq t\leq v). \tag{2.2}$$

Then, for any stopping time $T\leq v$, the process stops at the point $Z(T)$ and the corresponding expected reward is $E\{r(Z(T))|z\}$. We must extend the definition of $r(z)$ to include points with $v=0$ and put suitable restrictions on the magnitude of the reward function to ensure that such expectations are finite. In some applications, $r(u,v)\to-\infty$ as $v\to 0$, so we demand that $T<v$ with probability 1. Let us assume that, for each $z\varepsilon H$,

$$f(z) = \sup_{T\leq v} E\{r(Z(T))|z\} \tag{2.3}$$

is finite, we seek an optimal stopping time which attains this value.

Fortunately, it turns out that the only stopping times we need consider are those defined by reference to open subsets of the space H. Let $A \subset H$ be open and consider, for each initial point $z \in A$, the random point Z_A where the process (2.2) first hits the boundary of A. The corresponding expectation is

$$f_A(z) = E\{r(Z_A)|z\}$$

and we can extend the definition by setting $f_A(z) = r(z)$ for $z \notin A$. This function has the property that

$$\frac{1}{2} \Sigma \frac{\partial^2 f_A}{\partial u_i^2} = \frac{\partial f_A}{\partial v}, \qquad (z \in A), \tag{2.4}$$

which is a consequence of the fundamental relation between Brownian motion and the classical heat equation.

Returning to the maximum expected reward, it can be shown that f(z) is continuous in z and, hence, $B = \{z \in H: f(z) > r(z)\}$ is an open set. In general, we know that $f(z) \geq r(z)$ and by the definition of f(z), there is a stopping time which is preferable to stopping immediately if and only if $z \in B$. This suggests that B is the optimal continuation region and that

$$f_B(z) = f(z). \tag{2.5}$$

The proof of this result is quite long and technical, but it is worth commenting on one of the main ideas. The stopping rule associated with any open set A can be improved by reducing it to the subset $A' = \{z \in A: f_A(z) > r(z)\}$. In fact,

$$f_{A'}(z) = \max\{f_A(z), r(z)\}.$$

Further, it can be shown that when two modified open sets A_1' and A_2' are combined,

$$f_{A_1' A_2'}(z) \geq \max\{f_{A_1'}(z), f_{A_2'}(z)\}.$$

This argument leads eventually to the identity

$$f_{\cup A'}(z) = \sup_{A \subset H}\{f_A(z)\}, \tag{2.6}$$

where the union includes every modified open subset of H. Finally, it is established that the supremum here is equivalent to that in (2.3) and also that $B = \cup A'$, so (2.5) holds.

This construction of the optimal continuation region B as a union of open sets is revealing, but the local properties of the function f are also useful. In view of (2.4) and (2.5) we know that

$$\frac{1}{2} \Sigma \frac{\partial^2 f}{\partial u_i^2} = \frac{\partial f}{\partial v}, \quad (z \varepsilon B). \tag{2.7}$$

One of the boundary conditions for this equation follows from the continuity of f and the fact that $f(z) = r(z)$ for $z \notin B$. However, the boundary is not specified in advance and we need further information in order to locate it. Another necessary condition involves continuity of the first partial derivatives of f on the boundary of B. This holds provided that the given function r has continuous derivatives there. Thus, we have a free boundary problem specified by equation (2.7), together with the boundary conditions:

$$f(z) = r(z), \quad \frac{\partial f}{\partial u_i} = \frac{\partial r}{\partial u_i}, \quad \frac{\partial f}{\partial v} = \frac{\partial r}{\partial v}, \quad (i = 1,2,\ldots,m). \tag{2.8}$$

Strictly speaking, the conditions on first derivatives may not hold at certain irregular points of the boundary, but these form a set of probability measure zero with respect to the Brownian motion. Another qualification is that the maximum expected reward function may not be the only solution to the free boundary problem. In practice, this can be checked by making sure that a formal solution cannot be improved by any local modification.

Consider the distribution of z_A for any initial point z in an open set A. It is obvious that $Z_A = (u_A, v_A)$ always lies below $z = (u,v)$ in the sense that $v_A < v$. Hence $f_A(z) = E\{r(Z_A)|z\}$ is completely determined by the values of the reward function on the boundary of A below z. On the other hand, $f_A(z)$ is uniquely determined in A as the solution of equation (2.4) with the appropriate boundary values. This can be used to obtain inner approximations to the optimal continuation region B.

There are many solutions of the heat equation and, in principle, any of them can be used to generate stopping times in the following way. Let $g(z)$ be a function with continuous partial derivatives such that $\frac{1}{2}\sum \frac{\partial^2 g}{\partial u_i^2} = \frac{\partial g}{\partial v}$ in H. Suppose that $A = \{z \varepsilon H: g(z) > r(z)\}$ is non-empty and suppose further that $g(z) = r(z)$ at every point on the boundary of A. Then the functions g and f_A have similar properties throughout the closure of A. It follows from the results quoted in the last paragraph that

$$g(z) = f_A(z), \qquad (z \varepsilon A).$$

Hence, $f_A(z) > r(z)$ at every point of A and $A' = A \subset B$.

The idea we shall rely on to construct outer approximations to B depends on finding a formal solution to an auxiliary stopping problem. Let r^* be another reward function and imagine that we have determined the corresponding maximum expected reward f^* and optimal continuation set B^*, by solving the free boundary problem specified by (2.7) and (2.8). Now suppose that there is an initial point $z_0 = (u_0, v_0) \notin B^*$ such that

$$f^*(z_0) = r^*(z_0) = r(z_0) \text{ and } r^*(z) \geq r(z) \tag{2.9}$$

at every point $z = (u,v)$ with $v < v_0$. Since the maximum expectations $f^*(z_0)$ and $f(z_0)$ depend only on the values of the corresponding reward functions at points below z_0, it follows that $f^*(z_0) \geq f(z_0)$. Hence, $f(z_0) \leq r(z_0)$, so the optimal policy for the original problem is to stop at z_0: $f(z_0) = r(z_0)$ and $z_0 \notin B$.

The comparison technique needs careful application, but it can sometimes be effective. We shall apply it to one-dimensional Brownian motion in two special cases.

Example (i) $r(u,v) = \max(u,0)e^{-1/v}$

Example (ii) $r(u,v) = |u| - \frac{1}{v}$.

Both these examples arise in the determination of indices for sequential decision problems. Example (i) will be considered in Section 4. We conclude this section by giving an outer approximation of the optimal stopping boundary for Example (ii). This special case has also been investigated by Ferebee (1981), who obtained the same approximation. Some further results from his paper and their relation to the buyer's problem will be outlined in Section 5.

The symmetry of the reward function in Example (ii) shows that $f(u,v) = f(-u,v)$ and it suggests an optimal continuation region of the form $B = \{z: |u|<\lambda(v)\}$, with boundary curves $u = \pm\lambda(v)$. In order to see this, consider two initial points (u,v) and (u',v). It is enough to verify that, if $u>0$ and $(u,v)\varepsilon B$, then $(u',v) \varepsilon B$ whenever $0\le u'<u$. If $(u,v)\varepsilon B$, there is a stopping time $T\le v$ such that

$$E\{r(u + W(T), v - T)\} > r(u,v). \tag{2.10}$$

But the difference between the expected rewards obtained from the same stopping time for the initial points (u',v) and (u,v) is

$$E\{|u' + W(T)| - |u + W(T)|\} \ge u' - u,$$

since $u'<u$. Hence,

$$E\{r(u' + W(T), v - T)\} \ge E\{r(u + W(T), v - T)\} + u' - u. \tag{2.11}$$

The inequalities (2.10) and (2.11) show that $f(u',v)>r(u',v)$ and $(u',v)\varepsilon B$, as required.

We now seek an upper bound for the function $\lambda(v)$. This will be found by solving an auxiliary problem with parameters which can be chosen so that the conditions (2.9) are satisfied. Consider the reward function

$$r^*(u,v) = |u| + a + bv^{1/2}, \tag{2.12}$$

where a and b are constants, b>0. This is much easier to deal with than Example (ii), because it is associated with a separable solution of the heat equation. Let $y = uv^{-1/2}$ and define

$$f^*(u,v) = a + cv^{1/2}\{\phi(y) + y(\Phi(y) - \tfrac{1}{2})\} \tag{2.13}$$

for $|y| < \gamma$, where

$$\phi(y) = (2\pi)^{-1/2}e^{-y^2/2} \quad , \quad \Phi(y) = \int_{-\infty}^{y} \phi(\eta)d\eta \ .$$

We complete the definition of f* by setting $f^*(u,v) = r^*(u,v)$ when $|u| \geq \gamma v^{1/2}$. It is easy to verify that f* is a solution of equation (2.7) in the region given by $|u| < \gamma v^{1/2}$ and we intend to choose positive constants c and γ so that the boundary conditions (2.8) also hold.

For positive values of u and y,

$$f^*(u,v) - r^*(u,v) = v^{1/2}[c\{\phi(y) + y(\Phi(y) - \tfrac{1}{2})\} -y -b].$$

We must ensure that the expression on the right has its minimum with respect to y at $y=\gamma$ and that its value there is zero. This will guarantee that $f^*(u,v) > r^*(u,v)$ when $|u| < \gamma v^{1/2}$ and also that we obtain a solution of the free boundary problem. It is a straightforward exercise to find appropriate values for the constants:

$$b = \phi(\gamma)\{\Phi(\gamma) - \tfrac{1}{2}\}^{-1}, \quad c = \{\Phi(\gamma) - \tfrac{1}{2}\}^{-1} . \tag{2.14}$$

The constant a does not affect matters here, but it will be useful when we compare the auxiliary problem with Example (ii). It is not difficult to check that equations (2.14) determine γ and then c in terms of b. Since the solution of the free boundary problem defined by these values is unique, (2.13) and $B^* = \{z: |u|<\gamma v^{1/2}\}$ together represent the optimal stopping rule for the reward function (2.12).

In order to compare this with Example (ii), we examine

$$r^*(u,v) - r(u,v) = a + bv^{1/2} + v^{-1}.$$

The expression attains its minimum at $v = v_0$, given by $v_0^3 = 4b^{-2}$. Hence, we define

$$b = 2v_0^{-3/2}, \quad a = -3v_0^{-1}, \tag{2.15}$$

so that $r^*(z) \geq r(z)$ whenever $v \leq v_0$, with equality when $v = v_0$. We can establish all the conditions (2.9) by arranging that $f^*(z_0) = r^*(z_0)$ and this can be done by choosing a point on the boundary of B^* : $u_0 = \gamma v_0^{1/2}$. Then it follows from (2.9) that $z_0 \notin B$. In other words, $\lambda(v_0) \leq \gamma v_0^{1/2}$, where γ is defined in terms of v_0 by (2.14) and (2.15). Finally, we note that $v_0 > 0$ is arbitrary. The upper bound is quite general and it can be expressed in the form

$$\lambda(v) \leq \gamma v^{1/2}, \tag{2.16}$$

where γ is the unique solution of the equation

$$2\Phi(\gamma) - 1 = \phi(\gamma)v^{3/2}.$$

It turns out that the inequality (2.16) provides a reasonably good approximation to the optimal stopping boundary, for all $v > 0$. In particular, as we shall see in Section 5, it yields the correct asymptotic form of $\lambda(v)$ both as $v \to 0$ and as $v \to \infty$.

III. BANDIT PROBLEMS

A multi-armed bandit process consists of k projects or arms and the decision maker must choose, at each instant of discrete time, which of these projects to work on. The problem is formulated by assuming that there are k homogeneous Markov processes with state variables $x_1, x_2, \ldots, x_k$. If project i is chosen at time t when its state is $x_i(t)$, then it yields an immediate expected reward $R_i(x_i(t))$ and moves to a new state $x_i(t+1)$, according to a prescribed transition law. The state of the other projects remains unchanged: $x_j(t+1) = x_j(t)$, $j \neq i$. We suppose there is a fixed discount factor β, $0<\beta<1$, and the aim of the decision maker is to maximize his expected total discounted reward. For simplicity, this will be denoted by $E\{\Sigma\beta^t R_i(x_i(t))\}$, but it must be emphasized that $i = i(t)$ here and that the choice will depend on the allocation rule as well as the history of the state variables up to that time. We write $x(t) = (x_1(t), x_2(t),\ldots,x_k(t))$ for the vector of state variables and define

$$g(x) = \sup E\{\sum_{t=0}^{\infty} \beta^t R_i(x_i(t))\,|\,x(0) = x\}. \tag{3.1}$$

This depends only on the initial state vector x and the supremum includes all possible allocation rules. It will be assumed in this sketch of the theory that g(x) is finite always.

The key result for discounted multi-armed bandits is that one can define certain indices $M_i(x_i)$, $i = 1,2,\ldots,k$, which determine an optimal policy in the following way: at each time $t = 0,1,2\ldots$, choose any project i such that

$$M_i(x_i(t)) = \max_{1\le j\le k} M_j(x_j(t)). \tag{3.2}$$

It will be explained how the functions $M_i(x_i)$ can be calculated by solving an optimal stopping problem for each of the corresponding Markov processes. In effect, the result means that the multi-armed bandit problem is reduced to k separate one-armed bandit problems. The first proof was obtained by Gittins and Jones (1974) and the theory has been extended since then by Gittins and others: see Gittins (1979). The original argument was rather long and delicate and there is now a much simpler proof due to Whittle (1980). The indices are not easy to evaluate, particularly when the discount factor β is near 1, and our investigation will be limited to seeking approximations.

We introduce an arbitrary fixed reward M as an alternative to the sequence of rewards generated by the given Markov processes. Thus, at any stage, the decision maker may choose to abandon all the projects and accept a terminal reward M. Let $G(x,M)$ be the maximum total discounted expectation for this extended system and let $G_i(x_i,M)$ denote the corresponding maximum expectation for the i-th project alone. In other words, $G_i(x_i,M)$ is associated with an optimal stopping problem for the process $\{x_i(t)\}$, starting at $x_i(0) = x_i$, where the reward for stopping at any time t is $\beta^t M$. For any value of M, the function $G_i(.,M)$ is uniquely determined as the solution of a dynamic programming equation.

$$G_i(x_i,M) = \max[M, R_i(x_i) + \beta E\{G_i(x_i(1),M) | x_i(0) = x_i\}], \quad (3.3)$$

where the expectation is taken with respect to the state $x_i(1)$ after one step of the underlying Markov process. For the vector system, $G(x,M)$ is the solution of a more complicated equation, because there is a choice between k alternatives at each stage as well as the possibility of stopping.

The approach in Whittle's paper relies on the behaviour of the maximum expectations $G_i(x_i,M)$ and $G(x,M)$ as functions of M. It is obvious that $G(x,M)$ is non-decreasing in M and also that $G(x,M) \geq M$. It can be shown further that, for any fixed initial state x, $G(x,M)$ is a convex function of M. For large positive M, the optimal decision is to stop immediately and $G(x,M) = M$, whereas $G(x,M) \to g(x)$ as $M \to -\infty$ since this removes any possible advantage which might be obtained by stopping at some time in the future. The functions $G_i(x_i,.)$ have similar behaviour and it is clear from (3.3) that we can associate with each value of M an optimal continuation region consisting of those states x_i for which $G_i(x_i,M) > M$. The Gittins index for project i in state x_i can be defined as

$$M_i(x_i) = \min\{M : G_i(x_i,M) = M\}. \quad (3.4)$$

It is helpful to imagine increasing the terminal reward until it reaches a level at which the decision maker finds it immaterial whether he continues the project or abandons it.

For the original multi-armed bandit problem, the optimal policy based on these indices is difficult to evaluate because of the complicating effect of switching projects when one component and then another emerges as the one with largest index. We know that $g(x) \leq G(x,M)$ and, in principle, $g(x)$ can be determined by integrating an identity established by Whittle:

$$\frac{\partial G(x,M)}{\partial M} = \prod_{i=1}^{k} \frac{\partial G_i(x_i,M)}{\partial M} . \tag{3.5}$$

This holds for almost all values of M under the conditions we have assumed and it leads to an elegant proof that the index policy is optimal. However, in practice, it requires extensive calculations to solve the equations (3.3) for the functions $G_i(x_i,M)$ even to compute the indices $M_i(x_i)$.

So far, we have assumed that each of the component processes has transitions $x_i(t) \to x_i(t+1)$ described by a law which is homogeneous in time and does not involve any control parameters. It is natural to consider the generalisation in which each component is itself a Markov decision process: Gittins refers to such bandit processes as superprocesses. Some limited progress has been made in this direction but, although there is no difficulty in extending the definition of the indices, in general, they do not determine an optimal policy. The buyer's problem is one example of superprocesses, because it has components, each of which corresponds to a stopping problem with a terminal reward depending on its state. Since the identity (3.5) does not hold in this case, we shall rely on the general inequalities:

$$\max_{1 \leq j \leq k} \{G_j(x_j,M) - M\} \leq G(x,M) - M \tag{3.6}$$

$$\leq \sum_{j=1}^{k} \{G_j(x_j,M) - M\}.$$

These remain valid when the components of the multi-armed bandit are superprocesses and also, with a suitable interpretation of infinite expectations, when there is no discounting of future rewards. The first of the inequalities in (3.6) is an immediate consequence of the definitions and the second can be established by using the inductive approach described in Whittle's paper: a detailed proof will be published later. In the remaining sections of this paper, they will be needed only to confirm that

$$g(x) \leq \max_{1 \leq j \leq k} M_j(x_j). \tag{3.7}$$

This follows from the fact that $g(x) \leq G(x,M)$, by setting $M = \max M_j(x_j)$ in (3.6).

IV. GITTINS INDICES FOR NORMAL PROCESSES

Suppose that, for each $i = 1,2,\ldots,k$, there is a normal distribution with unknown mean θ_i and given variance σ_i^2. A-priori, θ_i is represented by the distribution $N(u_i,v_i)$. We define a project for each i regarding the corresponding observations with mean θ_i as a sequence of rewards. Thus, project i has initial state $x_i = (u_i,v_i)$ and a reward function given by $R_i(x_i) = u_i$. In order to simplify the notation, let us consider a single project with initial state (u,v) and write $\Gamma(u,v,M)$ instead of $G_i(u_i,v_i,M)$, omitting the subscript. Observations from the underlying distribution $N(\theta,\sigma^2)$ will generate a Markov process $\{(u(t), v(t)), t = 0,1,\ldots\}$ with transitions defined by relations (2.1). Hence, equation (3.3) for the maximum expected discounted reward takes the form:

$$\Gamma(u,v,M) = \max[M, u+\beta E\{\Gamma(u+\delta u, v-\delta v, M)\}], \tag{4.1}$$

where δu has the distribution $N(0,\delta v)$ and $\delta v = v^2(v+\sigma^2)^{-1}$. We are mainly interested in the Gittins index defined by

$$M(u,v) = \min\{M : \Gamma(u,v,M) = M\}. \tag{4.2}$$

We shall investigate a continuous version of the stopping problem (4.1). Notice that, in the discrete Markov process defined above, the variance decreases according to the equation

$$v(t) = \sigma^2 v(\sigma^2 + vt)^{-1}, \quad v(0) = v,$$

and, in general, u(t) has the conditional distribution N(u,v-v(t)), given that u(0) = u. When t and v(t) are treated as continuous parameters, the corresponding process is $\{(u + W(v-v(t)),v(t)\}$, where $\{W(s),s\geq 0\}$ is a standard Brownian motion. Consider the maximum expectation defined by

$$h(u,v,M) = \sup_{T\geq 0} E\int_0^T \{u + W(v-v(t))\ e^{-\alpha t}dt + e^{-\alpha T}M, \tag{4.3}$$

where the discount rate α is chosen so that $\beta = e^{-\alpha}$. The effect of discounting is slightly different here but, apart from this, $\Gamma(u,v,M)$ is the expectation obtained in (4.3) by restricting the supremum to discrete stopping times T. It is easily established that, in general,

$$\Gamma(u,v,M) \leq \alpha(1-e^{-\alpha})^{-1}h(u,v,M\alpha^{-1}(1-e^{-\alpha})). \tag{4.4}$$

The continuous stopping problem determines an index

$$\mu(u,v) = \min\{M : h(u,v,M) = M\} \tag{4.5}$$

and it follows from (4.4) that

$$M(u,v) \leq \alpha(1-e^{-\alpha})^{-1}\mu(u,v). \tag{4.6}$$

The discrete steps v(t) - v(t+1) are all relatively small if the initial value v is small compared with σ^2, since $(v(t)-v(t+1))/v(t) \leq v/\sigma^2$. In this case, (4.6) might be useful as an approximation.

The next step is to indicate a transformation which reduces the stopping problem (4.3) to the standard form considered in Section II. In fact, it is equivalent to Example (i). The rest of our investigation will be concerned with the optimal continuation region $B = \{(u,v) : u \leq \psi(v)\}$ for this example, because it turns out that

$$\mu(u,v) = \alpha^{-1}\{u + \alpha^{1/2}\sigma\psi(v\alpha^{-1}\sigma^{-2})\}. \tag{4.7}$$

Hence, (4.6) means that

$$M(u,v) \leq (1-e^{-\alpha})^{-1}\{u + \alpha^{1/2}\sigma\psi(v\alpha^{-1}\sigma^{-2})\}. \tag{4.8}$$

The required transformation can be found by examining the local properties of the function h, which are similar to those described in Section 2, so only the most important will be mentioned. The optimal continuation region for problem (4.3) is defined by the inequality $h(u,v,M)>M$ and the differential equation which holds there is

$$\frac{1}{2}\frac{\partial^2 h}{\partial u^2} = \frac{\partial h}{\partial v} + \frac{\sigma^2}{v^2}(\alpha h - u).$$

The last term arises partly from the effect of discounting, but also because there is a reward u per unit time for continuing the process. However, the equation can be transformed to the heat equation by defining

$$f(u,v) = \alpha^{1/2}\sigma^{-1}e^{-1/v}\{h(\alpha M-\alpha^{1/2}\sigma u,\alpha\sigma^2 v,M) + \alpha^{-1/2}\sigma u-M\}. \tag{4.9}$$

Since $h \geq M$ always holds, we have $f(u,v) \geq ue^{-1/v}$. It is a straightforward, but rather technical exercise to check that f is the maximum expected reward function for a standard problem with terminal rewards given by $r'(u,v) = ue^{-1/v}$. Notice that $r'(u,v) < 0$ in the quadrant $A = \{(u,v) : u<0, v>0\}$ and it converges to zero on the boundary of A. Thus, $f(u,v) \geq f_A(u,v) = 0$ in A and it follows that f can also be defined as the maximum expectation associated with the reward function

$$r(u,v) = \max(u,0)e^{-1/v}. \tag{4.10}$$

An argument similar to that used in Section 2 for Example (ii) can be applied here to establish a simple form for the optimal continuation region: $B = \{(u,v) : u<\psi(v)\}$, where $\psi(v)>0$. Then (4.9) shows that $f(u,v)>r'(u,v)$ if and only if

$$h(\alpha M-\alpha^{1/2}\sigma u, \alpha\sigma^2 v, M) > M,$$

so this is equivalent to the inequality $u<\psi(v)$. Finally, a simple change of variables leads to the conclusion that $h(u,v,M)>M$ if and only if

$$\alpha M < u + \alpha^{1/2}\sigma\psi(v\alpha^{-1}\sigma^{-2}),$$

and this proves equation (4.7).

We can now establish some properties of the optimal boundary curve $u = \psi(v)$ for Example (i).

Theorem 1

The function ψ *has the following properties;*

(a) $\psi(v)/\sqrt{v}$ *is non-decreasing in* v,

(b) $\psi(v)\leq v/\sqrt{2}$ *for all* $v > 0$,

(c) $\psi(v)/v\to 1/\sqrt{2}$ *as* $v \to 0$.

Proof. (a) We already know that $\psi(v)>0$ for $v>0$. Consider an initial state (u,v) and suppose that $0<u<\psi(v)$. Since $(u,v)\in B$, there is a stopping time $T\leq v$ for the Brownian motion $\{W(t)\}$ such that, in the notation of Section 2,

$$E\{r(u + W(T), v-T)\} > r(u,v) > 0. \tag{4.11}$$

The path leading from the initial to the final state consists of points $(u + W(t), v-t)$, $0\leq t\leq T$. We now use the fact that, for any constant $c>0$, the process $\{W'(t')\}$ defined by $W'(t') = cW(t'c^{-2})$ is also a standard Brownian motion. Consider shifting the above path to the points $(cu + W'(t'), c^2v - t')$, $0 \le t' \le c^2T$, so that it leads from the initial state (cu, c^2v) to $(cu + cW(T), c^2v - c^2T)$. Thus, we have a (1:1) correspondence between the paths from two initial states and the stopping time T determines the distribution of the final state in both cases. In particular,

$$f(cu, c^2v) \ge E\{r(cu + cW(T), c^2v - c^2T)\}. \tag{4.12}$$

We can use (4.10) to evaluate the ratio

$$\frac{r(cu + cW(T), c^2v - c^2T)}{r(cu, c^2v)} = \frac{\max(u + W(T), 0)}{u} \cdot \exp\{-Tc^{-2}v^{-1}(v-T)^{-1}\}.$$

This shows that, if $c>1$,

$$\frac{r(cu + cW(T), c^2v - c^2T)}{r(cu, c^2v)} \ge \frac{r(u + W(T), v - T)}{r(u,v)}.$$

When we consider expectations with respect to the distribution of T and W(T), and apply (4.11) and (4.12), we obtain the result that $f(cu, c^2v) > r(cu, c^2v)$. We have shown that, if $c>1$ and $0<u<\psi(v)$, then $cu<\psi(c^2v)$. In other words $\psi(c^2v) \ge c\psi(v)$ whenever $c>1$, which is equivalent to the required property (a).

(b) In order to establish this upper bound, we use the comparison technique based on conditions (2.9). Fix an initial state $z_0 = (u_0, v_0)$ with $u_0 = v_0/\sqrt{2}$ and consider an auxiliary problem with reward function

$$r^*(u,v) = \exp\{au + \tfrac{1}{2}a^2v + b\}.$$

Notice that, for any choice of the constants a and b, r^* is a solution of the heat equation everywhere in the (u,v) plane. This implies that, for any open set A, $f_A^*(z) = r^*(z)$ for all z and it follows that the optimal continuation region B^* is empty. Hence $f^*(z_0) = r^*(z_0)$. We need to prove that $\psi(v_0) \le v_0/\sqrt{2}$ and, according to the method established in Section 2, it will be enough to arrange that $r^*(z_0) = r(z_0)$ and, in general,

$r^*(z) \geq r(z)$. Let $a = \sqrt{2}/v_0$ and choose the constant b so that $r^*(u_0,v_0) = r(u_0,v_0)$. It is clear from (4.10) that $r^*(u,v) > r(u,v) = 0$ if $u \leq 0$ and when $u>0$, we have

$$\frac{r^*(u,v)}{r(u,v)} = \frac{u_0}{u} \exp\left\{\frac{\sqrt{2}(u-u_0)}{v_0} + \frac{v}{v_0^2} + \frac{1}{v} - \frac{2}{v_0}\right\}.$$

We are left with an easy exercise to minimize this quantity with respect to positive values of u and v. It leads to the result that $r^*(u,v) \geq r(u,v)$, with equality at $u = u_0$, $v = v_0$, so we obtain the required conclusion.

(c) We need a lower bound on $\psi(v)$ and one can be established by considering the function

$$g(u,v) = c \exp\{2^{1/2}b^{-1}u + b^{-2}(v-b)\} - c,$$

where $c = 2^{-1/2}b \exp\{-1-b^{-1}\}$ and b is an arbitrary positive constant. This is a solution of the heat equation similar to that used in part (b). It will eventually lead to a lower bound on $\psi(v_0)$ for a suitable choice of $v_0>b$ and, in fact, the result stated below in (4.15) is quite general. We are interested in a bounded subset A of the region in which $g(u,v)>r(u,v)$.

Suppose first that $u<0$. Then $r(u,v) = 0$ and $g(u,v)>0$ whenever $u > -2^{-1/2}b^{-1}(v-b)$. One of the boundaries of A will be defined by the line $u = -2^{-1/2}b^{-1}(v-b)$, for $v \geq b$. From now on, we restrict attention to points with $u \geq 0$, $v \geq b$ and examine the difference $K = g-r$:

$$K(u,v) = c \exp\{2^{1/2}b^{-1}u + b^{-2}(v-b)\} - c - u \exp\{-v^{-1}\}.$$

This is a strictly convex function of u and its minimum with respect to u occurs when

$$u = 2^{-1/2} b\{1 - v(b^{-1} - v^{-1})^2\}, \tag{4.13}$$

as can be seen by setting $\frac{\partial K}{\partial u} = 0$ and by using the definition of c. We also find that

$$L(v) = \min_u K(u,v) = 2^{-1/2} bv(b^{-1} - v^{-1})^2 \exp\{-v^{-1}\} - c.$$

It is easily verified that $L(b)<0$ and $L(v)$ is strictly increasing for $v>b$. The equation $L(v) = 0$ can be expressed in the form

$$v(b^{-1} - v^{-1})^2 = \exp\{v^{-1} - b^{-1} - 1\} \tag{4.14}$$

and this has a unique solution $v = v_0 > b$. We define $u = u_0$ when $v = v_0$ in (4.13) and note that $K(u_0,v_0) = 0$ and $u_0>0$.

It is clear that $K(0,v)\geq 0$ for $v\geq b$. We also know that $K(u,v) = L(v)\leq 0$ along the curve given by (4.13), for $b\leq v\leq v_0$, and it is easy to check that this lies in the positive quadrant. Hence, there is a curve $u = \chi(v)$ defined for $b\leq v\leq v_0$ on which $K(u,v) = 0$ and $g(u,v)>r(u,v)$ at every point of the open set

$$A = \{(u,v) : -2^{-1/2} b^{-1}(v-b)<u<\chi(v),\quad b<v<v_0\}.$$

According to the results quoted in Section 2, we have $f_A(u,v) = g(u,v)$ in A and hence $f(u,v)>r(u,v)$ at every point of this set. It follows by considering a neighbourhood of the boundary point (u_0,v_0) that $\chi(v_0)\leq\psi(v_0)$.

All that remains is to tidy up the notation by introducing a new parameter $\xi = b^{-1} - v_0^{-1}$. Then b, v_0 and $u_0 = \chi(v_0)$ can be expressed in terms of ξ, by using equations (4.13) and (4.14). We find that

$$\begin{aligned} b &= \{\xi + \xi^2 \exp(1+\xi)\}^{-1}, \quad v_0 = \xi^{-2} \exp(-1-\xi), \\ \chi(v_0) &= 2^{-1/2}\{\xi+\xi^2 \exp(1+\xi)\}^{-1} \{1 - \exp(-1-\xi)\}. \end{aligned} \tag{4.15}$$

As ξ increases through the range $(0,\infty)$, v_0 decreases over the same range and $\chi(v_0)$ provides a lower bound for $\psi(v_0)$ in each case. In particular, the bound is effective when $\xi\to\infty$ and $v_0\to 0$, in which case $\chi(v_0)/v_0\to 1/\sqrt{2}$. Hence,

$$\liminf_{v\to 0} \psi(v)/v \geq 1/\sqrt{2}$$

and this, together with the result of part (b), completes the proof.

Remarks. It is worth mentioning, without proof, another lower bound on $\psi(v)$, which turns out to be better than that given by (4.15) when v is large. The idea behind it is to use a fixed stopping time $T = \tau v$ at the initial state (u,v), where τ is a constant, $0<\tau<1$. The expected reward exceeds $r(u,v)$ for suitable values of u and one obtains a lower bound on $\psi(v)$. In particular, by letting $\tau\to 1$, it can be shown that

$$\liminf_{v\to\infty} \psi(v)\{2v \log v\}^{-1/2} \geq 1. \tag{4.16}$$

This complements part (a) of Theorem 1, by showing that $\psi(v)/\sqrt{v}\to\infty$ as $v\to\infty$.

I am grateful to Herman Chernoff and John Petkau for their computations of the optimal boundary curve for Example (i). In particular, the computations suggested the limit in part (c) of the theorem, before a proof was discovered. At the other extreme, computed values of $\psi(v)\ \{2v \log v\}^{-1/2}$ increase very slowly towards 1 as $v\to\infty$, in agreement with the asymptotic form suggested by (4.16). This example provided a rather severe test of the general programme developed by Chernoff and Petkau for solving optimal stopping problems, mainly because of the behaviour of the factor $\exp\{-v^{-1}\}$ in the reward function. In particular, the reward is very close to $\bar{r}(u,v) = \max(u,0)$ when v is large. For this alternative reward function, the optimal stopping time is

$T = v$ for every initial state (u,v) with $v>0$. In view of this, it is not surprising that the precise asymptotic behaviour of $\psi(v)$ as $v\to\infty$ is difficult to establish.

Finally, we turn to some implications of our results for the multi-armed bandit problem described at the beginning of this section. Project i has initial state (u_i,v_i) and variance parameter σ_i^2 so, according to (4.8), the Gittins index satisfies.

$$M(u_i,v_i) \leq (1-e^{-\alpha})^{-1}\{u_i + \alpha^{1/2}\sigma_i\psi(v_i\alpha^{-1}\sigma_i^{-2})\}.$$

Hence, by part (b) of the theorem,

$$M(u_i,v_i) \leq (1-e^{-\alpha})^{-1}\{u_i + (2\alpha)^{-1/2}v_i\sigma_i^{-1}\}.$$

We know that the index policy attains the maximum expected discounted reward, for any initial states, and the inequality (3.7) shows that this expectation cannot exceed

$$\max_{1\leq i\leq k} M(u_i,v_i) \leq (1-e^{-\alpha})^{-1} \max_{1\leq i\leq k}\{u_i + (2\alpha)^{-1/2}v_i\sigma_i^{-1}\}. \tag{4.17}$$

On the other hand, the expectation from the i-th project alone is $u_i\{1 + \beta + \beta^2 + \ldots\} = (1-\beta)^{-1}u_i$, where $\beta = e^{-\alpha}$. It follows that the maximum which can be achieved by restricting attention throughout to the best single project is $(1-e^{-\alpha})^{-1} \max\{u_i\}$. Thus, (4.17) means that the gain in expectation obtained by using the optimal policy is at most

$$(2\alpha)^{-1/2}(1-e^{-\alpha})^{-1} \max\{v_i\sigma_i^{-1}\}.$$

Of course, this quantity may be relatively large compared with $(1-e^{-\alpha})^{-1} \max\{u_i\}$, particularly if β is near 1 and α is small.

Another feature of the index policy is its sensitivity to the values of the parameters. It has been established by Kelly (1981) for projects involving Bernoulli trials that, when the discount factor β increases to 1, the index policy has a limit which is not a good rule for a long sequence of decisions. It is intuitively clear that a similar result holds for normal processes but, unfortunately, we are only in a position to derive it for the continuous version of the multi-armed bandit and then only when the projects have a common variance parameter: $\sigma_i^2 = \sigma^2$, $i = 1,2,\ldots,k$. In this case, the indices are given by (4.7), which shows that

$$\alpha\{\mu(u_i,v_i) - \mu(u_j,v_j)\} = u_i - u_j + \alpha^{1/2}\sigma\{\psi(v_i\alpha^{-1}\sigma^{-2}) - \psi(v_j\alpha^{-1}\sigma^{-2})\}.$$

Now suppose that $v_i > v_j$ for $j \neq i$. Then, part (a) of the theorem shows that $\psi(v_j\alpha^{-1}\sigma^{-2}) \leq (v_j/v_i)^{1/2}\psi(v_i\alpha^{-1}\sigma^{-2})$ and we have

$$\alpha\{\mu(u_i,v_i) - \mu(u_j,v_j)\} \geq u_i - u_j + (v_i^{1/2} - v_j^{1/2})\psi(v_i\alpha^{-1}\sigma^{-2})(v_i\alpha^{-1}\sigma^{-2})^{-1/2}.$$

It follows from (4.16) that the difference $\mu(u_i,v_i) - \mu(u_j,v_j)$, which depends on α, must become positive as α decreases to zero. In other words, for any fixed initial states with $v_i > v_j$ for some i and all $j \neq i$, $\mu(u_i,v_i)$ is the largest index when α is sufficiently small. This implies a limiting form of the index policy which selects the project having the largest posterior variance and this applies at every stage. Hence, it must lead to a situation in which the frequency of selection is the same for all the projects. This is not a good policy, because it cannot respond to any information collected about the unknown mean rewards $\theta_1, \theta_2, \ldots, \theta_k$.

V. THE BUYER'S PROBLEM

A general formulation of this problem has been investigated in a recent thesis by Bergman (1981). He proved that an index policy is optimal under rather special conditions: the buyer is confronted with a choice of one from an infinite collection of batches and it is assumed that he has independent prior information on the quality of each batch, such that the initial values of all the indices are the same. However, it is intuitively clear that index policies will be reasonably good under much more general conditions and one of the aims here is to suggest why, by sketching a proof of Bergman's result for a particular version of the buyer's problem. The other main purpose in this section is to explain the connection between his work and an application of the comparison technique carried out, quite independently, by Ferebee (1981). The model is based on the normal processes described in Section IV, with a modified structure of decisions and rewards. As before, we shall need several transformations and a continuous approximation of the underlying stopping problem, so only the main ideas will be included.

In the first place, suppose there are k batches of similar goods and let θ_i be the unknown quality of batch i, having a prior distribution $N(u_i,v_i)$. The cost of the batch is K_i and the difference $u_i - K_i$ represents the expected net profit to the buyer if he makes this particular choice immediately. The problem is to find an optimal strategy for sampling at cost $c>0$ per observation from some or all of the batches, until it is decided to stop and buy one of them. We suppose that c and the variance σ^2 of the observations are the same for each batch, in order to avoid technical complications when we let $k\to\infty$.

We must deal with a single batch, in order to describe the index, so the subscript i will be omitted for the moment. As before, $M(u,v)$ is defined as the smallest value of the alternative terminal reward M such that $\Gamma(u,v,M) = M$ and the function Γ is determined by solving a stopping problem:

$$\Gamma(u,v,M) = \max\{M,\ u - K,\ - c + E\ \{\Gamma(u+\delta u, v-\delta v, M)\}. \qquad (5.1)$$

This is similar to (4.1) except that there are two distinct ways of stopping, one of which corresponds to buying the batch and this will be preferred in any state where $\Gamma(u,v,M) = u-K$. The action of taking a new observation leads to a new state $(u + \delta u, v - \delta v)$ at cost c, where δu has the distribution $N(0,\delta v)$ and $\delta v = v^2(v + \sigma^2)^{-1}$.

In effect, K and M can be eliminated and we can reduce (5.1) to a standard form by defining

$$G(u,v) = 2\Gamma(u + K + M, v, M) - u - 2M. \tag{5.2}$$

Then $G(u,v) = G(-u,v)$, because the function satisfies the relation

$$G(u,v) = \max[|u|, -2c + E\{G(u + \delta u, v - \delta v)\}]. \tag{5.3}$$

Further, by examining this transformation carefully, one can deduce that

$$M(u,v) = u - K + L(v), \quad L(v) = \min\{u: G(u,v) = u\}. \tag{5.4}$$

As we shall see later, the stopping problem implicit in (5.3) is a discrete analogue of Example (ii). The argument based on (2.10) and (2.11) can be used again here to show that $G(u,v) > |u|$ if $|u| < L(v)$ and $G(u,v) = |u|$ if $|u| \geq L(v)$. When we consider the relation between this problem and Example (ii), it will be shown that, in general,

$$L(v) \leq a\lambda(va^{-2}), \quad a = (2c\sigma^2)^{1/3}, \tag{5.5}$$

where λ is the function describing the symmetric boundaries of the optimal continuation region for the continuous version of the problem.

A minor complication occurs when v is near zero, since $L(v) = 0$ and the solution of (5.3) is given by $G(u,v) = |u|$. This trivial solution can be checked by noting that it is valid provided that $E\{|u + \delta u|\} \leq 2c + |u|$ for all u and that a necessary and sufficient condition for this is $E\{|\delta u|\} \leq 2c$.

Because of the distribution of δu, we obtain a condition on v and it turns out that $L(v) = 0$ if and only if $v \le b$, where $b = \pi c^2\{1 + 2\pi^{-1}c^{-2}\sigma^2\}^{1/2}$.

We now return to (5.1) and consider an optimal stopping rule for the discrete Markov process $\{(u(t),v(t))\}$ described at the beginning of Section IV. Suppose the artificial terminal reward is fixed at the level $M_0 = M(u(0),v(0))$ and let $v(0)>b$ so that $L(v(0))>0$. The initial state is a boundary point of the optimal continuation region: $u(0) = M_0 + K - L(v(0))$. This means that the maximum expected net reward $\Gamma(u(0),v(0),M_0) = M_0$ can be attained either by stopping immediately or by taking an observation and, at times $t = 1,2,\ldots$, using the following

Decision rule: take another observation if $M_0 + K - L(v(t)) < u(t) < M_0 + K + L(v(t))$, *stop and buy the batch if* $u(t) \ge M_0 + K + L(v(t))$, *or stop and reject the batch in favour of the terminal reward* M_0 *if* $u(t) \le M_0 + K - L(v(t))$.

It needs a little patience to check, by using (5.1)...(5.4), that this sequential decision rule is equivalent to acting at each stage according to which of the terms on the right of (5.1) gives the maximum. In fact, the stopping time T implicit in the above rule can also be defined as the first $t \ge 1$ such that $\Gamma(u(t),v(t),M_0) = \max\{M_0,u(t)-K\}$. The decision procedure is also an index policy in the sense that it leads to rejection of the batch as soon as $M(u(t), v(t))$ falls below its initial level M_0.

Consider the states (u,v) with $v>b$, such that $M(u,v) = M_0$. According to (5.4), they lie on the curve given by $u = M_0 + K - L(v)$. We now associate with each of these states a stopping time $T = T(u,v)$, by using the above decision rule with $u(0) = u$, $v(0) = v$. There is also a corresponding acceptance probability $p = p(u,v)$ for the batch:

$$p = P\{u(T) \geq M_0 + K + L(v(T))\}$$

Since the decision rule is optimal, the expected net reward is $\Gamma(u,v,M_0) = M_0$ and this can also be evaluated as

$$pE\{u(T) - K|u(T) \geq M_0 + K + L(v(T))\} + (1-p)M_0 - cE(T).$$

It follows that

$$pE\{u(T) - K|u(T) \geq M_0 + K + L(v(T))\} - cE(T) = pM_0. \quad (5.6)$$

Note that the expectations here, as well as p, depend on the initial state (u,v). We shall need to apply the equation for the initial states of an infinite sequence of batches and it will simplify matters to arrange that the expectations remain bounded and that p(u,v) is bounded away from zero. This can be done by restricting attention to any bounded seqment of the curve:

$$u = M_0 + K - L(v), \quad b < v \leq d,$$

where d is a positive constant. For example, since $T \geq 1$, $p(u,v) \geq P\{u(1) \geq M_0 + K + L(v(1))\} = P\{\delta u \geq L(v) + L(v - \delta v)\}$ and δu has the distribution $N(0,\delta v)$ with $\delta v = v^2(v+\sigma^2)^{-1}$. Then the inequality (5.5) can be used to show that p(u,v) is bounded away from zero.

We can now establish Bergman's result. Suppose we have an infinite sequence of batches, each with the same initial index:

$$M(u_i,v_i) = M_0, \quad i = 1,2,\ldots . \quad (5.7)$$

Suppose further that $v_i \leq d$ always holds, for a suitable constant d. Let g be the maximum expected net reward in this situation. It follows from the inequality (3.7), by letting $k \to \infty$, that $g \leq M_0$. In fact, $g = M_0$ and this can be proved by finding a strategy which attains the expectation M_0 and is therefore optimal.

Let us assume first that $v_i \leq b$ for some $i \geq 1$. Then (5.7) means that $u_i = M_0 + K_i - L(v_i)$ and, since $L(v_i) = 0$, we know that $u_i - K_i = M_0$. Hence the net reward M_0 can be attained by a decision to buy batch i immediately and there is no better sequential decision procedure. We may assume from now that $b < v_i \leq d$, for each $i \geq 1$. Consider the following

Strategy: apply the decision rule to batch 1 and, if it is rejected, repeat the procedure with batch 2 and so on until one of the batches is accepted.

We can evaluate this strategy by using equation (5.6) repeatedly. Let $T_i = T(u_i, v_i)$, $p_i = p(u_i, v_i)$ and write $q_i = 1-p_i$. Now define N as the random number of batches up to and including the accepted one. It has the distribution given by $P(N=j) = q_1 q_2 \cdots q_{j-1} p_j$ and $P(N \geq j) = q_1 q_2 \cdots q_{j-1}$ for $j \geq 1$. Since the sequence $\{p_i\}$ is bounded away from zero and the random variables $\{T_i\}$ have bounded means, the expected net reward can be calculated without difficulty. It is

$$E\{u(T_N) - K_N\} - cE\{T_1 + T_2 + \ldots + T_N\}$$

$$= \sum_{j=1}^{\infty} P(N = j)\, E\{u(T_j) - K_j \mid N = j\} - c \sum_{j=1}^{\infty} P(N \geq j) E(T_j)$$

$$= \sum_{j=1}^{\infty} q_1 q_2 \cdots q_{j-1} [p_j E\{u(T_j) - K_j \mid N = j\} - cE(T_j)].$$

Equation (5.6) shows that the final series reduces to $\Sigma q_1 q_2 \cdots q_{j-1}\, p_j M_0 = M_0$ and this confirms that our strategy is optimal.

It still remains to indicate the link between the discrete stopping problem (5.3) and Example (ii), in order to justify the inequality (5.5). Let $h(u,v)$ represent the maximum expected net reward analogous to $G(u,v)$, when the process $\{(u(t)), v(t))\}$ starting at $u(0) = u$, $v(0) = v$, is no longer restricted to discrete stopping times. Clearly

$$G(u,v) \leq h(u,v), \tag{5.8}$$

provided that the cost of sampling the continuous process always matches the amount 2c incurred for each observation in (5.3). This can be arranged by defining the continuous sampling cost as $2c\sigma^2 v^{-2}$ per unit decrease in v. The reward for stopping is $|u|$ so, in general, $h(u,v) \geq |u|$. In the optimal continuation region $h(u,v) > |u|$ and it turns out that

$$\frac{1}{2}\frac{\partial^2 h}{\partial u^2} = \frac{\partial h}{\partial v} + \frac{2c\sigma^2}{v^2},$$

where the last term arises from the cost of sampling. This continuous stopping problem is easily reduced to the standard form of Section 2 by defining

$$f(u,v) = a^{-1}h(au,a^2 v) - v^{-1}, \tag{5.9}$$

with $a = (2c\sigma^2)^{1/3}$. Then it can be verified that f is the maximum expected reward function for a standard problem with terminal rewards given by

$$r(u,v) = |u| - v^{-1}.$$

We have already demonstrated, in Section 2, that this example has an optimal continuation region of the form $B = \{(u,v) : |u|<\lambda(v)\}$ and the condition $f(u,v)>r(u,v)$ which holds there is equivalent to $h(au,a^2v)>|au|$, by (5.9). It follows that $h(u,v)>|u|$ if and only if $|u|<a\lambda(va^{-2})$ and then the general inequality (5.8) implies that (5.5) holds:

$$L(v) \leq a\lambda(va^{-2}).$$

Remarks. This result, together with (5.4), gives no more than a rough indication of how the index policy might be applied to the buyer's problem. An exact specification of the policy would require extensive computations to solve the discrete stopping problem (5.3) and determine the function L. All we have is a crude upper bound obtained by combining the above inequality with the one established in Section II for the function λ. In fact, (2.16) provides a good approximation to $\lambda(v)$ both as $v\to 0$ and as $v\to\infty$. This claim is based on Ferebee's paper (1981), which establishes a lower bound on $\lambda(v)$ as well as the properties reproduced earlier in this paper. His inequality can be expressed, after minor changes of notation, in terms of two parameters $\eta,\xi>0$, where $\xi = \{2\Phi(\eta)-1\}/\phi(\eta)$. Here ϕ and Φ are the probability density and distribution function of the standard normal distribution. For each $\eta>0$, the corresponding $v = v(\eta)$ and a lower bound on $\lambda(v)$ are given by:

$$v = \{(2\eta + (1+\eta^2)\xi/4\}^{2/3}, \qquad \lambda(v) \geq v^{-1}\eta^2(2+\eta\xi)/4.$$

The implications are not immediately clear from these formulae, but it is easy to determine the asymptotic form of the bound as η,ξ and v approach 0 or ∞ together. Since the upper bound given by (2.16) has a similar asymptotic form at both extremes, it follows that

$$\lambda(v)/v^2 \to \frac{1}{2} \text{ as } v \to 0,$$

$$\lambda(v)/\sqrt{3v \log v} \to 1 \text{ as } v \to \infty.$$

VI. REFERENCES

Bather, J.A. (1962). "Bayes Procedures for Deciding the Sign of a Normal Mean." *Proc. Cambridge Philos. Soc. 58*, 599-620.

Bather, J.A. (1970). "Optimal Stopping Problems for Brownian Motion." *Adv. Appl. Prob. 2*, 259-286.

Bather, J.A. and Chernoff, H. (1966). "Sequential Decisions in the Control of a Spaceship." *Proc. Fifth Berkeley Symposium 3*, 181-207.

Bather, J.A. and Chernoff, H. (1967). "Sequential Decisions in the Control of a Spaceship (Finite Fuel)." *J. Appl. Prob. 4*, 584-604.

Bergman, S. (1981). Ph.D. Thesis, Statistics Dept., Yale.

Breakwell, J.V. and Chernoff, H. (1964). "Sequential Tests for the Mean of a Normal Distribution II (Large t). *Ann. Math. Statist. 35*, 162-173.

Chernoff, H. (1961). "Sequential Tests for the Mean of a Normal Distribution." *Proc. Fourth Berkeley Symposium 1*, 79-91.

Chernoff, H. (1965). "Sequential Tests for the Mean of a Normal Distribution III (Small t)". *Ann. Math. Statist. 36*, 28-54.

Chernoff, H. (1965). "Sequential Tests for the Mean of a Normal Distribution IV (Discrete Case)." *Ann. Math. Statist 36*, 55-68.

Ferebee, B. (1981). "Testing a New Drug Against a Standard, the Diffusion Approximation." Technical Report, SFB 123, Univ. Heidelberg.

Gittins, J.C. (1979). "Bandit Processes and Dynamic Allocation Indices (with discussion)." *J.R. Statist: Soc. B, 41*, 148-164.

Gittins, J.C. and Jones, D.M. (1974). "A Dynamic Allocation Index for the Sequential Design of Experiments." *Progress in Statistics* (J. Gani, ed.), Amsterdam: North Holland, 241-266.

Karatzas, I. (1982). "A Class of Singular Stochastic Control Problems". To appear, *Adv. Appl. Prob.*

Kelly, F.P. "Multi-armed Bandits with Discount Factor Near One: the Bernoulli Case." *Ann. Statist. 9*, 987-1001.

Robbins, H. (1952). "Some Aspects of the Sequential Design of Experiments." *Bull. Amer. Math. Soc. 58*, 527-535.

Whittle, P. (1980). "Multi-armed Bandits and the Gittins Index." *J.R. Statist. Soc. B, 42*, 143-149.

SEQUENTIAL DESIGN OF COMPARATIVE CLINICAL TRIALS

T. L. Lai
Herbert Robbins[1]

Department of Mathematical Statistics
Columbia University
New York, New York

David Siegmund[2]

Department of Statistics
Stanford University
Stanford, California

I. INTRODUCTION

Suppose that two treatments A and B of unknown efficacy are available and that a large number N of patients are to be treated. The trial consists of pairwise allocation of treatments to n pairs of patients; then a conclusion is made as to which treatment appears to be superior, so that the remaining N - 2n patients not on trial will be given the apparently superior treatment based on the results of the first n pairs.

In 1963, Anscombe *[1]* proposed a decision-theoretic approach to the problem of choosing n when N is given. Let z_i denote the difference in response between the patient receiving treatment A and the patient receiving treatment B in the ith trial pair.

[1] *Research supported by the National Institutes of Health and the National Science Foundation.*
[2] *Research supported by Office of Naval Research Contract N00014-77-C-0306 (NR-042-373).*

ISBN 0-12-589320-5

Assume that $z_1, z_2, \ldots$ are independently and normally distributed with unknown mean δ and known variance σ^2. If the trial period consists of n pairs and the remaining N - 2n patients are given treatment A or B according to whether $s_n > 0$ or $s_n < 0$, where $s_n = z_1 + \ldots + z_n$, then the total number of patients given the inferior treatment is $n + (N - 2n)I_{(s_n<0)}$ if $\delta > 0$ and $n + (N - 2n)I_{(s_n>0)}$ if $\delta < 0$, where $I_{(\cdot)}$ denotes the indicator function of the event in question. For any stopping rule T for determining n, the regret $R(\delta,T)$ is defined to be the expected total difference in response between the ideal procedure, which would assign all N patients to the superior treatment, and the procedure determined by the stopping rule T:

$$R(\delta,T) = \delta\, E_\delta\{T + (N - 2T)I_{(s_T<0)}\} \qquad \text{if } \delta > 0,$$

$$= |\delta|\, E_\delta\{T + (N - 2T)I_{(s_T>0)}\} \quad \text{if } \delta < 0.$$

In ignorance of $|\delta|$, Anscombe assumes a prior distribution G on δ and chooses a stopping rule T to minimize the Bayes risk $\int_{\infty}^{\infty} R(\delta,T)\, dG(\delta)$.

Since σ is known, we shall assume without loss of generality that $\sigma = 1$. A particular choice of G, considered by Anscombe, is the flat prior $dG(\delta) = d\delta$. Although the Bayes risk with respect to the flat prior is infinite for every stopping rule $T \geq 1$, the posterior risk $r(k,s)$ given that $s_k = s$ and that stopping occurs at stage k is well defined and is given by

$$r(k,s) = |s| + k^{-1/2} N\{\phi(s/k^{1/2}) - k^{-1/2}|s|\Phi(-|s|/k^{1/2})\},$$

where ϕ denotes the density function and Φ the distribution function of the standard normal distribution. For fixed k, $r(k,s)$ has its minimum at two s values defined by

$$1 - \Phi(|s|/k^{1/2}) = k/N. \tag{1}$$

Anscombe suggested using (1) to define a stopping rule

$$T_A = \inf\{k: 1 - \Phi(|s_k|/k^{1/2}) \leq k/N\}. \tag{2}$$

Noting that $r(k) = \inf_s r(k,s)$ is increasing with k for $k \leq .27\,N$, he conjectured that the stopping boundary of the Bayes rule is quite close to that given by (1), at least for smaller values of k.

The extent to which Ancombe's conjecture holds has recently been studied by Chernoff and Petkau *[3]* and by Lai, Levin, Robbins, and Siegmund *[7]*. There is a substantial difference between the stopping boundary of the Bayes rule and Anscombe's rule, although the two are indeed very close for k/N near 0. On the other hand, the numerical and asymptotic results in *[3]* and *[7]* show that the performance of Anscombe's rule is surprisingly close to that of Bayes rule.

The Bayes rule for an arbitrary normal prior G has been studied by Chernoff and Petkau *[3]*. Their results on the Bayes risks show that while Anscombe's rule is remarkably close to optimal, a later procedure proposed by Begg and Mehta *[2]* is a relatively poor competitor. Another rule, introduced by Lai, Levin, Robbins, and Siegmund was shown in *[7]* to have frequentist and Bayes properties comparable to Anscombe's rule and the flat prior Bayes rule. This rule is a sequential version of the optimal fixed sample size procedure when $|\delta|$ is known. If $|\delta|$ is known, then the optimal fixed sample size n minimizing $R(\delta,n)$ is obtained by solving the equation $\frac{\partial}{\partial n} R(\delta,n) = 0$, where we treat n as a continuous variable. This defines n implicitly by the equation

$$g(|\delta|n^{1/2}) = N/(2n), \tag{3}$$

where by definition,

$$g(x) = (2\Phi(x) - 1)/x\phi(x) + 1 \quad \text{if } x > 0,$$
$$= 3 \quad \text{if } x = 0.$$

Since $|\delta|$ is unknown in practice, (3) suggests the stopping rule

$$T^* = \inf\{k: g(|s_k|/k^{1/2}) \geq N/(2k)\}. \tag{5}$$

The similar behavior of the risk functions for the rules T_A, T^* and the flat prior Bayes rule led us to consider in [7] a more general class of stopping rules for which we stated without proof some theorems concerning their asymptotic optimality. In Section 2 below we establish some properties of these stopping rules, and apply the results in Section 3 to prove the asymptotic optimality theorems announced in [7]. The rule of Begg and Mehta [2] and some other rules proposed earlier by Colton [4] do not belong to this class. Their asymptotic properties are studied in Section 4, where we also discuss why these rules are asymptotically suboptimal.

II. THE CONTINUOUS-TIME PROBLEM AND A CLASS OF STOPPING RULES

Let $\theta = \delta N^{1/2}$ and let $w(t)$, $t \geq 0$, denote the Wiener process with drift coefficient θ. Given any continuous function $f: (0,\frac{1}{2}] \to [0,\infty)$, define the stopping rules

$$T(f,N) = \inf\{k \leq N/2: |s_k| \geq N^{1/2} f(k/N)\} \ (\inf \phi = N/2), \tag{6}$$

and

$$\tau_{f,N} = \inf\{t: t = k/N \text{ for some } 1 \leq k \leq N/2 \text{ and } |w(t)| \geq f(t)\} \ (\inf \phi = \tfrac{1}{2}). \tag{7}$$

Since the sequences $\{s_k\}$ and $\{N^{1/2} w(k/N)\}$ have the same distribution, $T(f,N)$ has the same distribution as $N\tau_{f,N}$, and the regret $R(\delta, T(f,N))$ can be expressed in terms of the process $w(t)$ as

$$R(\delta, T(f,N)) = N^{1/2} \rho(\theta, \tau_{f,N}), \tag{8}$$

where we define for any stopping rule $\tau(\leq \frac{1}{2})$ for $w(t)$

$$\begin{aligned} \rho(\theta,\tau) &= \theta\, E\{\tau + (1 - 2\tau) I_{(w(\tau)<0)} | \theta\} \quad \text{if } \theta > 0, \\ &= |\theta|\, E\{\tau + (1 - 2\tau) I_{(w(\tau)>0)} | \theta\} \text{ if } \theta < 0. \end{aligned} \tag{9}$$

Here and in the sequel, we use $P(\cdot|\theta)$ to denote the probability measure under which $w(t)$ has drift coefficient θ and use $E(\cdot|\theta)$ to denote expectation with respect to $P(\cdot|\theta)$.

As $N \to \infty$, $\tau_{f,N}$ converges with probability 1 to

$$\tau(f) = \inf\{t \varepsilon (0,\tfrac{1}{2}]: \quad |w(t)| \geq f(t)\}. \tag{10}$$

Hence, for large N, we can approximate the original stopping problem involving $\{s_k: 1 \leq k \leq N/2\}$ by the continuous-time stopping problem involving $\{w(t): 0 < t \leq \frac{1}{2}\}$ after we standardize the original problem by the following scale changes:

$$t = k/N, \quad w(t) = s_k/N^{1/2}, \quad \theta = \delta N^{1/2}.$$

In connection with the Bayes rule with respect to the flat prior for δ, Chernoff and Petkau [3] considered the corresponding continuous-time problem of minimizing the Bayes risk $\int_{-\infty}^{\infty} \rho(\theta,\tau)d\theta$ among all stopping rules $\tau \leq \frac{1}{2}$ on the process $w(t)$. The minimizing τ is of the form $\tau(f_B)$ in (10), and they obtained the following asymptotic expansion of the Bayes boundary $f_B(t)$ as $t \to 0$:

$$f_B(t) = \{2t(\log \frac{1}{t} - \frac{1}{2} \log \log \frac{1}{t} - \frac{1}{2} \log 16\pi + o(1))\}^{1/2}. \tag{11}$$

Anscombe's rule T_A can be written in the form (6) with $f(t)$ equal to

$$f_A(t) = t^{1/2}\, \Phi^{-1}(1 - t),$$

and the boundary $f_A(t)$ has the following asymptotic expansion as $t \to 0$:

$$f_A(t) = \{2t(\log \frac{1}{t} - \frac{1}{2} \log \log \frac{1}{t} - \frac{1}{2} \log 4\pi + o(1))\}^{1/2}. \quad (12)$$

Likewise, the rule T* is also of the form (6) with f(t) equal to

$$f^*(t) = t^{1/2}\, g^{-1}(1/2t),$$

where g is defined in (4). Moreover, as $t \to 0$,

$$f^*(t) = \{2t(\log \frac{1}{t} + \frac{1}{2} \log \log \frac{1}{t} - \frac{1}{2} \log 4\pi + o(1))\}^{1/2}. \quad (13)$$

Let $\mathcal{C}$ denote the set of all bounded functions $f\colon (0,\frac{1}{2}] \to [0,\infty)$ satisfying the following two conditions:

$$f(t) \sim (2t \log \frac{1}{t})^{1/2} \quad \text{as } t \to 0, \quad (14)$$

and there exist $\eta < 3/2$ and $t_0 > 0$ such that

$$f(t) \geq \{2t(\log \frac{1}{t} - \eta \log \log \frac{1}{t})\}^{1/2} \text{ for all } 0 < t \leq t_0. \quad (15)$$

In view of (11), (12) and (13), the boundaries f^*, f_A and f_B all belong to $\mathcal{C}$. In the following two lemmas, we study certain basic properties of the stopping rules $\tau(f)$ and $T(f,N)$ when f belongs to $\mathcal{C}$.

Lemma 1

Let $f\colon (0,\frac{1}{2}] \to [0,\infty)$ be a bounded function.

(i) Suppose that f satisfies (14). Then as $\theta \to \infty$,

$$E(\tau(f)|\theta) \sim 2\theta^{-2} \log \theta^2. \quad (16)$$

Moreover, letting $\lambda(f) = \sup\{t\varepsilon(0,\frac{1}{2}] : |w(t)| \leq f(t)\}$, we also have

$$E(\lambda(f)|\theta) \sim 2\theta^{-2} \log \theta^2. \quad (17)$$

(ii) Suppose that there exist real numbers η and γ such that as $t \to 0$

$$f(t) = \{2t(\log \frac{1}{t} - \eta \log \log \frac{1}{t} - \gamma + o(1))\}^{1/2}. \tag{18}$$

Then as $\theta \to \infty$,

$$P\{w(\tau(f)) < 0|\theta\} \sim \frac{e^{\gamma}\ (\log \theta^2)^{\eta-1/2}}{2\pi^{1/2}\ \theta^2} \tag{19}$$

Proof. Let $w_0(t)$ denote the standard Brownian motion with zero drift. Then under $P(\cdot|\theta)$, $w(t) = w_0(t) + \theta t$. The proof of (i) is straightforward and is omitted. (ii) can be proved by using a modification of the likelihood ratio argument given in [8]. Here we give an alternative proof by applying a recent result of Jennen and Lerche [6] to show that for all $0 < t_0 \le \frac{1}{2}$ and λ, as $\theta \to \infty$,

$$P[w_0(t) < -\theta t - \{2t(\log \frac{1}{t} - \eta \log \log \frac{1}{t} - \lambda)\}^{1/2}$$
$$\text{for some } 0 < t \le t_0] \sim \frac{1}{2} e^{\lambda}\ \pi^{-1/2}\ \theta^{-2}(\log \theta^2)^{\eta-1/2}. \tag{20}$$

Since $w_0(t)$ has the same distribution as $-tw_0(1/t)$, it suffices to show that

$$P[w_0(s) > \theta + \{2s(\log s - \eta \log \log s - \lambda)\}^{1/2} \text{ for some}$$
$$s \ge 1/t_0] \sim \frac{1}{2} e^{\lambda}\ \pi^{-1/2}\ \theta^{-2}(\log \theta^2)^{\eta-1/2}. \tag{21}$$

Let $\psi(s) = \{2s(\log s - \eta \log \log s - \lambda)\}^{1/2}$. Then by Theorem 2 of [6], as $\theta \to \infty$, the probability in (21) is asymptotically equivalent to the integral

$$(2\pi)^{-1/2} \int_{1/t_0}^{\infty} s^{-3/2}(\theta + \psi(s) - s\psi'(s))e^{-(\theta+\psi(s))^2/2s}\, ds.$$

Setting $s = (\theta^2 \log \theta^2)/t$ in the above integral, it is easy to see that the integral is asymptotically equivalent to

$$\frac{1}{2} e^{\lambda} \pi^{-1/2} \theta^{-2} (\log \theta^2)^{\eta-1/2} \int_0^{t_0\theta^2 \log \theta^2} \exp(-(2t)^{1/2}) dt,$$

giving the desired conclusion (21). By a standard argument, it can be shown that (19) follows from (20).

In view of (11), (12) and (13), the boundaries f*, f_A and f_B all satisfy (18). Therefore, applying Lemma 1 (ii) to these boundaries, we obtain that as $\theta \to \infty$,

$$P\{w(\tau(f^*)) < 0|\theta\} \sim \theta^{-2} (\log \theta^2)^{-1},$$

$$P\{w(\tau(f_A)) < 0|\theta\} \sim \theta^{-2},$$

$$P\{w(\tau(f_B)) < 0|\theta\} \sim 2\theta^{-2}.$$

We now apply Lemma 1 to prove the following.

Lemma 2

Let $f \varepsilon C$. Then as $N \to \infty$,

$$P_\delta\{s_{T(f,N)} < 0\} = o((N\delta^2)^{-1} \log(N\delta^2)) \tag{22}$$

uniformly in $|\delta| \geq a_N N^{-1/2}$, and

$$E_\delta\, T(f,N) \sim 2\delta^{-2} \log(N\delta^2) \tag{23}$$

uniformly in $b_N(\log N)^{1/2} \geq |\delta| \geq a_N N^{-1/2}$, where a_N and b_N are arbitrary sequences of positive numbers such that $\lim_{N\to\infty} a_N = \infty$ and $\lim_{N\to\infty} b_N = 0$.

Proof. Let $\theta = N\delta^{1/2}$. Since $EN\tau(f) \leq ET(f,N) \leq E\{N\lambda(f) + 1\}$, we obtain (23) from Lemma 1 (i). Moreover, since

$$P_\delta\{s_{T(f,N)} < 0\} \leq P\{w(t) < -f(t) \text{ for some } 0 < t \leq \tfrac{1}{2}|\theta\}$$

and since f satisfies (15) with $\eta < 3/2$, we obtain (22) from (20).

III. ASYMPTOTIC OPTIMALITY OF THE CLASS $\mathcal{C}$ OF STOPPING BOUNDARIES

From Lemma 2, we obtain the following asymptotic behavior of the regret $R(\delta,T(f,N))$ when $f\varepsilon\mathcal{C}$.

Corollary 1. Let $f\varepsilon\mathcal{C}$. Then as $N \to \infty$,

$$R(\delta,\ T(f,n)) \sim (2 \log N\delta^2)/|\delta| \tag{24}$$

uniformly in $a_N\ N^{-1/2} \leq |\delta| \leq b_N(\log N)^{1/2}$, where a_N and b_N are arbitrary sequences of positive numbers such that $\lim_{N\to\infty} a_N = \infty$ and $\lim_{N\to\infty} b_N = 0$.

Suppose that $|\delta| \neq 0$ is known and consider the optimal fixed sample size n minimizing $R(\delta,n)$. For this minimizing value of n, say $n^* = n^*(|\delta|)$, it follows from (3) and (4) that

$$R(\delta,n^*) \sim (2 \log N)/|\delta| \qquad \text{as } N \to \infty. \tag{25}$$

Comparison of (25) with (24) shows that for every given $\delta \neq 0$, stopping rules of the form $T(f,N)$ with $f\varepsilon\mathcal{C}$ are asymptotically (as $N \to \infty$) as efficient as the optimal fixed sample size n^* which assumes $|\delta|$ known. The following theorm shows that for every given $\delta \neq 0$, the order of magnitude of the regret $R(\delta,T(f,N))$ given by (24) is asymptotically minimal.

Theorem 1. Let $a > 1$ and let δ_N be a sequence of positive constants such that $\delta_N \to 0$ and $\log \delta_N^{-1} = o\ (\log N)$ as $N \to \infty$. For every $N \geq 2$, let $\mathcal{J}_N$ denote the class of stopping rules $T \leq N/2$ such that $R(\delta_N,T) + R(-\delta_N,T) \leq (\log N)^a/\delta_N$. Let $f\varepsilon\mathcal{C}$. Then $T(f,N)\varepsilon\mathcal{J}_N$ for all large N in view of (24), and for every fixed $\delta \neq 0$.

$$\inf_{T\varepsilon\mathcal{J}_N} R(\delta,T) \sim (2 \log N)/|\delta| \sim R(\delta,\ T(f,N) \qquad \text{as } N \to \infty. \tag{26}$$

Proof. Let $T\epsilon J_N$. Since $R(\delta_N,T) \leq (\log N)^a/\delta_N$,

$$N\delta_N P_{\delta_N}\{s_T < 0\} \leq R(\delta_N,T) + \delta_N E_{\delta_N} T$$

$$\leq 2 R(\delta_N,T) \leq 2(\log N)^a/\delta_N.$$

Therefore

$$P_{\delta_N}\{s_T < 0\} \leq 2(\log N)^a/(N\delta_N^2), \tag{27a}$$

and likewise

$$P_{-\delta_N}\{s_T > 0\} \leq 2(\log N)^a/(N\delta_N^2). \tag{27b}$$

By Hoeffding's lower bound (cf. (1.4) of [5]) on the expected value of any stopping rule T satisfying the probability constraints (27a) and (27b),

$$E_\delta T \geq 2(|\delta| + \delta_N)^{-2}\{[\tfrac{1}{2}\delta_N^2(|\delta| + \delta_N)^{-2} + \log(N\delta_N^2/4(\log N)^a]^{1/2} - 2^{-1/2}\delta_N(|\delta| + \delta_N)^{-1}\}^2. \tag{28}$$

Since $R(\delta,T) \geq |\delta|E_\delta T$, $\delta_N \to 0$ and $\log \delta_N^{-1} = o(\log N)$, we obtain from (28) that for fixed $\delta \neq 0$,

$$\inf_{T\epsilon J_N} R(\delta,T) \geq (2 + o(1))|\delta|^{-1} \log N.$$

Combining this result with Corollary 1 then proves (26).

Stopping rules of the form $T(f,N)$ with $f\epsilon C$ are also asymptotically optimal from a Bayesian point of view. This is the content of

Theorem 2. Let G be a distribution function on the real line such that G' is positive and continuous in some neighborhood of the origin and $\int_{-\infty}^{\infty}|\delta|dG(\delta) < \infty$. Let $f\epsilon C$. Then as $N \to \infty$,

$$\int_{-\infty}^{\infty} R(\delta,T(f,N))dG(\delta) \sim G'(0)(\log N)^2 \sim \inf_T \int_{-\infty}^{\infty} R(\delta,T)dG(\delta),$$

where $\inf_T$ denotes infimum over all stopping rules $T \leq N/2$.

Proof. We first show that as $N \to \infty$,

$$\inf_T \int_{-\infty}^{\infty} R(\delta,T)dG(\delta) \geq (G'(0) + o(1))(\log N)^2. \tag{29}$$

Given $0 < \varepsilon < 1$, we can choose $\eta > 0$ such that $G'(x) \geq \varepsilon G'(0)$ for all $|x| \leq \eta$. Obviously we need only consider stopping rules T satisfying

$$\int_{-\infty}^{\infty} R(\delta,T)dG(\delta) \leq 2G'(0)(\log N)^2. \tag{30}$$

Let $0 < \gamma < \frac{1}{2}$ and $\eta \geq \delta > 0$. Since

$$NxP_x\{s_T < 0\} \leq R(x,T) + xE_xT \leq 2R(x,T)$$

for $x > 0$, it follows from (30) that

$$\begin{aligned} 4G'(0)(\log N)^2 &\geq 2\int_{\gamma\delta/2}^{\gamma\delta} R(x,T)G'(x)dx \\ &\geq \varepsilon G'(0) \int_{\gamma\delta/2}^{\gamma\delta} NxP_x\{s_T < 0\}dx \\ &\geq \frac{3}{8}\varepsilon N\gamma^2\delta^2 G'(0)(\inf_{\gamma\delta/2 \leq x \leq \gamma\delta} P_x\{s_T < 0\}). \end{aligned}$$

Therefore there exists x such that $\frac{1}{2}\gamma\delta \leq x \leq \gamma\delta$ and

$$P_x\{s_T < 0\} \leq 11(\log N)^2/(\varepsilon\gamma^2 N\delta^2). \tag{31a}$$

Likewise there exists y such that $-\gamma\delta \leq y \leq -\frac{1}{2}\gamma\delta$ and

$$P_y\{s_T > 0\} \leq 11(\log N)^2/(\varepsilon\gamma^2 N\delta^2). \tag{31b}$$

Hence, by Hoeffding's lower bound [5] on the expected value of any stopping rule satisfying the probability constraints (31a) and (31b),

$$\begin{aligned} E_\delta T \geq 2\{(1 + \gamma)\delta\}^{-2} \{[\log(\varepsilon\gamma^2 N\delta^2/22(\log N)^2)]^{1/2} \\ - 2^{-1/2}\gamma(1 + \tfrac{1}{2}\gamma)^{-1}\}^2. \end{aligned} \tag{32}$$

Similarly, the lower bound in (32) also holds for $E_{-\delta}T$. Therefore, as $N \to \infty$,

$$\inf\{\int_{-\infty}^{\infty} |\delta| E_\delta T \, dG(\delta) : T \text{ satisfies } (30)\}$$

$$\geq \frac{(2 + o(1)) \, \varepsilon \, G'(0)}{(1+\gamma)^2} \int_{N^{-1/2+\gamma} \leq |\delta| \leq \eta} |\delta|^{-1} \log(N\delta^2) d\delta$$

$$\sim \varepsilon G'(0)(1 - 4\gamma^2)(\log N)^2/(1+\gamma)^2.$$

Letting $\varepsilon \to 1$ and $\gamma \to 0$, we then obtain (29).

It remains to show that for every $f \varepsilon C$,

$$\int_{-\infty}^{\infty} R(\delta, T(f,N)) \, dG(\delta) \sim G'(0)(\log N)^2. \tag{33}$$

Let $a_N = (\log N)^{1/3}$. Then by Corollary 1, as $N \to \infty$,

$$\int_{a_N N^{-1/2} \leq |\delta| \leq a_N^{-1}} R(\delta, T(f,N)) dG(\delta)$$

$$\sim 2G'(0) \int_{a_N N^{-1/2} \leq |\delta| \leq a_N^{-1}} |\delta|^{-1} \log(N\delta^2) d\delta \tag{34}$$

$$\sim G'(0)(\log N)^2.$$

Moreover, by Corollary 1,

$$\int_{a_N^{-1} \leq |\delta| \leq 1} R(\delta, T(f,N)) dG(\delta) \leq (2 + o(1)) a_N \log N \tag{35}$$

$$= o((\log N)^2).$$

Since $R(\delta, T(f,N)) \leq N|\delta|$, it follows that

$$\int_{|\delta| \leq a_N N^{-1/2}} R(\delta, T(f,N)) dG(\delta)$$

$$\sim G'(0) \int_{|\delta| \leq a_N N^{-1/2}} R(\delta, T(f,N)) d\delta \tag{36}$$

$$\leq NG'(0) \int_{|\delta| \leq a_N N^{-1/2}} |\delta| \, d\delta = a_N^2 \, G'(0) = o((\log N)^2).$$

As shown in the proof of Lemma 2 (where, in particular, we made use of the inequality $ET(f,n) \leq NE\lambda(f) + 1$), as $N \to \infty$,

$$R(\delta, T(f,N)) \leq |\delta| + (2 + o(1))|\delta|^{-1}(\log N + \log \delta^2)$$

uniformly in $|\delta| \geq a_N N^{-1/2}$. Therefore

$$\int_{|\delta|\geq 1} R(\delta, T(f,N))dG(\delta) \leq (2 + o(1))\log N. \tag{37}$$

From (34), (35), (36), and (37), the desired conclusion (33) follows.

IV. SOME ASYMPTOTICALLY SUBOPTIMAL STOPPING RULES

The conditions (14) and (15) defining the class $\mathcal{C}$ of stopping boundaries are in some sense minimal for rules of the form $T(f,N)$ to be asymptotically optimal, as shown by the following.

Theorem 3. Let f be a bounded nonnegative function on $(0, \frac{1}{2}]$.

(i) Suppose that $\limsup_{t \to 0} f(t)/(t \log \frac{1}{t})^{1/2} = \gamma < 2^{1/2}$. Then for every $\delta \neq 0$,

$$R(\delta, T(f,N)) \geq N^{1-\gamma^2/2 + o(1)} \quad \text{as } N \to \infty. \tag{38}$$

Moreover, as $N \to \infty$ and $\delta \to 0$ such that $N^{1/2}|\delta| \to \infty$,

$$R(\delta, T(f,N)) \geq |\delta|^{-1}(N\delta^2)^{1-\gamma^2/2 + o(1)}. \tag{39}$$

(ii) Suppose that $\liminf_{t \to 0} f(t)/(t \log \frac{1}{t})^{1/2} = \lambda > 2^{1/2}$. Then for every $\delta \neq 0$,

$$R(\delta, T(f,N)) \geq (\lambda^2|\delta|^{-1} + o(1)) \log N \quad \text{as } N \to \infty. \tag{40}$$

Moreover, as $N \to \infty$ and $\delta \to 0$ such that $N^{1/2}|\delta| \to \infty$,

$$R(\delta, T(f,N)) \geq (\lambda^2 + o(1))|\delta|^{-1} \log(N\delta^2). \tag{41}$$

(iii) Suppose that $f(t) \leq \{2t[\log \frac{1}{t} - (\eta + o(1)) \log\log \frac{1}{t}]\}^{1/2}$ as $t \to 0$ for some $\eta > 3/2$. Then as $N \to \infty$ and $\delta \to 0$ such that $\delta \log N \to 0$ and $N^{1/2}|\delta| \to \infty$,

$$R(\delta, T(f,N)) \geq |\delta|^{-1}(\log N\delta^2)^{\eta-1/2+o(1)}. \tag{42}$$

Proof. To prove (38), letting $T = T(f,N)$, we note that for $\delta > 0$,

$$P_\delta\{T = 1, s_T < 0\} = P_\delta\{z_1 < -N^{1/2}f(1/N)\}$$

$$\geq \exp\{-(\tfrac{1}{2}\gamma^2 + o(1))\log N\},$$

and therefore

$$E_\delta\{(N - 2T)I_{(s_T<0)}\} \geq N^{1-\frac{1}{2}\gamma^2+o(1)} \quad \text{as } N \to \infty.$$

To prove (39), we make use of a modification of the likelihood ratio argument given in [8] to obtain that for $\delta > 0$,

$$P_\delta\{T \leq \delta^{-2}, s_T < 0\}$$

$$= \int_{-\infty}^{\infty} (2\pi)^{-1/2}\{\int_{(T\leq\delta^{-2},\ s_T<0)} T^{1/2} e^{-(s_T-\delta T)^2/2T} dP_\mu\}d\mu$$

$$\geq (2\pi)^{-1/2}\int_{-2\gamma\delta(\log N\delta^2)^{1/2}}^{-\gamma\delta(\log N\delta^2)^{1/2}} \{\int_{(T\leq\delta^{-2},s_T<0)} T^{1/2} e^{-(s_T-\delta T)^2/2T} dP_\mu\}d\mu \tag{43}$$

$$\geq \exp\{-(\tfrac{1}{2}\gamma^2 + o(1))\log N\delta^2\}$$

as $N \to \infty$ and $\delta \to 0$ such that $N^{1/2}|\delta| \to \infty$. Since $\delta^{-2} = o(N)$, it then follows that

$$|\delta|\ E_\delta\{(N- T)I_{(s_T<0)}\} \geq |\delta|^{-1}(N\delta^2)^{1-\gamma^2/2+o(1)}.$$

To prove (ii), a standard argument as in the proof of Lemma 2 shows that as $N \to \infty$,

$$E_\delta\, T(f,N) \geq (\lambda^2 + o(1))\delta^{-2} \log(N\delta^2),$$

either for fixed $\delta \neq 0$ or for $\delta \to 0$ such that $N^{1/2}|\delta| \to \infty$.

The proof of (iii) makes use of a similar argument as in (43). As $N \to \infty$ and δ:0 such that $N\delta^2 \to \infty$ and $\delta \log N\delta^2 \to 0$,

$$P_\delta \{T \leq (\delta^2 \log N\delta^2)^{-1},\ s_T < 0\}$$

$$\geq (2\pi)^{-1/2} \int_{-2\delta \log N\delta^2}^{-2^{1/2}\delta \log N\delta^2} [\int_{\{T\leq(\delta^2 \log N\delta^2)^{-1}, s_T<0\}} T^{1/2} e^{-(s_T-\delta T)^2/2T}\, dP_\mu] d\mu$$

$$\geq (N\delta^2)^{-1/2} (\log N\delta^2)^{\eta-1/2+o(1)},$$

and therefore the desired conclusion (42) holds.

Assuming a prior distribution on δ, Begg and Mehta *[2]* proposed to stop sampling when no fixed additional time of observation would reduce the posterior risk. If we assume a flat prior and allow stopping at non-integral times, then the Begg - Mehta rule is of the form T(f,N), where

$$f(t) = \{2t(\log \frac{1}{t} - \frac{5}{2} \log \log \frac{1}{t} - \frac{1}{2} \log 4\pi - 4 + o(1))\}^{1/2} \qquad (44)$$

As $t \to 0$ (cf. [3], p. 131). Thus, this rule is an example of the asymptotically suboptimal rules of Theorem 3 (iii) with $\eta = 5/2$. Moreover, if G is a distribution function on the real line such that G' is positive and continuous in some neighborhood of the origin and $\int_{-\infty}^{\infty} |\delta| dG(\delta) < \infty$, then the argument of Theorem 2 and Theorem 3 (iii) show that the integrated risk $\int_{-\infty}^{\infty} R(\delta, T(f,N))dG(\delta)$ is of the order of magnitude of $(\log N)^3$ (instead of $(\log N)^2$ in Theorem 2) when f satisfies (44).

Colton *[4]* proposed to use horizontal stopping boundaries based on truncated sequential probability ratio tests. These stopping rules are of the form

$$T_C = \inf\{k \le N/2 : |s_k| \ge C\}. \quad (45)$$

Ignoring the effects of truncation and overshoot, Colton applied Wald's approximations to the error probabilities and expected sample size for the sequential probability ratio test to obtain the approximation

$$R(\delta, T_C) \doteq \frac{N|\delta|}{e^{2C|\delta|}+1} + C\left(\frac{e^{2C|\delta|}-1}{e^{2C|\delta|}+1}\right)^2 . \quad (46)$$

He suggested using the minimax value $C = .4131\ N^{1/2}$ for the approximate risk function given by (46). Such stopping rule can therefore be written in the form $T(f_d,N)$, where

$$f_d(t) = d \text{ for } 0 < t \le \frac{1}{2}, \quad (47)$$

and $d = .4131$ for Colton's minimax rule. The boundary f_d in (47) is an example of Theorem 3 (ii) with $\lambda = \infty$. Moreover, as $N \to \infty$,

$$R(\delta, T(f_d,N)) \sim dN^{1/2}$$

uniformly in $a_N N^{-1/2} \le |\delta| \le b_N N^{1/2}$, where a_N and b_N are arbitrary sequences of positive numbers such that $\lim_{N\to\infty} a_N = \infty$ and $\lim_{N\to\infty} b_N = 0$. Therefore the Bayes risk of $T(f_d,N)$ with respect to any prior distribution G on δ such that $\int_{-\infty}^{\infty} |\delta|\ dG(\delta) < \infty$ is of the order of magnitude of $N^{1/2}$.

Colton's minimax rule chooses C of the order of magnitude of $N^{1/2}$. On the other hand, if $|\delta|$ is known, then the value of C which minimizes the right hand side of (46) satisfies the asymptotic relation

$$C \sim \frac{1}{2}|\delta|^{-1} \log(N\delta^2) \quad \text{as } N \to \infty \quad \text{and } N\delta^2 \to \infty. \quad (48)$$

Although $|\delta|$ is unknown in practice, we can try to estimate it sequentially, as we have done before in the case of the rule T^*.

Hence, in ignorance of $|\delta|$, (45) and (48) suggest the stopping rule

$$\tilde{T} = \inf\{k \leq N/2 \ : \ s_k^2/k \geq \frac{1}{2} \log(N/k)\}. \tag{49}$$

The rule $\tilde{T}$, however, is an example of the asymptotically sub-optimal rules of Theorem 3(i) with $\gamma = 2^{-1/2}$.

Colton also considered Bayesian choice of C with respect to a prior normal distribution on δ. More generally, letting h be a bounded continuous density function such that $h(0) > 0$ and $\int_{-\infty}^{\infty} |\delta|\ h(\delta)d\delta < \infty$, it can be shown that the value of C which minimizes the integral

$$\int_{-\infty}^{\infty} \{\frac{N|\delta|}{e^{2C|\delta|}+1} + C(\frac{e^{2C|\delta|}-1}{e^{2C|\delta|}+1})^2\}h(\delta)d\delta$$

satisfies the asymptotic relation

$$C^3 \sim 2Nh(0) \int_{-\infty}^{\infty} x^2 e^{2|x|}(1 + e^{2|x|})^{-2}dx. \tag{50}$$

With this choice of C, the Bayes risk of the stopping rule T_C is of the order of magnitude of $N^{1/3}$, which is much larger than the order of $(\log N)^2$ for the rules $T(f,N)$ with $f \in \mathcal{C}$.

V. REFERENCES

[1] Anscombe, F.J. (1963). "Sequential Medical Trials." *J. Amer. Statist. Assoc. 58,* 365-383.

[2] Begg, C.B. and Mehta, C.R. (1979). "Sequential Analysis of Comparative Clinical Trials." *Biometrika 66,* 97-103.

[3] Chernoff, H. and Petkau, A.J. (1981). "Sequential Medical Trials Involving Paired Data." *Biometrika 68,* 119-132.

[4] Colton, T. (1963). "A Model for Selecting One of Two Medical Treatments." *J. Amer. Statist. Assoc. 58,* 388-400.

[5] Hoeffding, W. (1960). "Lower Bounds for the Expected Sample Size and Average Risk of a Sequential Procedure." *Ann. Math. Statist. 31,* 352-368.

[6] Jennen, C. and Lerche, H.R. (1981). "First Exit Densities of Brownian Motion Through One-Sided Moving Boundaries." *Z. Wahrscheinlichkeitstheorie verw. Gebiete 55,* 133-148.

[7] Lai, T.L., Levin, B., Robbins, H., and Siegmund, D. (1980). "Sequential Medical Trials." *Proc. Nat. Acad. Sci. U.S.A. 77,* 3135-3138.

[8] Lai, T.L. and Siegmund, D. (1977). "A Nonlinear Renewal Theory with Applications to Sequential Analysis I. *Ann. Statist.* 5, 946-954.

THE CRAMÉR-RAO INEQUALITY HOLDS ALMOST EVERYWHERE

Gordon Simons[1]

Department of Statistics
University of North Carolina
Chapel Hill, North Carolina

Michael Woodroofe[2]

Department of Statistics
University of Michigan
Ann Arbor, Michigan

I. INTRODUCTION

In its simplest form, the Cramer-Rao inequality asserts: under regularity conditions, the variance of an unbiased estimator of a parametric function is at least the square of the derivative of that function divided by the Fisher information in the sample for all values of the unknown parameter. Unfortunately, the regularity conditions may be severe. In early work, the regularity conditions placed restrictions on the estimator itself, which is undesirable if the goal is to obtain an inequality which must be satisfied by all estimators. Subsequently, Hannan and Fabian (1977) have avoided this problem by requiring that the likelihood ratios be differentiable in mean square.

[1] *Research supported by the National Science Foundation under MCS-8100748*
[2] *Research supported by the National Science Foundation under MCS-8101897*

ISBN 0-12-589320-5

In the sequential case, Wolfowitz (1947) has shown that the Cramér-Rao inequality holds with the information in the sample equal to the information in a single observation times the expected sample size. His regularity conditions seem especially severe. Moreover, Simons (1980), while expressing the opinion that Wolfowitz' conditions are too restrictive, has shown that some conditions are necessary by providing an example of an unbiased sequential estimator of a normal mean for which the Cramér-Rao inequality is violated at a single point.

Here it is shown that the Cramér-Rao inequality holds for almost every value of the parameter (Lebesgue) under the sole condition that information be definable; and even the definition of information, equation (6), is more general than previous ones. This result is detailed in Theorem 1 and Corollaries 1 and 2 in Section II. In Section III, an a.e. version of the Cramér-Rao inequality is established for sequential estimators, under the condition that the information in a single observation be positive. See Theorem 2 and Corollaries 3 and 4.

Similar results have been previously announced by Ibragimov and Has'minskii (1972). In particular, the interesting inequality in Lemma 1 may be found there. Ibragimov and Has'minskii did not include proofs of their results, however, and imposed stronger conditions than those required here. Moreover, the method of proof used here differs from that apparently intended by the earlier authors. By using approximate limits and derivatives, in place of real ones, the present proofs avoid requiring that the densities be absolutely continuous in the parameter or that the variance of the estimator be locally integrable. Connections between the present paper, that of Ibragimov and Has'minskii (1972), and Pitman's (1979) monograph are discussed in more detail in Section IV.

While the original motivation for studying the Cramér-Rao inequality--finding minimum variance unbiased estimators--no longer seems compelling, new applications have been found. Blythe and Roberts (1972) survey uses of the Cramér-Rao and related inequalities in admissibility proofs; and Alvo (1977) has recently used the sequential Cramér-Rao inequality to obtain lower bounds on the Bayes risk of a sequential estimator. Alvo's application only required the Cramér-Rao inequality to hold a.e.; and a.e. versions of the Cramér-Rao inequality may be applicable to prove almost admissibility.

II. THE FIXED SAMPLE SIZE CASE

Let $(\mathcal{X},\mathcal{B})$ denote a measurable space; let Ω denote an open subinterval of $(-\infty,\infty)$; and let F_ω, $\omega \in \Omega$, be a dominated family of probability measures on $\mathcal{B}$, say

$$dF_\omega = f_\omega d\mu, \qquad \omega \in \Omega , \tag{1}$$

where μ is a sigma-finite measure on $\mathcal{B}$. Denote the affinity between F_θ and F_ω by $\rho(\theta,\omega)$ and the Hellinger distance of F_θ from F_ω by $h(\theta,\omega)$, so that

$$\begin{aligned} h^2(\theta,\omega) &= \int (\sqrt{f_\omega} - \sqrt{f_\theta})^2 d\mu = 2 - 2 \int \sqrt{(f_\theta f_\omega)}\, d\mu \\ &= 2(1 - \rho(\theta,\omega)), \quad \theta \quad \omega \in \Omega. \end{aligned} \tag{2}$$

It is assumed throughout that the sigma algebra $\mathcal{B}$ is countably generated[3] and that the family F_ω, $\omega \in \Omega$, is measurable in the sense that $F_\omega\{B\}$ depends measurably on ω for each $B \in \mathcal{B}$. It follows that there are versions of the densities for which $f_\omega(x)$ is jointly measurable in (ω,x). See, for example, Mackey (1949).

Next, let γ denote an (unbiasedly) estimable, real valued (finite) function on Ω, and let $\hat{\gamma}$ denote an unbiased estimator. Thus,

[3] *This represents no essential loss of generality, since one may always restrict attention to a sufficient sigma algebra which is countably generated. See Neveu (1965, p. 124).*

$$\int \hat{\gamma}\, dF = \gamma(\omega), \qquad \omega \in \Omega. \tag{3}$$

Of course, (3) includes the assumption that $\hat{\gamma}$ be integrable with respect to F_ω for each $\omega \in \Omega$. The variance of $\hat{\gamma}$ is denoted by

$$\sigma^2(\omega) = \int [\hat{\gamma}(x) - \gamma(\omega)]^2 dF_\omega(x) \leq \infty, \qquad \omega \in \Omega. \tag{4}$$

Lemma 1

For all $\theta, \omega \in \Omega$ *with* $h(\theta,\omega) > 0$,

$$\frac{1}{2}[\sigma^2(\theta) + \sigma^2(\omega)] \geq \frac{[\gamma(\omega) - \gamma(\theta)]^2}{4h^2(\theta,\omega)} - \frac{1}{4}[\gamma(\omega) - \gamma(\theta)]^2. \tag{5}$$

Proof. For $\theta, \omega \in \Omega$ and $-\infty < c < \infty$,

$$\begin{aligned}[\gamma(\omega) - \gamma(\theta)]^2 &= \{\int(\hat{\gamma} - c)(f_\omega - f_\theta)d\mu\}^2 \\ &= \{\int(\hat{\gamma} - c)(\sqrt{f_\omega} + \sqrt{f_\theta})(\sqrt{f_\omega} - \sqrt{f_\theta})d\mu\}^2 \\ &\leq \int(\hat{\gamma} - c)^2(\sqrt{f_\omega} + \sqrt{f_\theta})^2 d\mu\, h^2(\theta,\omega) \\ &\leq \{2\int(\hat{\gamma} - c)^2 f_\omega d\mu + 2\int(\hat{\gamma} - c)^2 f_\theta d\mu\}h^2(\theta,\omega).\end{aligned}$$

by Schwarz' inequality and some simple algebra. When $c = [\gamma(\omega) + \gamma(\theta)]/2$, the term in braces is $2\sigma^2(\theta) + 2\sigma^2(\omega) + [\gamma(\omega) - \gamma(\theta)]^2$; and the lemma follows by substitution.

Lemma 1 may be exploited in several ways. The one detailed below leads to quite general results. Some other ways are explored in Section IV.

There is natural interest in letting $\omega \to \theta$ in (5). In this process, it is convenient to use approximate limits, as defined by Saks (1937, pp. 218-220), for example. Recall that a point x is a point of density (respectively, dispersion) of a measurable set $B \subset (-\infty,\infty)$ iff $\lim m(BI)/m(I) = 1$ (respectively, 0) as $I \downarrow x$ through open intervals, where m denotes Lebesgue measure. Then a.e. $x \in B$ is a point of density of B. If g is an extended real valued, measurable function on $(-\infty,\infty)$ and if $-\infty < x < \infty$, then ap lim $\sup_{y \to x}$ g(y) is defined to be the infimum of the set of z for which x is a point of dispersion of {y : g(y) > z}; and ap lim $\inf_{y \to x}$ g(y) is defined similarly. If two extended real valued functions g and h are equal on a measurable set $B \subset (-\infty, \infty)$, then ap lim $\sup_{y \to x}$ g(y) = ap lim $\sup_{y \to x}$ h(y) and ap lim $\inf_{y \to x}$ g(x) = ap lim $\inf_{y \to x}$ h(y) for every point of density x of B. In particular, these relations hold for a.e. $x \in B$. See Saks (1937, p. 219). If ap lim $\inf_{y \to x}$ g(y) = ap lim $\sup_{y \to x}$ g(y), possibly infinite, then g is said to have an approximate limit as $y \to x$, and the common value is denoted by ap $\lim_{y \to x}$ g(y). It may be shown that ap $\lim_{y \to x}$ g(x) = L iff there is a measurable set B which has x as a point of density and through which $\lim_{y \to x}$ g(x/ = L in the usual sense. See Geman and Horowitz (1980). A function g is said to be approximately continuous at x iff g(x) = ap $\lim_{y \to x}$ g(y), possibly infinite. Approximate limits enjoy many of the properties of ordinary limits. In particular, the following is needed: if g has an approximate limit at x and if h is any extended real valued measurable function, the ap lim $\sup_{y \to x}$ g(y)h(y) = [ap $\lim_{y \to x}$ g(y)][ap lim $\sup_{y \to x}$ h(y)] provided that the latter produce is not $0 \cdot \infty$. It is clear that lim inf $\leq$ ap lim inf $\leq$ ap lim sup $\leq$ lim sup; so, the approximate limits reduce to ordinary ones, when the latter exist.

The following important theorem is proved by Saks (1937, p. 132): a measurable, real valued (finite) function g is approximately continuous at a.e. x, $- < x < \infty$. In fact, the requirement that g be finite may be dropped, by considering arctan(g).

Using approximate limits, one may define approximate derivatives. A measurable, real valued function g is said to be approximately differentiable at x iff g'(x) = ap $\lim_{y \to x}$ [g(y) - g(x)]/(y - x) exists. It is known that any measurable, real valued function g is approximately differentiable at a.e. x for which ap lim $\sup_{y \to x}$ $|g(y) - g(x)|/|y - x| < \infty$. See Saks (1937, pp. 295-297).

Upper and lower information numbers are defined by

$$i^*(\theta) = \operatorname{ap} \limsup_{\omega \to \theta} 4h^2(\theta,\omega)/(\theta - \omega)^2 \tag{6a}$$

and

$$i_*(\theta) = \operatorname{ap} \liminf_{\omega \to \theta} 4h^2(\theta,\omega)/(\theta - \omega)^2 , \tag{6b}$$

and when $i^*(\theta) = i_*(\theta)$, the common value is called the information and is denoted by $i(\theta)$. These definitions are not as arbitrary as might appear. If the square roots of the densities $s_\omega = \sqrt{f_\omega}$ are differentiable in mean square at $\omega = \theta$, then $\dot{s}_\theta = \frac{1}{2} \dot{f}_\theta/\sqrt{f_\theta}$, where $\dot{f}_\theta$ is the derivative in the mean, and

$$i(\theta) = \lim_{\omega \to \theta} \frac{4h^2(\theta,\omega)}{(\omega-\theta)^2} = \int 4\dot{s}_\theta^2 d\mu , \tag{7}$$

which is equivalent to the conventional definition of Fisher information, under additional smoothness conditions. Conversely, it is shown in Appendix B that (7) holds with $\dot{s}_\theta$ equal to an approximate mean square derivative at a.e. θ for which ap lim $\sup_{\omega \to \theta}$ $h^2(\omega,\theta)/(\omega - \theta)^2 < \infty$. Thus, $i(\theta)$ provides a natural extension of the definition of Fisher information. In fact, Pitman (1979) argues that the conventional definition $\int f'^2_\theta/f_\theta d\mu$ is only appropriate when s_ω, $\omega \in \Omega$, are differentiable in mean square. See Section IV.

Lemma 2

Let γ be a real valued, estimatle function on Ω; let $\hat{\gamma}$ be an unbiased estimator; and let σ^2 denote the variance of $\hat{\gamma}$. Then γ is approximately differentiable at a.e. θ for which $\sigma^2(\theta) < \infty$ and $i^*(\theta) < \infty$.

Proof. Since σ^2 and γ are approximately continuous a.e.,

$$\operatorname{ap}\limsup_{\omega \to \theta}\left[\frac{\gamma(\omega) - \gamma(\theta)}{\omega - \theta}\right]^2 \le \operatorname{ap}\limsup_{\omega \to \theta}\left\{\frac{1}{2}\sigma^2(\theta) + \frac{1}{2}\sigma^2(\omega)\right\}\frac{4h^2(\theta,\omega)}{(\omega-\theta)^2}$$

$$= \sigma^2(\theta)i^*(\theta) < \infty,$$

for a.e. $\theta \in \Omega$ for which $\sigma^2(\theta) < \infty$ and $i^*(\theta) < \infty$.

Theorem 1. Let γ be a real valued, estimable function on Ω; and let Ω_0 be the set of θ at which γ is approximately differentiable. If $\hat{\gamma}$ is an unbiased estimator then

$$\sigma^2(\theta) \ge \gamma'(\theta)^2/i_*(\theta), \qquad \text{for a.e. } \theta \in \Omega_0,$$

where γ' denotes the approximate derivative of γ, σ^2 denotes the variance of $\hat{\gamma}$, and 0/0 is to be interpreted as 0.

Proof. Since σ^2 is approximately continuous a.e.,

$$\sigma^2(\theta) = \operatorname{ap}\limsup_{\omega \to \theta} \frac{1}{2}[\sigma^2(\theta) + \sigma^2(\omega)]$$

$$\ge \operatorname{ap}\limsup_{\omega \to \theta}\left[\frac{\gamma(\omega) - \gamma(\theta)}{\omega - \theta}\right]^2 \Big/ 4\left[\frac{h(\theta,\omega)}{\omega - \theta}\right]^2 = \gamma'(\theta)^2/i_*(\theta)$$

for a.e. $\theta \in \Omega_0$.

Corollary 1. Suppose that $i(\theta) = i_*(\theta) = i^*(\theta) < \infty$ for a.e. $\theta \in \Omega$. Then

$$\sigma^2(\theta) \ge \gamma'(\theta)^2/i(\theta) \qquad \text{for a.e. } \theta \in \Omega. \tag{8}$$

where 0/0 = 0 and $\gamma'(\theta)^2$ is to be interpreted as ∞ for $\theta \notin \Omega_0$.

The corollary follows directly from the theorem and Lemma 2.

Theorem 1 may be applied to samples of arbitrary size. Let F_ω, $\omega \in \Omega$, be as in (1); let $n \ge 1$ be an integer, and let

ρ_n, h_n, i^*_n, and i_{n*} denote affinity, Hellinger distance, and upper and lower information for the family $F^n_\omega = F_\omega \times \cdots \times F_\omega$, $\omega \in \Omega$, of product measures. Then

$$\rho_n(\theta,\omega) = \rho(\theta,\omega)^n, \qquad \theta,\ \omega \in \Omega,$$

$$h^2_n(\theta,\omega) = 2[1 - \rho(\theta,\omega)^n], \qquad \theta,\ \omega \in \Omega, \tag{9}$$

$$i^*_n(\theta) = ni^*(\theta), \text{ and } i_{n*}(\theta) = ni_*(\theta), \qquad \theta \in \Omega.$$

From these simple remarks, it is easy to deduce the following corollary.

Corollary 2. Suppose that $i(\theta)$ exists and is finite for a.e. $\theta \in \Omega$; let $X_1,\ldots,X_n$ be i.i.d. with common distribution F_ω for some unknown $\omega \in \Omega$ and let $\hat{\gamma}_n = \hat{\gamma}_n(X_1,\ldots,X_n)$ be an unbiased estimator of a real valued function γ. Then

$$\sigma^2_n(\theta) \underset{\text{def}}{=} E_\theta[(\hat{\gamma}_n - \gamma(\theta))^2] \geq \gamma'(\theta)^2/ni(\theta) \quad \text{for a.e. } \theta \in \Omega,$$

where $0/0 = 0$ and $\gamma'(\theta)^2 = \infty$ for $\theta \notin \Omega_0$.

Example 1. Let $X = (-\infty,\infty)$ and let f be a density (w.r.t. Lebesgue measure) for which $s = \sqrt{f}$ is absolutely continuous and

$$0 < i_f = \int_{-\infty}^{\infty} 4s'^2 dx < \infty;$$

and let

$$f_\theta(x) = f(x - \theta), \qquad -\infty < x,\ \theta < \infty.$$

Then Corollary 2 is applicable, with $i(\theta) = i_f$, $-\infty < \theta < \infty$, even if the support of f is a finite interval. In the latter case f_ω/f_θ is not integrable for any choice of $\theta \neq \omega$; so, the results of Fabian and Hannan (1977) are not applicable, although a suggested extension may be. Likewise, the general results of Chapman and Robbins (1951) are not applicable, nor is the recent result of Pitman (1979, p. 34).

A specific example is provided by letting $f(x) = \frac{3}{2}(1 - |x|)^2$ for $-1 \le x \le 1$ and zero for other values of x, in which case $i_f = 12$.

Example 2. Here is an example of a smooth family of densities for which the Cramér-Rao inequality fails at a single point. Let $a > 0$; let ϕ denote the standard normal density; let $f_0 = \phi$; and let

$$f_\theta(x) = \frac{a}{a+\theta^2}\phi(x - \theta) + \frac{\theta^2}{a+\theta^2}\phi(x - \frac{1}{\theta})$$

for $-\infty < x < \infty$ and $\theta \ne 0$. Thus, f_θ is a weighted average of the $N(\theta,1)$ density and the $N(\frac{1}{\theta},1)$ density for $\theta \ne 0$. For small $|\theta| > 0$, the latter is much more informative than the former, but much less weight is attached to it. Let $\hat{\gamma}(x) = x$, $-\infty < x < \infty$, and let $\gamma(\theta) = E_\theta(X)$, $-\infty < \theta < \infty$. Then straightforward calculations yield

$$\gamma'(0) = 1 + \frac{1}{a}, \quad \sigma^2(0) = 1, \text{ and } i(0) = 1 + \frac{4}{a};$$

and it is easily checked that $\sigma^2(\theta) < \gamma'(0)^2/i(0)$ if $a < \frac{1}{2}$.

There are two interesting features to this example. First, the square roots $s_\theta = \sqrt{f_\theta}$ are not differentiable in mean square at $\theta = 0$. In fact, $\int(\dot{f}_0^2/f_0)dx = 1$, so that the conventional and extended definitions of information differ. Second, the variance of γ is not continuous at $\theta = 0$; for $\sigma^2(\theta) = (a + 1)(a + \theta^4)/(a + \theta^2)^2$ and $\lim_{\theta\to 0} \sigma^2(\theta) = 1 + \frac{1}{a}$.

Example 3. Example 2 can be extended to produce countably infinite violation sets as follows. Let N be any countable subset of the real line (dense or not). In Appendix A, a function α is defined on $(-\infty,\infty)$ with the properties (i) $\alpha(\theta) = 0$, $\theta \in N$, and (ii) $\alpha'(\theta) = 1$, $\theta \in N$. Let

$$f_\theta(x) = \begin{cases} \phi(x-\theta) & : \quad \alpha(\theta) = 0 \\ \dfrac{a}{a+\alpha^2(\theta)}\,\phi(x-\theta) + \dfrac{\alpha^2(\theta)}{a+\alpha^2(\theta)}\,\phi\left[x-\theta-\dfrac{1}{\alpha(\theta)}\right] & \\ & : \quad \alpha(\theta) \neq 0, \end{cases}$$

$-\infty < x < \infty$; and let $\hat{\gamma}(x) = x$ and $\gamma(\theta) = E_\theta(X)$ for $-\infty < x, \theta < \infty$. Then on N, $\gamma'(\theta) = 1 + a^{-1}$, $\sigma^2(\theta) = 1$, and $i(\theta) = 1 + 4a^{-1}$; so the Cramér-Rao inequality is violated on N when $0 < a < \frac{1}{2}$. The information $i(\theta)$ and $\gamma'(\theta)$ are defined, as well, for a.e. θ not in N.

III. THE SEQUENTIAL CASE

Let F_ω, $\omega \in \Omega$, be as in (1). Further, let $X_1, X_2, \ldots$ be i.i.d. with common distribution F_ω under a probability measure P_ω for each $\omega \in \Omega$; let $A_n = \sigma(X_1, \ldots, X_n)$ be the sigma-algebra generated by $X_1, \ldots, X_n$ for $n \geq 1$; and let t be a finite stopping time with respect to A_n, $n \geq 1$. Thus it is assumed that $\{t = n\} \in A_n$ for all $n \geq 1$, and $P_\omega\{t < \infty\} = 1$ for all $\omega \in \Omega$. Let A_t be the sigma-algebra of events which are determined by time t, the class of events A for which $A\{t = n\} \in A_n$ for all $n \geq 1$. It is no restriction to assume that the measure μ which dominates the family F_ω, $\omega \in \Omega$, is a probability measure, and that there is a probability measure Q under which $X_1, X_2, \ldots$ are i.i.d. with common distribution μ. Let P_ω^t denote the restriction of P_ω to A_t for $\omega \in \Omega$, and let Q^t be the corresponding restriction of Q. Then the measures P_ω^t, $\omega \in \Omega$, are dominated by Q^t with densities (likelihood ratios)

$$g_\omega(X_1, \ldots, X_t) = \frac{dP_\omega^t}{dQ^t} = \prod_{i=1}^{t} f_\omega(X_i), \qquad \omega \in \Omega;$$

on $t < \infty$. When $t = \infty$ (which may happen with positive Q-probability), all of the above likelihood ratios are zero. Hellinger distance and information for the family P_ω^t, $\omega \in \Omega$, are denoted by $H(\theta,\omega)$ and $I(\theta)$ for $\theta, \omega \in \Omega$.

For $\theta, \omega \in \Omega$, let

$$Y_k = Y_k(\theta,\omega) = \sqrt{[f_\theta(X_k) f_\omega(X_k)]}, \qquad k \geq 1,$$

$$Z_n = Z_n(\theta,\omega) = \prod_{k=1}^{n} Y_k(\theta,\omega), \qquad n \geq 1,$$

$$Z_0 = 1,$$

and

$$C(\theta,\omega) = \sum_{k=1}^{\infty} \int_{t \geq k} Z_{k-1}\, dQ, \qquad \theta, \omega \in \Omega.$$

Lemma 3

$$H^2(\theta,\omega) = h^2(\theta,\omega) C(\theta,\omega), \qquad \theta \neq \omega.$$

Proof. Since $h^2(\theta,\omega) = 2E(1 - Y_1)$ and $H^2(\theta,\omega) = 2E(1 - Z_t 1_{t<\infty})$, where "E" refers to expectation under Q, it must be shown that $E(1 - Z_t 1_{t<\infty}) = E(1 - Y_1)\, C(\theta,\omega)$. Let $t \wedge m = \min(t,m)$ for $m \geq 1$; then

$$E(1 - Z_{t\wedge m}) = \sum_{n=1}^{m} \int_{t\wedge m=n} (1 - Z_n)\, dQ$$

$$= \sum_{n=1}^{m} \sum_{k=1}^{n} \int_{t\wedge m=n} Z_{k-1}(1 - Y_k)\, dQ$$

$$= \sum_{k=1}^{m} \int_{t\geq k} Z_{k-1}(1 - Y_k)\, dQ$$

$$= E(1 - Y_1) \sum_{k=1}^{m} \int_{t\geq k} Z_{k-1}\, dQ,$$

for $m \geq 1$ by the independence of Y_k and $\{t \geq k\}$. As $m \to \infty$, the latter sum converges up to $C(\theta,\omega)$, while $E(1 - Z_{t\wedge m}) \to E(1 - Z_t 1_{t<\infty})$, since

$$\left(\int_{t>m} Z_m\, dQ\right)^2 \leq P_\theta(t > m)\, P_\omega(t > m) \to 0 \quad \text{as } m \to \infty,$$

by Schwarz' inequality.

Lemma 4

$$\text{ap}\lim_{\omega\to\theta} h(\theta,\omega) = 0 \qquad \textit{for a.e. } \theta \in \Omega.$$

Proof. Since the sigma algebra $\mathcal{B}$ is assumed to be countably generated, the square roots $s_\omega = \sqrt{f_\omega}$, $\omega \in \Omega$, define a measurable s mapping from $\mathcal{X}$ into the separable Hilbert space $L^2(\mathcal{X},\mathcal{B},\mu)$. See, for example, Cambanis (1975). Observe that $||s_\omega - s_\theta|| = h(\theta,\omega)$ for all ω, $\theta \in \Omega$. By Lusin's Theorem[4], for each $\varepsilon > 0$, there is a closed set $K \subset \Omega$ for which $m(\Omega - K) < \varepsilon$ and the restriction of s to K is strongly continuous. It follows easily that $\text{ap}\lim_{\omega\to\theta} h(\theta,\omega) = 0$ for a.e. $\theta \in K$ (for all θ which are points of density of K); and the theorem then follows, since $\varepsilon > 0$ was arbitrary.

Lemma 5

$$C(\theta,\omega) \leq \sqrt{E_\theta(t)\, E_\omega(t)} \qquad \theta,\ \omega \in \Omega, \tag{10}$$

$$\sum_{k=1}^{N} P_\theta(t \geq k) \leq C(\theta,\omega) + N^{\frac{3}{2}}\, h(\theta,\omega) \qquad \theta,\ \omega \in \Omega,\ N \geq 1, \tag{11}$$

$$\text{ap}\lim_{\omega\to\theta} C(\theta,\omega) = E_\theta(t) \qquad \textit{for a.e. } \theta \in \Omega. \tag{12}$$

Proof. Inequality (10) is Schwarz' inequality applied to the definition of $C(\theta,\omega)$. Inequality (11) depends upon Schwarz' inequality as well; in fact,

[4] *This form of Lusin's Theorem holds for any measurable function from the real line into a separable metric space. See, for example, Dinculeanu (1967, p. 334).*

$$\sum_{k=1}^{N} P_\theta(t \geq k) - C(\theta,\omega)$$

$$\leq \sum_{k=1}^{N} \int_{t \geq k} \prod_{j=1}^{k-1} f_\theta^{\frac{1}{2}}(X_j)\left[\prod_{j=1}^{k-1} f_\theta^{\frac{1}{2}}(X_j) - \prod_{j=1}^{k-1} f_\omega^{\frac{1}{2}}(X_j)\right] dQ$$

$$\leq \prod_{k=1}^{N} \{P_\theta(t \geq k) \cdot \int\left[\prod_{j=1}^{k-1} f_\theta^{\frac{1}{2}}(X_j) - \prod_{j=1}^{k-1} f_\omega^{\frac{1}{2}}(X_j)\right]^2 dQ\}^{\frac{1}{2}}$$

$$\leq \{N \sum_{k=1}^{N} (k-1)\, h^2(\theta,\omega)\}^{\frac{1}{2}} \leq N^{\frac{3}{2}} h(\theta,\omega).$$

The inequality ap lim $\sup_{\omega\to\theta} C(\theta,\omega) \leq E_\theta(t)$ follows from (10) for a.e. $\theta \in \Omega$; and the inequality ap lim $\inf_{\omega\to\theta} C(\theta,\omega) \geq E_\theta(t)$ follows from (11) and Lemma 4, by letting $\omega \to \theta$ and $N \to \infty$ in that order.

Theorem 2. Whenever the products make sense (i.e., not $0 \cdot \infty$),

$$I^*(\theta) = E_\theta(t)i^*(\theta) \text{ and } I_*(\theta) = E_\theta(t)i_*(\theta) \text{ for a.e. } \theta\in\Omega, \quad (13)$$

$$I^*(\theta) \geq E_\theta(t)i^*(\theta) \text{ and } I_*(\theta) \geq E_\theta(t)i_*(\theta) \text{ for all } \theta\in\Omega. \quad (14)$$

Proof. The equalities in (13) follow directly from Lemma 3, (12), and the definitions of upper and lower information. For (14), observe that $C(\theta,\omega) \geq 1$ for all θ, $\omega \in \Omega$, so that $H(\theta,\omega) \geq \delta$ whenever $h(\theta,\omega) \geq \delta > 0$. Thus

$$\frac{H^2(\theta,\omega)}{(\omega-\theta)^2} = C(\theta,\omega)\,\frac{h^2(\theta,\omega)}{(\omega-\theta)^2}$$

$$\geq \min\{\left[\sum_{k=1}^{N} P_\theta(t \geq k) - N^{\frac{3}{2}}\delta\right]\frac{h^2(\theta,\omega)}{(\omega-\theta)^2},\ \frac{\delta^2}{(\omega-\theta)^2}\}$$

for ω, $\theta \in \Omega$, $\delta > 0$, and $N \geq 1$. The inequalities (14) now follow by taking approximate upper and lower limits as $\omega \to \theta$, letting $\delta \downarrow 0$, and letting $N \to \infty$ in that order.

Corollary 3. Suppose that $i(\theta)$ exists and that either $i(\theta) > 0$ or $E_\theta(t) < \infty$ for a.e. θ. Then

$$I(\theta) = E_\theta(t)\, i(\theta) \qquad \text{for a.e. } \theta \in \Omega.$$

The corollary follows directly from (13). When $i(\theta) = 0$ and $E_\theta(t) = \infty$, $I(\theta)$ may be 0, finite, or infinite.

Example 4. Suppose that F_θ is the normal distribution with mean $\mu = |\theta|^\alpha$ and unit variance for $-\infty < \theta < \infty$, where $\alpha > 1$. Then $i(\theta) = \alpha^2|\theta|^{2\alpha-2}$, $-\infty < \theta < \infty$, and $i(0) = 0$. Next, let

$$t = \inf(n \geq 1: S_n > 0),$$

where $S_n = X_1 + \ldots + X_n$, $n \geq 1$. Then $t < \infty$ w.p. 1 (P_θ) for all θ, $E_\theta(t) < \infty$ for $\theta \neq 0$, and $E_0(t) = \infty$. Now

$$H^2(0,\theta) = 2\int(1 - e^{\frac{1}{2}\mu S_t - \frac{1}{4}\mu^2 t})\, dP_0$$

$$= 2\int(1 - e^{-\frac{1}{4}\mu^2 t})\, dP_0 + 2\int(1 - e^{\frac{1}{2}\mu S_t})e^{-\frac{1}{4}\mu^2 t}\, dP_0$$

$$= A(\mu) + B(\mu), \qquad \text{say.}$$

It is easily seen that $\mu^{-1}B(\mu) \to E_0(S_t)$ as $\theta \to 0$; and $E_0(S_t) = 1/\sqrt{2}$ by Theorem 1 of Feller (1966, p. 575). For A write

$$A(\mu) = 2(1 - e^{-\frac{1}{4}\mu^2}) \sum_{n=0}^{\infty} e^{-\frac{1}{4}n\mu^2} P_0(t > n).$$

According to Feller (1966, p. 398),

$$P_0(t > n) = \binom{2n}{n} 2^{-2n} \sim \frac{1}{\sqrt{\pi n}} \qquad \text{as } n \to \infty.$$

So

$$\mu^{-1} A(\mu) \approx \frac{1}{2}\mu \sum_{n=0}^{\infty} e^{-\frac{1}{4}n\mu^2} \frac{1}{\sqrt{\pi n}}$$

$$\to \frac{1}{2\sqrt{\pi}} \int_0^{\infty} x^{-\frac{1}{2}} e^{-\frac{1}{4}x} dx = 1.$$

Thus

$$\lim_{\theta \to 0} \frac{1}{\mu} H^2(0,\theta) = 1 - \frac{1}{\sqrt{2}};$$

and $I(0)$ is zero, finite and positive, or infinite accordingly as $\alpha > 2$, $\alpha = 2$, or $1 < \alpha < 2$.

We do not know whether (13) extends, in general, to "all θ". We have some evidence that it does hold for all θ, but no proof. In this regard, observe that one cannot expect (12) to hold "for all θ", although it may hold whenever $\text{ap}\lim_{\omega \to \theta} h(\theta,\omega) = 0$.

Significantly, (13) is strong enough, for an immediate application of Theorem 1 and Corollary 1.

Corollary 4. Suppose that $i(\theta)$ *exists and* $0 < i(\theta) < \infty$ *for a.e.* $\theta \in \Omega$. *Let* γ *be a real valued function on* Ω*; let* t *be a finite stopping time; and let* $\hat{\gamma}$ *be an* A_t*-measurable unbiased estimator of* γ. *Then the variance of* $\hat{\gamma}$ *(denoted* $\sigma^2(\theta)$*) satisfies*

$$\sigma^2(\theta) \geq \gamma'(\theta)^2/\{E_\theta(t) \cdot i(\theta)\} \qquad \text{for a.e. } \theta \in \Omega.$$

Here $0/0 = 0$ *and* $\gamma'(\theta)^2$ *can be interpreted as infinity whenever* $\gamma'(\theta)$ *fails to exist as an approximate derivative.*

IV. REMARKS

The definition of information given here differs from the conventional one. In the case that the densities $f_\omega(x)$, $\omega \in \Omega$, are differentiable with respect to ω for a.e. x (μ), Pitman (1979) explores a closely related difference.[5] He defines the information and sensitivity of f_ω, $\omega \in \Omega$, by

$$i_0(\theta) = \int (f_\theta'^2/f_\theta)\, d\mu$$

and

$$i_1(\theta) = \lim_{\omega \to \theta} h^2(\theta,\omega)/(\omega - \theta)^2,$$

respectively, provided that the limit exists in the latter case, and shows that $i_0(\theta) \leq i_1(\theta)$ for all θ for which $i_1(\theta)$ is defined with equality iff $s_\omega = \sqrt{f_\omega}$ are differentiable in mean square at $\omega = \theta$. Moreover, Le Cam (1970) has shown that s_ω are differentiable in mean square at a.e. θ for which $\limsup h^2(\theta,\omega)/(\omega - \theta)^2 < \infty$ as $\omega \to \theta$, so that the two quantities cannot be too different when $i_1(\theta) < \infty$ for all $\theta \in \Omega$. Based on an example for which $i_1(\theta) = \infty$ and $i_0(\theta) = 1$ for all $\theta \in \Omega$, Pitman (1979, p. 39) concludes that sensitivity, not information, is the quantity to use in variance bounds.

Of course, Pitman's sensitivity differs from our definition of information only in the sense in which the limit is taken. In this regard, observe that using approximate limits has two advantages: $i(\theta)$ exists whenever $i_1(\theta)$ does, and $i(\theta)$ transforms naturally to $I(\theta) = E_\theta(t)i(\theta)$ in the sequential case. It is easy to construct examples in which $i(\theta)$ exists for a.e. θ, but $i_1(\theta)$ exists for no θ. We do not know to what extent $i_1(\theta)$ transforms to $I_1(\theta) = E_\theta(t)i_1(\theta)$ in the sequential case, but suspect that additional conditions are necessary.

[5] *Actually, Pitman considers a slightly more general model.*

The simplest way to exploit Lemma 4 is to let $\omega \to \theta$ in (4). If $i_1(\theta)$ exists and γ is differentiable at θ, then this operation yields

$$\frac{1}{2}[\sigma^2(\theta) + \liminf_{\omega \to \theta} \sigma^2(\omega)] \geq \gamma'(\theta)^2/i_1(\theta)$$

for the given θ; and if the conditions hold for all $\theta \in \Omega$, then the conclusion holds for all $\theta \in \Omega$ (cf Chernoff (1958, Lemma 1)). Lusin's theorem shows that $\liminf_{\omega\to\theta} \sigma^2(\omega) \leq \sigma^2(\theta)$ for a.e. θ to yield $\sigma^2(\theta) \geq \gamma'(\theta)^2/i_1(\theta)$ for a.e. θ, provided i_1 and γ' exist. The use of approximate limits yields sharper, more general results.

Suppose next that $s_\omega(x) = \sqrt{f_\omega}(x)$ are absolutely continuous in ω for a.e. x (μ); extend the definition of i_0 by letting

$$i_0(\omega) = \int 4\dot{s}_\omega^2 \, d\mu \qquad \omega \in \Omega;$$

and suppose that i_0 is locally integrable with respect to $\omega \in \Omega$. Let i_1 be as above. Then, for $\theta, \omega \in \Omega$,

$$h^2(\theta,\omega) = \int_\theta^\omega \int_\theta^\omega [\int \dot{s}_\xi \dot{s}_\eta \, d\mu] d\eta \, d\xi \leq \frac{1}{4} \int_\theta^\omega \int_\theta^\omega \sqrt{i_0(\xi)} \sqrt{i_0(\eta)} \, d\eta \, d\xi,$$

so that $i_1(\theta) = i_0(\theta) = i(\theta)$ for a.e. $\theta \in \Omega$. Similarly, if $\hat{\gamma}$ is an unbiased estimator of a real valued function γ, and if both $i_0(\omega)$ and the variance $\sigma^2(\omega)$ of $\hat{\gamma}$ are locally integrable, then γ must be absolutely continuous with derivative

$$\dot{\gamma}(\theta) = \int \hat{\gamma} \cdot 2 s_\theta \dot{s}_\theta \, d\mu \qquad \theta \in \Omega,$$

in which case the Cramér-Rao inequality (8) follows for a.e. θ either by letting $\omega \to \theta$ in (4) or directly from Schwarz' inequality. This approach is pursued by Ibragimov and Has'minskii (1972). Of course, this approach places conditions on the estimator.

V. ACKNOWLEDGMENTS

The authors gratefully acknowledge helpful discussions of aspects of this paper with Stamatis Cambanis, Vidyadhar Mandrakar, Karl Petersen, Doraiswamy Ramachandran, the receipt of some helpful references from Donald Geman, and some helpful correspondence from Andrew Bruckner. Acknowledgement is also due Robert Berk, who convinced one of the authors several years ago that a satisfactory description of the Cramér-Rao inequality in the sequential setting did not exist, and that the methods for improvement that had been developed for the fixed sample size setting were not fully adequate for the sequential setting.

VI. APPENDIX A

The following construction is required for Example 3.

Theorem 3. For any fixed countable subset N *of the real line, there exists a real-valued measurable function* α *of a real variable with the properties that on* N, *(i)* $\alpha(x) = 0$, *and (ii)* $\alpha'(x)$ *exists, at least as an approximate derivative, and* $\alpha'(x) = 1$.

Proof. The assertion is straightforward when N is finite, so we will assume N is (countably) infinite with members $r_1, r_2, \ldots$. For each real x, let n(x) be chosen so as to make $2^n |x - r_n|$ minimum with respect to $n \geq 1$. Using a variation of the Borel Cantelli lemma, one finds that

$$\sum_{k=1}^{\infty} m\{x: 2^k |x - r_k| \leq M\} = \sum_{k=1}^{\infty} 2^{1-k} M < \infty, \qquad M > 0,$$

and thus $2^k |x - r_k| \to \infty$ as $k \to \infty$ for a.e. x (m = Lebesgue measure). Consequently, n(x) is well-defined, in a unique manner, for a.e. x. Clearly, it is well-defined on N, and specifically $n(x) = k$ when $x = r_k$. Let $\alpha(x) = (x - r_{n(x)})$ for $-\infty < x < \infty$. Then, clearly $\alpha(x) = 0$ for $x \in N$. To show that the approximate derivative of α is one at a given $r_k \in N$, it suffices to show that

$$\lim \frac{1}{m(I)} m[x \in I: n(x) \neq k] = 0$$

as $I \downarrow r_k$. Let $A_n = [x: n(x) = n]$ for $n = 1,2,\ldots$. Then it suffices to show that

$$\lim \frac{1}{m(I)} \sum_{n \neq k}^{\infty} m(IA_n) = 0 \tag{15}$$

as $I \downarrow r_k$; and, since $A_n \cap I$ is empty for all sufficiently small I for each fixed n, the summation in (15) may be changed from all $n \neq k$ to $n \geq n_0$ for any fixed $n_0 > k$.

The proof of (15) depends on the simple observation that, for $n \geq k$, $A_n \subset [x: 2^n|x - r_n| \leq 2^k|x - r_k|]$. Suppose that $I = (r_k - a, r_k + b)$, where $a,b > 0$. Then $|x - r_k| \leq a \vee b$ for $x \in I$, so that

$$m(A_n \cap I) \subset m[x: |x - r_n| \leq 2^{-n+k}(a \vee b)] \leq (\tfrac{1}{2})^{n-k}(a + b)$$

for $n \geq n_0 > k$. It follows that for sufficiently small a and b, (15) does not exceed $\sum_{k=n_0}^{\infty} (\frac{1}{2})^{n-k}$, which may be made arbitrarily small by taking n_0 sufficiently large.

Whenever $x \in N$ is a limit point of N, $\alpha'(x)$ exists only as an approximate derivative. Tolstoff (1938) has shown that a function which is approximately differentiable at every point of an interval is actually differentiable (in the classical sense) except on some (closed) nowhere dense subset of the interval. Thus α must have points of non-approximate differentiability in any interval on which N is dense. This reasoning applies no matter how α is constructed so as to satisfy (i) and (ii). For instance, when N is the set of rational numbers, then no matter how α is constructed, there must always be a dense set of irrational points at which α is not approximately differentiable. On the other hand, the function α can always (regardless of N) be constructed so as to be approximately differentiable a.e. Indeed, the function α constructed in the proof of Theorem 3 is approximately differentiable at a.e. x, with $\alpha'(x) = 1$. This is a simple consequence of the fact that the integer-valued (measurable) function n, defined therein, is necessarily approximately continuous a.e.

This discussion has relevance for Example 3, since the function γ (which is being estimated without bias) is approximately differentiable, and the information i is defined, finite and positive, at all points $\theta \in \Omega$ for which $\alpha'(\theta)$ is defined. In summary, $\gamma'(\theta)$ and $i(\theta) \in (0,\infty)$ are defined for every $\theta \in N$ and, at least, for a.e. other θ in Ω.

Andrew Bruckner has provided us with a proof that the countability assumption in Theorem 3 is absolutely essential.

VII. APPENDIX B

Here, the well-known result that a real-valued function g of a real variable is approximately differentiable at a.e. x for which ap lim $\sup_{y\to\infty} |g(y) - g(x)|/|y - x| < \infty$ is extended, in Theorem 4 below, to the setting of a Hilbert space-valued function of a real variable. As noted in the proof of Lemma 4, $s_\theta = \sqrt{f_\theta}$ can be viewed as an element of a separable Hilbert space, and $||s(\omega) - s(\theta)|| = h(\theta,\omega)$. Theorem 4 permits us to conclude that the approximate derivative in mean square $\dot{s}_\theta$ of s_θ, and the information $i(\theta)$ are defined for a.e. θ for which the upper information $i^*(\theta)$ is finite. Moreover, for such θ, $i(\theta) = \int 4\dot{s}_\theta^2 \, d\mu$.

Theorem 4. Let H be a separable Hilbert space with norm $||\cdot||$, let s be a measurable mapping from a real interval Ω into H, and let Ω_0 denote the subset of points $\theta \in \Omega$ for which

$$\text{ap lim sup}_{\omega\to\theta} \left|\left|\frac{s(\omega) - s(\theta)}{\omega - \theta}\right|\right| < \infty. \tag{16}$$

Then s *possesses an approximate strong* (Fréchet) *derivative for a.e.* $\theta \in \Omega_0$. *I.e., for a.e.* $\theta \in \Omega_0$*; there is an element* $\dot{s}(\theta) \in H$ *for which*

$$\text{ap lim}_{\omega\to\theta} \left|\left|\frac{s(\omega) - s(\theta)}{\omega - \theta} - \dot{s}(\theta)\right|\right| = 0. \tag{17}$$

Our proof of this theorem uses techniques from Sak's (1937, pp. 295-297) proof for the real-valued version and LeCam's (1970, Theorem 1) proof of an analogous result involving limits rather than approximate limits.

Proof. It is easily seen that Ω_0 is a measurable set; and without loss of generality, Ω can be taken to be bounded. Let Ω_1 be the set of $\theta\in\Omega_0$ at which the approximate derivative exists; then it suffices to show for arbitrary preassigned $\eta > 0$, that $m(\Omega_0 - \Omega_1) \leq 2\eta$. By Lusin's Theorem, there exists a closed $K \subset \Omega_0$ for which $m(\Omega_0 - K) \leq \eta$ and the restriction of s to K is strongly continuous.

Let $E(\theta,n) = \{\omega\in K: ||s(\omega) - s(\theta)|| \leq n|\omega - \theta|\}$ for $\theta \in K$, $n \geq 1$; and for each $n \geq 1$, let K_n be the set of $\theta \in K$ for which $m(I \cap E(\theta,n)) \geq \frac{3}{4}m(I)$ for all open intervals I for which $\theta \in I$ and $m(I) \leq 1/n$. Then each K_n is a closed set by Lemma 6 below; and $K_n \uparrow K$ by (16) since $K \subset \Omega_0$. Let $\Omega_2 = K_{n_0}$ for some n_0 such that $m(k - K_{n_0}) \leq \eta$, so that $m(\Omega_0 - \Omega_2) \leq 2\eta$.

It will now be shown that s satisfies a uniform Lipschitz condition on Ω_2:

$$||s(\theta_2) - s(\theta_1)|| \leq C|\theta_2 - \theta_1|, \quad \theta_1, \theta_2\in\Omega_2, \tag{18}$$

for some constant C. Since the restriction of s to Ω_2 is continuous, and therefore bounded, it suffices to verify (18) when $|\theta_2 - \theta_1|$ is small, say $\leq \frac{1}{2n_0}$. If $\theta_1, \theta_2 \varepsilon \Omega_2$, $\theta_1 < \theta_2$ and $\theta_2 - \theta_1 \leq \frac{1}{2n_0}$, then, by Lemma 7 below, there is an $\omega \varepsilon K$ for which $\theta_1 < \omega < \theta_2$ and $\omega \varepsilon E(\theta_1, n_0) \cap E(\theta_2, n_0)$. Hence

$$||s(\theta_2) - s(\theta_1)|| \leq ||s(\theta_2) - s(\omega)|| + ||s(\omega) - s(\theta_1)||$$

$$\leq n_0(\theta_2 - \omega) + n_0(\omega - \theta_1) = n_0|\theta_2 - \theta_1|,$$

which establishes (18) for some C.

By means of linear interpolation and a simple limiting operation, a function t can be defined on all of Ω which agrees with s on Ω_2 and satisfies

$$||t(\theta_2) - t(\theta_1)|| \leq C|\theta_2 - \theta_1|, \qquad \theta_1, \theta_2 \varepsilon \Omega. \tag{19}$$

The Lipschitz condition for t implies that the functions $(t(\cdot), e)$, $e \varepsilon H$, are absolutely continuous. Thus t is weakly differentiable a.e., and the weak derivative $\dot{t}$ satisfies the integral condition

$$\left(\frac{t(\omega) - t(\theta)}{\omega - \theta} - \dot{t}(\theta), e\right) = \frac{1}{\omega - \theta} \int_\theta^\omega (t(s) - t(\theta), e)ds,$$

$$e \varepsilon H, \ \theta \varepsilon \Omega,$$

which implies

$$\left|\left|\frac{t(\omega) - t(\theta)}{\omega - \theta} - \dot{t}(\theta)\right|\right| \leq \left|\frac{1}{\omega - \theta}\right| \int_\theta^\omega ||\dot{t}(s) - \dot{t}(\theta)||ds, \quad \theta \varepsilon \Omega.$$

The latter tends to zero as $\omega \to \theta$ for a.e. $\theta \varepsilon \Omega$ by Lusin's Theorem. Thus (17) holds for a.e. $\theta \varepsilon \Omega_2$ with $\dot{s}(\theta)$ set equal to $\dot{t}(\theta)$.

The following two lemmas complete the proof of the theorem.

Lemma 6.

Each K_n, $n \leq 1$, *is a closed set.*

Proof. Fix $n \geq 1$; and let θ_k, $k \geq 1$, be a convergent sequence from K_n, say $\theta_k \to \theta$. If $\omega \in E(\theta_k, n)$ for all but a finite number of k, then

$$||s(\omega) - s(\theta)|| = \lim_{k\to\infty} ||s(\omega) - s(\theta_k)||$$

$$\leq \lim_{k\to\infty} n|\omega - \theta_k| = n|\omega - \theta|;$$

so $\omega \in E(\theta,n)$, i.e. $E(\theta,n) \supset \liminf_{k\to\infty} E(\theta_k,n)$.
If I is an open interval for which $\theta \in I$ and $m(I) \leq 1/n$, then $\theta_k \in I$ for all large k; so

$$m[I \cap E(\theta,n)] \geq \liminf_{k\to\infty} m[I \cap E(\theta_k,n)] \geq \frac{3}{4}m(I).$$

That is, $\theta \in K_n$.

Lemma 7

If θ_1, $\theta_2 \varepsilon \Omega_2$ *and* $0 < \theta_2 - \theta_1 \leq \frac{1}{2n_0}$, *then there is an* ω *for which* $\theta_1 < \omega < \theta_2$ *and* $\omega \in E(\theta_1,n) \cap E(\theta_2,n)$

Proof. Let $\varepsilon = \theta_2 - \theta_1$, let $I = (\theta_1 - \frac{1}{4}\varepsilon, \theta_2 + \frac{1}{4}\varepsilon)$, and observe that $m(I) = \frac{3}{2}\varepsilon \leq 1/n_0$. Let $F_i = I \cap E(\theta_i, n_0)$ $i = 1,2$. Then

$$m(I - [F_1 \cap F_2 \cap (\theta_1,\theta_2)]) \leq m(I - F_1) + m(I - F_2) + m(I - (\theta_1,\theta_2))$$

$$\leq \frac{1}{4}m(I) + \frac{1}{4}m(I) + \frac{1}{2}\varepsilon = \frac{5}{4}\varepsilon.$$

Thus $m(F_1 \cap F_2 \cap (\theta_1,\theta_2)) \geq \frac{1}{4}\varepsilon > 0$, and the desired conclusion follows.

We have, in fact, essentially generalized "Whitney's Theorem" (1951) from the setting of a real valued function to that of a separable Hilbert space-valued function: *A measurable, separable Hilbert space-valued function* $s(\theta)$ *defined on an interval* Ω *is approximately strongly differentiable a.e. in a measurable set*

Ω_0 *if and only if to each* $\varepsilon > 0$ *corresponds a closed set* K *and a strongly continuously differentiable function* t, *defined on* Ω *and taking values in the Hilbert space, such that* $m(\Omega_0 - K) < \varepsilon$ *and* s = t *on* K. (Showing that t can be chosen so as to have a *continuous* derivative requires some additional "fiddling" with the t we have found in the proof of Theorem 4. Cf., Lemma 1 of Whitney (1951)).

VIII. REFERENCES

Alvo, M. (1977). "Bayesian Sequential Estimates." *Ann. Statist. 5,* 955-968.

Blythe, C.R. and Roberts, D. (1972). "On Inequalities of Cramér-Rao Type and Admissibility Proofs." *Proc. Sixth Berk. Symp. 1,* 17-30.

Bruckner, A. (1978). "Differentiation of Real Functions". *Lecture Notes in Mathematics,* 659, Springer-Verlag.

Cambanis, S. (1975). "The Measurability of a Stochastic Process of Second Order and its Linear Space." *Proc. Amer. Math. Soc. 47,* 467-475.

Chapman, D. and Robbins, H. (1951). "Minimum Variance Estimation Without Regularity Assumptions." *Ann. Math. Statist. 22,* 581-586.

Chernoff, H. (1956). "Large Sample Theory, Parametric Case". *Ann. Math. Statist. 27,* 1-22.

Dinculeanu, N. (1967). *Vector Measures,* Pergamon Press.

Fabian, V. and Hannan, J. (1977). "On the Cramér-Rao Inequality." *Ann. Statist. 5,* 197-205.

Feller, W. (1966). *An Introduction to Probability Theory and its Applications 2,* Wiley, New York.

Geman, D. and Horowitz, J. (1980). "Occupation Densities". *Ann. Prob. 8,* 1-67.

Ibragimov, I.A. and Has'minskii, R. (1972). "Information Inequalities and Supereffective Estimates." *Selected Translations in Probability and Statistics 13,* 821-824.

Le Cam, L. (1970). "On the Assumptions Used to Prove Asymptotic Normality of Maximum Likelihood Estimates." *Ann. Math. Statist. 41,* 802-827.

Mackey, G. (1949). "A Theorem of Stone and von Neumann." *Duke Math. J. 16,* 313-326.

Neveu, J. (1965). *Mathematical Foundations of the Calculus of Probability,* Holden-Day.

Pitman, E.S.G. (1979). *Some Basic Theory for Statistical Inference,* Chapman and Hall.

Saks, S. (1937). *Theory of the Integral,* Dover.

Simons, G. (1980). "Sequential Estimators and the Cramér-Rao Lower Bound." *J. Stat. Planning and Inf. 4,* 67-74.

Toltoff, G. (1938). "Sur la Dérivée Approximate Exacte". *Rec. Math. (Mat. Sbornik), 4* (46), 499-504.

Whitney, H. (1951). "On Totally Differentiable and Smooth Functions". *Pacific J. Math. 1,* 143-154.

Wolfowitz, J. (1947). "The Efficiency of Sequential Estimates and Wald's Equation for Sequential Processes." *Ann. Math. Statist. 18,* 215-230.

A TWO-SAMPLE SEQUENTIAL TEST FOR SHIFT WITH ONE SAMPLE SIZE FIXED IN ADVANCE

Paul Switzer

Department of Statistics
Stanford University
Stanford, California

I. INTRODUCTION

Suppose we have a sample of m observations of a real-valued random variable taken at some point in the past. We now plan to take a new sample of size n in order to test whether the shift in the mean from then until now is positive or negative. If n is fixed we have a standard two-sample shift problem. However, if it is possible to take our new sample in a <u>sequential</u> manner, then there may be procedures which afford an average savings in the sample-size needed to attain specified error levels.

The characteristic feature of this setup which distinguishes it from the usual two-sample setup is that we do not begin taking the second sample until the first sample is completely in. Apart from the two-points-in-time application mentioned above, a similar situation arises in comparing two widely separated geographic points when it is not practical to travel to each of the points more than once. Another application might involve comparing a treatment with a control where all the control data is already in hand and it is not possible to obtain

ISBN 0-12-589320-5

more. When we are in the fortunate position that neither of the two samples has yet been collected, we can make the choice of m (the fixed size of the first sample) part of the problem.

Section 2 proposes a sequential sampling plan for the second sample appropriate to the situation described above. In Section 3 the operating characteristic function of the proposed procedure is derived under the simplest normal shift model with common specified variance for the two samples using a Wald-type approximation. It is shown there that the OC function has a simple closed form, suitable for calculation, for certain combinations of model parameters. Section 4 discusses the behavior of the mean sample size of the second sample. Section 5 compares the mean sample size of the sequential procedure with completely fixed sample-size procedures which attain the same error probabilities. Section 6 considers the question of optimizing the fixed size of the first sample and also makes comparisons with sequential procedures which sample both populations simultaneously. Sections 3, 4, 5, and 6 all contain calculated examples along with appropriate simulations to check on the accuracy of the Wald approximations.

In an earlier version, this paper appeared as a 1970 Stanford University Technical Report *[4]* under Herman Chernoff's Department of Defense research contract. Subsequently, two related papers appeared in 1977. One by Choi *[3]* uses Theorem 1 of Section 2 and extensions thereof for a similar problem although no use was made of the series representation of Theorem 2. The second paper by Wolfe *[6]* also considers a two-sample problem with one sample size fixed and one sample size random, although Wolfe's procedure differs completely from those of Switzer and Choi. This seems to be the extent to date of work on such two sample problems.

Existing sequential treatments of two-sample problems, unrestricted in the above manner, usually require alternating sampling from each of the two distributions. For example, in *[2]* we can find a sequential treatment of observations taken in pairs, for a variety of model specifications. Such treatments are inapplicable to our problem at least insofar as only one of the sample-sizes may be random.

II. PROBLEM SPECIFICATION AND THE SUGGESTED PROCEDURE

Let $X_1, X_2, \ldots$ be independent replications of a random variable X which is normally distributed with unknown mean μ_x and variance $\sigma^2 = 1$; we will call this the "old data" sequence. Let $Y_1, Y_2, \ldots$ be independent replications of a random variable Y which is normally distributed with mean $\mu_x + d$ and variance $\sigma^2 = 1$; we will call this the "new data" sequence. The shift d is also unknown and the Y's are independent of the X's. Our job is to decide whether $d \geq 0$ or $d < 0$ on the basis of finite data sequences.

The old data sequence is stopped at a predetermined m. [Later, in Section 5, we make the choice of m part of the problem.] The new data sequence is stopped after n observations, where n may depend on $X_1, X_2, \ldots, X_m$ and on $Y_1, Y_2, \ldots$.

The suggested sequential stopping rule for n is to continue sampling the Y's until

$$|\bar{Y}_n - \bar{X}_m| \geq c/n. \tag{1}$$

The terminal decision rule declares the shift d to be positive or negative according to whether the final difference $\bar{Y}_n - \bar{X}_m$ is positive or negative. The constant $c > 0$ is chosen to provide specified error probabilities as will be later explained.

This procedure mimics the Wald one-sample SPRT [2] that would be appropriate if μ_x were known. Specifically, $\bar{X}_m$ would be replaced by μ_x, and c would be approximately $c_0 = [\log_e(1 - \alpha)/\alpha]/2|d|$ in order to achieve error rate α at values of the shift parameter $\pm d$. With μ_x unknown, c decreases to c_0 as m gets larger.

III. CALCULATION OF ERROR PROBABILITIES

It is convenient to express the continuation region of the suggested stopping rule in the following way:

$$n(\overline{X}_m - \mu_x) - c < \sum_{i=1}^{n} Z_i < n(\overline{X}_m - \mu_x) + c,$$

where $Z_i = (Y_i - \mu_x)$ and the Z_i are i.i.d. N(d,1) independent of $\overline{X}_m$. If we consider plotting ΣZ_i against n, then the continuation region is bounded by the pair of parallel lines with common slope $S = (\overline{X}_m - \mu_x)$ and with intercepts $\pm c$. The form of the procedure when viewed this way differs from Wald's one-sample SPRT (for the mean of Z) only in that the slope S of the boundary is a random variable in our case, whereas in Wald's problem the boundary lines have fixed slope. However, since ΣZ_i is independent of S we can immediately use Wald's approximations to obtain an OC-function conditional on S. Specifically,

$$\text{Prob}_d[\text{decide } d < 0 | S] = [1 + e^{2c(d - S)}]^{-1}. \tag{2}$$

This is a specialization of formula (7:21) in (5) to the case of intercepts of opposite sign and equal magnitude. The approximation involved is one of ignoring "excess", i.e., assuming the sampling terminates right on the boundary. Of course (2) cannot be computed because S involves the unknown quantity μ_x. However, the distribution of S is $N(0,1/m)$ and is free of any unknown quantities so that an unconditional OC function $\alpha(d)$ can indeed be computed. Specifically, we now have

Theorem 1

$$\text{Prob}_d[\text{decide } d < 0] \equiv \alpha(d)$$

$$= E[1 + e^{2c(d - S)}]^{-1} \text{ where } S \sim N(0, 1/m)$$

$$= E[1 + e^{AV}]^{-1} \text{ where } V \sim N(1, 1/B^2) \tag{3}$$

$$\text{and } A = 2cd, \quad B^2 = md^2.$$

Thus the OC function, up to Wald's approximation, may be expressed in terms of the pair of numbers A,B. The OC function may be interpreted as the error probability for $d > 0$ and as the complement of the error probability for $d < 0$.

A remarkably simplified expression for $\alpha(d)$ can be obtained for lattice values of the argument d.

Theorem 2

If the argument $d = I\cdot(2c/m)$ for a positive integer I, then the Wald approximation for the error probability $\alpha(d)$ as given in Theorem 1 may be represented by the finite sum

$$\alpha(d) = \phi(B) \sum_{k=1}^{2I-1} (-1)^{k+1} M(kB/I - B) \tag{4}$$

where ϕ is the standard Gaussian density function, Φ is the standard Gaussian distribution function, $M(t)$ is the Mill's ratio $\Phi(-t)/\phi(t)$, and $B^2 = md^2$, as before.

The proof of Theorem 2 depends on first obtaining an alternating absolutely convergent infinite series representation of $\alpha(d)$ for all d. We state the preliminary result as a *Lemma:* for all d,

$$\alpha(d) = \Phi(-B) + \phi(B) \sum_{k=1}^{\infty} (-1)^{k+1}\{M(kA/B - B) - M(kA/B + B)\} \tag{5}$$

where $A = 2cd$ and $B^2 = md^2$. The proof of the lemma is given in the Appendix. There is a connection between formula (5) and a similar expression given by Anderson [1] for a one-sample truncated sequential procedure; this connection is described briefly at the end of this section.

The lemma is itself useful for computation for a general argument d. If the series (5) is truncated after r terms then the residual is bounded by the first neglected term. By noting that $M(t) = t^{-1} + O(t^{-3})$ it can be shown that the residual is $O(r^{-2})$.

Proof of Theorem 2. The requirement that the argument $\bar{d} = I\cdot(2c/m)$ is equivalent to $A = B^2/I$ for positive integer I. By direct substitution,

$$\sum_{k=1}^{\infty} (-1)^{k+1} \{M(kA/B - B) - M(kA/B + B)\}$$

$$= \sum_{k=1}^{\infty} (-1)^{k+1} \{M(kB/I - B) - M(kB/I + B)\}$$

$$= \sum_{k=1}^{2I} (-1)^{k+1} M(kB/I - B) +$$

$$\sum_{k=2I+1}^{\infty} (-1)^{k+1}\{M(kB/I - B) - M([k - 2I]B/I + B)\}.$$

(The individual summands vanish in the second summation; the rearrangement of terms in the sums is justified because the original sum is absolutely convergent.)

$$= \sum_{k=1}^{2I-1} (-1)^{k+1} M(kB/I - B) - M(B).$$

Substituting the above into expression (5) for $\alpha(d)$ and noting that $\Phi(B) \cdot M(B) = \Phi(-B)$ completes the proof.

Table 1 displays values of the error probability α for a few selected values of the new parameters A and B. The entries in rows labeled I were computed using (4) or (5) based on the Wald-type approximation. It is apparent from the table that the error probability is somewhat sensitive to the fixed sample-size m of the old data.

TABLE I. Error Probability $\alpha(d)$ for the Sequential Procedure (1) with Width Constant c and with Old Sample Size m; $A \equiv 2cd$, $B^2 \equiv md^2$

		B = 1.5	B = 1.75	B = 2.00	B = 2.25	B = ∞
A = 3.00	I	.130[a]	.110	.097	.087	.047
	II	.123[b]	.109	.097	.086	
	III	.110[c]	.094	.075	.063	
A = 4.00	I	.106	.084	.068	.056	.018
	II	.095	.076	.061	.053	
	III	.101	.076	.056	.044	
A = 5.00	I	.093	.069	.052	.041	.0067
	II	.083	.059	.044	.034	
	III	.092	.063	.042	.030	
A = 6.00	I	.086	.061	.043	.032	.0025
	II	.073	.047	.040	.024	
	III	.083	.055	.036	.024	
A = ∞	I	.063	.040	.023	.012	0

[a] Calculated using Wald-type approximation and Theorem 2.
[b] Based on Monte Carlo simulation with $d = 0.25\sigma$.
[c] Based on Monte Carlo simulation with $d = 0.50\sigma$.

As a check on the accuracy of the Wald-type approximation of the OC function and its equivalent series representation, our suggested sequential sampling and decision procedure was computer-simulated 1000 times, each time with an independently chosen $\bar{X}_m$ value. We took $d = 0.25\sigma$. The parameter values A and B were varied within a single simulation so that the resulting Monte Carlo estimates of α are not independent for different A and B. These M-C estimates are shown in rows II of Table I. Individually, the standard error of a table entry is $[\alpha(1 - \alpha)/1000]^{1/2}$ or roughly between .005 and .01 in the range of the table. These standard errors are rather large, so it would be wise to regard the M-C estimates with care.

A comparison of rows I and II of Table I would indicate that the Wald-type approximation overestimates the error probability by 10% - 30%. Also, it should be remembered that the Wald-type approximation requires only the specification of $B^2 = md^2$ and not d and m separately. The exact calculation of $\alpha(d)$ would require separate specifications of d and m, so a second independent series of 1000 simulation was carried out with $d = 0.50\sigma$. By thus doubling d, the number of new data observations is reduced by about a factor of 4 so we might expect to have a poorer approximation of $\alpha(d)$. Rows III of Table I give the M-C estimates of $\alpha(d)$ with $d = 0.50\sigma$; they are sometimes closer to the calculated row I figures than are the row II figures, and sometimes they are not. Recall that the sampling variability in these M-C estimates is not negligible.

Two limiting cases of the OC function, $\alpha(d)$, for fixed d, are readily obtained and easily interpreted:

$$\lim_{c\to\infty} \alpha(d) = \lim_{A\to\infty} \alpha(d) = \Phi(-B) \tag{6}$$

$$\lim_{m\to\infty} \alpha(d) = [1 + e^A]^{-1} \tag{7}$$

where, as before, $A = 2cd$ and $B^2 = md^2$ in terms of the original constants of the problem.

As we broaden the continuation region for new data sampling by increasing c, we will decrease the error probability $\alpha(d)$ for fixed $d > 0$. There is, however, a positive lower bound for α, determined by the old data sample-size m. This limit (6) is obtained by noting that the summands in the expression (5) become individually negligible for every k as $A \to \infty$ (equivalently $c \to \infty$). This limit corresponds to the one-sample case where the new data mean μ_y becomes known exactly and we want to test whether the old data mean is greater or less than μ_y based on a sample of fixed size m. This limiting case is shown in the last row in Table I.

There is also a positive lower bound for $\alpha(d)$ if c is held fixed and the old data sample-size m is allowed to go to infinity. Then the old data mean μ_x becomes known exactly. If we replace $\bar{X}_m$ by μ_x in the recommended new data sequential procedure then it reduces to Wald's one-sample SPRT for testing whether the new data mean is greater or less than the known constant μ_x. The relation between α and c in this one-sample problem is given by Wald's approximation $\alpha = [1 + e^A]^{-1}$, where A = 2cd as before. This limiting case (7) is the last column in Table I. The limit (7) could also be obtained by letting $m \to \infty$ in expression (3), since $[1 + e^{AV}]^{-1}$ is a bounded and continuous function of V and $V \underset{p}{\to} 1$.

We conclude this section on error probabilities by noting a formal correspondence between our expression (5) and error probabilities calculated by Anderson in [1] for a *truncated one-sample* sequential testing problem. The nature of this correspondence can be made clear by referring to his section 4.1 and 4.3.

Specifically, suppose $Y_1, Y_2, \ldots$ were i.i.d. $N(d,1)$ and we are testing d > 0 vs. d < 0 using the Wald SPRT with continuation region

$$n|\bar{Y}_n| < (1 - n/m)c.$$

This region is bounded by nonparellel lines which meet at n = m, and the sampling is therefore truncated at n = m. Compare this with the formulation (1), where c and m are as before. The error probability for this truncated one-sample procedure is exactly (up to Wald approximation) the error probability of the two-sample untruncated procedure (1), although derived quite differently.

This correspondence permits the use of Anderson's general results to adjust error probabilities when the two-sample procedure is modified to have nonparallel boundaries with truncation. For example, the simple case where procedure (1) is truncated at $n = T$, say, corresponds to the one-sample procedure truncated at $n = m \cdot T/(T + m)$. Hence, by Anderson's formula (4.74), the truncation at T causes an adjustment in the error probability requiring the summands of expression (5) to be multiplied by $\exp(-k^2A^2/2d^2T)$. The correspondence does not carry over to the calculation of average sample sizes.

IV. AVERAGE NEW DATA SAMPLE-SIZE

When we use the sequential sampling procedure of Section 2, the number of new data observations n we will need is a random variable. The randomness is imparted by the old data as well as the new data. The expected value of n, taken over all this randomness will depend on the value of $d = \mu_y - \mu_x$ but fortunately not on the nuisance parameter μ_x itself. We use the symbol $E_d(n)$. Of course, the expected new data sample-size will also depend on the old data sample-size m and on the width constant c of the sequential procedure, but these dependences are suppressed in the notation.

An approximation for $E_d(n)$ can be obtained if we use Wald's approach by ignoring excess. In Section 3 it was shown that the two-sample procedure is equivalent to a one-sample SPRT with observations drawn from $N(d,1)$, where the parallel straight line sampling boundaries have random slope S, and where S is $N(0, 1/m)$ independent of the observations. For a fixed value of S, the standard Wald approximation for the expected sample-size is (see [5], p. 123)

$$\{c - 2c\cdot[1 + e^{2c(d-S)}]^{-1}\}/(d - S). \tag{8}$$

[For $S = d$, the above expression is defined by continuity to be c^2.] The expectation of (8) with respect to the distribution of S will then provide a representation for $E_d(n)$; or alternatively we have, in the notation of Theorem 1,

Theorem 3

Ignoring "excess",

$$d^2 E_d(n) = A \cdot E\{\tfrac{1}{2} - (1 + e^{AV})^{-1}\}/V \tag{9}$$

where $V \sim N(1, 1/B^2)$, $A = 2cd$ and $B^2 = md^2$.

It is interesting to note that this approximation for $d^2 E_d(n)$ does not require that the parameter d be specified explicitly, although this would be required for an exact calculation of $d^2 E_d(n)$.

Formula (9) implies that the average new data sample-size is largest when d = 0. This is not surprising because d = 0 is the boundary between the two hypothesis sets; the proof uses the fact that $\{\frac{1}{2} - (1 + e^{AV})^{-1}\}/V$ is a decreasing function of $|V|$ which has monotone likelihood ratio in d^2.

It may also be shown that for d = 0 the average new data sample-size $E_0(n)$ *increases* with m (the old data sample-size) to its limiting value c^2. This can be shown by putting d = 0 in (9) and making the substitution $U = m^{1/2}V$. Then U will have a distribution not depending on m and the integrand will be a monotonically increasing function of m for each value of U.

When the old data fixed sample size m is large, one obtains the unsurprising result that

$$\lim_{m \to \infty} d^2 E_d(n) = A\{\tfrac{1}{2} - [1 + e^{A}]^{-1}\}. \tag{10}$$

It follows because, as $m \to \infty$, the random variable V of (9) tends in probability to 1. As expected, this is exactly the Wald approximation for the one-sample SPRT one would use if the old data mean μ_x were known without error. When, in addition, the width parameter c of the continuation region is large (corresponding to small error probabilities) then the term $[1 + e^{A}]^{-1}$ becomes negligible if d is strictly positive, and we get the following approximation for large m and c:

$$d^2 \, E_d(n) \approx \frac{1}{2} A, \qquad d \neq 0 \tag{11}$$

or $E_d(n) \approx c/d$ in terms of the original constants.

This remarkably simple approximation indicates that the expected size of the new data sample (for large m and c) is proportional to the width of the continuation region, inversely proportional to the size of the shift parameter d, and independent of the old data sample size m. For the case d near 0, the upper bound $c^2 > E_0(n)$ can be used as an approximation when m is large.

Table II (rows labeled I) gives values of $d^2 \, E_d(n)$ for selected values of the parameters A and B. These are the same parameter values used in Table I for the error probability calculations. The entries were computed from formula (9) by numerical methods. It should be remembered that the formula itself is based on the Wald approximation which has been shown in [2] to actually be a lower bound. In contrast to the error probabilities in Table 1 which were sensitive to uncertainty in the old data mean, we see that the expected new data sample-size remains virtually constant for given A within the range of the table. Indeed, the simple approximation $d^2 \, E_d(n) \sim \frac{1}{2} A$ given in (11) does moderately well throughout the whole table.

As a check on the usefulness of the Wald approximations (9), the 1000 independent simulations referred to in Section 3 were used also to obtain Monte Carlo estimates of $d^2 \, E_d(n)$ and their standard errors. Two independent sets of estimates were obtained, one for $d = 0.25\sigma$ and one for $d = 0.50\sigma$, since in

TABLE II. Average Sample Sizes for the Sequential Procedure (1). The Tabled Quantity is $d^2 \cdot E_d(n)$; $A \equiv 2cd$ and $B^2 \equiv md^2$

		$B = 1.50$	$B = 1.75$	$B = 2.00$	$B = 2.25$	$B = \infty$
	I	1.4	1.4	1.4	1.4	1.36
$A = 3.00$	II	1.6	1.6	1.6	1.6	
	III	1.9	1.9	1.9	1.8	
	I	2.2	2.1	2.1	2.1	1.93
$A = 4.00$	II	2.4	2.3	2.3	2.3	
	III	2.6	2.6	2.6	2.6	
	I	2.8	2.8	2.8	2.8	2.47
$A = 5.00$	II	3.1	3.1	3.0	3.0	
	III	3.4	3.4	3.4	3.3	
	I	3.4	3.4	3.3	3.3	3.00
$A = 6.00$	II	3.9	3.8	3.7	3.6	
	III	4.4	4.2	4.1	4.0	

reality $d^2\ E_d(n)$ may vary somewhat with d even after A and B are specified. These Monte Carlo estimates are shown in rows II and III of Table II and are to be compared with each other and with the approximations of rows I.

The approximations appear to be about 10% below the estimates for $d = 0.25\sigma$ and about 20% below the $d = 0.50\sigma$ estimates. As noted earlier in connection with the error probability, the approximations will do better when we expect a longer sequence of observations, i.e., when d is smaller. The estimated standard errors of the above mentioned sample-size estimates ranged from about 2% - 4%.

While the standard errors seem small, the estimated standard deviations of $d^2\ E_d(n)$ are $\sqrt{1000}$ or about 30 times greater.

This means that they are anywhere from approximation 50% - 100% of the mean sample-size and that the sample-size distribution is markedly skewed. Even with low mean sample-sizes there might still be appreciable probabilities of long sampling sequences. Truncated sampling procedures were discussed briefly at the end of Section 3.

V. COMPARISON WITH A FIXED NUMBER OF NEW DATA OBSERVATIONS

For a given old data sample-size m, the elementary normal theory specifies a unique fixed new data sample-size n_F, which yields a specified α error probability for specified values of the shift parameter d, viz.

$$n_F = mz_\alpha^2/(md^2 - z_\alpha^2) \text{ where } \Phi(z_\alpha) = \alpha, \tag{12}$$

or rather the next largest integer. *[*For a given m, there is a lower bound to the error probability no matter how large we take n_F. This is the same lower α-bound given for the sequential procedure in (6).*]*

Perhaps everyone expects to find that n_F exceeds $E_d(n)$. The sample-size comparisons are exhibited in Table III for values of α ranging from 0.03 to 0.10. Rows I and II were found by interpolation in Table I and Table II: row I entries are based on Theorems 1, 2, and 3; row II entries are based on the 1000 simulated Monte Carlo experiments using $d = 0.25\sigma$. Row III was computed using formula (12). Dashes in the table indicate entries which were not computed.

Inspection of Table III shows that rows I and II are nearly identical (as one hopes). But rows III indicate that enormous savings involving factors of 6 and higher in new data sampling effort - on the average - can be effected by using the recommended sequential procedure. The savings are highest when α is smallest. Typically, the fixed sample-size is about double the expectation of the random sample-size.

TABLE III. Comparison of Fixed and Random Sample Sizes for the New Data Sample. Tabled Values are $d^2 E_d(n)$ for Rows I and II, and $d^2 n_F$ for Row III. $B^2 = md^2$

		B = 1.50	B = 1.75	B = 2.00	B = 2.25	B = ∞
α = 0.03	I	∞	∞	-	3.4	1.6
	II	∞	∞	-	3.3	-
	III(fixed)	∞	∞	30.0	11.8	3.5
α = 0.04	I	∞	∞	-	2.8	1.5
	II	∞	∞	3.7	2.8	-
	III(fixed)	∞	∞	13.0	7.8	3.1
α = 0.05	I	∞	-	2.9	2.4	1.3
	II	∞	3.7	2.8	2.4	-
	III(fixed)	∞	22.5	8.3	5.8	2.7
α = 0.06	I	∞	3.4	2.5	2.0	1.1
	II	∞	3.1	2.3	2.1	-
	III(fixed)	∞	11.6	6.1	4.7	2.4
α = 0.07	I	-	2.8	2.1	1.8	1.1
	II	-	2.6	2.1	1.9	-
	III(fixed)	60.0	7.4	4.7	3.8	2.2
α = 0.08	I	-	2.3	1.8	1.6	1.0
	II	3.3	2.2	1.9	1.7	-
	III	16.0	5.5	3.9	3.2	2.0
α = 0.09	I	3.1	1.9	1.6	-	1.0
	II	2.7	2.0	-	-	-
	III(fixed)	9.0	4.3	3.3	2.8	1.8
α = 0.10	I	2.4	1.7	-	-	0.9
	II	2.3	1.8	-	-	-
	III(fixed)	5.9	3.5	2.8	2.4	1.6

It may be tempting to simply ignore the uncertainty in the old data average $\bar{X}_m$ and calculate error probabilities and average new data sample-sizes using the standard results from the theory of the one sample SPRT. This corresponds to the B = ∞ column of Table III. The entries in that column were calculated using the limiting form $(\frac{1}{2} - \alpha)\ln(1 - \alpha)/\alpha$ for rows I, and the limiting form z_α^2 for rows III. However, ignoring the old data uncertainty could be very misleading. For example, suppose we want an error probability $\alpha = 0.08$ at $d = 0.25\sigma$. Table III indicates that by ignoring the $\bar{X}_m$ uncertainty one has $B = \infty$ and an expected new data sample size $E_{0.25}(n) = (1.0)/(0.25)^2 = 16$; but if m were 81 then $B = 2.25$ and the expected new data sample-size should really be $E_{0.25}(n) = (1.7)/(0.25)^2 = 27.2$. This effect is even more pronounced for smaller α values. In general, it appears that very large old data sample-sizes are needed before we can safely ignore the variation in the old data sample average.

VI. OPTIMIZATION OF THE SEQUENTIAL PROCEDURE WHEN THE "OLD" DATA IS STILL TO BE COLLECTED; COMPARISON WITH BEST FIXED SAMPLE-SIZE PROCEDURES AND PROCEDURES FOR SAMPLING IN PAIRS

Suppose we are in the fortunate position of being able to choose the old data sample-size m in advance. For any m greater than z_α^2/d^2 we can achieve an error probability α for a shift d using a sequential sampling plan for the new data of the type we have been discussing. As Table III well illustrates, a larger m leads to a smaller $E_d(n)$. We would like to minimize $\nu = m + E_d(n)$ for a chosen α, i.e., minimize the expected total sample-size for fixed d.

A general approach to this minimization problem seems to lead to great difficulties. An early conjecture that $m = E_d\{n\}$ yields minimum ν, i.e., equal sample-sizes, was easily proved false in general. A glance at Table III for $\alpha = 0.08$ suggests that $d^2\nu = md^2 + d^2 \cdot E_d(n)$ has a local minimum of approximately 5.3 near $md^2 = 3$. Unreported refinements of this table strongly suggest that this is a global minimum and that we can come within 10% of the minimum using md^2 values extending over most of the interval from 2 to 4. In Table IV, columns I and II show the approximate optimized value of md^2 and the approximate minimum value of $md^2 + d^2 \cdot E_d(n)$ for α ranging from 0.03 to 0.10.

It is interesting to compare the minimum expected total sample-size ν with the corresponding total $m + n_F$ when the best *fixed* sample-size procedure is used. It follows that $m + n_F$ is smallest for given α when $md^2 = d^2 n_F$, i.e., equal sample sizes. The corresponding minimum value of $d^2(m + n_F)$ is $4z_\alpha^2$. These minimum values are shown in column III of Table IV and can be compared directly with the corresponding minimum average total sample-sizes of column II for the recommended sequential procedure. In so doing we note that failure to sample the new data sequentially will increase the new plus old sample-size by about 50% - 75%, a moderate increase.

The more common two-sample sequential sampling procedure takes the observations in pairs and uses an SPRT based on the sequence of differences. Strictly speaking, this is not a competitor to the sequential procedure of Section 2 because we have stipulated that sampling of the "old" data must be completed before "new" data sampling can begin. Nevertheless it seems interesting to compare the average total sample-size $E_d(n_p)$ of the sampling-in-pairs procedure with the other sample-size results in Table IV. Values of $d^2 \cdot E_d(n_p)$ are listed in column IV of the table and were computed using the Wald approximation for such a procedure [which is really a lower bound]:

$$d^2 \cdot E_d(n_p) = 2(1 - \alpha) \ln \frac{(1 - \alpha)}{\alpha} .$$

Standard results for the SPRT based on differences can be found in *[2]*, for example. Within the range of the table, it would be fair to conclude that not being able to sample in pairs increases the average sample-size by 25% - 30%.

TABLE IV. Optimization of the Size m of the old Data Sample for Fixed Shift d

	I	II	III	IV
	Optimized md^2	min $d^2[m+E_d(n)]$	min $d^2[m+n_F]$	$d^2E_d(n_p)$ (Sampling in pairs)
$\alpha = .03$	4.5	8.3	14.1	6.5
$\alpha = .04$	4.0	7.6	12.2	5.9
$\alpha = .05$	4.0	6.8	10.8	5.3
$\alpha = .06$	3.5	6.0	9.7	4.8
$\alpha = .07$	3.5	5.6	8.7	4.4
$\alpha = .08$	3.0	5.3	7.9	4.1
$\alpha = .09$	2.5	5.0	7.2	3.8
$\alpha = .10$	2.5	4.6	6.6	3.5

VII. APPENDIX

Proof of Lemma of Section 3. Begin with the expression (3) for $\alpha(d)$,

$$[1 + e^{AV}]^{-1} = \sum_{k=0}^{\infty} (-e^{AV})^k \quad \text{for all } V < 0$$

$$= e^{-AV} \sum_{k=0}^{\infty} (-e^{-AV})^k \quad \text{for all } V > 0 \tag{13}$$

$$= \frac{1}{2}[1 - \text{sgn}(V)] - \text{sgn}(V) \sum_{k=1}^{\infty} (-e^{A|V|})^k$$

$$\text{for all } V \neq 0.$$

The series (13) converges absolutely and uniformly in any V region which excludes a neighborhood of V = 0. We can integrate the individual terms of the infinite sum with respect to the normal distribution of V, excluding such a neighborhood, and get

$$\int_{|V|B>\varepsilon>0} \operatorname{sgn}(V)e^{-kA|V|}\, d\Phi(VB - B) \tag{14}$$
$$= e^{-\varepsilon kA/B}\{\phi(B - \varepsilon)M(\varepsilon - B + kA/B) - \phi(B + \varepsilon)M(\varepsilon + B + kA/B)\}.$$

Summing the above expression over k, one obtains another infinite sum which converges to

$$\int_{|V|B>\varepsilon} [1 + e^{AV}]^{-1}\, d\Phi(VB - B)$$

for any $\varepsilon > 0$ by virtue of the previously mentioned uniformity.

Considering the integrated series as a function of ε, we see that the convergence of the integrated series is absolute and uniform on every finite interval $0 \leq \varepsilon \leq \varepsilon_0$ including $\varepsilon = 0$. This follows because (i) it can be shown that (14) achieves its largest value at $\varepsilon = 0$ for all sufficiently large k, and (ii) the integrated series is absolutely convergent at $\varepsilon = 0$. Since the terms of the integrated series are continuous functions of ε, their infinite sum is right continuous at $\varepsilon = 0$ and its value at $\varepsilon = 0$ is $\int_{-\infty}^{\infty} [1 + e^{AV}]^{-1}\, d\Phi(VB - B)$, i.e., $\alpha(d)$ as given by (3). The RHS of (5) is precisely the integrated series when $\varepsilon = 0$, and this completes the proof.

VIII. REFERENCES

[1] Anderson, T.W. (1960). "A Modification of the Sequential Probability Ratio Test to Reduce the Sample Size." *Ann. Math. Statist. 31*, 165-197.

[2] Bechhoffer, R.E., Kiefer, J., and Sobel, M. (1968). *Sequential Identification and Ranking Procedures*, University of Chicago Press.

[3] Choi, S.C. (1977). "Two-Sample Sequential Test of Normal Distribution when one of the Sample Sizes is Fixed." *Comp. and Maths. with Appls. 3*, 125-129.

[4] Switzer, P. (1970). "A Two-Sample Sequential Test with One Sample-Size Fixed." Stanford University, Department of Statistics, Technical Report No. 158.

[5] Wald, A. (1947). *Sequential Analysis,* John Wiley and Sons, Inc., New York.

[6] Wolfe, D.A. (1977). "On a Class of Partially Sequential Two-Sample Test Procedures." *J. Amer. Statist. Assn. 72,* 202-205.

ON SEQUENTIAL RANK TESTS

Michael Woodroofe[1]

Department of Statistics
University of Michigan
Ann Arbor, Michigan

I. INTRODUCTION

Let F and G be continuous distribution functions and let $X_1, X_2, \ldots$ and $Y_1, Y_2, \ldots$ be independent random variables for which $X_1, X_2, \ldots$ have common distribution function F and $Y_1, Y_2, \ldots$ have common distribution function G. In the context of nonparametric tests of the hypothesis H_0: F = G, there is interest in statistics of the form

$$T = n \int J(F_m, G_n) dF_m,$$

where

$$F_m(x) = (1/m) \sum_{i=1}^{m} I_{(-\infty,x]}(X_i), \quad -\infty < x < \infty,$$

and (1)

$$G_n(y) = (1/n) \sum_{j=1}^{n} I_{(-\infty,y]}(Y_j), \quad -\infty < y < \infty,$$

denote the empirical distribution functions of $X_1, \ldots, X_m$ and $Y_1, \ldots, Y_n$ for $m \geq 1$ and $n \geq 1$, and J is a function on the square

[1] *Research supported by the National Science Foundation under MCS-8101897.*

ISBN 0-12-589320-5

$[0,1] \times [0,1]$. The fundamental result of Chernoff and Savage (1958) established the asymptotic normality of T, suitably rescaled, as m and $n \to \infty$ with m/n bounded away for 0 and ∞, under modest conditions on the function J. One of the most interesting extensions of the Chernoff-Savage Theorem is that of Lai (1975). In the special case that m = n, his results show that $T = T_n$ may be written in the form

$$T = S_n + \xi_n,$$

where S_n, $n \geq 1$, is a random walk, which is described in more detail in Section 3, and

$$n^{-\alpha} \cdot \xi_n \to 0 \qquad \text{w.p.1. as } n \to 0,$$

for certain values of $\alpha > 0$, depending on the function J. For example, a law of the iterated logarithm for $T = T_n$ follows easily from Lai's theorem.

For applications to sequential analysis, one needs to know when the ξ_n, $n \geq 1$, are slowly changing in the following sense: there is a β for which

$$\frac{1}{2} < \beta < 1,$$

and

$$\max_{0 \leq k \leq n^\beta} |\xi_{n+k} - \xi_n| \to_p 0, \qquad \text{as } n \to \infty,$$

where $\to_p$ denotes convergence in probability. In fact, this is the major condition imposed in Lai and Siegmund's (1977) non-linear renewal theory. Conditions under which ξ_n, $n \geq 1$, are slowly changing are developed here.

The paper proceeds as follows: Section 2 reviews some properties of empirical distribution functions; Section 3 presents the main theorem and its proof; and Section 4 develops an application to the sequential probability ratio tests of Savage and Sethuraman (1966) and Sethuraman (1970).

II. PRELIMINARIES

In this section, some properties of the sample distribution function are reviewed. The notations and standing assumptions of the introduction are used throughout. Thus, F and G denote continuous distribution functions; $X_1, X_2, \ldots, Y_1, Y_2, \ldots$ denote independent random variables for which $X_1, X_2, \ldots$ have common distribution function F and $Y_1, Y_2, \ldots$ have common distribution function G; and F_n and G_n denote the empirical distributions functions of $X_1, \ldots, X_n$ and $Y_1, \ldots, Y_n$, as in (1). In addition, let

$$H = \frac{1}{2}F + \frac{1}{2}G \text{ and } H_n = \frac{1}{2}F_n + \frac{1}{2}G_n, \qquad n \geq 1.$$

Then

$$F \leq 2H \geq G \text{ and } F_n \leq 2H_n \geq G_n, \qquad n \geq 1.$$

Proposition 1

Let $\varepsilon_n = \log^4 n/n$, $n \geq 2$. Then

$$\limsup_{n \to \infty} \sup_{\varepsilon_n < F < 1 - \varepsilon_n} \frac{\sqrt{n}\cdot|F_n - F|}{\sqrt{[2F(1-F)\log\log n]}} = 1 \qquad \text{w.p.1.}$$

Proposition 1 follows directly from Theorem 5 of Csorgo and Revesz (1974) by applying to the latter the random variables $F(X_1)$ and $1 - F(X_1)$.

Let $\log_1 x = \max\{1, \log x\}$, $x > 0$.

Corollary 1. Let $0 < \alpha < 1$ and let

$$D_n = \sqrt{[\frac{n}{\log_1 n}]}\,\frac{|F_n - F|}{\sqrt{[H(1-H)]}} \qquad n \geq 1.$$

Then

$$D = \sup_{n \geq 1} \sup_{n^{-\alpha} < H < 1 - n^{-\alpha}} D_n < \infty \qquad \text{w.p.1.}$$

Proof. Let $I_{n1} = \{x: F(x) \leq \varepsilon_n, n^{-\alpha} < H(x) < 1 - n^{-\alpha}\}$ and $I_{n2} = \{x: F(x) \geq 1 - \varepsilon_n, n^{-\alpha} < H(x) < 1 - n^{-\alpha}\}$. Then it suffices to show that $\sup_{n \geq 1} \sup_{I_{n1}} D_n$ and $\sup_{n \geq 1} \sup_{I_{n2}} D_n$ are finite w.p.1. The two arguments are similar, so only the first is detailed. Let $x_n = \inf\{x: F(x) \geq \varepsilon_n\}$. Then, for all $x \in I_{n1}$

$$D_n \leq \sqrt{\left[\frac{2n^{1+\alpha}}{\log n}\right]}\,[F_n(x_n) + \varepsilon_n], \qquad n \geq 3;$$

and

$$P\{F_n(x_n) > 2\sqrt{\left[\frac{\log n}{n^{1+\alpha}}\right]} \leq P\{nF_n(x_n) > 2n^{\frac{1}{2}(1-\alpha)}\}, \quad n \geq 3,$$

which is summable by a simple application of Berstein's inequality. Thus, $P\{\sup_{I_{n1}}, D_n > 3, \text{ i.o.}\} = 0$ by the Borel Cantelli Lemmas.

The second result was used by Lai (1975, expression (2.5)).

Proposition 2

If $c > 0$, $\delta > 1$, and $0 < \alpha < 1$, then there are positive constants C and λ for which

$$P\{\sup_{F \geq cn^{-\alpha}} \frac{F}{F_n} > \delta \text{ or } \sup_{1-F \geq cn^{-\alpha}} \frac{1-F}{1-F_n} > \delta\} \tag{2}$$

$$\leq C \exp[-\lambda n^{\frac{1}{2}(1-\alpha)}]$$

for all $n \geq 1$.

Corollary 2. Let $0 < \alpha < 1$ and let B_n be the event

$$B_n = \{H_k(1 - H_k) > \frac{1}{11} H(1 - H) \text{ on } n^{-\alpha} < H < 1 - n^{-\alpha},$$

$$\forall k \geq n\}, \quad n \geq 1.$$

Then

$$P(B_n) \to 1, \qquad \text{as } n \to \infty.$$

Proof. Let $A_{n,1}$ be the event whose probability is bounded in (2) with $c = 1$ and $\delta = 9/8$; let $A_{n,2}$ be the same event with F_n and F replaced by G_n and G; and let $A_n = A_{n,1} \cup A_{n,2}$, $n \geq 1$. Then it suffices to show that $B_n' \subset \cup_{k=n}^{\infty} A_k$. Consider a value of x for which $F(x) \geq G(x)$ and $n^{-\alpha} \leq H(x) \leq \frac{1}{2}$. Then $F(x) \geq n^{-\alpha} \geq k^{-\alpha}$ for all $k \geq n \geq 1$. Thus, if $\cup_{k=n}^{\infty} A_k$ does not occur, then

$$H_k(1 - H_k) \geq \frac{2}{9}H_k \geq \frac{1}{9}F_k \geq \frac{1}{11}F \geq \frac{1}{11}H(1 - H)$$

for all $k \geq n$. The other three cases--$G \leq F$ and $n^{-\alpha} < H \leq \frac{1}{2}$, $F \geq G$ and $\frac{1}{2} \leq H < 1 - n^{-\alpha}$, and $G \geq F$ and $\frac{1}{2} \leq H \leq 1 - n^{-\alpha}$ may be analyzed similarly to complete the proof.

III. THE MAIN THEOREM

As in the introduction, let F and G denote continuous distribution functions; let $X_1, X_2, \ldots, Y_1, Y_2, \ldots$ be independent random variables for which $X_1, X_2, \ldots$ have common distribution function F and $Y_1, Y_2, \ldots$ have common distribution function G; and let

$$T_n = n \int J(F_n, G_n)\, dF_n,$$

where F_n and G_n denote the empirical distribution functions of $X_1, \ldots, X_n$ and $Y_1, \ldots, Y_n$ for $n \geq 1$ and J satisfies the following condition. <u>Condition C</u>: J is a real valued function on the square $[0,1] \times [0,1]$; J is twice continuously differentiable on the interior $(0,1) \times (0,1)$; and there are constants C and δ for which

$$0 < C < \infty \quad \text{and} \quad 0 < \delta < \frac{1}{4}$$

and

$$|J_{20}(x,y)| + |J_{11}(x,y)| + |J_{02}(x,y)| \le C\{\frac{1}{z(1-z)}\}^{2+\delta}$$

for

$$0 < x < 1 \quad \text{and } 0 < y < 1 \text{ where } \quad z = \frac{1}{2}(x + y)$$

and

$$J_{ij}(x,y) = \frac{\partial^{i+j}}{\partial x^i \partial y^j} J(x,y) \qquad i + j \le 2.$$

The condition implies the existence of another constant, also denoted by C, for which

$$|J_{10}(x,y)| + |J_{01}(x,y)| \le C\{\frac{1}{z(1-z)}\}^{1+\delta}$$

and

$$|J(x,y)| \le C\{\frac{1}{z(1-z)}\}^{\delta}.$$

Thus, J_{10} and J_{01} admit continuous extensions to all points of the square except for (0,0) and (1,1); and J must be continuous at all points except (0,0) and (1,1). No assumption of continuity at (0,0) and (1,1) is made.

To describe the decomposition $T_n = S_n + \xi_n$, $n \ge 1$, it is convenient to let

$$\Delta_{n,0} = J(F_n,G_n) - J(F,G)$$

$$\Delta_{n,1} = J_{10}(F,G)(F_n - F) + J_{01}(F,G)(G_n - G)$$

and

$$\Delta_{n,2} = \Delta_{n,0} - \Delta_{n,1}, \qquad n \ge 1.$$

Then

$$T_n = S_n + \xi_n,$$

where

$$S_n = n \int J(F,G)\,dF_n + n \int \Delta_{n,1}\,dF$$

and

$$\xi_n = n \int \Delta_{n,1}(dF_n - dF) + n \int \Delta_{n,2}\,dF_n, \qquad n \geq 1.$$

Clearly, S_n, $n \geq 1$, is a random walk. In fact, letting x_0 be a value with $0 < H(x_0) < 1$,

$$S_n = \sum_{i=1}^{n} J[F(X_i),G(X_i)] - \sum_{i=1}^{n} [u_{10}(X_i) - \int u_{10}\,dF]$$

$$- \sum_{i=1}^{n} [u_{01}(Y_i) - \int u_{01}\,dG],$$

where

$$u_{10}(x) = \int_{x_0}^{x} J_{10}[F(z),G(z)]\,dF(z)$$

and

$$u_{01}(x) = \int_{x_0}^{x} J_{01}[F(z),G(z)]\,dF(z), \quad \text{for } 0 < H(x) < 1.$$

Theorem 1

Suppose that condition C is satisfied. If $0 < \beta < 1 - 2\delta$

$$\max_{k \leq n^{\beta}} |\xi_{n+k} - \xi_n| \to_p 0, \qquad n \to \infty. \tag{3}$$

The proof is given in Lemmas 1-4 below. To begin, fix a value of β for which $0 < \beta < 1 - 2\delta$ and let α be a value for which

$$1 - \tfrac{1}{2}\delta < \alpha < 1 \quad \text{and} \quad 0 < \beta < (1 - 2\delta)\alpha.$$

Write

$$m = [n^{\beta}], \quad \max_k = \max_{0 \leq k \leq m},$$

and

$$I_n = \{z\colon n^{-\alpha} < H(z) < 1 - n^{-\alpha}\}, \qquad n \geq 1.$$

Lemma 1. As $n \to \infty$,

$$\max_k \left| (n+k) \int_{I_n'} \Delta_{n+k,0} dF_{n+k} - n\int_{I_n'} \Delta_{n,0} dF_n \right| \to_p 0, \tag{4a}$$

$$\max_k \left| (n+k) \int_{I_n'} \Delta_{n+k,1} dF_{n+k} - n\int_{I_n'} \Delta_{n,1} dF_n \right| \to_p 0, \tag{4b}$$

$$\max_k \left| (n+k) \int_{I_n'} \Delta_{n+k,1} dF - n\int_{I_n'} \Delta_{n,1} dF \right| \to_p 0, \tag{4c}$$

and

$$\max_k \left| (n+k) \int_{I_n'} \Delta_{n+k,2} dF_{n+k} - n\int_{I_n'} \Delta_{n,2} dF_n \right| \to_p 0. \tag{4d}$$

Proof. In all four cases, one may write

$$\int_{I_n'} = \int_{H \leq n^{-\alpha}} + \int_{H \geq 1-n^{-\alpha}}, \qquad n \geq 1, \tag{5}$$

and the analyses of the two integrals on the right side of (5) are similar. Only the analyses of $\int_{H \leq n^{-\alpha}}$ are given.

Let B_n be the event

$$B_n = \{H(X_{n+i}) > n^{-\alpha} \text{ and } H(Y_{n+i}) > n^{-\alpha}, \text{ for all } i \leq m\}.$$

Then

$$P(B_n') \leq mP\{F(X_1) < 2n^{-\alpha} \text{ or } G(Y_1) < 2n^{-\alpha}\} \leq 4mn^{-\alpha} \to 0$$

as $n \to \infty$, since $F \leq 2H \geq G$ and $\beta < \alpha$. Observe that B_n implies

$$F_{n+k} = \frac{n}{n+k} F_n \quad \text{and} \quad G_{n+k} = \frac{n}{n+k} G_n$$

on $H \leq n^{-\alpha}$ for all $k \leq m$.

To prove (4a), let $\varepsilon(n,k)$ be the term on the left side of (4a), but with I_n' replaced by $\{H \leq n^{-\alpha}\}$ for $k \leq m$. Then

$$\varepsilon(n,k) I_{B_n} \leq n\int_{H \leq n^{-\alpha}} \left| \Delta_{n+k,0} - \Delta_{n,0} \right| dF_n, \; 1 \leq k \leq m, \; n \geq 1.$$

Now

$$\left| \Delta_{n+k,0} - \Delta_{n,0} \right| = \left| J(F_{n+k}, G_{n+k}) - J(F_n, G_n) \right|$$

$$\leq |J_{10}^*| \, | F_{n+k} - F_n | + |J_{01}^*| \, | G_{n+k} - G_n |$$

where J^*_{10} and J^*_{01} denote J_{10} and J_{01} evaluated at intermediate points between F_{n+k} and F_n and G_{n+k} and G_n for $1 \leq k \leq m$ and $n \geq 1$. Thus, there is a constant C for which

$$|\Delta_{n+k,0} - \Delta_{n,0}|I_{B_n} \leq C\left(\frac{1}{H_n}\right)^{1+\delta} \left[\left(\frac{k}{n+k}\right)(F_n + G_n)\right]$$

$$\leq \frac{4Ck}{n}\left[F_n^{-\delta} + F_n^{-1-\delta}G_n\right], \quad 1 \leq k \leq m, \quad n \geq 1.$$

Let $X_{n1},\ldots,X_{nn}$ denote the order statistics of $X_1,\ldots,X_n$; and let $K_n = \#\{i \leq n\colon H(X_i) \leq n^{-\alpha}\}$, so that K_n has the binomial distribution with parameters n and $p_n \leq 2n^{-\alpha}$ for $n \geq 1$. Then

$$\max_k \varepsilon(n,k) I_{B_n} \leq \frac{4Cm}{n}\sum_{i=1}^{K_n}\left(\frac{n}{i}\right)^{\delta} + \frac{4Cm}{n}\sum_{i=1}^{K_n}\left(\frac{n}{i}\right)^{1+\delta} G_n(X_{ni})$$

$$= \varepsilon_1(n) + \varepsilon_2(n), \quad \text{say, for } n \geq 1.$$

Now,

$$\varepsilon_1(n) \leq C'm\left(\frac{K_n}{n}\right)^{1-\delta} = O_p[n^{\beta-(1-\delta)\alpha}] = o_p(1), \text{ as } n \to \infty;$$

and

$$E[\varepsilon_2(n)|X_1,\ldots,X_n] = \frac{4Cm}{n}\sum_{i=1}^{K_n}\left(\frac{n}{i}\right)^{1+\delta} G(X_{ni})$$

$$\leq \frac{8Cm}{n^{1+\alpha}}\sum_{i=1}^{K_n}\left(\frac{n}{i}\right)^{1+\delta}$$

$$\leq 8Cn^{\beta+\delta-\alpha}\sum_{i=1}^{\infty}\left(\frac{1}{i}\right)^{1+\delta} \to 0, \quad \text{as } n \to \infty$$

for some constant C'. So, $\varepsilon_2(n) \to_p 0$ as $n \to \infty$, by Markov's inequality and the bounded convergence theorem. It follows that $\max_k \varepsilon(n,k) \to_p 0$ as $n \to \infty$, completing the discussion of (4a). See (5).

The analysis of (4b) is similar. Let $\varepsilon(n,k)$ be the term on the left side of (4b) with I_n' replaced by $\{H \le n^{-\alpha}\}$ for $1 \le k \le m$ and $n \ge 1$. Then, as above,

$$\varepsilon(n,k) I_{B_n} \le n \int_{H \le n^{-\alpha}} |\Delta_{n+k,1} - \Delta_{n,1}| \, dF_n$$

$$\le Cm \int_{H \le n^{-\alpha}} \left(\frac{1}{F}\right)^{1+\delta} (F_n + G_n) \, dF_n$$

for $1 \le k \le m$ and $n \ge 1$ for some constant C. Let $U_{ni} = F(X_{ni})$, $1 \le i \le n$, where $X_{n1}, \ldots, X_{nn}$ are the order statistics of $X_1, \ldots, X_n$, and let $V_n = nU_{n1}$, $n \ge 1$. Then V_n has a limiting exponential distribution, so that $1/V_n$ is stochastically bounded. So.

$$\max_k \varepsilon(n,k) I_{B_n} \le \frac{Cm}{n} \sum_{i=1}^{K_n} \left(\frac{1}{U_{ni}}\right)^{1+\delta} [(\tfrac{i}{n}) + G_n(X_{ni})]$$

$$\le Cmn^{\delta} \left(\frac{1}{V_n}\right)^{1+\delta} \sum_{i=1}^{K_n} [(\tfrac{i}{n}) + G_n(X_{ni})].$$

As above, the latter is of order n^q, where $q = \beta + \delta + 1 - 2\alpha < 0$, and so it tends to zero in probability as $n \to \infty$. This completes the discussion of (4b).

Observe that (4d) follows directly from (4a) and (4b).

Finally, let $\varepsilon(n,k)$ denote the left side of (4c) with I_n' replaced by $\{H < n^{-\alpha}$, for $1 \le k \le m$ and $n \ge 1$. Then

$$\varepsilon(n,k) = \varepsilon_1(n,k) + \varepsilon_2(n,k), \quad 1 \le k \le m, \ n \ge 1$$

where

$$\varepsilon_1(n,k) = (n + k) \int_{H \le n^{-\alpha}} [\Delta_{n+k,1} - \Delta_{n,1}] \, dF$$

and

$$\varepsilon_2(n,k) = k \int_{H \le n^{-\alpha}} \Delta_{n,1} \, dF, \quad 1 \le k \le m, \ n \ge 1.$$

The proof that $\max_k |\varepsilon_1(n,k)| \to_p 0$ as $n \to \infty$ is similar to, but slightly simpler than, the proofs of (4a) and (4b). The details are left to the reader. Finally, $\max_k |\varepsilon_2(n,k) = |\varepsilon_2(n,m)|$, and for some C

$$E[|\varepsilon_2(n,m)|] \leq m\int_{H \leq n^{-\alpha}} E|\Delta_{n,1}|\, dF$$

$$\leq Cm\int_{H \leq n^{-\alpha}} (\tfrac{1}{H})^{1+\delta} E[|F_n - F| + |G_n - G|]\, dF$$

$$\leq \frac{2Cm}{\sqrt{n}} \int_{H \leq n^{-\alpha}} (\tfrac{1}{H})^{1+\delta} \sqrt{H}\, dH \to 0, \text{ as } n \to \infty,$$

to complete the proof of the lemma.

Below, estimates are needed for integrals of the form

$$I_n = \int_{I_n} \{\frac{1}{H(1-H)}\}^{1+\gamma} dF_n \text{ and } E(I_n) = \int_{I_n} \{\frac{1}{H(1-H)}\}^{1+\gamma} dF,$$

where $\gamma > 0$. The second of these two integrals is easily bounded. One finds

$$E(I_n) = 0(n^{\alpha\gamma}) \text{ and } I_n = 0_p(n^{\alpha\gamma}), \text{ as } n \to \infty. \tag{6}$$

To continue with the proof of the theorem, let

$$\xi_1(n,k) = (n + k) \int_{I_n} \Delta_{n+k,1}(dF_{n+k} - dF)$$
$$- n \int_{I_n} \Delta_{n,1}(dF_n - dF)$$

and

$$\xi_2(n,k) = (n + k) \int_{I_n} \Delta_{n+k,2}\, dF_{n+k} - n \int_{I_n} \Delta_{n,2}\, dF_n$$

for $k,n \geq 1$. Then it suffices to show that $\max_k |\xi_1(n,k)| \to_p 0$ and $\max_k |\xi_2(n,k)| \to_p 0$ as $n \to \infty$. In the proofs of these relations, it is convenient to let $F_{n,k}$ and $G_{n,k}$ denote the empirical distribution functions of $X_{n+1},\ldots,X_{n+k}$, and $Y_{n+1},\ldots,Y_{n+k}$. Thus

$$F_{n,k}(z) = (1/k) \sum_{i=1}^{k} I_{(-\infty,z]}(X_{n+i}), \qquad -\infty < z < \infty,$$

and

$$G_{n,k}(z) = (1/k) \sum_{i=1}^{k} I_{(-\infty,z]}(Y_{n+i}), \qquad -\infty < z < \infty,$$

for $k,n \geq 1$. Then the processes $F_n(z)$, $-\infty < z < \infty$, $G_n(z)$, $-\infty < z < \infty$, $F_{n,k}(z)$, $-\infty < z < \infty$, $k \geq 1$, and $G_{n,k}(z)$, $-\infty < z < \infty$, $k \geq 1$, are independent for each fixed n; and, for example,

$$F_{n+k} = \left(\frac{n}{n+k}\right) F_n + \left(\frac{k}{n+k}\right) F_{n,k} \qquad \text{for } k,n \geq 1. \tag{7}$$

Lemma 2. $\max_k |\xi_2(n,k)| \to_p 0$, as $n \to \infty$.

Proof. In view of (7), we may write

$$\xi_2(n,k) = k\int_{I_n} \Delta_{n+k,2}\, dF_{n,k} + n\int_{I_n} [\Delta_{n+k,2} - \Delta_{n,2}]\, dF_n$$

$$= \varepsilon_1(n,k) + \varepsilon_2(n,k), \quad \text{say,}$$

for $k,n \geq 1$. These two terms are considered separately. First, letting J^*_{ij} denote J_{ij} evaluated at intermediate points between F_{n+k} and F and between G_{n+k} and G, one may write

$$\varepsilon_1(n,k) = \tfrac{1}{2}k\int_{I_n} J^*_{20}(F_{n+k} - F)^2\, dF_{n,k}$$

$$+ k\int_{I_n} J^*_{11}(F_{n+k} - F)(G_{n+k} - G)\, dF_{n,k}$$

$$+ \tfrac{1}{2}k\int_{I_n} J^*_{02}(G_{n+k} - G)^2\, dF_{n,k}$$

$$= \varepsilon_{11}(n,k) + \varepsilon_{12}(n,k) + \varepsilon_{13}(n,k), \quad \text{say,}$$

if $0 < H_n < 1$ on I_n. The analyses of these terms are similar, so only that of the first is detailed. Let B_n denote the event

$$B_n = \{H_k(1 - H_k) > \tfrac{1}{11}H(1 - H) \text{ on } n^{-\alpha} < H < 1 - n^{-\alpha}, \tag{8}$$

$$\forall k \geq n\}, \quad n \geq 1.$$

Then

$$P(B_n) \to 1, \qquad \text{as } n \to \infty,$$

by the Corollary to Proposition 2; and B_n implies that $|J^*_{ij}| \leq C[1/H(1-H)]^{2+\delta}$ on $n^{-\alpha} < H < 1 - n^{-\alpha}$ for all $k \geq n$ and $i + j = 2$ for some constant C. Let

$$D = \sup_n \sup_{I_n} \sqrt{(\frac{n}{\log_1 n})}\{\frac{|F_n - F| + |G_n - G|}{\sqrt{[H(1-H)]}}\}. \tag{9}$$

Then $D < \infty$ w.p.1. by the Corollary to Proposition 1. It follows that there is a constant C for which

$$\max_k |\varepsilon_{11}(n,k)| I_{B_n} \leq \frac{CD^2 m\log n}{n} \int_{I_n} \{\frac{1}{H(1-H)}\}^{1+\delta} dF_{n,m}$$
$$= O_p[n^{\beta-1+\alpha\delta} \log n] = o_p(1), \text{ as } n \to \infty, \tag{10}$$

by relation (6). Thus, $\max_k |\varepsilon_{11}(n,k)| \to_p 0$ as $n \to \infty$. Similar arguments show that $\max_k |\varepsilon_{12}(n,k)| \to_p 0$ and $\max_k |\varepsilon_{13}(n,k)| \to_p 0$ as $n \to \infty$. So, $\max_k |\varepsilon_1(n,k)| \to_p 0$ as $n \to \infty$.

Next, consider $\varepsilon_2(n,k)$. If $0 < H_n < 1$, on I_n, then

$$\varepsilon_2(n,k) = \varepsilon_{21}(n,k) + \ldots + \varepsilon_{25}(n,k),$$

where

$$\varepsilon_{21}(n,k) = n\int_{I_n} [J_{10}(F_n,G_n) - J_{10}(F,G)](F_{n+k} - F_n)\, dF_n,$$

$$\varepsilon_{22}(n,k) = n\int_{I_n} [J_{01}(F_n,G_n) - J_{01}(F,G)](G_{n+k} - G_n)\, dF_n,$$

$$\varepsilon_{23}(n,k) = \frac{1}{2}n\int_{I_n} J^*_{20}(F_{n+k} - F_n)^2\, dF_n,$$

$$\varepsilon_{24}(n,k) = n\int_{I_n} J^*_{11}(F_{n+k} - F_n)(G_{n+k} - G)\, dF_n,$$

$$\varepsilon_{25}(n,k) = \frac{1}{2}n\int_{I_n} J^*_{02}(G_{n+k} - G_n)^2\, dF_n,$$

and J^*_{ij} denotes J_{ij} evaluated at intermediate points between F_{n+k} and F_n and between G_{n+k} and G_n for $n,k \geq 1$.

To estimate $\varepsilon_{21}(n,k)$, write $F_{n+k} - F_n = k/(n+k)[(F_{n,k} - F) - (F_n - F)]$ and

$$\varepsilon_{21}(n,k) = (\frac{n}{n+k})[\varepsilon_{211}(n,k) + \varepsilon_{212}(n,k)]$$

with

$$\varepsilon_{211}(n,k) = k\int_{I_n} [J_{10}(F_n,G_n) - J_{10}(F,G)](F_{n,k} - F)\, dF_n$$

and

$$\varepsilon_{212}(n,k) = k\int_{I_n} [J_{10}(F_n,G_n) - J_{10}(F,G)](F_n - F)\, dF_n$$

for $n,k \geq 1$. Let B_n be as in (8) and D as in (9). Then, for some C,

$$\max_k |\varepsilon_{212}(n,k)| I_{B_n} \leq \frac{CD^2 m \log n}{n} \int I_n \{\frac{1}{H(1-H)}\}^{1+\delta}\, dF_n,$$

which tends to zero in probability by (6), as in (10). Next, for fixed n, $\varepsilon_{211}(n,k)$ is a martingale with respect to $A_{n,k} = \sigma\{Y_1,\ldots,Y_n,X_1,\ldots,X_{n+k}\}$, $k \geq 1$. So, for $c > 0$,

$$P\{\max_k |\varepsilon_{211}(n,k)| > c \mid A_{n,0}\} \leq c^{-2} E[\varepsilon_{211}(n,m)^2 \mid A_{n,0}] .$$

It suffices to show that the right side approaches zero. Let $u_n = J(F_n,G_n) - J(F,G)$, $n \geq 1$. Then $|u_n| \leq CD\sqrt{[(\log_1 n)/n]}\{1/H(1-H)\}^{\frac{3}{2}+\delta}$ on I_n for $n \geq 1$; and the conditional expectation is

$$2m \iint_{x \leq y,\ x,y \in I_n} u_n(x) u_n(y) F(x)(1-F(y))\, dF_n(x)\, dF_n(y),$$

which is of order $n^q \log n$ as $n \to \infty$ with $q = \beta - 1 + 2\alpha\delta$, by an argument which is similar to (6). This completes the discussion of ε_{21}, and the analysis of ε_{22} is virtually identical.

Next, consider ε_{23}. One has

$$|\varepsilon_{23}(n,k)| \leq \varepsilon_{231}(n,k) + \varepsilon_{232}(n),$$

where

$$\varepsilon_{231}(n,k) = \frac{k^2}{n} \int_{I_n} J_{20}^*(F_{n,k} - F)^2\, dF_n$$

and

$$\varepsilon_{232}(n) = \frac{m^2}{n} \int_{I_n} J^*_{20}(F_n - F)^2 \, dF_n$$

for $n,k \geq 1$. As above, $\varepsilon_{232}(n) I_{B_n} \leq CD^2 (m/n)^2 \log_1 n \int_{I_n} \{1/H(1 - H)\}^{1+\delta} dF_n = o_p(1)$ as $n \to \infty$ for some constants C. Also, as above, $\varepsilon_{231}(n,k)$ is a sub-martingale with respect to $A_{n,k} = \sigma\{Y_1, \ldots, Y_n, X_., \ldots, X_{n+k}\}$, $k \geq 1$, for each fixed n, and

$$E\{\varepsilon_{231}(n,m) \mid A_{n,0}] \, I_{B_n} \leq \frac{Cm}{n} \int_{I_n} \{1/H(1 - H)\}^{1+\delta} dF_n = o_p(1)$$

as $n \to \infty$. The analyses of ε_{24} and ε_{25} are similar.

Next consider $\xi_1(n,k)$. Since $(n + k)F_{n+k} = nF_n + kF_{n,k}$,

$$\xi_1(n,k) = \xi_{11}(n,k) + \xi_{12}(n,k)$$

with

$$\xi_{11}(n,k) = (n + k) \int_{I_n} [\Delta_{n+k,1} - \Delta_{n,1}](dF_{n+k} - dF)$$

and

$$\xi_{12}(n,k) = k\int_{I_n} \Delta_{n,1}(dF_{n,k} - dF) \quad \text{for } k,n \geq 1.$$

Lemma 3.

$$\max_k |\xi_{12}(n,k)| \to_p 0, \qquad \text{as } n \to \infty .$$

Proof. First observe that $\xi_{12}(n,k)$, $k \geq 1$, is a martingale with respect to the sigma-algebras $A_{n,k} = \sigma\{Y_1, \ldots, Y_n, X_1, \ldots, X_{n+k}\}$, $k \geq 1$, for each fixed $n \geq 1$. Thus, $p\{\max_k |\varepsilon_{12}(n,k)| > c\} \leq c^{-2} E[\xi_{12}(n,m)^2]$ for $c > 0$ and $n \geq 1$ by the martingale inequality. So, it suffices to show that $E[\xi_{12}(n,m)^2] \to 0$ as $n \to \infty$. Now, for some constants C and C',

$$\begin{aligned} E[\xi_{12}(n,m)^2] &\leq m\int_{I_n} E(\Delta_{n,1}^2) \, dF \\ &\leq \frac{Cm}{n} \int_{I_n} [J_{10}(F,G)^2 F(1 - F) \\ &\quad + J_{01}(F,G)^2 G(1 - G)] \, dF \\ &\leq \frac{C'm}{n} \int_{I_n} \left\{\frac{1}{H(1 - H)}\right\}^{1+2\delta} dH \end{aligned}$$

$$= O[n^{\beta-1+2\alpha\delta}] = o(1)$$

as $n \to \infty$ by (6), completing the proof of Lemma 3.

Finally, consider $\xi_{11}(n,k)$.

Lemma 4.

$$\max_k |\xi_{11}(n,k)| \to_p 0, \qquad \text{as } n \to \infty.$$

Proof. By definition of $\Delta_{n,1}$, one may write

$$\xi_{11}(n,k) = \varepsilon_1(n,k) + \varepsilon_2(n,k) \qquad k,n \geq 1,$$

where

$$\varepsilon_1(n,k) = (n + k) \int_{I_n} J_{10}(F,G)(F_{n+k} - F_n)(dF_{n+k} - dF)$$

and

$$\varepsilon_2(n,k) = (n + k) \int_{I_n} J_{01}(F,G)(G_{n+k} - G_n)(dF_{n+k} - dF),$$

$$k,n \geq 1.$$

These two terms are considered separately. First, since $F_{n+k} - F = (F_{n+k} - F_n) + (F_n - F)$ and $(n + k)F_{n+k} = kF_{n,k} + nF_n$, one may write

$$\varepsilon_1(n,k) = \varepsilon_{11}(n,k) + \varepsilon_{12}(n,k) - \varepsilon_{13}(n,k),$$

where

$$\varepsilon_{11}(n,k) = \frac{k^2}{n+k} \int_{I_n} J_{10}(F,G)(F_{n,k} - F_n)(dF_{n,k} - dF_n),$$

$$\varepsilon_{12}(n,k) = k \int_{I_n} J_{10}(F,G)(F_{n,k} - F)(dF_n - dF)$$

and

$$\varepsilon_{13}(n,k) = k \int_{I_n} J_{10}(F,G)(F_n - F)(dF_n - dF), \qquad k,n \geq 1.$$

Clearly, $\max_k |\varepsilon_{13}(n,k)| = |\varepsilon_{13}(n,m)|$; and Lai's (1975) result shows that $\varepsilon_{13}(n,m) \to 0$ w.p.1. as $n \to \infty$. See the analysis of Q_{1n} on page 839 of Lai's article.

To estimate $\varepsilon_{12}(n,k)$, let a_n and b_n be the endpoints of $I_n = \{x: n^{-\alpha} < H(x) < 1 - n^{-\alpha}\}$; let c me a median of H; and let

$$J_n(y) = \begin{cases} \int_{(c,y]} J_{10}(F,G)(dF_n - dF) & : y > c \\ \int_{(y,c]} -J_{10}(F,G)(dF_n - dF) & : y \leq c . \end{cases}$$

Then

$$\varepsilon_{12}(n,k) = k \int_{I_n} (F_{n,k} - F)\, dJ_n$$

$$= k(F_{n,k} - F)\, J_n(y) \Big|_{y=a_n}^{b_n} - k\int_{I_n} J_n(dF_{n,k} - dF),$$

w.p.1,

for $k,n \geq 1$. Thus, $\varepsilon_{12}(n,k)$, $k \geq 1$, is a martingale with respect to $A_{n,k} = \sigma\{X_1,\ldots,X_{n+k}\}$, $k \geq 1$, for each fixed $n \geq 1$; and it suffices to estimate $E[\varepsilon_{12}(n,m)^2]$, as in the proof of Lemma 3. Towards this end, an estimate for $E[J_n^2(y)]$ is obtained first. Let $A_y = (y,c]$ for $y \leq c$ and $A_y = (c,y]$ for $y > c$. Then there are constants C and C' for which

$$E[J_n^2(y)] \leq \frac{1}{n} \int_{A_y} J_{10}(F,G)^2\, dF$$

$$\leq \frac{C}{n} \int_{A_y} \left\{\frac{1}{H(1-H)}\right\}^{2+2\delta} dH$$

$$\leq \frac{C'}{n} \left\{\frac{1}{H(1-H)}\right\}^{1+2\delta} \Big|_y , \qquad y \in I_n,$$

Thus,

$$E\{m^2[\int_{I_n} J_n(dF_{n,m} - dF)]^2\} \leq m \int_{I_n} E[J_n^2(y)]\, dF(y)$$

$$\leq \frac{C'm}{n} \int_{I_n} \left\{\frac{1}{H(1-H)}\right\}^{1+2\delta} dH \rightarrow 0$$

as $n \rightarrow \infty$, by (6). A similar, simpler argument shows that $E\{m^2(F_{n,m} - F)^2\, J_n^2\big|_{a_n}^{b_n}\} \rightarrow 0$ as $n \rightarrow \infty$. Thus, $\max_k |\varepsilon_{12}(n,k)| \rightarrow 0$ as $n \rightarrow \infty$.

The analysis of $\varepsilon_{11}(n,k)$ is similar, if more tedious. Integration by parts shows that

$$\varepsilon_{11}(n,k) = \frac{1}{2}\left(\frac{k^2}{n+k}\right)(F_{n,k} - F_n)^2 J_{10}(F,G)^2 \Big|_{a_n}^{b_n}$$

$$- \frac{1}{2}\left(\frac{k^2}{n+k}\right) \int_{I_n} (F_{n,k} - F_n)^2 J_{20}(F,G)\, dF$$

$$- \frac{1}{2}\left(\frac{k^2}{n+k}\right) \int_{I_n} (F_{n,k} - F_n)^2 J_{11}(F,G)\, dG$$

$$+ \frac{k^2}{2n(n+k)} \int_{I_n} J_{10}(F,G)\, dF_n$$

$$+ \frac{k}{2(n+k)} \int_{I_n} J_{10}(F,G)\, dF_{n,k}$$

$$= \varepsilon_{111}(n,k) + \ldots + \varepsilon_{115}(n,k), \quad \text{say, for } k,n \geq 1.$$

After bounding $1/(n+k)$ by $1/n$, k by m, and $kF_{n,k}$ by $mF_{n,m}$ in $\varepsilon_{114}(n,k)$ and $\varepsilon_{115}(n,k)$, it follows directly from (6) that $\max_k |\varepsilon_{114}(n,k)| \to_p 0$ and $\max_k |\varepsilon_{115}(n,k)| \to_p 0$ as $n \to \infty$. Next, as in the proof of Lemma 3, $-(n+k)\varepsilon_{113}(n,k)/n$ is a non-negative submartingale with respect to $A_{n,k} = \sigma\{X_1,\ldots,X_{n+k}\}$, $k \geq 1$, and there is a constant C for which

$$E[-\varepsilon_{113}(n,m)] \leq \frac{Cm}{n} \int_{I_n} \left\{\frac{1}{H(1-H)}\right\}^{1+\delta} dG,$$

which approaches zero as $n \to \infty$ by (6). Thus, $\max_k |\varepsilon_{113}(n,k)| \to 0$ in probability as $n \to \infty$; and similar arguments show that $\max_k |\varepsilon_{112}(n,k)| \to 0 \leftarrow \max_k |\varepsilon_{111}(n,k)|$ in probability to complete the analysis of ε_1.

It remains to consider $\varepsilon_2(n,k)$. Letting $G_{n,k}$ denote the empirical distribution function of $Y_{n+1},\ldots,Y_{n+k}$ for $k,n \geq 1$, one may write $(n+k)[G_{n+k} - G_n] = k[G_{n,k} - G_n]$ and

$$\varepsilon_2(n,k) = \varepsilon_{21}(n,k) - \varepsilon_{22}(n,k) \qquad \text{for } k,n \geq 1,$$

where

$$\varepsilon_{21}(n,k) = k \int_{I_n} J_{01}(F,G)(G_{n,k} - G)(dF_{n+k} - dF)$$

and

$$\varepsilon_{22}(n,k) = k \int_{I_n} J_{01}(F,G)(G_n - G)(dF_{n+k} - dF), \quad k,n \geq 1.$$

Now $(n + k)\varepsilon_{21}(n,k)$, $k \geq 1$, is a martingale with respect to $A_{n,k} = \sigma\{X_1,\ldots,X_{n+k},Y_1,\ldots,Y_{n+k}\}$, $k \geq 1$, for each fixed n;

$$\begin{aligned} E[\varepsilon_{21}(n,m)^2] &\leq \left(\frac{m^2}{n+m}\right) \int_{I_n} J_{01}(F,G)^2 \, E[(G_{n,m} - G)^2] \, dF \\ &\leq \frac{Cm}{n} \int_{I_n} \left\{\frac{1}{H(1-H)}\right\}^{2+2\delta} G(1-G) \, dF \end{aligned} \tag{12}$$

for some constant C; and the last line in (12) tends to zero as $n \to \infty$ by (6). So, $\max_k |\varepsilon_{21}(n,k)| \to_p 0$ as $n \to \infty$. Similarly, $|(n + k)\varepsilon_{22}(n,k)|$, $k \geq 1$, is a submartingale with respect to $A_{n,k} = \sigma\{Y_1,\ldots,Y_n,X_1,\ldots,X_{n+k}\}$, $k \geq 1$, for each fixed n, and

$$E[\varepsilon_{22}(n,m)^2] \leq \left(\frac{m}{n}\right) \int_{I_n} J_{01}(F,G)^2 \, G(1-G) \, dF,$$

which tends to zero as $n \to \infty$ by (6). So $\max_k |\varepsilon_{22}(n,k)| \to_p 0$ as $n \to \infty$. So, $\max_k |\varepsilon_2(n,k)| \to_p 0$ as $n \to \infty$ to complete the proof of the lemma and of the theorem.

Remark 1

In the context of Theorem 1, Lai's result asserts that $n^{-\alpha}\xi_n \to 0$ for all $\alpha < 1 - \delta$, provided that Condition C holds with $\delta < \frac{1}{2}$. It is not clear whether (3) holds for $\beta < 1 - \delta$, if condition C is satisfied with $\delta < \frac{1}{2}$. A search for counter examples produced none.

Remark 2

Theorem 1 requires that the function J not depend on n. This requirement may be circumvented as follows. Suppose that

$$T_n^* = n \int J_n(F_n, G_n)\, dF_n$$

where J_n, $n \geq 1$, are functions on the square $[0,1] \times [0,1]$. Suppose further that J_n may be written in the form

$$J_n = J + \frac{1}{n}K + \varepsilon_n$$

where J satisfies condition C for some δ, $0 < \delta < \frac{1}{4}$,

$$\overline{K} = \int K(F,G)\, dF < \infty$$

$$\int K(F_n, G_n)\, dF_n \to \int K(F,G)\, dF \quad \text{w.p.1}$$

and

$$\sum_{i=1}^{n} \max_{1 \leq j \leq n} \left| \varepsilon\left(\frac{i}{n}, \frac{j}{n}\right) \right| \to 0 \qquad \text{as } n \to \infty$$

Then one may write $T_n^* = S_n + \xi_n^*$, $n \geq 1$, where S_n, $n \geq 1$, is the random walk constructed as in Theorem 1 with the given J and ξ_n^*, $n \geq 1$, satisfies (3). In fact, $\xi_n^* = \xi_n + \overline{K} + o(1)$ w.p.1 as $n \to \infty$, where ξ_n, $n \geq 1$, are as in Theorem 1.

Remark 3

Lai and Siegmund (1979), consider processes of the form $S_n + \xi_n$, $n \geq 1$, where ξ_n, $n \geq 1$, satisfy conditions related to (3) and $\xi_n - c_n$ converge in distribution for suitably chosen constants c_n, $n \geq 1$. They obtain asymptotic expansions for the expected value of the first passage times $t_a = \inf\{n \geq 1: S_n + \xi_n > a\}$ as $a \to \infty$. The hypothesis of Theorem 1 do not imply existence of constants c_n, $n \geq 1$, for which $\xi_n - c_n$ has a limiting distribution.

IV. A SEQUENTIAL PROBABILITY RATIO TEST

Recall that F and G denote continuous distribution functions and that $X_1, X_2, \ldots, Y_1, Y_2, \ldots$ denote independent random variables for which $X_1, X_2, \ldots$ have common distribution function F and $Y_1, Y_2, \ldots$ have common distribution function G. Suppose that (X_1, Y_1), $X_2, Y_2) \ldots$ are observed sequentially and let

$$R_{n,1}, \ldots, R_{n,2n} = \text{ranks of } X_1, \ldots, X_n, \; Y_1, \ldots, Y_n,$$

$$\mathcal{A}_n = \sigma\{R_{n,1}, \ldots, R_{n,2n}\}, \qquad n \geq 1$$

and

$$\mathcal{A}_\infty = \sigma\{ \bigcup_{n=1}^{\infty} \mathcal{A}_n \}.$$

Thus, $R_{n,1}, \ldots, R_{n,2n}$ are the ranks of the observations available at time n. Next, let $0 < \theta < \infty$ and $\theta \neq 1$ and consider a sequential probability ratio test of the hypotheses

$$G = F \quad \text{vs.} \quad G = F^\theta,$$

based on the sequence $R_n = (R_{n,1}, \ldots, R_{n,2n})$, $n \geq 1$. If $G = F^\omega$ for some $\omega, 0 < \omega < \infty$, then the restriction of the probability measure to $\mathcal{A}_\infty$ depends only on ω. Denote this restriction by P_ω and let

$$P_\omega^n = P_\omega | \mathcal{A}_n$$

denote the restriction of P_ω to $\mathcal{A}_n$ for $n \geq 1$ and $0 < \omega < \infty$. Then the sequential probability ratio test uses the sequence $L_n = dP_\theta^n / dP_1^n$, $n \geq 1$, of likelihood ratios. The latter have been computed by Savage (1956) as

$$L_n = (2n)! \, \theta^n / \prod_{k=1}^{n} \{[nF_n(X_k) + \theta n G_n(X_k)][nF_n(Y_k) + \theta n G_n(Y_k)]\}, \tag{13}$$

where F_n and G_n denote the empirical distribution functions of $X_1, \ldots, X_n$ and $Y_1, \ldots, Y_n$ for each $n \geq 1$. Let

$$\ell_n = \log L_n, \qquad n \geq 1,$$

and

$$t = \inf\{n \geq 1\colon \ell_n < -b \text{ or } \ell_n > a\},$$

where $a,b > 0$. The sequential probability ratio test of $\omega = 1$ vs. $\omega = \theta$ takes t observations and decides in favor of $\omega = \theta$ if and only if $\ell_t > b$. That $t < \infty$ w.p.1. under both hypotheses was shown by Savage and Sethuraman (1966). In fact, t is exponentially bounded under P_ω for all ω, $0 < \omega < \infty$, by the results of Savage and Sethuraman (1966) and Lai (1975). Here the error probabilities

$$\alpha = P_1\{\ell_t > a\} \quad \text{and} \quad \beta = P_\theta\{\ell_t < -b\} \tag{14}$$

are of primary interest. Wald's (1947), pp. 40-42) approximations assert that for any choice of $a,b > 0$,

$$\alpha \approx (e^b - 1)/(e^c - 1) \quad \text{and} \quad \beta \approx (e^a - 1)/(e^c - 1),$$

where $c = a + b$; but these approximations may overestimate α and β substantially. Refined approximations may be obtained from Lai and Siegmund's (1977) non-linear renewal theorem.

To develop refined approximations, one must study the sequence ℓ_n, $n \geq 1$, in more detail. By (13),

$$\ell_n = n[\log 4\theta - 2] - T_{n,1} - T_{n,2} + c_n, \qquad n \geq 1,$$

where

$$c_n = \log(2n)! - \log n^{2n} - 2n[\log 2 - 1], \quad n \geq 1,$$

$$T_{n,1} = n\int J[F_n, G_n]dF_n, \text{ and } T_{n,2} = n\int J[F_n, G_n]dG_n$$

with

$$J(x,y) = \begin{cases} \log(x + \theta y) & : 0 \leq x,y \leq 1 \text{ and } x + y > 0 \\ 0 & : \text{otherwise.} \end{cases}$$

Observe that $c_n = \frac{1}{2}\log 4\pi n + o(1)$ as $n \to \infty$ by Stirling's Formula. It is easily seen that J satisfies condition C for any δ, $0 < \delta < \frac{1}{4}$. Thus, by Theorem 1, $T_{n,i} = S_{n,i} + \xi_{n,i}$, where

$S_{n,i}$, $n \geq 1$, is a random walk and $\xi_{n,i}$ satisfies (3) for all β, $0 < \beta < 1$, for $i = 1,2$. Combining these two results with some simple algebra, shows that

$$\ell_n = S_n + \xi_n,$$

where

$$S_n = n[\log 4\theta - 2] - \sum_{k=1}^{n} \{J[F(X_k),G(X_k)] + J[F(Y_k),G(Y_k)]\}$$

$$- \sum_{k=1}^{n} \{[u(X_k) - \int udF] + [v(Y_k) - \int vdG]\}, \quad n \geq 1,$$

with

$$u(x) = \int_x^\infty J_{10}(F,G)(dF + dG)$$

and

$$v(y) = \int_y^\infty J_{01}(F,G)(dF + dG), \qquad -\infty < y < \infty,$$

and

$$\max_{k \leq n^\beta} |\xi_{n+k} - \xi_n| \to_p 0 \quad \text{as } n \to \infty,$$

for every β, $0 < \beta < 1$. Of course, S_n, $n \geq 1$, is a random walk under P_ω for all ω, $0 < \omega < \infty$. Savage and Sethuraman (1966) have computed the drift $\mu_\omega = E_\omega(S_1)$ of the random walk as

$$\mu_\omega = 2 \log[\frac{2\sqrt{\theta}}{1 + \theta}] + (\theta - 1)(\omega - 1)\int_0^1 \frac{s^{\omega-1}}{1 + \theta s^{\omega-1}}\, ds$$

and shown

$$\mu_1 < 0 < \mu_\theta.$$

Thus the random walk drifts to $-\infty$ under P_1 and drifts to ∞ under P_θ. Let

$$\tau+ = \inf\{n \geq 1: S_n > 0\}$$

and

$$\tau- = \inf\{n \geq 1: S_n < 0\}.$$

Then $\tau^- < \infty$ w.p.1 (P_1) and $\tau^+ < \infty$ w.p.o (P_θ), so that $S_{\tau-}$ is well defined under P_1 and $S_{\tau+}$ is well defined under P_θ. Next let

$$H^+\{dr\} = (1/E_\theta(S_{\tau+}))P_\theta\{S_{\tau+} > r\}dr$$

and

$$H^-\{dr\} = (1/E_1|S_{\tau-}|)P_1\{|S_{\tau-}| > r\}dr, \qquad r > 0.$$

Then H^+ is the asymptotic distribution of residual waiting time for the random walk S_n, $n \geq 1$, under P_θ; and H^- is the asymptotic distribution of residual waiting time for the random walk $-S_n$, $n \geq 1$, under P_1. See, for example, Woodroofe (1982, Ch. 2). Finally, let

$$\gamma^+ = \int_0^\infty e^{-r}H^+\{dr\}$$

and

$$\gamma^- = \int_0^\infty e^{-r}H^-\{dr\}.$$

Theorem 2

Define α and β by (14). Then, as $a,b \to \infty$,

$$\alpha \sim \gamma^+e^{-a} \qquad \text{and} \qquad \beta \sim \gamma^-e^{-b}.$$

The proof is essentially the same as that of Theorem 3.1 of Woodroofe (1982). The only difference is that the non-linear renewal theorem is used to determine the asymptotic distribution of the excess over the boundary. Theorem 1 and Theorem 2 of Lai (1975) are used to verify that the conditions imposed in the non-linear renewal theorem are satisfied.

It is possible to compute the Laplace transform γ^- in closed form. When G = F simple algebra shows that

$$S_n = Z_1 + \ldots + Z_n, \qquad n \geq 1,$$

where

$$Z_k = a[\log G(Y_k) - \log F(X_k)] - c, \qquad k \geq 1,$$

with

$$a = \left(\frac{\theta - 1}{\theta + 1}\right) \qquad \text{and} \qquad c = -2 \log\left[\frac{2\sqrt{\theta}}{1 + \theta}\right].$$

Thus, each Z_k is a linear function of a standard bilateral exponential random variable, and each Z_k has an exponential right tail. When this observation is combined with the factorization theorem, one finds that

$$\gamma^- = \{2 - (2 + c)e^{-c}\}/\{c[1 + \sqrt{(1 - e^{-c})}]\},$$

The computation of γ^+ presents more difficulties, and will not be pursued here.

Remark 4

Postulating that G is a specific Lehmann alternative may seem unrealistic, but assuming that G is some Lehmann alternative may be a reasonable surrogate for alternatives in which Y_1 is either stochastically larger or stochastically smaller than X_1. If one postulates that $G = F^\theta$ for some $\theta > 0$, then one may estimate θ and perform repeated likelihood ratio rests of the null hypothesis $\theta = 1$. The analysis of repeated likelihood ratio tests is more complicated than that of the S.P.R.T.; but the hardest part of the analysis is to show that terms of the form ξ_n, $n \geq 1$, in Theorem 1, are slowly changing.

Properties of the likelihood function, based on ranks and Lehmann alternatives, will be reported elsewhere.

V. REFERENCES

Chernoff, H. and Savage, I.R. (1958). "Asymptotic Normality and efficiency of Certain Non Parametric Test Statistics". *Ann. Math. Stat. 29*, 972-994.

Csörgö, M. and Revesz, P. (1974). "Some Notes on the Empirical Distribution Function and the Quantile Process." In *Strong Laws of Invariance Principle,* Mathematics Department, Carleton University.

Lai, T.L. (1975). "On Chernoff Savage Statistics and Sequential Rank Tests." *Ann. Statist. 3,* 825-845.

Lai, T.L. and Siegmund, D. (1977). "A Non-Linear Renewal Theory With Applications to Sequential Analysis I." *Ann. Statist. 5,* 946-954.

Lai, T.L. and Siegmund, D. (1979). "A Non-Linear Renewal Theory With Applications to Sequential Analysis II." *Ann. Statist. 7,* 60-76.

Savage, I.R. (1956). "Contributions to the Theory of Rank Order Statistics: the Two Sample Case." *Ann. Math. Statist. 27,* 590-616.

Savage, I.R. and Sethuraman, J. (1966). "Stopping Time of a Rank Order Sequential Test Based on Lehmann Alternatives." *Ann. Math. Statist. 37,* 1154-1160.

Sethuraman, J. (1970). "Stopping Time of a Rank Order Sequential Test Based on Lehmann Alternatives, II." *Ann. Math. Statist. 41,* 1322-1333.

Wald, A. (1947). *Sequential Analysis,* John Wiley and Sons.

Woodroofe, M. (1982). *Non-Linear Renewal Theory in Sequential Analysis.* S.I.A.M.

II. OPTIMIZATION INCLUDING CONTROL THEORY

A NON-HOMOGENEOUS MARKOV MODEL OF A CHAIN-LETTER SCHEME

P. K. Bhattacharya[1]

Division of Statistics
University of California at Davis

J. L. Gastwirth[2]

Department of Statistics
George Washington University
Washington, D. C.

I. INTRODUCTION

In spite of Federal and state laws concerning deceptive business practices such as false promise and misrepresentation, variations of chain letters and pyramid schemes continue to be offered to the public. In 1979 one of the authors was contacted by the State Attorney of Illinois to see whether a previous model (Gastwirth [4]), which emphasized how dependent a participant's potential earnings are to the time of entry in the chain, could be adapted to a chain letter being circulated at that time.

The letter being sold had 6 names on it and a purchaser paid \$500 to the person selling the letter and \$500 to the person at the top of the list. The purchaser then re-did the list by

[1,2] *Research supported by National Science Foundation Grants MCS-81-01976 and MCS-80-05872.*

ISBN 0-12-589320-5

crossing off the person at the top and placing his name at the bottom, renumbered the list and had the right to sell 2 copies of the letter. The process continues and the promoters indicated that eventually there will be 32 people selling 64 letters with the participant's name at the top of the list so that the participant should ultimately receive $32,000. The promoters emphasize that a participant recovers his original investment as soon as he sells his two letters after which he can only make a profit. In court, the promoters asserted that due to this feature many persons would re-enter the system by purchasing another letter so that the usual argument that the pool of participants is finite does not apply to their letter.

In Section II we describe a Markov model of the process which keeps track of the number of participants who have sold 0, 1 or 2 letters and allows those persons who have sold two letters to re-join the system.

The role of the model in the legal proceedings is to demonstrate that the promises made by the promoters are inherently deceptive and constitute mistrepresentation because it is virtually impossible for the majority of participants to make anywhere near the large sums indicated by the promotional material. In fact, many participants will lose all or part of their entrance fee. Since a participant's earnings depends on the time of entrance, defendant promoters often can produce a few "winners" to testify that they did well. The probabilistic approach enables one to show that the proportion of winners is small and that most participants will lose money.

The basic properties of the process and its limiting diffusion approximation are developed in Section II and the asymptotic distributions of the stopping time and the number of participants with no letter sold at that time are derived in Section III. Using these results we show that

1. Even when re-joining is allowed the process must stop after a finite number of letters have been sold. Moreover, every participant in the system at this time, including re-joiners, will not have recouped their payment for the letters they are currently selling.

2. Assuming that the number of potential participants is n (large), with high probability, the process terminates when approximately 1.62n letters have been sold. From this it follows that most participants cannot receive large monetary profits as at least 2n letters need to be sold just for the participants to recoup their initial purchase price.

3. Moreover, if all persons who sell their 2 letters re-join, then at the termination of the process approximately 61.8% of the participating population (n) will be attempting to sell 2 letters and will be $1000 behind (the purchase price of their current letters), while 38.2% still will have 1 letter to sell and will be $500 behind. Thus in order to make a profit, one has to receive money by reaching the top of a letter sold to a future participant.

4. An *upper* bound to the expected number of letters sold by a participant and his subsequent recruits is obtained. From the bounds which depend on the time of entry into the chain, it follows that the substantial majority of participants (80%) have a *less* than 10% probability of reaping the large financial gains promised by the promoters.

An appendix to the paper presents the derivation of the diffusion which approximates the Markov chain model except at its very early stages. For this we treat the chain conditionally from a certain stage onwards. The transition properties of this conditional chain present two difficulties when the state variable is appropriately normalized: the increments neither have bounded conditional expectations, nor are their conditional variances bounded away from zero. For this reason, standard results of Gikman and Skorokhod [5] or Strook and Varadhan [6] are not directly applicable here. These difficulties are overcome by modifying the chain beyond constant boundaries and then showing that the probability of the chain moving outside these boundaries is sufficiently small so as to make the effect of this modification negligible. A suitable probability inequality for the constant boundary crossing event is obtained for this purpose, using a "rebounding effect" in the transition probabilities of the chain. It is hoped that this result may be of use in Markov models for other phenomenae exhibiting this behavior.

II. A MARKOV MODEL OF THE RECRUITMENT PROCESS AND ITS DIFFUSION APPROXIMATION

At the kth stage (time k) of the recruitment process, i.e., when k letters have been sold, let U_k, V_k and W_k denote respectively the number of participants who have sold 0 or 1 or 2 letters. At this time, the $U_k + V_k$ participants who have not used up their quota of selling 2 letters are competing for the next sale. Assuming that each of them has an equal chance to sell the next letter, the process $\{(U_k,V_k,W_k),k=0,1,2,\ldots\}$ becomes a Markov chain with transition probabilities

$$P[(U_{k+1},V_{k+1},W_{k+1}) =$$

$$= (u,v,w) + (0,1,0) \mid (U_k,V_k,W_k) = (u,v,w)] = u/(u+v)$$

$$P[(U_{k+1},V_{k+1},W_{k+1}) =$$

$$= (u,v,w) + (1,-1,1) \mid (U_k,V_k,W_k) = (u,v,w)] = v/(u+v), \tag{2.1}$$

and initial state $(U_0,V_0,W_0) = (c,0,0)$, where $c \geq 1$ is the number of persons who iniated the sales.

By (2.1), $U_k+V_k+W_k$ increases by 1 at each step and $U_{k+1}-U_k = W_{k+1}-W_k$, so that

$$W_k = U_k - c,\ V_k = k + 2c - 2U_k \quad \text{with probability 1} \tag{2.2}$$

in view of the initial condition. Thus the process is completely described by the Markov chain $\{U_k,\ k=0,1,2,\ldots\}$ starting at $U_0 = c$ and having transition probabilities

$$P[U_{k+1}=u\,|\,U_k=u] = 1-P[U_{k+1}=u+1\,|\,U_k=u] = u/(k+2c-u). \tag{2.3}$$

Before analyzing the Markov chain in depth, note that the non-decreasing property of $\{U_k\}$ already implies that some participants will lose their entire investment when the process terminates. Moreover, even if those who sell their 2 letters rejoin as new recruits, the process must terminate in less than 2n steps in a population of n potential participants. This is because the number of persons selling letters must be less than n for the process to continue, i.e., $U_k + V_k < n$, while $V_k \geq 0$. By (2.2), these inequalities become

$$k-n+c < U_k \leq \frac{1}{2}k+c. \tag{2.4}$$

Hence the process cannot continue beyond 2n-2c steps and consequently, only a very small proportion of the participants can possibly gain large amounts as promised by the promoters.

From now on, we consider the chan $\{U_k, 0\leq k\leq 2n\}$ in a population of n individuals with the assumption that all those who sell their 2 letters rejoin. The process terminates at the smallest k for which the first inequality in (2.4) is violated. Thus the stopping time is

$$K_n = min\{k\,|\,U_k\leq k-n+2c\}. \tag{2.5}$$

The asymptotic distribution of K_n and U_{K_n} as $n \to \infty$ will be obtained from the weak convergence property of a suitably normalized version of $\{U_k\}$. For this normalization, it is essential to separate out the long term trend in $\{U_k\}$, and a linear trend is seen to be consistent with the following heuristic argument. By (2.3),

$$E[U_{k+1}-U_k|U_k] = \frac{k+2c-2U_k}{k+2c-U_k} \underset{-}{\sim} \frac{k-2U_k}{k-U_k} \underset{-}{\sim} \frac{1-2\alpha}{1-\alpha}$$

for large k, where $\alpha \underset{-}{\sim} U_k/k$ represents the rate of growth of $\{U_k\}$. On the other hand, the rate of growth should equal the average increase per step. This leads to the equation

$$(1-\alpha)/(1-\alpha) = \alpha \text{ or } \alpha^2 - 3\alpha+1 = 0, \tag{2.6}$$

of which the solution $\alpha = (3 - \sqrt{5})/2 = .382$ in (0,1) is appropriate for our purpose.

We now center U_k by subtracting the initial value c and the linear trend k to arrive at

$$X_k = U_k-c-k\alpha, \quad 0 \leq k \leq 2n \tag{2.7}$$

which is a Markov chain with $X_0 \equiv 0$ and transition probabilities obtained by rewriting (2.3) as

$$P[X_{k+1}-x=-\alpha|X_k=x] = \frac{k\alpha+x+c}{k(1-\alpha)-x+c} = 1-\alpha-h_k(x)$$

$$P[X_{k+1}-x=1-\alpha|X_k=x] = \alpha + h_k(x) , \tag{2.8}$$

where

$$h_k(x) = 1-\alpha- \frac{k\alpha+x+c}{k(1-\alpha)-x+c} = \frac{(2-\alpha)x+c\alpha}{x-\{k(1-\alpha)+c\}} , \tag{2.9}$$

using (2.6). Finally, we normalize both the state variable and the time in $\{X_k\}$ to arrive at the continuous time process $\{\xi_n(t),0\leq t\leq 2\}$ obtained by extending

$$\{\xi_n(k/n) = X_k/\sqrt{n}, \quad k = 0,1,\ldots, 2n\}$$

by linear interpolation, i.e.,

$$\xi_n(t)=n^{-\frac{1}{2}}[([nt]+1-nt)X_{[nt]}+(nt-[nt])X_{[nt]+1}], 0\leq t\leq 2 \quad . \quad (2.10)$$

To avoid unnecessary complications we shall treat nt as an integer and work with the simpler formula

$$\xi_n(t) = n^{-\frac{1}{2}}X_{nt}, \; 0 \leq t \leq 2. \qquad (2.10a)$$

The limiting behavior of $\{\xi_n(t)\}$ as $n \to \infty$ is understood in terms of the conditional mean and variance of its rate of change from time t to t + 1/n. By (2.10a) the conditional behavior of $\xi_n(t+1/n)-x$ given $\xi_n(t)=x$ is the same as that of $n^{-\frac{1}{2}}(X_{nt+1} - \sqrt{n}\,x)$ given $X_{nt} = \sqrt{n}\,x$. Using (2.8) and (2.9), we thus have

$$\begin{aligned} m_n(t,x) &= nE[\xi_n(t+1/n)-x \,|\, \xi_n(t)=x] \\ &= n^{\frac{1}{2}}E[X_{nt+1} - \sqrt{n}\,x \,|\, X_{nt} = \sqrt{n}\,x] \\ &= n^{\frac{1}{2}}h_{nt}(\sqrt{n}\,x)\;, \qquad (2.11) \end{aligned}$$

$$\begin{aligned} v_n(t,x) &= n\,Var[\xi_n(t+1/n)-x \,|\, \xi_n(t) = x] \\ &= Var[X_{nt+1}-\sqrt{n}\,x \,|\, X_{nt} = \sqrt{n}\,x] \\ &= \alpha(1-\alpha) + (1-2\alpha)h_{nt}(\sqrt{n}\,x). \qquad (2.12) \end{aligned}$$

For large n,

$$h_{nt}(\sqrt{n}\,x) = \frac{(2-\alpha)\sqrt{n}\,x + c\alpha}{\sqrt{n}\,x-\{nt(1-\alpha)+c\}} = n^{-\frac{1}{2}}[-(\frac{2-\alpha}{1-\alpha})\,\frac{x}{t} + o(1)], \qquad (2.13)$$

so that (2.11) and (2.12) approximately become

$$m_n(t,x) \simeq -(\frac{2-\alpha}{1-\alpha})\,\frac{x}{t}\,, \qquad v_n(t,x) \simeq \alpha(1-\alpha).$$

This suggests that $\xi_n(t)$ tends to a diffusion with these infinitesimal properties. However, there are obvious difficulties near t = 0 and even for t bounded away from 0, the o(1) term in (2.13) being non-uniform in x causes problems. These technical difficulties are overcome in the appendix where a full justification is given for the following theorem.

Theorem 2.1. For $\varepsilon > 0$, the conditional process $\{\xi_n(t), \varepsilon \le t \le 2\}$ given $\xi_n(\varepsilon) = x$ converges weakly to a diffusion $\{\xi(t), \varepsilon \le t \le 2\}$ described by the stochastic differential equation

$$d\xi(t) = -(\beta_1/t)\xi(t)dt + \sqrt{\beta_2}\, dW(t), \quad \varepsilon \le t \le 2 \tag{2.14}$$

starting at $\xi(\varepsilon) = x$, where

$$\beta_1 = (2-\alpha)/(1-\alpha), \ \beta_2 = \alpha(1-\alpha)$$

and W(t) is a standard Wiener process. Moreover, this convergence is uniform for $|x| \le L$.

An explicit solution of (2.14) is obtained by the following standard result.

Lemma 2.2. The solution of the stochastic differential equation

$$d\xi(t) = [A(t)\xi(t) + a(t)]dt + b(t)dW(t), \ t \le t \le T$$

starting at $\xi(t_0) = x_0$ is given by

$$\xi(t) = \psi(t)\left[x_0 + \int_{t_0}^{t} \frac{a(s)}{\psi(s)}\, ds + \int_{t_0}^{t} \frac{b(s)}{\psi(s)}\, dW(s)\right],$$

$$t_0 \le t \le T$$

where $\psi(t)$ is the solution of the deterministic equation

$$x'(t) = A(t)x(t) \text{ with } x(t_0) = 1.$$

Proof. See Arnold [2], pages 129-130.

For (2.14), the coefficients are $A(t) = -\beta_1/t$, $a(t) = 0$, $b(t) = \sqrt{\beta_2}$, and the solution of $x'(t) = -(\beta_1/t)x(t)$ starting at $x(\varepsilon) = 1$ is $\psi(t) = (\varepsilon/t)^{\beta_1}$. We thus have

Theorem 2.3. For $0 < \varepsilon < t \leq 2$, the conditional distribution of $\xi_n(t)$ given $\xi_n(\varepsilon) = x$ converges to that of

$$\xi(t) = (\varepsilon/t)^{\beta_1} x + \sqrt{\beta_2} \int_\varepsilon^t (s/t)^{\beta_1} dW(s),$$

the convergence being uniform for $|x| \leq L$.

Corollary 2.4. In the conditional distribution given $\xi_n(\varepsilon) = x$,

$$\xi_n(t) \xrightarrow{L} N\left(\left(\frac{\varepsilon}{t}\right)^{\beta_1} x, \frac{\beta_2 t}{2\beta_1+1}\left\{1-\left(\frac{\varepsilon}{t}\right)^{2\beta_1+1}\right\}\right) \text{ uniformly in } |x| \leq L.$$

To obtain an unconditional convergence from Corollary 2.4, observe that

(i) by Lemma A.1 we can make $P[|\xi_n(\varepsilon)| > M]$ smaller than an arbitrary $\delta > 0$ for all ε in $[0,2]$ by making M sufficiently large, but not depending on ε;

(ii) as $\varepsilon \downarrow 0$, the conditional mean $(\varepsilon/t)^{\beta_1} x$ of $\xi_n(t)$ tends to 0 uniformly in $|x| \leq M$ and the conditional variance tends to $\beta_2 t/(2\beta_1+1)$; and

(iii) the convergence in Corollary 2.4 is uniform in $|x| \leq M$.

We have thus proved

Theorem 2.5. $\xi_n(t) \xrightarrow{L} N(0, \beta_2 t/(2\beta_1+1))$.

In terms of the original variables U_k, Theorem 2.5 means that for large values of k, U_k is approximate normal with

$$\text{mean} \qquad \alpha k = .382k \qquad \text{and}$$
$$\text{variance} \qquad \beta_2(2\beta_1+1)^{-1}k = .0379k.$$

Equivalently, for large k, the fraction $k^{-1}U_k$ of participants (including repeaters) who have not sold *any* letter is asymptotically normal with mean .382 and s.d. $.195/\sqrt{k}$, i.e., approximately 38.2% of all participants will *not sell any letter*.

III. TERMINATION OF THE RECRUITMENT PROCESS

The stopping time K_n of the recruitment process is given by (2.5). Equivalently, in terms of $X_k = U_k - k\alpha - c$,

$$K_n = \min\{k | X_k \leq k(1-\alpha)-n+c\}.$$

In this section, asymptotic properties of K_n and U_{K_n} as $n \to \infty$, will be examined.

Lemma 3.1. $K_n/n \xrightarrow{p} (1-\alpha)^{-1}$ as $n \to \infty$. Hence $\lim_{n\to\infty} E(K_n/n) = (1-\alpha)^{-1}$.

Proof. For arbitrary $\delta > 0$, $K_n/n > a = (1-\alpha)^{-1} + \delta$ implies

$$X_{na} > na(1-\alpha)-n+c = n(1-\alpha)\delta+c.$$

On the other hand, since the boundary $k(1-\alpha)-n+c$ is an increasing function of k, $K_n/n < b = (1-\alpha)^{-1}-\delta$ implies

$$\min_{1\leq k\leq nb} X_k \leq nb(1-\alpha)-n+c = -n(1-\alpha)\delta+c.$$

Hence

$$P[|K_n/n-(1-\alpha)^{-1}|>\delta]\leq P[X_{na}\geq n(1-\alpha)\delta+c]+P[\min_{1\leq k\leq nb} X_k\leq -n(1-\alpha)+c]$$

$$<2P[\max_{1\leq k\leq na} |X_k|\geq n(1-\alpha)\delta-c]$$

which tends to 0 as $n \to \infty$ by Lemma A.1. This proves the first assertion which in turn implies the second assertion because $1 \leq K_n/n \leq 2$ with probability 1.

Theorem 3.2. $X_{K_n}/\sqrt{n} \xrightarrow{L} N(0,\beta_2(2\beta_1+1)^{-1}(1-\alpha)^{-1})$ as $n \to \infty$.

Proof. Since $K_n/\{n(1-\alpha)^{-1}\} \xrightarrow{p} 1$, the asymptotic distribution of $X_{K_n}/\sqrt{n}$ will be the same as that of $X_{n(1-\alpha)^{-1}}/\sqrt{n}$ by Theorem 1 of Anscombe [1] subject to the verification of his condition of uniform continuity in probability, viz.

$$\lim_{n\to\infty} P[\max_{(1-\delta)n\leq k\leq (1+\delta)n} |X_k-X_n| > \sqrt{n}\ \varepsilon] = 0.$$

for each $\varepsilon > 0$ and some $\delta = \delta(\varepsilon) > 0$. Theorem 2.5 will then imply the result. Now

$$P[\max_{(1-\delta)n\leq k\leq (1+\delta)n} |X_k-X_n|>\sqrt{n}\ \varepsilon]$$

$$\leq P[\max_{(1-\delta)n\leq k\leq (1+\delta)n} |X_k-X_{(1-\delta)n}|\geq \tfrac{1}{2}\sqrt{n}\varepsilon]$$

$$=EP[\max_{1-\delta\leq t\leq 1+\delta} |\xi_n(t)-\xi_n(1-\delta)|\geq\varepsilon/2|\xi_n(1-\delta)]$$

$$\to EP[\max_{1-\delta\leq t\leq +\delta} |\{(\tfrac{1-\delta}{t})^{\beta_1}-1\}\xi(1-\delta)$$

$$+\sqrt{\beta_2}\ t^{-\beta_1}\int_{1-\delta}^{t} s^{\beta_1}dW(s)|\geq\varepsilon/2|\xi(1-\delta)]$$

by Theorem 2.3, where $\xi(1-\delta)$ is $N(0,\beta_2(1-\delta)/(2\beta_1+1))$ by Theorem 2.5. Thus

$$\lim_{n\to\infty} P[\max_{(1-\delta)n\leq k\leq(1+\delta)n} |X_k-X_n|>\sqrt{n}\,\varepsilon]$$

$$\leq P[|\xi(1-\delta)|)\geq \frac{1}{4}\varepsilon\{1-(\frac{1-\delta}{1+\delta})^{\beta_1}\}^{-1}]$$

$$+P[\max_{1-\delta\leq t\leq 1+\delta} |\int_{1-\delta}^{t} s^{\beta_1} dW(s)|\geq\varepsilon\cdot\frac{(1-\delta)^{\beta_1}}{4\sqrt{\beta_2}}] ,$$

of which the first term obviously $\to 0$ as $\delta\downarrow 0$ and by Gikhman and Skorokhod [5], page 393, the second term is bounded by

$$16\beta_2\varepsilon^{-2}(1-\delta)^{-2\beta_1}\int_{1-\delta}^{1+\delta} s^{2\beta_1} ds \to 0 \text{ as } \delta\downarrow 0.$$

This concludes the proof.

Using an argument similar to one used by Bhattacharya and Mallik [3], the asymptotic distribution of K_n is obtained from that of X_{K_n}.

Theorem 3.3. $n^{-\frac{1}{2}}[K_n-n(1-\alpha)^{-1}] \xrightarrow{L} N(0,\beta_2(2\beta_1+1)^{-1}(1-\alpha)^{-3})$ as $n\to\infty$.

Proof. By definition of K_n,

$$X_{K_n} \leq K_n(1-\alpha)-n+c,\ X_{K_n-1} > (K_n-1)(1-\alpha)-n+c,$$

so that

$$n^{-\frac{1}{2}}(1-\alpha)^{-1}(X_{K_n}-c) \leq n^{-\frac{1}{2}}[K_n-n(1-\alpha)^{-1}]$$

$$< n^{-\frac{1}{2}}\{1+(1-\alpha)^{-1}(X_{K_n-1}-c)\}.$$

Since the two extremes of the above inequality differ by at most $n^{-\frac{1}{2}}(1-\alpha)^{-1}$, it follows that

$$n^{-\frac{1}{2}}[K_n - n(1-\alpha)^{-1}] = n^{-\frac{1}{2}}(1-\alpha)^{-1}(X_{K_n} - c) + o_p(1)$$

$$= n^{-\frac{1}{2}}(1-\alpha)^{-1} X_{K_n} + o_p(1),$$

and the desired result follows from Theorem 3.2.

The remark made in paragraph 2 of *Introduction* is now justified.

IV. A BOUND FOR THE SIZE OF THE SUB-CHAIN STARTED BY THE kth PARTICIPANT

As a participant's ultimate earnings depend on the success of his recruits and their subsequent recruits in selling their letters, the number of descendants of the kth recruit is of interest. Notice that the number of descendants of a participant is at least as large as the number of letters ever sold with the participant's name on this (as the original participant's name is dropped once one of his descendants sells a letter with his name at the top of the list) and is much larger than the number of letters sold with his name at the top of the list. In this section, a bound on the average number of descendants is obtained and used in the standard Markov inequality to show that the majority of participants have less than a 1% chance of having 2 + 4 + ... + 64 = 126 descendants which is necessary but not sufficient for them to receive money from 64 future entrants.

For the arguments in this section, at time k fix attention on a participant who has not sold any letter, call him (k,0), and another participant who has sold one letter, call him (k,1), assuming $V_k \geq 1$. Let ξ_k and η_k denote respectively the number of future descendants of (k,0) and (k,1) respectively and define

$\mu_k(u) = E[\xi_k|U_k = u]$, $\nu_k(u) = E[\eta_k|U_k = u]$ for $c \leq u \leq \frac{1}{2}(k-1) + c$. This covers all cases where $U_k \geq c$ and $V_k \geq 1$. For $U_k = u = \frac{1}{2}k + c$ and $V_k = 0$, which is possible only for k even, let

$$\mu_k(\tfrac{1}{2}k+c) = \{E[K_n|U_k = \tfrac{1}{2}k+c]-k\}/(\tfrac{1}{2}k+c), \quad \nu_k(\tfrac{1}{2}k+c) = 0,$$

K_n being the stopping time of the process. Then $\mu_k(u)$ is the conditional expectation of the number of descendants of the new recruit at time k, given $U_k = u$. The total number of recruits from time k is K_n-k, which is the same as the total number of descendants of those U_k who have not sold any letter and those $k + 2c - 2U_k$ who have sold one each. Thus

$$U_k\mu_k(U_k) + (k+2c-2U_k)\nu_k(U_k) = E[K_n|U_k]-k. \tag{4.1}$$

Starting from k, let $k+\tau$ denote the time at which participants (k,0) and (k,1) between them sell their first letter and let $\tau = K_n-k$ if neither can sell a letter before termination. Define

$$\mu_k(u|\ell) = E[\xi_k|U_k = u,\tau=\ell], \quad \nu_k(u|\ell)=E[\eta_k|U_k=u,\tau=\ell].$$

Since the participants (k,0) and (k,1) each has conditional probability 1/2 of selling the first letter between them, it follows that

$$\mu_k(u|\ell)=E[\tfrac{1}{2}\{1+\mu_{k+\ell}(U_{k+\ell})+\nu_{k+\ell}(U_{k+\ell})\}+ \tfrac{1}{2}\mu_{k+\ell}(U_{k+\ell})|U_k=u,\tau=\ell],$$

$$\nu_k(u|\ell)=E[\tfrac{1}{2}\nu_{k+\ell}(U_{k+\ell}) + \tfrac{1}{2}\{1+\mu_{k+\ell}(U_{k+\ell})\}|U_k = u,\tau= \ell]$$

for $1 \leq \ell \leq K_n-k-1$, and

$$\mu_k(u|K_n-k) = \nu_k(u|K_n-k) = 0,$$

so that

$$\mu_k(u|\ell)-\nu_k(u|\ell) = \begin{cases} \frac{1}{2} E[\mu_{k+\ell}(U_{k+\ell})|U_k=u,\tau=\ell], & 1\leq\ell\leq K_n-k-1 \\ 0, & \ell = K_n-k \end{cases}$$

$$\leq \frac{1}{2}\mu_k(u), \qquad 1 \leq \ell \leq K_n-k, \tag{4.2}$$

as $\mu_{k+\ell}(U_{k+\ell}) \leq \mu_k(U_k)$ because at time $k + \ell$ there are more participants and fewer opportunities for sale before termination than at time k. Thus

$$\mu_k(u)-\nu_k(u) = \sum_{\ell=1}^{K_n-k} P[\tau=\ell|U_k=u]\{\mu_k(u|\ell)-\nu_k(u|\ell)\} \leq \frac{1}{2}\mu_k(u). \tag{4.3}$$

On the other hand, it follows from (4.2) that $\mu_k(u|\ell)-\nu_k(u|\ell)\geq 0$ for all ℓ, so that

$$\mu_k(u)-\nu_k(u) \geq 0. \tag{4.4}$$

Combining (4.3) and (4.4) we have

$$\frac{1}{2}\mu_k(U_k) \leq \nu_k(U_k) \leq \mu_k(U_k). \tag{4.5}$$

Substituting the bounds for $\nu_k(U_k)$ provided by (4.5) in (4.1), we have

$$\frac{E[K_n|U_k] -k}{k-U_k+2c} \leq \mu_k(U_k) \leq \frac{2\{E[K_n|U_k]-k\}}{k+2c} . \tag{4.6}$$

Note that if $V_k = 0$ so that U_k takes on its largest possible value $(k+2c)/2$, then the expected number of descendants of any of the current participants is the expected number of future participants divided by $(k+2c)/2$, which is the upper bound in (4.6).

In order to calculate the bounds in (4.6), we need an expression for $E[K_n|U_k]$. It can be shown as in Lemma 3.1 that for every $\varepsilon > 0$,

$$\lim_{n\to\infty} P[|K_n/n-(1-\alpha)^{-1}| > \varepsilon | U_k=k\alpha+\sqrt{n}\, x] = o$$

uniformly for $|x| \leq M$. Since $1 \leq K_n/n \leq 2$ with probability 1, this implies

$$\lim_{n\to\infty} E[K_n/n | U_k=k\alpha + \sqrt{n}\, x] = (1-\alpha)^{-1}$$

uniformly for $|x| \leq M$. Furthermore, $P[n^{-\frac{1}{2}}|U_k-k\alpha| \leq M]$ can be made arbitrarily close to 1 for large n if M is made large. Letting $k = n\beta$, we now have

$$k^{-1}E[K_n|U_k] = \beta^{-1}E[K_n/n|U_k] = (1-\alpha)^{-1}\beta^{-1}+o_p(1).$$

Substituting this in (4.6) and remembering

$$k^{-1}U_k = \alpha + O_p(n^{-\frac{1}{2}}),$$

we have

$$(1-\alpha)^{-1}[(1-\alpha)^{-1}\beta^{-1}-1] \leq \mu_k(U_k) \leq 2[(1-\alpha)^{-1}\beta^{-1}-1]$$

with high probability when n is large.

Using the right side of (4.7) in the Markov inequality, an upper bound for $P(\xi_k \geq x)$ is obtained, where ξ_k is the number of descendants of the kth recruit. Some numerical results are presented in Table 1.

TABLE I. Bounds for the Expected Value and Exceedance Probabilities of the Number of Descendants ξ_k of the k-th Participant.

$\beta = k/n$	$\beta(1-\alpha)$	$E(\xi_k)$	$P(\xi_k \geq 7)$	$P(\xi_k \geq 126)$
.1618	.100	18	1	.143
.3236	.200	8	1	.063
.5388	.333	4	.571	.032
.809	.500	.4722	.067	.004

Although these bounds are quite generous, they indicate how strongly a participant's probability of earning money depends on the time he enters the process. Moreover, they show a person entering when the process has run only one-tenth of its course (i.e., $\beta(1-\alpha) = .1$) has less than a one-in-seven chance of having 126 descendants and his chance of receiving money from 64 persons is even smaller. For persons entering at the half-way point, this probability is less than 0.4% and they have less than a one-in-fifteen chance of having 7 descendants which is essential to make a profit.

V. MONTE CARLO SIMULATION OF THE PROCESS

The mathematical results obtained in the previous sections were based, in large part, on the limiting diffusion process to which the chain letter Markov process converges as the number of letters sold increases. In order to assess the accuracy of these large sample results, we conducted a computer simulation of the process assuming a moderate sized ($n = 500$) population of eligible participants. The simulation enables us to obtain a better estimate of how *small* the probability a new participant has of realizing a large sum of money by reaching the top of letters sold to future entrants, once the chain reaches *ten percent* of its ultimate size. In only 6 percent of the simulations did such a participant receive $5,000 (10 payments) or more while in over *half* of the simulations they did not receive any money by being at the top of the list of a letter by a future entrant. The simulations showed that persons entering at a later stage have a much smaller chance of earning a profit.

The Monte Carlo simulation consisted of 200 replications, 100 with each of two different initial random numbers (seeds). Before simulating the entire process we first assessed the accuracy of the mean and standard deviation of the limiting normal distribution for the fraction, $k^{-1}U_k$, without putting a limit on the number of eligible participants. Starting with one person, i.e., the initial state (U,V,W) = (1,0,0) we determined U_k and $k^{-1}U_k$ when k reached 809 and 2,000, respectively. The number 809 was selected as it is approximately the expected stopping time of the process (Lemma 3.1) when n = 500. The results in Table 2 agree with the theoretical ones.

TABLE II. Accuracy of the Parameters of the Normal Approximation for the Fraction of Participants with No Recruits, $k^{-1}U_k$.

	k = 809		k = 2,000	
	Mean	*Std. Dev.*	*Mean*	*Std. Dev.*
Theory	.382	.0068	.382	.0044
Simul. I	.384	.0072	.383	.0048
Simul. II	.382	.0079	.381	.0045

As the promoters of the Illinois chain claimed that once participants recouped their initial cost by selling two letters they would rejoin the system, we next assumed this. In a population of 500 eligible participants, Lemma 3.1 asserts that a process should stop after about 809 persons (including re-entrants) purchase letters. Theorem 3.3 indicates that the standard deviation of the stopping time should be approximately

$$\sqrt{\frac{\beta_2}{(2\beta_1+1)} \; \frac{1}{(1-\alpha)^3}} \; \sqrt{500} = 8.95.$$

Similarly, Theorem 3.2 and Lemma 3.1 show that the fraction $k^{-1}U_k$ of all participants (including re-entrants) with 2 letters to sell at the random time (K_n) when the process stops is approximately normally distributed with mean .382 and STD .0068. The results of the simulation, presented in Table 3, confirm the theory, showing that the effect of the random stopping on the distribution of $k^{-1}U_k$ is small for our sample size.

TABLE III. Stopping Time and Fraction of Participants Having 2 Letters to Sell, $k^{-1}U_k$, When the Process Stops.

	Stopping Time (K_n)		Fraction, $k^{-1}U_k$	
	Expectation	*STD.*	*Expectation*	*STD.*
Theory	809.0	8.953	.3820	.0068
Simul. I	811.0	9.402	.3807	.0072
Simul. II	807.5	9.007	.3834	.0074

The simulation of the entire chain letter system is quite involved as one must keep track of the number of letters in circulation with the kth participant's name on them indexed by his place on each list. Therefore, we describe the sub-chain initiated by the kth participant by a 12 component vector. The first 6 components describe the number of letters being sold by sub-chain members with 2 letters to sell indexed by the kth participant's place, i.e., $(m_1, m_2, \ldots, m_6)$ means that there are a total of $\sum_{i=1}^{6} m_i$ persons in the sub-chain with 2 letters to sell and the kth participant's name is in the ith place on m_i of them. Similarly, the next 6 components $(n_1, \ldots, n_6)$ index the number of letters being sold by sub-chain members with one letter to sell by the kth participant's place on the list. Whenever one of the (m_1+n_1) letters with the kth participant's name at the top of the list is sold, he receives a payoff ($500) and the new recruit is *not* added to the sub-chain as the kth participant is no longer on the letter.

The transitions were simulated under the assumption that all current participants were equally likely to recruit the next one. The simulation first decided whether a member of the sub-chain recruited the next participant giving the sub-chain a probability of success equal to the fraction of all participants who are members of the sub-chain. When the sub-chain was successful, the letter purchased by the new entrant was randomly selected from the available ones and the appropriate transition in the $(\vec{m},\vec{n})$ vector made.

In order to examine how dependent a participant's earnings are to the time of entry into the process, we assumed that the kth entrant was either the 81st, 162nd or 405th participant to join, i.e., the process was approximately at 10, 20 or 50 percent of its ultimate size. Finally, we assumed the process was at its equilibrium position, $U_{k-1} = \alpha(k-1)$, when the kth entrant purchased his letter. Thus, the simulations of the payoffs received by the 81st entrant started with the main chain in state (31, 18, 31) and the sub-chain at (1,0,0), so U = 32 and there are 50 active participants.

In Table 4 we present the results of the simulation which clearly demonstrate how small the probability a typical participant has of receiving money by reaching the top of the list of letters sold to future purchasers. There was no simulation in which a participant joining when the process was half of its ultimate size, received any money in this manner. Participants entering when the process is only 10% of its ultimate size earned big money ($5,000 or more) in only 6% of the simulations, while they received no money in about half of them.

TABLE IV. Distribution of the Number of Payoffs to a Participant by Time of Entry

Payoffs	*0*	*1*	*2*	*3-5*	*6-9*	*10+*	*AVG.*	*STD.*
Enter at 10%								
Seed 1	*56*	*7*	*9*	*14*	*10*	*4*	*2.21*	*4.26*
Seed 2	*51*	*8*	*7*	*17*	*9*	*8*	*2.86*	*4.87*
Enter at 20%								
Seed 1	*80*	*10*	*5*	*5*	*0*	*0*	*.38*	*.92*
Seed 2	*86*	*6*	*4*	*2*	*2*	*0*	*.37*	*1.28*
Enter at 50%								
Seed 1	*100*	*0*	*0*	*0*	*0*	*0*	*0*	*0*
Seed 2	*100*	*0*	*0*	*0*	*0*	*0*	*0*	*0*

Since the entire eligible population are in the system when it stops, they all are behind by either \$500 or \$1,000 depending on whether they still have 1 or 2 letters to sell and approximately $\alpha(1-\alpha)^{-1} = 61.8\%$ of them will be in the latter category. Thus, the vast majority of participants will lose money as they would have to receive at least 1 or 2 payoffs just to break even and the results in Table IV indicate that this is very unlikely.

VI. ACKNOWLEDGMENT

The authors are happy to acknowledge the assistance of Pat Hammick who conducted the computer simulation in a most efficient and helpful manner.

APPENDIX

The weak convergence property of $\{\xi_n(t), \varepsilon \leq t \leq 2\}$ stated in Theorem 2.1 will be established here. Our weak convergence arguments will depend critically on a Kolmogorov-type probability bound for a class of Markov chains defined below.

Definition. A Markov chain $\{Z_k, k=0,1,2,\ldots\}$ with $Z_0 \equiv 0$ and transition probabilities

$$P[Z_{k+1} - z = -\alpha \mid Z_k = z] = 1 - \alpha - g_k(z \mid a)$$

$$P[Z_{k+1} - z = -\alpha \mid Z_k = z] = \alpha + g_k(z \mid a)$$

where $0 < \alpha < 1$ and $g_k(z|a) <, =$ or > 0 according as $z >, =$ or $<$ a will be called a *Bernoulli random walk with drift towards a*.

The rebounding property inherent in the transition probabilities of such a chain tends to move it down or up respectively as it goes above or below the level a. The chain $\{X_k, 0 \leq k \leq 2n\}$ starting at $X_0 \equiv 0$ and with transition probabilities (2.8) enjoys this property with drift towards

$$a_0 = -c\alpha/(2-\alpha).$$

The *conditional chain*

$\{Y_k = X_{n\varepsilon+k} - X_{n\varepsilon},\ 0 \le k \le (2-\varepsilon)n\}$ *given* $X_{n\varepsilon} = \sqrt{n}\, x_0$ starting at

$Y_0 \equiv 0$ and with transition probabilites

$$P_{n\varepsilon\sqrt{n}\,x_0}[Y_{k+1} - y = -\alpha \mid Y_k = y] = 1-\alpha-h_{n\varepsilon+k}(\sqrt{n}\, x_0 + y)$$

$$P_{n\varepsilon\sqrt{n}\,x_0}[Y_{k+1} - y = 1-\alpha \mid Y_k = y] = \alpha + h_{n\varepsilon+k}(\sqrt{n}\, x_0 + y), \tag{A.2}$$

where $P_{n\varepsilon\sqrt{n}\,x_0}$ denotes the conditional probability distribution given $X_{n\varepsilon} = \sqrt{n}\, x_0$ and $h_{n\varepsilon+k}(\sqrt{n}\, x_0 + y)$ is defined according to (2.9), also enjoys this property with

$$a_n(x_0) = -[c\alpha/(2-\alpha) + \sqrt{n}\, x_0/\sqrt{1-\varepsilon}\,].$$

We now obtain the probability bound in the following lemma.

Lemma A.1. Let $\{X_i, i=0,1,2,\ldots\}$ be a Bernoulli random walk with drift towards a and let $\{X_i^*, i=0,1,2,\ldots\}$ be a corresponding Bernoulli random walk without drift, i.e.,

$X_i^* = \sum_{j=0}^{i} Y_j$ where $Y_0 \equiv 0$ and $Y_1, Y_2, \ldots$ are iid, taking values $1-\alpha$ and $-\alpha$ with probabilities α and $1-\alpha$ respectively. Then for $\lambda > 1 + |a|$,

$$P[\max_{0 \le i \le k} |X_i| \ge \lambda] \le P[\max_{0 \le i \le k} |X_i^*| \ge \tfrac{1}{2}\{\lambda - (\ +|a|)\}]$$

$$\le 4k\alpha(1-\alpha)/\{\lambda - (1+|a|)\}^2 .$$

Proof. Since a is fixed, there is no confusion in writing $g_i(z) = g_i(z|a)$. We prove the first inequality by representing the two events in a common probability space and establishing an inclusion relation between them in that space. To find such a representation try to write $X_i = \xi_i + \eta_i$ where $\{\xi_i\}$ has the same law as the random walk $\{X_i^*\}$ and the increments $\Delta\eta_i = \eta_{i+1}-\eta_i$ of the $\{\eta_i\}$, whose conditional distribution given $(\xi_1,\eta_1),\ldots,(\xi_i,\eta_i)$ depends only $\xi_i+\eta_i$ by the Markov property of X_i, are constructed in such a manner as to provide the necessary drift adjustment on $\{\xi_i\}$. For this we need to choose the distribution of $\Delta\eta_i$ depending on $\xi_i + \eta_i$ as follows. If $X_i = \xi_i + \eta_i > a$, then keeping $\Delta\eta_i \equiv 0$ when $\Delta\xi_i = \xi_{i+1}-\xi_i = -\alpha$ and allowing $\Delta\eta_i$ to be 0 or -1 with appropriate probabilities when $\Delta\xi_i = 1-\alpha$ will provide the necessary negative drift in $\Delta X_i = \Delta\xi_i+\Delta\eta_i$. On the other hand if $X_i = \xi_i + \eta_i < a$, then keeping $\Delta\eta_i \equiv 0$ when $\Delta\xi_i = 1-\alpha$ and allowing $\Delta\eta_i$ to be 0 or +1 with appropriate probabilities when $\Delta\xi_i = -\alpha$ will provide the necessary positive drift in ΔX_i. Specifically, consider the bivariate Markov chain $\{(\xi_i,\eta_i),\ i = 0, 1, 2, \ldots\}$ with $(\xi_0,\eta_0) \equiv (0,0)$ and transition probabilities

$$P[\xi_{i+1}-x = 1-\alpha \mid (\xi_i,\eta_i) = (x,y)]$$

$$= 1-P[\xi_{i+1}-x = -\alpha \mid (\xi_i,\eta_i) = (x,y)] = \alpha;$$

$$P[\eta_{i+1}-y=-1 \mid (\xi_i,\eta_i) = (x,y),\ \xi_{i+1}=1-\alpha]$$

$$= 1-P[\eta_{i+1}-y=0 \mid (\xi_i,\eta_i) = (x,y),\ \xi_{i+1}=1-\alpha]$$

$$= -I_{(a,\infty)}(x+y)g_i(x+y)/\alpha,$$

$$P[\eta_{i+1}-y=+1 \mid (\xi_i,\eta_i) = (x,y),\ \xi_{i+1} = -\alpha]$$

$$= 1-P[\eta_{i+1}-y=0 \mid (\xi_i,\eta_i) = (x,y),\xi_{i+1}=-\alpha]$$

$$= I_{(-\infty,a)}(x+y)/(1-\alpha).$$

We shall now show that

(i) $\{\xi_i\}$ and $\{X_i^*\}$ have identical joint distributions, and

(ii) $\{\xi_i+\eta_i\}$ and $\{X_i\}$ have identical joint distributions.

By this representation, both $\{X_i\}$ and $\{X_i^*\}$ are embedded on $\{(\xi_i,\eta_i)\}$, so the first inequality of (A.3) will be proved by demonstrating that

$$\max_{0\le i\le k} |\xi_i+\eta_i| \ge \lambda \Rightarrow \max_{0\le i\le k} |\xi_i| \ge \frac{1}{2}\{\lambda-(1+|a|)\}. \tag{A.4}$$

The second inequality of (A.3) will then follow by the Kolmogorov Kolmogrov-inequality applied to the random walk $\{X_i^*\}$.

From the description of $\{(\xi_i,\eta_i)\}$, (i) is obvious. To prove (ii), fix i and $(z_1,\ldots,z_i)$ arbitrarily, let S_i denote the set of all paths $\omega_i = ((0,0),(x_1,y_1),\ldots,(x_i,y_i))$ until time i which satisfy $x_j+y_j=z_j$, $1\le j\le i$, and let $f(\omega_i|S_i)$ denote the conditional distribution given S_i. Then

$$P[(\xi_{i+1}+\eta_{i+1})-(\xi_i+\eta_i)=-\alpha|S_i]$$

$$= \Sigma_{S_i} f(\omega_i|S_i)\{P[\xi_{i+1}-\xi_i=1-\alpha,\eta_{i+1}-\eta_i=-1|\omega_i] +$$

$$+P[\xi_{i+1}-\xi_i=-\alpha,\eta_{i+1}-\eta_i=0|\omega_i]\}$$

$$=\Sigma_{S_i} f(\omega_i|S_i)\{-\alpha\cdot I_{(a,\infty)}(z_i)g_i(z_i)/\alpha +$$

$$+ (1-\alpha)(1-I_{(-\infty,a)}(z_i)g_i(z_i)/(1-\alpha))\}$$

$$= 1-\alpha-g_i(z_i),$$

since $\Sigma_{S_i} f(\omega_i|S_i) = 1$, and

$$P[(\xi_{i+1}+\eta_{i+1})-(\xi_i+\eta_i) = 1-\alpha|S_i] = \alpha+g_i(z_i)$$

follows simiarly. Thus (ii) is verified.

We now demonstrate that the implication (A.4) holds. Consider a sample sequence ω for which $\max_{0\le i\le k} [\xi_i(\omega)+\eta_i(\omega)]\ge\lambda$. Let $\nu = \nu(\omega) \le k$ denote the smallest i for which $\xi_i+\eta_i \ge \lambda$. If $\eta_\nu \le \frac{1}{2}[\lambda+(1+|a|)]$, then $\xi_\nu \ge \frac{1}{2}[\lambda-(1+|a|)]$ ensuring (A.4); so suppose $\eta_\nu > \frac{1}{2}[\lambda+(1+|a|)]$. There exists $i < \nu$ such that $\eta_i = \eta_\nu - 1$ and $\eta_{i+1} - \eta_i = +1$. Let

$$\tau = \tau(\omega) = \max\{i|i<\nu(\omega),\eta_i(\omega)=\eta_{\nu(\omega)}(\omega)-1,\ \eta_{i+1}(\omega)-\eta_i(\omega)=+1\}$$

But $\eta_{i+1}-\eta_i$ can be +1 only if $\xi_i+\eta_i < a$. We thus have

$$\xi_\tau < -\eta_\tau+a = -\eta_\nu+1+a<-\frac{1}{2}[\lambda+(1+|a|)]+1+|a|$$
$$= -\frac{1}{2}[\lambda-(1+|a|)],$$

establishing (A.4). Similarly, $\min_{0\le i\le k} [\xi_i+\eta_i] \le -\lambda$ entails the same implication and the lemma is proved.

Consider the conditional chain $\{Y_k\}$ given $X_{n\varepsilon} = \sqrt{n}\, x_0$ given by (A.1) and (A.2) and let

$$\eta_n(t) = \eta_{n,nt} = Y_{nt}/\sqrt{n}, \qquad 0\le t \le 2-\varepsilon. \tag{A.5}$$

Comparing (A.5) with the defining formula (2.10a) for $\xi_n(t)$, we see that the conditional distribution of $\{\xi_n(t),\varepsilon\le t\le 2\}$ given $\xi_n(\varepsilon) = x_0$ is the same as the conditional distribution of $\{x_0+\eta_n(t-\varepsilon),\ \varepsilon\le t\le 2\}$ given $X_{n_\varepsilon} = \sqrt{n}\, x_0$:

$$\{\xi_n(t),\varepsilon\le t\le 2|\xi_n(\varepsilon)=x_0\} \overset{L}{=} \{x_0+\eta_n(t-\varepsilon),\varepsilon\le t\le 2|X_{n\varepsilon}=\sqrt{n}\, x_0\}.$$

The proof of Theorem 2.1 will, therefore, be accomplished by establishing the weak convergence of $\{\eta_n(t),\ 0\le t\le 2-\varepsilon\}$ under $P_{n\varepsilon\sqrt{n}\, x_0}$. To this end we examine the infinitesimal properties of this conditional process, viz.

$$a_n(k/n,y) = n\, E_{n\varepsilon\sqrt{n}\, x_0}[I_{\{|\eta_{n,k+1}-y|\leq 1\}}(\eta_{n,k+1}-y)\,|\,\eta_{nk}=y]$$

$$= \sqrt{n}\, E_{n\varepsilon\sqrt{n}\, x_0}[I_{\{|Y_{k+1}-\sqrt{n}\, y|\leq\sqrt{n}\}}(Y_{k+1}-\sqrt{n}y)\,|\,Y_k=\sqrt{n}y]$$

$$= \sqrt{n}\, h_{n\varepsilon+k}(\sqrt{n}\,(x_0+y)) \tag{A.7}$$

and

$$\sigma_n^2(k/n,y) = n\, E_{n\varepsilon\sqrt{n}\, x_0}[I_{\{|\eta_{n,k+1}-y|\leq\ \}}(\eta_{n,k+1}-y)^2\,|\,\eta_{nk}=y]$$

$$= E_{n\varepsilon\sqrt{n}\, x_0}\, I_{\{|Y_{k+1}-\sqrt{n}y\}\leq\sqrt{n}\}}(Y_{k+1}-\sqrt{n}y)^2\,|\,Y_k=\sqrt{n}\, y$$

$$= \alpha(1-\alpha)+(1-2\alpha)h_{n\varepsilon+k}(\sqrt{n}(x_0+y)), \tag{A.8}$$

using the transition probabilities under $P_{n\varepsilon\sqrt{n}\, x_0}$ given by (A.2) to calculate the corresponding expectations $E_{n\varepsilon\sqrt{n}\, x_0}$ and remembering that $I_{\{|Y_{k+1}-Y_k|\leq\sqrt{n}\}} \equiv 1$.

The limiting form of $h_{n\varepsilon+k}(\sqrt{n}\,(x_0+y))$ in (2.13) shows that for large n, $a_n(k/n,y)$ and $\sigma_n^2(k/n,y)$ are approximated by $a(k/n,y)$ and $\sigma^2(k/n,y)$ respectively, where

$$a(t,y) = -\left(\frac{2-\alpha}{1-\alpha}\right)\cdot\frac{x_0+y}{\varepsilon+t}, \tag{A.9}$$

$$\sigma^2(t,y) = \alpha(1-\alpha). \tag{A.10}$$

However, these convergences are uniform only on compact sets of y. For this reason we modify the $\{Y_k\}$ chain, as follows. Let

$$\phi_M(y) = \begin{cases} -M, y<-M \\ y, |y|\leq M, \quad h^{(M)}_{n\varepsilon+k}(\sqrt{n}\ x_0+y)=h_{n\varepsilon+k}(\sqrt{n}\ x_0+\sqrt{n}\ \phi_M(y/\sqrt{n})). \\ M, y>M \end{cases}$$

Consider a Markov chain $\{Y_k^{(M)}, 0\le k\le (2-^2)n\}$ with $Y_0^{(M)} \equiv 0$ and transition probabilities obtained by replacing $h_{n\varepsilon+k}(\sqrt{n}\, x_0+y)$ by $h_{n\varepsilon+k}^{(M)}(\sqrt{n}\, x_0+y)$ in (A.2). Since both $\{Y_k^{(M)}\}$and $\{Y_k\}$ start at $Y_0^{(M)} = Y_0 = 0$ and have the same transition probabilities when both processes lie between $[-\sqrt{n}\, M, \sqrt{n}\, M]$, we can and do assume that they are defined on the same probability space and that

$$Y_k^{(M)} \equiv Y_k \quad \text{for} \quad k \le \min\{j \,\big|\, |Y_j| \ge \sqrt{n}\, M\}.$$

Now let

$$\eta_n^{(M)}(t) = \eta_{n,nt}^{(M)} = Y_{nt}^{(M)}/\sqrt{n}, \qquad 0\le t \le 2-\varepsilon.$$

Then $\eta_n^{(M)}(\cdot) \equiv \eta_n(\cdot)$ on the set $\{\sup_{0\le t\le 2-\varepsilon} |\eta_n(t)| \le M\}$.

We now examine the weak convergence of $\{\eta_n^{(M)}(t)\}$ by approximating its infinitesimal properties analogous to (A.7) and (A.8), viz.

$$a_n^{(M)}(k/n,y) = \sqrt{n}\, h_{n\varepsilon+k}^{(M)}(\sqrt{n}\,(x_0+y))$$

$$\sigma_n^{(M)^2}(k/n,y) = \alpha(1-\alpha) + (1-2\alpha)h_{n\varepsilon+k}^{(M)}(\sqrt{n}\,(x_0+y))$$

with $a^{(M)}(k/n,y)$ and $\sigma^{(M)^2}(k/n,y$ respectively, where

$$a^{(M)}(t,y) = a(t,\phi_M(y)) = -\left(\frac{2-\alpha}{1-\alpha}\right) \cdot \frac{x_0+\phi_M(y)}{\varepsilon+t}, \tag{A.11}$$

$$\sigma^{(M)^2}(t,y) = \sigma^2(t,\phi_M(y)) = \alpha(1-\alpha) \tag{A.12}$$

are the truncated versions of $a(t,y)$ and $\sigma^2(t,y)$ given by (A.9) and (A.10).

Lemma A.2. For each L and M, there is a constant B such that for sufficiently large n,

$$|a_n^{(M)}(k/n,y)-a^{(M)}(k/n,y)| \le B/\sqrt{n} ,$$

$$|\sigma_n^{(M)^2}(k/n,y)-\sigma^{(M)^2}(k/n,y)| \le \frac{1-2\alpha}{\sqrt{n}}\left[\left(\frac{2-\alpha}{1-\alpha}\right)\left(\frac{L+M}{\varepsilon}\right) + \frac{B}{\sqrt{n}}\right]$$

holds for all $|x_0| \le L$ and $0 \le k \le (2-\varepsilon)n$.

Proof.

$$|a_n^{(M)}(k/n,y)-a^{(M)}(k/n,y)|$$

$$= |\sqrt{n}\, h_{n\varepsilon+k}^{(M)}(\sqrt{n}\,(x_0+y))-a^{(M)}(k/n,y)|$$

$$= \frac{1}{\sqrt{n}}\left|\frac{(2-\alpha)(x_0+\phi_M(y))^2+c\alpha(1-\alpha)(\varepsilon+kn^{-1})-n\;c(2-\alpha)(x_0+\phi_M(y))}{(1-\alpha)(\varepsilon+kn^{-1})\{(1-\alpha)(\varepsilon+kn^{-1})-n^{-\frac{1}{2}}(x_0+\phi_M(y))+n^{-1}c\}}\right|$$

$$\le B/\sqrt{n} ,$$

where $B = B_1/B_2$ is given by

$$B_1 = (2-\alpha)(L^2+M^2) + 2c\alpha(1-\alpha)+c(2-\alpha)(L+M),$$

$$B_2 = (1-\alpha)\varepsilon/2.$$

The second bound follows immediately from the first.

Since $|Y_{k+1}^{(M)}-Y_k^{(M)}| < 1$ with probability 1, and since

$$\lim_{n\to\infty} \sup_{0\le k\le(2-\varepsilon)n,\,y\varepsilon\mathbb{R}} |a_n^{(M)}(k/n,y)-a^{(M)}(k/n,y)| = 0 \text{ and}$$

$$\lim_{n\to\infty} \sup_{0\le k\le(2-\varepsilon)n,\,y\varepsilon\mathbb{R}} |\sigma_n^{(M)^2}(k/n,y)-\sigma^{(M)^2}(k/n,y)| = 0$$

by virtue of Lemma A.2, the conditional process $\{\eta_n^{(M)}(t),\ 0\underline{<}t\underline{<}2-\varepsilon\}$ given $X_{n\varepsilon} = \sqrt{n}\, x_0$ satisfies all conditions of Theorem 10.3 of Strook and Varadhan [6] and we have the following theorem.

Theorem A.3.

$\{\eta_n^{(M)}(t),\ 0\leq t\leq 2-\varepsilon \mid X_{n\varepsilon} = \sqrt{n}\, x_0\} \overset{w}{\to} \{\eta^{(M)}(t),\ 0\leq t\leq 2-\varepsilon\}$ where $\eta^{(M)}(t)$ is the solution of the stochastic differential equation

$$\eta^{(M)}(t) = \int_0^t a^{(M)}(s,\eta^{(M)}(s))ds + \int_0^t \sigma^{(M)}(s,\eta^{(M)}(s))dW(s) \tag{A.13}$$

starting at $\eta^{(M)}(0) = 0$, where $a^{(M)}(t,y)$ and ${\sigma^{(M)}}^2(t,y)$ are given by (A.11) and (A.12), and W(s) is a standard Wiener process. Moreover, the convergence is uniform for $|x_0| \leq L$.

From this, we now have

Theorem A.4.

$\{\eta_n(t),\ 0\leq t\leq 2-\varepsilon \mid X_{n\varepsilon}=\sqrt{n}\, x_0\} \overset{w}{\to} \{\eta(t),\ 0\leq t\leq 2-\varepsilon\}$ uniformly for $|x_0|\leq L$, where $\eta(t)$ is the solution of the stochastic differential equation obtained by replacing $a^{(M)}(t,y)$ and ${\sigma^{(M)}}^2(t,y)$ in (A.13) by $a(t,y)$ and $\sigma^2(t,y)$ respectively, given in (A.9) and (A.10).

Proof. The processes $\eta_n^{(M)}(\cdot)$ and $\eta_n(\cdot)$ are defined on the same probability space and are identical until $\eta_n(\cdot)$ moves outside -M,M for the first time. Hence the events $\sup_{0\leq t\leq 2-\varepsilon} |\eta_n(t)| \leq M$ and $\sup_{0\leq t\leq 2-\varepsilon} |\eta_n^{(M)}(t)| \leq M$ are identical.

Denote this event by S_n. Using Lemma A.1 with $a_n(x_0) = -[c\alpha/(2-\alpha) + \sqrt{n}\, x_0/\sqrt{1-\varepsilon}]$,

$$P_{n\varepsilon\sqrt{n}\,x_0}[\sup_{0\le t\le 2-\varepsilon}|\eta_n(t)|>M] =$$

$$= P_{n\varepsilon\sqrt{n}\,x_0}[\max_{0\le k\le(2-\varepsilon)n}|Y_k| >\sqrt{n}\,M]$$

$$\le 4n(2-\varepsilon)\alpha(1-\alpha)/\{\sqrt{n}\,M-(1+|a_N(x_0)|)\}^2$$

$$\le 2/\{M-L-n^{-\frac{1}{2}}(c+1)\}^2$$

for $M > L+c+1$. Hence $P_{n\varepsilon\sqrt{n}\,x_0}(S_n^c)$ can be made smaller than an arbitrary $\delta > 0$ for all n and $|x_0| \le L$ by choosing M large enough. Moreover, let $\eta(t)$ by governed by the same Wiener process $W(\cdot)$ which gives $\eta^{(M)}(t)$. Since the coefficients $a^{(M)}(t,y)$ and $\sigma^{(M)}(t,y)$ of $\eta^{(M)}(t)$ are the same as the coefficients $a(t,y)$ and $\sigma(t,y)$ of $\eta(t)$ for $|y| \le M$, it follows that $\eta(\cdot)$ and $\eta^{(M)}(\cdot)$ defined on the same probability space are identical until $\eta^{(M)}(\cdot)$ moves outside $[-M,M]$ for the first time. Hence the events $\sup_{0\le t\le 2-\varepsilon}|\eta(t)| \le M$ and $\sup_{0\le t\le 2-\varepsilon}|\eta^{(M)}(t)|\le M$ are identical. Denote this event by S. By the weak convergence of Theorem A.3,

$$P(S^c) < P_{n\varepsilon\sqrt{n}\,x_0}(S_n^c) + \delta < 2\delta .$$

We now consider a bounded continuous function f on $C[0,2-\varepsilon]$ and let $\sup_\eta|f(\eta)| \le B$. Then for large n,

$$|E_{n\varepsilon\sqrt{n}\,x_0}f(\eta_n)-Ef(\eta)| \le (6B+1)\delta,$$

because

$$\left|E_{n\varepsilon\sqrt{n}\, x_0} f(\eta_n^{(M)}) - E_{n\varepsilon\sqrt{n}\, x_0} f(\eta_n)\right| \leq \int_{S_n^c} \left|f(\eta_n^{(M)} - f(\eta_n)\right| dP_{n\varepsilon\sqrt{n}\, x_0} \leq 2B\delta,$$

$$\left|Ef(\eta^{(M)}) - Ef(\eta)\right| \leq \int_{S^c} f(\eta^{(M)}) - f(\eta)\left| dP \leq 2B \cdot 2\delta,\right.$$

and by the weak convergence of Theorem A.3,

$$\left|E_{n\varepsilon\sqrt{n}\, x_0} f(\eta_n^{(M)}) - Ef(\eta^{(M)})\right| < \delta \text{ for large } n .$$

This completes the proof.

Rewriting Theorem A.4 in view of (A.6), we have

$$\{\xi_n(t), \varepsilon \leq t \leq 2 \mid \xi_n(\varepsilon) = x_0\} \overset{W}{\rightarrow} \{x_0 + \eta(t-\varepsilon), \varepsilon \leq t \leq 2\} .$$

Call $\xi(t) = x_0 + \eta(t-\varepsilon)$. Then

$$\begin{aligned} \xi(t) &= x_0 - \left(\frac{2-\alpha}{1-\alpha}\right) \int_0^{t-\varepsilon} \frac{x_0 + \eta(s)}{s+\varepsilon}\, ds + \sqrt{\alpha(1-\alpha)} \int_0^{t-\varepsilon} dW(s) \\ &= x_0 - \left(\frac{2-\alpha}{1-\alpha}\right) \int_0^{t-\varepsilon} \frac{\xi(s+\varepsilon)}{s+\varepsilon}\, ds + \sqrt{\alpha(1-\alpha)} \int_0^{t-\varepsilon} dW(s) \\ &= x_0 - \left(\frac{2-\alpha}{1-\alpha}\right) \int_\varepsilon^{t} \frac{\xi(s)}{s}\, ds + \sqrt{\alpha(1-\alpha)} \int_\varepsilon^{t} dW(s), \end{aligned}$$

and Theorem 2.1 is proved.

REFERENCES

[1] Anscombe, F. J. (1952). "Large-Sample Theory of Sequential Estimation," *Proc. Cambridge Phil. Soc.*, *48*, 600-607.

[2] Arnold, L. (1974). *Stochastic Differential Equations: Theory and Applications*. Wiley.

[3] Bhattacharya, P. K. and Mallik, A. (1973). "Asymptotic Normality of Stopping Times of Some Sequential Procedures," *Ann. Statist.*, *1*, 1203-1209.

[4] Gastwirth, J. L. (1977). "A Probability Model of a Pyramid Scheme," *The American Statistician, 31,* 79-82.
[5] Gikhman, I. I. and Skorokhod, A. V. (1969). *Introduction to the Theory of Random Processes.* Saunders.
[6] Strook, D. W. and Varadhan, S. R. S. (1969). "Diffusion Processes with Continuous Coefficients II. *Comm. Pure and Appl. Math., 22,* 479-530.

SET-VALUED PARAMETERS AND SET-VALUED STATISTICS

Morris H. DeGroot[1]
William F. Eddy[2]

Department of Statistics
Carnegie-Mellon University
Pittsburgh, Pennsylvania

I. INTRODUCTION

The problem of statistical inference for a random sample from a uniform distribution usually evokes in the mind of a statistician the vision of a problem that has been thoroughly worked out and that contains few, if any, remaining challenges. That vision is incorrect, however. Even a uniform distribution on a fairly simple subset of the real line can involve unexpected complexities, and for uniform distributions in higher dimensions or abstract spaces these complexities can be quite intricate.

Consider, as an example, the problem of determining a minimal sufficient statistic for a random sample from the uniform distribution supported by the union of two intervals on the real line $[0,\theta]$ and $[2\theta,3\theta]$, where $\theta > 0$ is an unknown parameter. Alternatively, one might consider the minimal sufficient statistic for a sample from the uniform distribution supported

[1] *Partially supported by the National Science Foundation under grant SES-7906386.*
[2] *Partially supported by an NSF Mathematical Sciences Post-doctoral Research Fellowship at Brown University.*

ISBN 0-12-589320-5

by the union of the two intervals $[0,\theta]$ and $[2\theta,1]$, where $0 < \theta < \frac{1}{2}$. (These two examples and some of the others mentioned here are discussed in detail in Section VI.) Generalizing, one might consider a uniform distribution supported by the union of three intervals $[0,\theta]$, $[2\theta,3\theta]$, and $[4\theta,5\theta]$. Also one might introduce another parameter and consider a uniform distribution supported by the union of $[0,\theta]$ and $[\lambda\theta,(\lambda+1)\theta]$ for $\theta > 0$ and $\lambda > 1$, or a uniform distribution supported by the union of $[0,\theta]$ and $[\lambda\theta,1]$ for $\lambda > 1$ and $0 < \theta < 1/\lambda$. In summary a minimal sufficient statistic for a family of uniform distributions of subsets of the line can be relatively complicated if the support of each distribution is not an interval. Standard discussions of sufficiency such as Huzurbazar (1976) do not seem to treat this situation.

In the plane one might seek a minimal sufficient statistic for a sample from a uniform distribution on a triangle where the triangle belongs to some family of congruent or similar triangles. One might consider a uniform distribution over the union of two or three such triangles or one might consider other geometric figures. The possibilities are literally countless.

In this paper we shall focus on problems of this type in which the primary purpose of a statistical experiment is to determine the boundaries of some unknown set. We will attempt to remain quite general. Our observations will take values in abstract spaces; the distributions over the unknown set will be arbitrary rather than just a uniform distribution; and we will consider several different sampling models for obtaining the observational data. In much of the statistical literature, the concept of completeness is directly linked to that of sufficiency [see, e.g., Fraser (1953 and 1957, p. 27ff) and Bell, Blackwell, and Breiman (1960)]. However, because of the generality of the parameter space in the models that we shall present, we do not have access to the standard techniques for establishing completeness and will not treat this topic in this paper.

In Section II we describe the different sampling models that we will consider, present the appropriate likelihood functions based on these methods, and define and study the set-valued statistics that will be used. In Section III, the minimal sufficient statistics for the different likelihood functions are derived. In Section IV, maximum likelihood estimators of unknown sets are developed based on these likelihood functions. In Section V, Bayesian methods are studied. Finally, in Section VI, minimal sufficient statistics and maximum likelihood estimators are worked out for several of the examples mentioned at the beginning of this paper.

II. PRELIMINARIES

Let $(\underline{X},\mathcal{X})$ be an abstract measure space with generic point x and let $\mathcal{S} = \{S\}$ be an arbitrary collection of subsets of $\underline{X}$. To avoid various circumlocutions concerning inner and outer measure, it is assumed that $\mathcal{S} \subseteq \mathcal{X}$. For $S_0 \varepsilon \mathcal{S}$, define

$$J(x) = I_{S_0}(x) = I(x\varepsilon S_0) = \begin{cases} 1 \text{ if } x\varepsilon S_0 \\ 0 \text{ otherwise.} \end{cases}$$

Suppose now that S_0 is unknown (a "parameter"), that x is the realization of a random variable X taking values in $\underline{X}$, and that the probability distribution of X depends on S_0. There are several possible forms of this dependence.

2.1 *Deterministic Model*

Choose fixed $z_1,\ldots,z_N$ in $\underline{X}$ and observe $J(z_1),\ldots,J(z_N)$. In other words, observe whether each of the specified points $z_1,\ldots,z_N$ lies inside or outside the unknown set S_0. This model is used in mineral exploration: specific locations are chosen by a non-random mechanism and test borings are made at those locations. The observation at each location is the presence or absence of the mineral of interest and the problem is to locate the ore body denoted by S_0. The likelihood function is

$$L[S|z_1,\ldots,z_N,\ J(z_1),\ldots,J(z_N)] \propto \prod_{i=1}^{N} [1-|I_S(z_i)-J(z_i)|]. \quad (2.1)$$

It is convenient to relable the data so that those observations with J=0 are denoted $y_1,\ldots,y_m$ and those observations with J=1 are denoted $x_1,\ldots,x_n$ where m+n=N. In this notation the likelihood (2.1) reduces to

$$L(S|x_1,\ldots,x_n,\ y_1,\ldots,y_m) \propto \prod_{i=1}^{n} I_S(x_i) \prod_{j=1}^{m} I_{\bar{S}}(y_j) \quad (2.2)$$

where $\bar{S}$ is the complement of S in $\underline{X}$.

2.2 *Shotgun Model*

Choose $z_1,\ldots,z_N$ according to a common distribution P on $\underline{X}$ and observe $J(z_1),\ldots,J(z_N)$. Again, the observations indicate whether each of the random points $z_1,\ldots,z_N$ lies inside or outside the unknown set S_0. This model has been studied by Hachtel, Meilijson, and Nádas (1981) in a problem in which $\mathcal{S}$ is the collection of compact convex sets in d-dimensional Euclidean space. Under this model the likelihood of S is again given by (2.1) because the distribution P does not depend on S_0. Again, when the data are relabelled $x_1,\ldots,x_m$ and $y_1,\ldots,y_n$ as above, the likelihood reduces to (2.2). Of course, if we can only observe n and m and not the individual points x_i and y_j, the likelihood is simply

$$L(S|n,m) \propto P(S)^n P(\bar{S})^m. \quad (2.3)$$

To improve the realism of this model, we introduce an additional complication into the data collection process. Let W be a fixed known subset of $\underline{X}$; again, for simplicity assume $W \varepsilon \mathcal{X}$. Suppose that the only data that can be observed are those in W;

the set W is a sampling window. Again a sample of size N is drawn from P on $\underline{X}$ but we only observe the points $z_1,\ldots,z_{N_0}$ that fall in W and $J(z_1),\ldots,J(z_{N_0})$. In this case the likelihood is simply

$$L[S|z_1,\ldots,z_{N_0},J(z_1),\ldots,J(z_{N_0}),W] \propto \prod_{i=1}^{N_0} [1-|I_S(z_i)-J(z_i)|]. \tag{2.4}$$

It is possible that N is unknown. Since N_0 has a binomial distribution with parameters N and P(W), and

$$\binom{N}{N_0} P(W)^{N_0} P(\overline{W})^{N-N_0} \propto \binom{N}{N_0} P(\overline{W})^N,$$

the joint likelihood of S and N is just

$$L[S,N|N_0,z_1,\ldots,z_{N_0}, J(z_1),\ldots,J(z_{N_0}),W] \propto \binom{N}{N_0} P(\overline{W})^N \prod_{i=1}^{N_0} [1-|I_S(z_i)-J(z_i)|]. \tag{2.5}$$

Again it is convenient to relabel the data so that the observations with J=0 are $y_1,\ldots,y_{m_0}$ and those with J=1 are $x_1,\ldots,x_{n_0}$, where $m_0 + n_0 = N_0$. Then the likelihood of S is simply

$$L(S|x_1,\ldots,x_{n_0}, y_1,\ldots,y_{m_0},W) \propto \prod_{i=1}^{n_0} I(x_i \varepsilon S\cap W) \prod_{j=1}^{m_0} I(y_j \varepsilon \overline{S}\cap W) \tag{2.6}$$

and the joint likelihood of S and N is just the product of (2.6) and the factor $\binom{N}{N_0} P(\overline{W})^N$.

It is, perhaps, more interesting to compute the joint likelihood of S, n, and m. If a sample of unknown size n is drawn from S in accordance with the restriction of P to S and a sample of unknown size m from $\overline{S}$ in accordance with the restriction of P to $\overline{S}$, then we find

$$L(S,n,m|x_1,\ldots,x_{n_0},y_1,\ldots,y_{m_0},W) \propto \binom{n}{n_0} P(S)^{-n} P(S\cap\overline{W})^{n-n_0} \binom{m}{m_0} P(\overline{S})^{-m} P(\overline{S}\cap\overline{W})^{m-m_0} \prod_{i=1}^{n_0} I(x_i \varepsilon S\cap W) \prod_{j=1}^{m_0} I(y_j \varepsilon \overline{S}\cap W). \quad (2.7)$$

On the other hand if we regard N as fixed and n and m as random but unobserved, then we must multiply by $\binom{N}{n} P(S)^n P(\overline{S})^m$. Thus, letting $n_1 = n - n_0$ and $m_1 = m - m_0$ be the number of unobserved x's and unobserved y's respectively, we get

$$L(S,n_1,m_1|x_1,\ldots,x_{n_0},y_1,\ldots,y_{m_0},W) \propto \binom{N}{n_0,m_0,n_1,m_1} P(S\cap\overline{W})^{n_1} P(\overline{S}\cap\overline{W})^{m_1} \cdot \prod_{i=1}^{n_0} I(x_i \varepsilon S\cap W) \prod_{j=1}^{m_0} I(y_j \varepsilon \overline{S}\cap W). \quad (2.8)$$

2.3 Inclusion - Exclusion Model

Suppose that P and Q are measures on $(\underline{X},X)$ which satisfy $0 < P(S) < \infty$ and $0 < Q(\overline{S}) < \infty$ for each $S\varepsilon S$. Let P_S and $Q_{\overline{S}}$ be the probability measures defined as follows:

$$P_S(\sigma) = P(\sigma)/P(S) \qquad \sigma\varepsilon X, \quad \sigma \subseteq S,$$

$$Q_{\overline{S}}(\sigma) = Q(\sigma)/Q(\overline{S}) \qquad \sigma\varepsilon X, \quad \sigma \subseteq \overline{S}.$$

Choose $x_1,\ldots,x_n$ independently according to a common distribution P_{S_0} and (independently) choose $y_1,\ldots,y_m$ independently according to a common distribution $Q_{\overline{S}_0}$. As special cases, we might have only an inclusion model (m=0) or only an exclusion model (n=0). A simple example of the inclusion-exclusion model arises when $\underline{X}$ is a bounded set and an experimenter can observe a random sample of n observations drawn from a uniform distribution on S_0 and an independent random sample of m observations drawn from a uniform distribution on $\bar{S}_0$. From these observations we wish to make inferences about the unknown set S_0.

In general, under an inclusion-exclusion sampling model, the likelihood of $S \varepsilon \mathcal{S}$ is given by [see (2.2) and (2.3)]

$$L(S|x_1,\ldots,x_n,\ y_1,\ldots,y_m) \propto \prod_{i=1}^{n}\left[\frac{I(x_i \varepsilon S)}{P(S)}\right]\prod_{j=1}^{m}\left[\frac{I(y_j \varepsilon \bar{S})}{Q(\bar{S})}\right]. \quad (2.9)$$

Now introduce the sampling window W as in Section 2.2 and again let $n_1 = n - n_0$ and $m_1 = m - m_0$ be the numbers of unobserved x's and unobserved y's. The likelihood is

$$L(S,n_1,m_1|x_1,\ldots,x_{n_0},\ y_1,\ldots,y_{m_0},W) \propto$$

$$\binom{n}{n_0} P(S)^{-n} P(S\cap\bar{W})^{n_1} \prod_{i=1}^{n_0} I(x_i \varepsilon S\cap W) \binom{m}{m_0} Q(\bar{S})^{-m} Q(\bar{S}\cap\bar{W})^{m_1}$$

$$\cdot \prod_{j=1}^{m_0} I(y_j \varepsilon \bar{S}\cap W). \quad (2.10)$$

2.4 Coverage and Support

Define the inclusion sets in $\mathcal{S}$ by

$$\mathcal{T}(z_1,\ldots,z_N) = \{S|J(z_i) = 1 \Rightarrow z_i \varepsilon S,\ i=1,\ldots,N\}.$$

In terms of x and y, this becomes

$$\mathcal{T}(x_1,\ldots,x_n,\ y_1,\ldots,y_m) = \mathcal{T}(x_1,\ldots,x_n) = \{S|x_i \varepsilon S,\ i=1,\ldots,n\}.$$

If n = 0, then $\mathcal{T} = \mathcal{S}$. Similarly, define the exclusion sets in $\mathcal{S}$ by

$$\mathcal{R}(z_1,\ldots,z_N) = \{S|J(z_i) = 0 \Rightarrow z_i \varepsilon \bar{S},\ i=1,\ldots,N\}$$

or

$$\mathcal{R}(x_1,\ldots,x_n,y_1,\ldots,y_m) = \mathcal{R}(y_1,\ldots,y_m) = \{S|y_j \varepsilon \bar{S},\ j=1,\ldots,m\}.$$

If m = 0, then $\mathcal{R} = \mathcal{S}$. Finally, define the inclusion-exclusion sets in $\mathcal{S}$ by

$$H(z_1,\dots,z_N) = H(x_1,\dots,x_n,\ y_1,\dots,y_m) = \{S \mid S\varepsilon\mathcal{T},\ S\varepsilon\mathcal{R}\}.$$

Note that the "true" S_0 is an element of $\mathcal{T}$, of $\mathcal{R}$, and of H.

We now introduce several set-valued statistics. The coverage of $z_1,\dots,z_N$ (which depends only on $x_1,\dots,x_n$) with respect to $\mathcal{S}$ is given by

$$T = \bigcap_{S\varepsilon\mathcal{T}} S.$$

The support of $z_1,\dots,z_N$ (which depends only on $y_1,\dots,y_m$) with respect to $\mathcal{S}$ is given by

$$R = \bigcup_{S\varepsilon\mathcal{R}} S.$$

The maximal coverage is given by

$$T^* = \bigcap_{S\varepsilon H} S$$

and the minimal support is given by

$$R^* = \bigcup_{S\varepsilon H} S.$$

It is obvious from these definitions that

$$\{x_1,\dots,x_n\} \subseteq T \subseteq T^* \subseteq S_0 \subseteq R^* \subseteq R \subseteq \overline{\{y_1,\dots,y_m\}}\ . \qquad (2.11)$$

Note that if $m = 0$ either by chance (the deterministic or the shotgun model) or by design (the inclusion-exclusion model), then $T \equiv T^*$. For certain collections $\mathcal{S}$, the event $T \equiv T^*$ may also occur when $m \neq 0$ but in general this relation does not hold. Also, when $m = 0$,

$$R = \bigcup_{S\varepsilon\mathcal{S}} S$$

is independent of the data $\{z_1,\dots,z_N\} = \{x_1,\dots,x_n\}$, and $R^* = \bigcup_{S\varepsilon\mathcal{T}} S$. Similar results hold when $n = 0$: $R \equiv R^*$, $T = \bigcap_{S\varepsilon\mathcal{S}} S$ does not depend on the data, and $T^* = \bigcap_{S\varepsilon\mathcal{R}} S$.

III. SUFFICIENCY

In the previous section we introduced several sets (statistics) which summarize the data under our various sampling models. In this section we will characterize their sufficiency properties. In a remarkably prescient paper, Fraser (1952) has studied one of these cases; see also Fraser (1966) and Smith (1956). In our terminology and notation, Fraser considered the inclusion-exclusion model with m = 0 and assumed that P belonged to a parametric family, say $\{P_\theta;\ \theta\varepsilon\Theta\}$, on $\underline{X}$. Fraser (1952, Theorem 1) showed that if $M(z_1,\ldots,z_N)$ is minimal sufficient for θ then the pair (M,T) is jointly minimal sufficient for the inclusion family $\Theta\times\mathcal{S}$. Following Fraser we could extend our shotgun model and our inclusion-exclusion model so that P (and Q) belong to parametric families; then using his technique we could display the joint minimal sufficient statistics. Because we are primarily interested in set-valued statistics here we will forgo this opportunity.

We begin by considering the three models without a sampling window.

Theorem 3.1. The statistics (T, R*) are jointly minimal sufficient for the likelihoods (2.2) and (2.9).*

Proof. The likelihoods (2.2) and (2.9) only depend on the data through the indicator functions

$$\prod_{i=1}^{n} I(x_i \varepsilon S)\ \prod_{j=1}^{m} I(y_j \varepsilon \overline{S}).$$

Thus, for sufficiency of (T*, R*), it is enough to show that for every $S\varepsilon\mathcal{S}$ and each $\underset{\sim}{x}$ and $\underset{\sim}{y}$, the relation

$$\{x_1,\ldots,x_n\} \subseteq S \subseteq \overline{\{y_1,\ldots,y_m\}}$$

is equivalent to

$$T^* \subseteq S \subseteq R^*.$$

This follows from the definitions; see (2.11). To see that (T^*, R^*) are jointly minimal sufficient, consider any other sufficient statistic A. From A it must be possible to construct $\mathcal{T}$ and $\mathcal{R}$ (by sufficiency). Once this is done, $\mathcal{H} = \mathcal{T} \cap \mathcal{R}$ can be determined and then T^* and R^* can be recovered. □

This proof shows that any support set T' and coverage set R' satisfying

$$\{x_1, \ldots, x_n\} \subseteq T' \subset T^* \text{ and } \{y_1, \ldots, y_m\} \subseteq \overline{R'} \subseteq \overline{R^*}$$

for all $\underset{\sim}{x}$ and $\underset{\sim}{y}$ must be jointly sufficient for $\mathcal{S}$.

We now introduce the sampling window W. Paralleling some definitions in Section 2.4, we define

$$\mathcal{T}_W = \{S \mid x_i \varepsilon S \cap W,\ i=1,\ldots,n_0\},$$

$$\mathcal{R}_W = \{S \mid y_j \varepsilon \overline{S} \cap W,\ j=1,\ldots,m_0\},$$

and

$$\mathcal{H}_W = \mathcal{T}_W \cap \mathcal{R}_W.$$

Also,

$$T_W = \bigcap_{S \varepsilon \mathcal{T}_W} S, \quad R_W = \bigcup_{S \varepsilon \mathcal{R}_W} S$$

and

$$T_W^* = \bigcap_{S \varepsilon \mathcal{H}_W} S, \quad R_W^* = \bigcup_{S \varepsilon \mathcal{H}_W} S.$$

Obviously [see (2.11)],

$$\{x_1, \ldots, x_{n_0}\} \subseteq T_W \subseteq T_W^* \subseteq T^* \subseteq S_0 \subseteq R^* \subseteq R_W^* \subseteq R_W \subseteq \{\overline{y_1, \ldots, y_{m_0}}\} \tag{3.1}$$

and

$$\{x_1, \ldots, x_{n_0}\} \subseteq T_W \subseteq T \subseteq T^* \subseteq S_0 \subseteq R^* \subseteq R \subseteq R_W \subseteq \{\overline{y_1, \ldots, y_{m_0}}\}.$$

Theorem 3.2. The statistics (T_W^*, R_W^*) *are jointly minimal sufficient for the likelihood* (2.6).

The proof is essentially the same as the proof of Theorem 3.1 and is omitted.

Theorem 3.3. The statistics (T_W^*, R_W^*, n_0, m_0) *are jointly minimal sufficient for the likelihoods* (2.7), (2.8), and (2.10).

Proof. Each of the likelihood functions (2.7), (2.8), and (2.10) depends on the data only through the indicator functions and the numbers of observations in the sets $S \cap W$, $S \cap \overline{W}$, $\overline{S} \cap W$, and $\overline{S} \cap \overline{W}$. Given these counts we can proceed as in the proof of Theorem 3.1. Note that by (3.1),

$$\{x_1, \ldots, x_{n_0}\} \subseteq S \subseteq \overline{\{y_1, \ldots, y_{m_0}\}}$$

if and only if

$$T_W^* \subseteq S \subseteq R_W^* .$$

Thus, sufficiency is established. Also, the collections $\mathcal{T}_W$ and $\mathcal{R}_W$ (and hence $\mathcal{H}_W$) can be recovered from any other sufficient statistic. Thus, minimal sufficiency is also established. □

IV. MAXIMUM LIKELIHOOD ESTIMATORS

In this section we examine the sets which maximize the likelihoods (2.2), (2.6), (2.7), (2.8), (2.9), and (2.10). Unfortunately, because of the generality of the models, it is not always possible to give an explicit representation for these estimators. Up to this point we have not made any restrictions on the sets in $\mathcal{S}$. However, consideration of the simplest examples shows that maximum likelihood estimators do not in general exist. Under the models of Section II, we may be sampling from both S_0 and $\overline{S}_0$. In every example that we have looked at, the non-existence of maximum likelihood estimators can be remedied by the simple device of extending $\mathcal{S}$ to include both open and closed sets. We will not address this point further, and henceforth assume that $\mathcal{S}$ has the appropriate properties.

4.1 Deterministic Model and Shotgun Model

It can be seen that the likelihoods (2.2) and (2.6) take only the values 0 and 1. Thus a maximum likelihood estimator (MLE) of S is any set, say $\hat{S}$, satisfying $L(\hat{S}) = 1$. For the likelihood (2.2) these sets are exactly those which satisfy

$$T^*(\underset{\sim}{x},\underset{\sim}{y}) \subseteq \hat{S} \subseteq R^*(\underset{\sim}{x},\underset{\sim}{y}).$$

For the likelihood (2.6) these sets satisfy

$$T_W^*(\underset{\sim}{x},\underset{\sim}{y}) \subseteq \hat{S} \subseteq R_W^*(\underset{\sim}{x},\underset{\sim}{y}).$$

4.2 Inclusion-Exclusion Model With Complementary Measures

We begin by noting that if $P(S) + Q(\bar{S}) = c < \infty$, for some constant c that does not depend on S, then the likelihood (2.9) is $P(S)^{-n}[c-P(S)]^{-m}$ for $S\varepsilon\mathcal{H}$ and it is 0 otherwise. To maximize this expression, suppose that m = 0 and observe that the value of the likelihood is $P(S)^{-n}$ if all the x's are in S and it is 0 otherwise. Thus an MLE of S (if it exists) is any set $\hat{S}$ satisfying $T^*(\underset{\sim}{x}) \subseteq \hat{S}$ which minimizes P. The probability P is increasing on $\mathcal{S}$ in the sense that if $S_1 \subseteq S_2$ and $S_1, S_2\varepsilon\mathcal{S}$, then $P(S_1) \leq P(S_2)$. Consequently, if $T^*\varepsilon\mathcal{S}$ then T^* is an MLE. Furthermore, if $\mathcal{S}$ is closed with respect to intersection, then $T^*\varepsilon\mathcal{S}$.

The case where n = 0 is similar. The likelihood becomes

$$[c-P(S)]^{-m} \prod_{j=1}^{m} I(y_j \varepsilon \bar{S}).$$

An MLE of S is any $\hat{S}$ satisfying $\hat{S} \subseteq R^*$ which maximizes $P(\hat{S})$. Since P is increasing on $\mathcal{S}$, if $R^*\varepsilon\mathcal{S}$ then R^* is an MLE. Furthermore, if $\mathcal{S}$ is closed with respect to union, then $R^*\varepsilon\mathcal{S}$.

Now we consider the case where $mn \neq 0$. Lince $L(s) = 0$ if $S \notin \mathcal{H}$, it is sufficient to maximize

$$L(P) = P^{-n}(c-P)^{-m}$$

for $P = P(S)$, $S \varepsilon \mathcal{H}$. Since L is a convex function of P, the maximum will occur at either the smallest or largest value of P on $\mathcal{H}$. To be explicit, define $\hat{P}_1$ and $\hat{S}_1 \varepsilon \mathcal{H}$ by

$$\hat{P}_1 = P(\hat{S}_1) = \inf_{S \varepsilon \mathcal{H}} P(S);$$

$\hat{S}_1$ may not exist or may not be unique. Similarly, define $\hat{P}_2$ and $\hat{S}_2 \varepsilon \mathcal{H}$ by

$$\hat{P}_2 = P(\hat{S}_2) = \sup_{S \varepsilon \mathcal{H}} P(S);$$

again $\hat{S}_2$ may not exist or may not be unique. Finally define

$$\hat{L} = \left[\frac{\hat{P}_1}{\hat{P}_2}\right]^n \left[\frac{1-\hat{P}_1}{1-\hat{P}_2}\right]^m .$$

An MLE of S (if one exists) is given by

$$\hat{S} = \begin{cases} \hat{S}_1 & \text{if } \hat{L} \leq 1 \\ \hat{S}_2 & \text{if } \hat{L} \geq 1 . \end{cases} \tag{4.1}$$

If R* and T* are elements of $\mathcal{H}$ then at least one of them maximizes L.

4.3 Inclusion-Exclusion Model for Arbitrary Measures

As in Section 4.2, the likelihood (2.9) is 0 if $S \notin \mathcal{H}$. Thus, we wish to maximize

$$L(P,Q) = P^{-n} (1-Q)^{-m} , \tag{4.2}$$

where $P = P(S)$, $Q = Q(S)$, and $S \varepsilon \mathcal{H}$. If either $m = 0$ or $n = 0$ then the problem reduces to a case in Section 4.2. Assume therefore that $mn \neq 0$ and define

$$P(\hat{S}_1) = \inf_{S \in H} P(S),$$

$$Q(\hat{S}_2) = \sup_{S \in H} Q(S).$$

Assuming that $\hat{S}_1$ and $\hat{S}_2$ exist, let

$$\hat{L}_1 = P(\hat{S}_1)^{-n}(1-Q(\hat{S}_1))^{-m},$$

$$\hat{L}_2 = P(\hat{S}_2)^{-n}(1-Q(\hat{S}_2))^{-m},$$

$$\hat{M} = \max(\hat{L}_1, \hat{L}_2).$$

If it happens that

$$L(P, Q) \leq \hat{M} \tag{4.3}$$

for all $S \in H$, then $\hat{S}_1$ or $\hat{S}_2$ will be an MLE of S according as $\hat{L}_1 = \hat{M}$ or $\hat{L}_2 = \hat{M}$. Substituting (4.2) into (4.3) we conclude that $\hat{S}$ is given by $\hat{S}_1$ or $\hat{S}_2$ if and only if

$$Q(S) \leq 1 - [\hat{M}\, P(S)^n]^{-1/m}$$

for all $S \in H$. A more general solution is not possibly analytically.

4.4 *Window Sampling*

We begin by considering the likelihood (2.7). Suppose first that it is known that m = 0 and let $A = S \cap W$ and $B = S \cap \overline{W}$, so that $A \cap B = \emptyset$ and $A \cup B = S$. We find $L(A, B, n|x_1, \ldots, x_{n_0}) \propto$

$$\binom{n}{n_0}[P(A) + P(B)]^{-n}\, P(B)^{n-n_0} \prod_{i=1}^{n_0} I(x_i \in A).$$

It is obvious that the set A should be chosen to minimize P(A). As in the previous two subsections one such choice is $\hat{A} = T^*_W$ if possible. Now

$$\frac{L(\hat{A}, B, n)}{L(\hat{A}, \emptyset, n_0)} = \binom{n}{n_0}\left[\frac{P(B)}{P(\hat{A}) + P(B)}\right]^n \left[\frac{P(\hat{A})}{P(B)}\right]^{n_0}. \tag{4.4}$$

We cannot guarantee that this ratio is less than one, but for each fixed B it is maximized when

$$n = \left\lceil \frac{n_0 \, P(B)}{P(\hat{A})} \right\rceil + n_0, \tag{4.5}$$

where as usual $\lceil x \rceil$ denotes the greatest integer $\leq x$. If $n_0 P(B)/P(\hat{A})$ is an integer, it is also maximized by $n_0 [P(B)/P(\hat{A})] + n_0 + 1$. We cannot proceed further analytically but must choose B to maximize (4.4) after n has been replaced by (4.5). A similar solution obtains when $n = 0$ and $m \neq 0$. When $mn \neq 0$, the MLE of S will presumably lie between the solutions found when $m = 0$ and when $n = 0$.

For the likelihood (2.8) we know somewhat less. The set S must be chosen, inside W, so that the products of indicators are 1. This is easy: Any S satisfying $T_W^* \subseteq S \cap W \subseteq R_W^*$ will do. Outside W, still less is known. For example, fix A, B and m_1. Then (2.8) is maximized by

$$\hat{n}_1 = \max\left\{ 0, \left\lceil \frac{(n_0+m_0+m_1+1)P(A)-1}{1-P(A)} \right\rceil \right\}.$$

This could be substituted back into (2.8) and A chosen to maximize the likelihood. However, analytic results are not possible. Finally, the analysis for the likelihood (2.10) is identical to that for (2.7) and is omitted.

V. BAYESIAN METHODS

We assume throughout this section that there exist densities (with respect to some dominating measure) on $\mathcal{S}$. Suppose that a prior density ξ on $\mathcal{S}$ is given. If the likelihood of $S \in \mathcal{S}$ is given by (2.2) then the posterior density for S is proportional to

$$\xi(S|x_1,\ldots,x_n,y_1,\ldots,y_m) = \xi(S) \prod_{i=1}^{n} I_S(x_i) \prod_{j=1}^{m} I_{\bar{S}}(y_j).$$

That is, the posterior density is proportional to the prior density truncated to those sets "allowed" by the data. In particular, the relative probability of two sets in $\mathcal{S}$ is unchanged by the data if both sets are "allowed". The identical anaysis applies to the likelihood (2.6).

A much more interesting case occurs under the likelihood (2.9). Suppose that P = Q and the prior density for S has the same general form as the likelihood [a "conjugate" prior; see e.g., DeGroot (1970), Chapter 9]:

$$\xi(S|S_1,S_2,\alpha,\beta) \propto P(S)^{-\alpha}P(\overline{S})^{-\beta}I(S_1\subseteq S\subseteq S_2). \tag{5.1}$$

The constants α and β and the sets S_1 and S_2 can be interpreted as hyperparameters. Since the likelihood (2.9) can be written as

$$P(S)^{-n}P(\overline{S})^{-m}I(T^*\subseteq S\subseteq R^*),$$

the posterior density is proportional to

$$\xi(S|S_1,S_2,\alpha,\beta,\underset{\sim}{x},\underset{\sim}{y}) = P(S)^{-n-\alpha}P(\overline{S})^{-m-\beta}I(S_1^*\subseteq S\subseteq S_2^*)$$

where $S_1^* = T^*\cup S_1$ and $S_2^* = R^*\cap S_2$.

As an example, suppose that $(\underline{X}, \mathcal{X})$ is some compact convex subset of the plane, say the unit disk, with the usual Borel field and $\mathcal{S}$ is the collection of all closed convex polygonal subsets of $\underline{X}$. Suppose for simplicity that m = 0 and, in (5.1), that β = 0 and $S_2 = \underline{X}$. Then

$$\xi(S|S_1,\alpha) = \xi(S|S_1,\alpha,k)\ p(k),$$

where p(k) is the prior probability that S has exactly k vertices. In this case, the collection of sets S with k vertices is finite-dimensional. The conjugate prior (5.1) has a Pareto form,

$$\xi(S|S_1,\alpha) = \alpha P(S_1)^{\alpha}P(S)^{-(\alpha+1)}I(S_1\subseteq S)\ p(k),$$

and together with the likelihood (2.9) yields the posterior density

$$\xi(S|S_1,\alpha,\underset{\sim}{x}) = (n+\alpha)P(S_1')^{n+\alpha}P(S)^{-(n+\alpha+1)}I(S_1'\subseteq S)p(k),$$

where S_1' is the convex hull of S_1 and $\underset{\sim}{x}$. Notice that conditionally on k, P(S) has a Pareto distribution. Hence,

$$E[P|S_1,\alpha,\underset{\sim}{x},k] = \frac{n+\alpha}{n+\alpha-1} P(S_1'),$$

which does not depend on k.

For the window likelihoods (2.7), (2.8), and (2.10) the situation is more complex. It is possible to take a prior density which is conjugate in P but not in the various counts (n, m, n_1, m_1). We shall not pursue these ideas here.

VI. EXAMPLES

In this section we shall discuss some of the examples presented in the introduction.

6.1 Uniform Distribution on $[0,\theta] \cup [2\theta,3\theta]$

Suppose that a random sample of size n is drawn from a uniform distribution having this form. This problem (and each of the problems mentioned in Section I) is an example of inclusion sampling. By Theorem 3.1, T = T* is minimal sufficient; the specific form of T* depends on the observed data. Let $x_1 < x_2 < \ldots < x_n$ denote the ordered observations. For each fixed θ there are three mutually exclusive and exhaustive possibilities:

a) $x_n \leq \theta$

b) $2\theta \leq x_1$ and $x_n \leq 3\theta$

c) There is at least one value of i (i=1,...,n-1) such that $x_i \leq \theta$ and $2\theta \leq x_{i+1}$.

Now consider the possibilities for θ given $\underset{\sim}{x}$. Suppose first that $x_1/x_n \geq 2/3$ and consider again the three cases just mentioned:

a) Any value of $\theta \ge x_n$ is possible.

b) Any value of θ such that $\frac{x_n}{3} \le \theta \le \frac{x_1}{2}$ is possible.

c) No value of θ is possible since

$$x_1 \le \theta,\ 2\theta \le x_n \Rightarrow 2x_1 \le x_n \Rightarrow \frac{x_1}{x_n} \le 1/2.$$

Now suppose that $1/2 < \frac{x_1}{x_n} < 2/3$ and consider the three cases:

a) Any $\theta \ge x_n$ is possible.

b) No value of θ is possible since

$$2\theta \le x_1, x_n \le 3\theta \Rightarrow \frac{x_n}{3} \le \frac{x_1}{2} \Rightarrow \frac{x_1}{x_n} \ge 2/3.$$

c) No θ is possible, as in the previous case (c).

Finally suppose that $0 < \frac{x_1}{x_n} \le 1/2$:

a) Any $\theta \ge x_n$ is possible.

b) No θ is possible, as in the previous case (b).

c) There is at most one value of i such that $2x_i \le x_{i+1}$ and $\frac{x_n}{x_{i+1}} \le 3/2$. To see this, suppose that these relations are satisfied for some value of i and, for some $j > i$, it is also true that $2x_j \le x_{j+1}$. Then

$$x_n \le \frac{3}{2} x_{i+1} \le \frac{3}{2} x_j \le \frac{3}{4} x_{j+1} \le \frac{3}{4} x_n,$$

a contradiction. Let I denote this unique value of i when it exists. Then any θ satisfying $x_I \le \theta \le \frac{1}{2}x_{I+1}$ is possible.

In summary, the minimal sufficient statistic is (A,B,C,D), where

$$A = x_n,$$

$$B = \begin{cases} x_1 & \text{if } \frac{x_1}{x_n} \ge \frac{2}{3}, \\ 0 & \text{otherwise,} \end{cases}$$

$$(C,D) = \begin{cases} (x_I, x_{I+1}) & \text{if } \frac{x_1}{x_n} \le 1/2,\ x_I \le 1/2 x_{I+1}, x_n \le \frac{3}{2} x_{I+1}, \\ (0,0) & \text{otherwise.} \end{cases}$$

The MLE of θ is the smallest allowable value of θ. That is,

$$\hat{\theta} = \begin{cases} \frac{x_n}{3} \text{ if } \frac{x_1}{x_n} \geq \frac{2}{3}, \\ x_n \text{ if } \frac{1}{2} < \frac{x_1}{x_n} < \frac{2}{3}, \\ x_I \text{ if } \frac{x_1}{x_n} \leq \frac{1}{2}, \ x_I \leq \frac{1}{2}x_{I+1}, \ x_n \leq \frac{3}{2}x_{I+1}, \\ x_n \text{ otherwise.} \end{cases}$$

The extension to the uniform distribution on $[0,\theta] \cup [\lambda\theta,(\lambda+1)\theta]$, where $\lambda > 1$ and $\theta > 0$, is easy. Let $0 = x_0 < x_1 < \ldots < x_n$ be the ordered data. For some i, $x_{i-1} < \frac{1}{2} x_n < x_i$. If i=1, then $T^* = [0,\frac{1}{2}x_n] \cup [x_1,x_n]$. If i=n, then $T^* = [0,x_{n-1}] \cup [\frac{1}{2}x_n,x_n]$. Otherwise, $T^* = [0,x_{i-1}] \cup [x_i,x_n]$.

6.2 *Uniform Distribution on* $[0,\theta] \cup [2\theta,1]$, $0 < \theta < \frac{1}{2}$

The easy way to do this problem is to think of $\overline{T^*}$ as the union of all sets of the form $(\theta,2\theta)$ which are contained in the complement of the data. Again let $0 = x_0 < x_1 < \ldots < x_n < x_{n+1} = 1$ be the ordered data. If $\frac{x_i}{x_{i+1}} < \frac{1}{2}$ for i taking the values $0 = i_1, i_2, \ldots, i_k = n$, then

$$T^* = \bigcup_{j=1}^{k-1} [x_{i_j+1}, x_{i_{j+1}}] \cup [0] \cup [1] .$$

If $x_i/x_{i+1} < \frac{1}{2}$ for i taking the values $0 = i_1, i_2, \ldots, i_k < n$, then

$$T^* = \bigcup_{j=1}^{k-1} [x_{i_j+1}, x_{i_{j+1}}] \cup [0] \cup [x_{i_k+1}, 1]. \tag{6.1}$$

If $x_i/x_{i+1} < \frac{1}{2}$ for i = 0 only, then the first term of (6.1) is empty and $T^* = [0] \cup [x_1,1]$. For the model which is uniform on the union of $[0,\theta]$ and $[\lambda\theta,1]$ for $\lambda > 1$ and $0 < \theta < 1/\lambda$, the set $\{x_1,\ldots,x_n\}$ is minimal sufficient; no reduction is achieved.

6.3 Convex Subsets of a Compact Set

Let $\underline{X}$ be a compact set in some Euclidean space equipped with the usual Borel field $\mathcal{X}$ and let $\mathcal{S}$ consist of all convex subsets of $\underline{X}$. For the various statistics introduced in Section 2.4 we have

a) $\underline{T} = T^* = C(\underset{\sim}{x})$ = the closed convex hull of $x_1,\ldots,x_n$

b) $\overline{R} = \{y_1,\ldots,y_n\}$

c) R^* = the open union of all convex sets which include $\underset{\sim}{x}$ and exclude $\underset{\sim}{y}$.

It is difficult to provide a simpler description of R^*; Hachtel, Meilijson and Nádas (1981) refer to $\overline{R}^*$ as the shade of T^* on $\overline{R}$. Some asymptotic properties of T^* and R^* are known; see Eddy (1980, 1982), Eddy and Gale (1981), and McClure and Vitale (1975). We conclude this paper with the following interesting property of R^*.

Proposition. The set R^* *is star-shaped from the set* T^* .

Proof. To establish this proposition we must show that if $r \in R^*$, $t \in T^*$, and $s = \alpha r + (1-\alpha)t$ for $0 \leq \alpha \leq 1$, then $s \in R^*$. If $r \in R^*$, then from the definition of R^* there is some $S \in \mathcal{H}$ such that $r \in S$; call it S^*. From the definition of T^*, $t \in T^*$ implies that $t \in S^* \in \mathcal{H}$. Since $r \in S^*$ and $t \in S^*$, and S^* is conves, it follows that $s \in S^*$. But $R^* \supseteq S^*$, so $s \in R^*$. Thus, R^* is star-shaped from T^*. □

REFERENCES

Bell, C.B., Blackwell, D., and Breiman L. (1960). "On the Completeness of Order Statistics." *Ann. Math. Statist. 31,* 794-797.

DeGroot, M.H. (1970). *Optimal Statistical Decisions,* McGraw-Hill Book Co., New York.

Eddy, W.F. (1980). "The Distribution of the Convex Hull of a Gaussian Sample." *J. Appl. Prob. 17,* 686-695.

Eddy, W.F. (1982). Laws of Large Numbers for Intersection and Union of Random Closed Sets. Technical Report No. 227, Department of Statistics, Carnegie-Mellon University.

Eddy, W.F. and Gale, J.D. (1981). "The Convex Hull of a Spherically Symmetric Sample." *Adv. Appl. Prob. 13,* 751-763.

Fraser, D.A.S. (1952). "Sufficient Statistics and Selection Depending on the Parameter." *Ann. Math. Statist. 23,* 417-425.

Fraser, D.A.S. (1953). "Completeness of Order Statistics." *Can. J. Math. 6,* 42-45.

Fraser, D.A.S. (1957). *Nonparametric Methods in Statistics,* Wiley, New York.

Fraser, D.A.S. (1966). "Sufficiency for Selection Models." *Sankhyā, Series A 28,* 329-334.

Hachtel, G.D., Meilijson, I. and Nadas, A. (1981). The Estimation of a Convex Subset of $\mathbb{R}^K$ and Its Probability Content. Technical Report RC 8666 (#37890) IBM T.J. Watson Research Center, Yorktown Heights, New York.

Huzurbazar, V.S. (1976). *Sufficient Statistics: Selected Contributions,* Marcel Dekker, Inc., New York.

McClure, D.E. and Vitale, R.A. (1975). "Polygonal Approximation of Plane Convex Bodies." *J. Math. Anal. Appl. 51,* 326-358.

Ripley, B.D. and Rason, J.P. (1977). "Finding the Edge of a Poisson Forest." *J. Appl. Prob. 14,* 483-491.

Smith, W.L. (1957). "A Note on Truncation and Sufficient Statistics." *Ann. Math. Statist. 28,* 247-252.

OPTIMAL SEQUENTIAL DECISIONS IN PROBLEMS INVOLVING MORE THAN ONE DECISION MAKER[1]

Morris H. DeGroot
Joseph B. Kadane

Department of Statistics
Carnegie-Mellon University
Pittsburgh, Pennsylvania

"What did the President know and when did he know it?" -- Senator Howard Baker, United States Senate hearings on Watergate and Related Activities, June, 1973.

I. INTRODUCTION

This paper is an initial exploration into the question of what information it is necessary for Bayesians to specify in order to play a sequential game against each other. We consider here the simplest sort of such a game, in which there are two players who take turns making decisions, although generalizations to other cases are discussed in the concluding section. While we have some interest in the form of the optimal strategies, our principal concern is the elicitation of opinion from the players. What would we need to know from each of them about their opinions and beliefs in order to advise them on their optimal actions?

[1] *This research was supported in part by the National Science Foundation under grant SES-7906386 and in part by the Office of Naval Research under contract 014-82-K-0622.*

ISBN 0-12-589320-5

The basic technique underlying our proofs is backward induction. Consequently we consider games with a fixed finite number of moves in which Player 1 has the last move, Player 2 the next-to-last move, etc. The kind of solution we consider is purely competitive, in the sense that each player is assumed to make at each stage the best decision for that player, minimizing his expected loss. It is assumed that the players do not have the institutional ability to make a binding agreement with enforcement that could alter their loss functions. However, nothing here forbids communication that might alter one player's opinions about the other's likely actions. We begin Section II by discussing a simple example that is completely deterministic in the sense that there are no stochastic variables or unknown parameters. In Section III we discuss a stochastic version of this problem in which each player has private information in order to give an idea of the kind of opinions that might be required. In Section IV we give the general case, which appears, surprisingly, to be simpler than the special case discussed in Section III. Section V gives our interpretation of this apparent paradox, and discusses extensions to problems involving more than two players, games in which the moves are made simultaneously by the two players rather than alternatingly or in which the order of play is not certain, and the relation of our findings to others in the literature.

II. A SIMPLE GAME

Consider a problem in which two players try to move an object on the real line in such a way that it ends up near certain targets that the players have. In particular, suppose that the target of Player 1 is x, the target of Player 2 is y, and the game consists of three stages such that Player 1 can move the object at the first and third stages, and Player 2 can move the object at the second stage. Thus, the game proceeds as follows;

The object begins at the position s_0. At the first stage, Player 1 can move the object by any amount u (postive or ne-ative), so that after this choice the object is at position $s_1 = s_0 + u$. At the second stage, Player 2 can move the object by any amount v, so that after this choice the object is at position $s_2 = s_1 + v = s_0 + u + v$. Finally, Player 1 has another choice w, so that the last stage of the system is $s_3 = s_2 + w$. We take Player 1's loss to be

$$L_1 = q(s_3 - x)^2 + u^2 + w^2 \tag{2.1}$$

and Player 2's loss to be

$$L_2 = r(s_3 - y)^2 + v^2, \tag{2.2}$$

where q and r are positive constants that are known to both players. Thus, the loss function of each player is the sum of a cost proportional to the squared distance of the final position of the object from the player's target and a quadratic cost of control or adjustment. It should be noted that this game is completely deterministic in the sense that it contains no exogenous chance elements. In fact, if the targets x and y are known to both players, the game contains no unknown or uncertain quantities whatsoever, other than the moves that the players might make.

In order to "solve" this game, we will adopt the following principle of optimality based on backward induction: At the final stage, player 1 will select a move that will minimize his expected loss given the outcomes of the first two stages. At the second stage, player 2 will select a move that will minimize his expected loss given the outcome of the first stage and given that at the last stage player 1 will follow the strategy just described. In general, in any game with alternating choices by the players, we assume that, at any stage, a player makes his decision to minimize his expected loss given the outcomes of the previous stages and given that at all later stages both players will follow this same rule.

The optimal strategies in our simple deterministic game are now easily derived.

Theorem 1

If x and y are known to both players, then the optimal decisions are

$$w = q(x-s_2)/(q+1) \tag{2.3}$$

$$v = (1-k)(m-s_1) \tag{2.4}$$

$$u = [(q+1)(x-s_0)+r(x-y)]/(q+1)[1+(q+1)/qk^2] \tag{2.5}$$

$$m = (q+1)y-qx$$

$$k = (q+1)^2/[r+(q+1)^2].$$

Proof. The solution is by backward induction. Thus, start with Player 1's last decision, the value of w. He knows that the object is at s_2 and that his target is x. At this time u will already have been determined. Then the function to be minimized with respect to w is

$$L_1 = q(s_2+w-x)^2 + u^2 + w^2.$$

Taking the derivative, we have

$$\frac{\partial L_1}{\partial w} = 2q(s_2+w-x) + 2w.$$

Setting the derivative equal to zero and solving, we obtain

$$w = q(x-s_2)/(q+1),$$

which is (2.3).

Next, we suppose that Player 2 is given s_1 and $w(s_2)$, and we now study the optimal choice of v. Hence Player 2 minimizes

$$L_2 = r(s_1+v+w-y)^2 + v^2.$$

Calculating the derivative, we have

$$\frac{\partial L_2}{\partial v} = 2r(s_1+v+w-y)(1 + \frac{dw}{dv}) + 2v.$$

Now

$$1 + \frac{dw}{dv} = 1 + \frac{d}{dv}\left[\frac{q(x-s_1-v)}{q+1}\right] = 1 - \frac{q}{q+1} = \frac{1}{q+1} .$$

Substituting for w and solving for v, we obtain

$$v = (1-k)(m-s_1)$$

which is (2.4).

Finally, we compute Player 1's optimal value for u, his first choice. Player 1 minimizes

$$L_1 = q(s_0+u+v+w-x)^2 + u^2 + w^2 .$$

Thus the optimal u satisfies

$$0 = \frac{dL_1}{du} = 2q(s_0+u+v+w-x)(1 + \frac{dv}{du} + \frac{dw}{du}) + 2u + 2w\frac{dw}{du} . \qquad (2.6)$$

Computing the needed derivatives, we obtain

$$\frac{dv}{du} = \frac{d}{du}\,[(y-s_0-u) + q(y-x)][1-k]$$
$$= k-1.$$

Similarly

$$\frac{dw}{du} = \frac{d}{du}\,\{\frac{q}{q+1}\,(x-s_0-v-u)\}$$
$$= -\frac{q}{q+1}\,(1 + \frac{dv}{du}) = -\,qk/(q+1).$$

Now the values for these derivatives, and (2.3) and (2.4) which involve u through s_1 and s_2, are substituted into (2.6). Solving the resulting equation for u, yields (2.5). □

It is worthwhile noting that even the apparently straightforward solution presented in Theorem 1 has some interesting and surprising features.

Theorem 2

Suppose that x and y are both known to both players. (i)

(i) Suppose that $y < s_0 < x$. Then $u > x - s_0$ if and only if

$$\frac{x-s_0}{x-y} < qk(1-k). \tag{2.7}$$

(ii) Suppose that $s_0 < x < y$. Then $u < 0$ if and only if

$$\frac{x-s_0}{y-x} < \frac{r}{1+q}. \tag{2.8}$$

Proof. Both parts of this theorem follow directly from the expression (2.5) for u and some straightforward algebra. □

The interpretations of the two parts of Theorem 2 are as follows. Part (i) presents a simple condition under which Player 1 will move the object from its initial position s_0 to a point s_1 *beyond* his target x, accepting the larger costs that attend this apparently wasteful move. Part (ii) presents a simple condition under which Player 1 will move the object in a direction away from his target x, again accepting the costs that attend this apparently counterproductive move.

It should be emphasized that these moves are not made by Player 1 in an attempt to fool Player 2 in any way, since both players have full information. The properties displayed in Theorem 2 make it clear that the solution in Theorem 1 will not provide a Pareto optimum for the players in the sense that if they could make an enforceable agreement, their choices would not be those derived here.

III. THE STOCHASTIC GAME

We shall now change the conditions of the game by assuming that neither player knows the other's target. Thus, Player 1 knows x but not y, and Player 2 knows y but not x. We continue to assume that both players know q and r. We refer to this version as a stochastic game, even though there are no exogenous stochastic elements, because each player's target is uncertain, and hence a random variable, to the other player, and it is possible for a player to gain information pertaining to the values about which he is uncertain by observing the moves of the other player.

Theorem 3

If Player 2 does not know x and Player 1 does not know y, and this condition is known to both players, then the optimal decisions are

$$w = q(x-s_2)/(q+1), \tag{3.1}$$

$$v = (1-k)[M(u)-s_1] \tag{3.2}$$

where $M(u) = (q+1)y-qE_2(x|u)$, and u is given as the implicit solution to the equation

$$0 = \frac{1}{2}\frac{d}{du}E_1\{K^2(u)\} + u(q+1)/q, \tag{3.3}$$

where $K(u) = (s_0+u)k-x+(1-k)M(u)$, under the assumption that the minimum of $E_1(L_1)$ over u can be found by setting the derivative with respect to u equal to zero, and there is a unique solution to this equation.

Proof. Player 1's situation, after s_2 is known and before the choice of w, is just as it is under the conditions of Theorem 1: He minimizes

$$L_1 = q(s_2+w-x)^2 + u^2 + w^2$$

with respect to w, and the minimum occurs at

$$w = q(x-s_2)/(q+1),$$

which is (3.1).

Player 2 does not know the value of x, but may have some information about it in the value of u chosen by Player 1.

Player 2's loss is

$$L_2 = r(s_1+v+w-y)^2 + v^2$$

$$= r[s_1 + v + \frac{q(x-s_1-v)}{q+1} - y]^2 + v^2$$

$$= [r/(q+1)^2](s_1+v-m)^2 + v^2 .$$

Taking the expectation of L_2 with respect to Player 2's uncertainty over x given u, yields

$$E_2(L_2) = [r/(q+1)^2]\{E_2(m^2|u) - 2(s_1+v)E_2(m|u)+(s_1+v)^2\}+v^2.$$

Recall that $M(u) = E_2(m|u)$. Then the optimal v satisfies

$$0 = \frac{d}{dv} E_2(L_2) = [r/(q+1)^2]\{-2M(u) + 2(s_1+v)\} + 2v.$$

Solving this equation for v yields (3.2).

Finally we address the question of the optimal choice of u for Player 1. Substituting for v and w, $L_1^* = ((q+1)/q)L_1$ can be written, after some tedious algebra, as

$$L_1^* = \{(s_0+u)k-x+(1-k)M(u)\}^2 + u^2(q+1)/q = K^2(u)+u^2(q+1)/q.$$

Now Player 1 is uncertain about both M(u) and y, and is supposed to have some joint prior distribution on them. Thus we minimize the expectation of L_1^* (which is equivalent to but more convenient than minimizing the expectation of L_1), namely

$$E_1(L_1^*) = E_1K^2(u) + u^2(q+1)/q.$$

Then, calculating the derivative with respect to u yields

$$0 = \frac{d}{du} E_1(L_1^*) = \frac{d}{du} E_1K^2(u) + 2u(q+1)/q.$$

Now formula 3.3 follows directly. □

It should be noted that the optimal w is the same in (3.1) as it was in the non-stochastic version of the game, since when Player 1 chooses w, he knows the location s_2 and his own target x. While he does not know Player 2's target y, such knowledge is irrelevant to his expected loss. The difference between the two versions shows up first in the choice of v.

In the stochastic version, Player 2, not knowing x, must consider his conditional distribution for m having observed u. Because of the squared-distance nature of L_2, the conditional expectation of this distribution suffices. Thus we find $M(u) = E_2(m|u)$ as an essential element in v. It can be seen from (2.4) and (3.2) that when Player 2 chooses v in the stochastic version, he simply replaces m in the optimal choice for the non-stochastic version with his expectation of m.

We must note the possibility that, although Player 2 is sure he knows m, he may be mistaken. This possibility would relate to a non-stochastic game in which Player 2 might "know" a wrong m. However in the non-stochastic game he could compute Player 1's optimal u as a function of m given in Theorem 1, and observe that his actual play was suboptimal.

A somewhat similar difficulty arises in the definition of M(u). One way of computing M(u) is to elicit a joint distribution for m and u from Player 2, and then compute the expectation of the conditional distribution of m for each given u. The principles of the subjectivist Bayesian viewpoint require us to allow Player 2 to hold any opinion about (m,u). Thus, in particular, Player 2 may regard some values of u as having not only zero probability but zero density. If he is sure such values of u will not occur, then the joint distribution of (m,u) does not entail, using only the rules of coherence (i.e., Bayes' Theorem) a value for M(u) at these values. Nonetheless we may elicit M(u) directly for those values. The conversation would be somewhat like asking a scientist for an opinion about the orbit of the moon if it really did turn out to be made of green cheese. One would have to say, "I know you think it impossible that the moon is made of green cheese. Nonetheless, if it were, what would you think about its orbit?". Hence we take M(u) to be defined for all real values of u, including those regarded as impossible by Player 2.

Knowing that Player 2's decision about v depends on M(u) imposes a difficult burden on Player 1 in his choice of u. He must have an opinion about the value of M(u) for each value of u that he might choose. Thus, when choosing u, Player 1 must take into account the conclusion M(u) he believes Player 2 will draw about Player 1's m based on the observation of Player 1's choice u. It can be seen from inspection of the functional equation (3.3) to be solved for the optimal u that Player 1's optimal choice is not obvious.

For loss functions other than the one that we have assumed in this paper, Player 1's choice would be even more complicated. In general, for each possible value of u he would need a probability distribution on Player 2's conditional distribution of m given u. This may be somewhat more difficult analytically, but it is not different in principle.

There are however, a few special cases of Theorem 3 that are worth noting. One of these is the situation in which $E_2(x|u)$ is linear in u and y.

Corollary 1. Suppose that $E_2(x|u) = a + bu + cy$, where a, b, and c are known to Player 2 but not to Player 1. Then Player 1's optimal u is u = C/D where

$$C = E_1\{[(1-k)qb-k][s_0k-x+(1-k)((q+1-qc)y-qa)]\}$$

and

$$D = (q+1)/q + E_1[(1-k)qb-k]^2$$

Proof. If

$$E_2(x|u) = a + bu + cy, \text{ then}$$

$$M(u) = (q+1-qc)y-qa-qbu.$$

Noting $M'(u) = -bq$, and substituting into (3.3), we get the required result. □

It is not hard to imagine that Player 2 might wish to take c = 0. In this case, the following special argument gives a reasonable value for a. Suppose that Player 2 thinks that if s_0 and x coincide, then Player 1 would choose u = 0. That is, if the object starts out exactly at Player 1's target, he might reasonably let it alone in his first move.

In this case,

$$s_0 = E_2(x|u) = a+b(0) = a.$$

Thus Player 2 may regard $a = s_0$ as a reasonable value but of course this is not required in any sense by the Bayesian structure.

Another special case worth noting because of its simplicity is described in the following result.

Corollary 2. Suppose that both x and $E_1(y)$ are known to Player 2, and that Player 1 is aware of this knowledge on the part of Player 2. Then the optimal choice of u is given by (2.5) with y replaced by $E_1(y)$.

Proof. The conditions of this corollary imply that $E_2(x|u) = x$ for all values of u. Thus, in the terminology of Corollary 1, Player 2 knows for certain that a = x and b = c = 0. When this knowledge is used, it can be shown after much straightforward algebra that the optimal u, as given in Corollary 1, reduces to (2.5) with y replaced by $E_1(y)$. □

IV. A MORE GENERAL GAME

Consider now a game with a total of N moves, where as before the two players take turns moving. In general, N could be either odd or even (in the example of the previous sections, N=3) and either player could move first. To be specific in the discussion here, we will assume that N is even, N = 2n, and that Player 2 goes first and makes every odd-numbered move, while Player 1 makes the last move and every even-numbered move.

It is convenient to number each player's moves from the end of the game. Accordingly, let e_1 denote Player 1's last move and, in general, for k=1,...,n, let e_k denote his k^{th} move from the end. Similarly, for k=1,...,n, let f_k denote Player 2's k^{th} move from the end. Thus if we count backward from the end of the game, the moves are e_1, f_1, e_2, $f_2,\ldots,e_n$, f_n. We suppose that Player i wishes to minimize the expected value of his loss $L_i(f_n,e_n,\ldots,f_1,e_1)$. The notation $\tilde{f}_i = (f_n,e_n,\ldots,e_{i+1},f_i)$ and $\tilde{e}_i = (f_n,e_n,\ldots,f_i,e_i)$ will be useful.

Player 1, making the last move, chooses e_1 to minimize $L_1(\tilde{e}_1)$ given $\tilde{f}_1$. The resulting loss is denoted $L_1'(f_1)$. Player 2, making his last move, chooses f_1 to minimize $E_2[L_2(\tilde{e}_1)|(\tilde{e}_2,f_1)]$, where the expectation is over his conditional distribution of e_1 given $\tilde{f}_1$. The resulting expected loss is denoted $L_2'(\tilde{e}_2)$.

In general Player 1, making his k^{th} move from the end (k=2,...,n), chooses e_k to minimize $E_1[L_1^{(k-1)}(\tilde{f}_{k-1})|(\tilde{f}_k,e_k)]$, where the expectation is over his conditional distribution of f_{k-1} given $\tilde{e}_k$. The resulting expected loss is $L_1^{(k)}(\tilde{f}_k)$. Similarly Player 2, making his k^{th} move from the end, chooses f_k to minimize $E_2[L_2^{(k-1)}(\tilde{e}_k)|(\tilde{e}_{k+1},f_k)]$ where the expectation is over his conditional distribution of e_k given $\tilde{f}_k$. The resulting expected loss is $L_2^{(k)}(\tilde{e}_{k+1})$.

This backward induction is general in the sense that it embraces any number of moves, and any opinions that each player might have about the other player. In fact, for each player to play optimally by this method, it is not necessary to assume that the other player is playing optimally according to some particular loss function. *Any* assumptions about the other player can be entertained in this general model.

V. DISCUSSION

Our approach in this paper is fundamentally different from the classic work in the theory of games, as developed originally by von Neumann and Morgenstern (1947) in that we explicitly assume that our players adopt a Bayesian, rather than a minimax, outlook. It is also different, however from other work on multi-agent Bayesian decision theory, such as Harsanyi (1968, 1977), Prescott and Townsend (1980), and Weerahandi and Zidek (1981), where the emphasis is on group decisions and "rational expectations". The emphasis in our work is on the specification of the type of opinion each player must have in order to make each of his non-cooperative moves in a subjectively optimal fashion.

The non-stochastic version of the example in Section II has forerunners in the work by Cyert and DeGroot (1970, 1977). One mystery is why the special game, involving only three moves, seems so much more complex in the stochastic version than does the general game. To some extent we believe that the simplification in the general case is more apparent than real, in that the required conditional expectations are not simple to think about. However, the special structure introduced by the example may in fact also make choices more complicated by introducing a special theory about the opponent's behavior that is particular in some respects and general in others.

Another lesson to come from this study is that a statement such as the assumption that "x and y are known to both players" actually stands for many statements: x and y are known to both players, both players know that the other knows x and y, both players know that both players know,... See "common knowledge" in Aumann (1976). Exactly who knows what, when, is critical to the analysis we are making.

Finally we remark that our analysis can be extended to more than two players, to games in which the order of the moves of the two players is not determined in advance and may be uncertain to the players, and to games in which simultaneous play is a possibility. In the latter connection, see Kadane and Larkey (1981, 1982).

The essence of all these decision-making problems is in the elicitation of the opinions of the players. The principal use of special structure may be to simplify this elicitation somewhat, although even for very simple cases, such as our three-move stochastic game, elicitation now seems quite difficult.

VI. REFERENCES

Aumann, R. (1976). "Agreeing to Disagree." *Ann. of Stat. 4,* 1236-1239.

Cyert, R.M. and DeGroot, M.H. (1970). "Multiperiod Decision Models with Alternating Choice as a Solution to the Duopoly Problem." *Quart. J. Econ. 84,* 410-429.

Cyert, R.M. and DeGroot, M.H. (1977). "Sequential Strategies in Dual Control Problems." *Theory and Decision 8,* 173-192.

Harsanyi, J.C. (1968). "Games with Incomplete Information Played by 'Bayesian' Players." *Management Sci. 14,* 159-182, 320-334, 481-502.

Harsanyi, J.C. (1977). *Rational Behavior and Bargaining Equilibrium in Games and Social Situations,* Cambridge University Press, New York.

Kadane, J.B. and Larkey, P.D. (1981). "The Confusion of Is and Ought in Game Theoretic Contexts." *Management Sci.,* to appear.

Kadane, J.B. and Larkey, P.D. (1982). "Subjective Probability and the Theory of Games." *Management Sci. 28,* 113-120.

Prescott, E.D. and Townsend, R.M. (1980). "Equilibrium under Uncertainty: Multiagent Statistical Decision Theory." *Bayesian Analysis in Econometrics and Statistics* (ed. by A. Zellner), 169-194, North-Holland Publishing Co., Amsterdam.

von Neumann, J. and Morgenstern, O. (1947). *Theory of Games and Economic Behavior,* 2nd ed., Princeton University Press, Princeton, N.J.

Weerahandi, S. and Zidek, J.V. (1981). "Multi-Bayesian Statistical Decision Theory." *J. Royal Statist. Soc., Series A 144,* 85-93.

AN AVERAGING METHOD FOR STOCHASTIC APPROXIMATIONS WITH DISCONTINUOUS DYNAMICS, CONSTRAINTS, AND STATE DEPENDENT NOISE[1]

Harold J. Kushner

Divisions of Applied Mathematics and Engineering
Lefschetz Center for Dynamical Systems
Brown University
Providence, Rhode Island

I. INTRODUCTION

Since the work of Robbins and Monro [1] and Kiefer and Wolfowitz [2], much attention has been devoted to the problems of stochastic approximation [3, 4]. In recent years, work on the recursive estimation algorithms in adaptive control and communication systems has both rekindled widespread interest and required results under rather different assymptions on the noise and dynamics than were used in the early years (see, e.g., [5-10]). Problems with constraints have been treated by similar methods.

A typical form of current interest is the following: The iterate sequence is given by

$$X_{n+1} = X_n + a_n h(X_n, \xi_n), \; X_n \in R^r, \text{ Euclidean r-space,} \tag{1.1}$$

$\sum a_n = \infty$, $0 < a_n \to 0$ as $n \to \infty$, and $h(\cdot,\cdot)$ is not necessarily continuous. For example, it might be an indicator function.

[1] *Work supported in part by the Air Force Office of Scientific Research under AFOSR 76-306D, in part by the National Science Foundation under NSF-Eng 77-12946-A02 and in part by the Office of Naval Research under N00014-76-C-0279-P0004.*

ISBN 0-12-589320-5

Also, the noise sequence $\{\xi_n\}$ might depend on $\{X_n\}$ in a complicated way. The references [5] - [10] contain a variety of techniques which are useful for proving w.p.1 or weak convergence for fairly general types of stochastic approximations, both with and without constraints. But they are not good enough to treat many problems where $h(\cdot,\cdot)$ is discontinuous or $\{\xi_n\}$ 'state dependent'.

In [11] and [12], averaging methods were used to get weak convergence results for suitably scaled stochastic difference equations, where the 'dynamical' term corresponding to our $h(\cdot,\cdot)$ might have the properties mentioned above. Here we adapt the method of [11] and [12] with that of [6] to develop a technique which is quite useful and versatile for the problems of interest. In a sense, we rely on the assumption that even if $h(\cdot,\cdot)$ is not smooth, expectations or conditional expectations of the types $Eh(\cdot,\xi_n)$ or $E[h(\cdot,\xi_n)|\xi_{n-1}, \xi_{n-2}, \ldots]$ are smooth functions of x. This situation occurs in many cases, as attested to by the examples in Section V. Results for both w.p.1 and weak convergence (see remark at end of Section IV) are available.

We also obtain analogous results for the following projection algorithm. Let $q_1(\cdot), \ldots, q_m(\cdot)$ denote continuously differentiable functions and define $G = \{x:q_i(x)\leq 0, i = 1, \ldots, m\}$. Let $\pi_G(y)$ denote any closest point in G to y. Then the algorithm is defined by

$$X_{n+1} = \pi_G(X_n + a_n h(X_n,\xi_n)) \tag{1.2}$$

In Section II, we treat the algorithm (1.1), and the projected algorithm is treated in Section III. A method for 'state dependent' noise $\{\xi_n\}$ is given in Section IV, and Section V contains two non-standard examples.

2. w.p.1 CONVERGENCE FOR (1.1)

Assumptions. Write δX_n for $X_{n+1} - X_n$, and let $\hat{C}_0^2$ denote the space of real valued continuous functions on R^r with compact support and continuous second partial derivatives. Let E_n denote expectation conditioned on $\{\xi_j, j < n\}$, and K will be used to denote a constant (its value might change from usage to usage).

One of the key difficulties is proving w.p.1 boundedness of $\{X_n\}$. For this we use a stability assumption on the differential equation $\dot{x} = \bar{h}(x)$, where $\bar{h}(x)$ is (*very loosely speaking*) $Eh(x,\xi_j)$. The boundedness argument uses a perturbed form of the Liapunov function $V(\cdot)$ for that differential equation, and various differences or derivatives of $V(\cdot)$ appear in the (mixing type) conditions. Owing to this, some of the conditions might seem at first glance a little unnatural, but they in fact are frequently readily verifiable. Theorem 1 and its conditions should be viewed as a prototype of a method which can be adapted to a wide variety of problems.

The assumptions are written such that (A1) - (A4) can be used for both bounded and unbounded $\{\xi_n\}$. With bounded $\{\xi_n\}$, we can let the $\alpha_{in} \equiv K$, all i,n. The (A6) and (A7) would not often hold as stated when $\{\xi_n\}$ is unbounded. The forms of (A6) and (A7) which are useful for the unbounded noise case depend on the particular form of $h(\cdot,\cdot)$ and it does not seem reasonable to try to get the most general form here. The unbounded case will be discussed after the theorem, and in the examples. Owing to our desire to have a proof which can be used with only minor changes when $\{\xi_n\}$ is unbounded, the details are a little more complicated than necessary.

A1. $\sum a_n^2 < \infty$, $\sum a_n = \infty$, $a_n > 0$, $\{a_{n+1}/a_n\}$ *is bounded.* $h(\cdot,\cdot)$ is *measurable and for each compact set Q there is a sequence*

$\{\alpha_{0n}\}$ *such that* $|h(x,\xi_n)| \leq \alpha_{0n}$ *for* $x \in Q$ *and* $\sum \alpha_{0n}^2 a_n^2 < \infty$ w.p.1.

A2. There is a twice continuously differentiable Liapunov function $V(\cdot) \geq 0$ *such that* $|V_{xx}(\cdot)|$ *is bounded,* $V(x) \to \infty$ *as* $|x| \to \infty$. *There are* $\varepsilon_0 > 0$ *and* $\lambda_0 < \infty$ *such that for* $x \notin Q_0 \equiv \{x: V(x) \leq \lambda_0\}$, *we have* $V_x'(x)\bar{h}(x) \leq -\varepsilon_0$, *where* $\bar{h}(\cdot)$ *is defined in* (A3).

A3. Let $\{\alpha_{1n}\}$ *denote a sequence of random variables such that* $\sum_n \alpha_{1n}^2 < \infty$, w.p.1. *There is a continuously differentiable* $\bar{h}(\cdot)$ *such that the limit defined (pointwise in x) by*

$$V_0(x,n) = \sum_{j=n}^{\infty} a_j V_x'(x) E_n [h(x,\xi_j) - \bar{h}(x)]$$

exists and (together with the partial sums) is bounded by $\alpha_{1n}(1+|V_x'(x)\bar{h}(x)|)$. *The bound also holds if* $V(\cdot)$ *is replaced by a continuously differentiable function with compact support* ($\{\alpha_{in}\}$ *can depend on the function).*

A4. There is a random sequence $\{\alpha_{in}\}$ *such that*

$$E_n|h(x,\xi_n)|^2 \leq \alpha_{2n}^2(1+|V_x'(x)\bar{h}(x)|)$$

and $a_n \alpha_{2n}^2 \to 0$ w.p.1 *as* $n \to \infty$.

A5. $|V_x'(x)\bar{h}(x)| \leq K(1+V(x))$.

A6. With $[\ ]_x$ *denoting gradient with respect to* x, *let*

$$\left|\sum_{j=n}^{\infty} a_j [V_x'(x)\{E_n h(x,\xi_j) - \bar{h}(x)\}]_x\right| \leq K a_n (1+|V_x'(x)\bar{h}(x)|^{\frac{1}{2}})$$

The inequality holds with $V(\cdot)$ *replaced by an arbitrary continuously differentiable function with compact support.*

A7. For $s \le 1$

$$E_n|V_x'(x+sa_nh(x,\xi_n))\bar{h}(x+sa_nh(x,\xi_n))| \le K(\;+|V_x'(x)\bar{h}(x)|)$$

Remark. As seen from the Examples in Section V, the assumptions include some hard and interesting cases. In a sense, the 'prototype' model for (A1) - (A7) is the case where $|h(\cdot,\cdot)|$ and $|\bar{h}(\cdot)|$ have at most a linear growth in $|x|$ as $|x| \to \infty$ and $V(\cdot)$ has a growth one order higher than that of $h(\cdot,\cdot)$ and $\bar{h}(\cdot)$. Then the bounds in the assumptions make sense, under various mixing type conditions on $\{\xi_n\}$. Here, $\{\xi_n\}$ is not treated as being explicitly 'state dependent.' In the state dependent case, we must take into account the way that $\{\xi_n\}$ evolves as a function of $\{X_n\}$, and use a slightly different form of $V_0(x,n)$. See Section IV and Example 2.

Theorem 1. Assume (A1) - (A7). *The sequence* $\{X_n\}$ *is bounded* w.p.1. *If* $V_x'(x)\bar{h}(x) \le 0$ *for all* x, *then* $X_n \to \Lambda_0 \equiv \{x: V_x'(x)\bar{h}(x) \le 0\}$ w.p.1. *Otherwise,* $\{X_n\}$ *converges* w.p.1 *to the largest bounded invariant set* S[2] *of*

$$\dot{x} = \bar{h}(x). \tag{2.1}$$

Remark. If $V_x'(x)\bar{h}(x)$ is not ≤ 0 for all x or if Λ_0 contains more than one point, then the limit set of $\{X_n\}$ might be more than one point, or even be a non-degenerate trajectory of (2.1). Such possibilities do exist in applications. But the theorem can be refined as follows. Let $x_0 \equiv x(t)$ be an asymptotically stable solution of (2.1) (in the sense of Liapunov) with domain of attraction $DA(x_0)$. There is a null set N such that if $\omega \notin N$

[2]*Let* S *denote a bounded* invariant set *of (2.1). Then for each* $x \in S$, *there is a trajectory* $x(\cdot)$ *of (2.1) contained in* S *for* $t \in (-\infty,\infty)$ *and* $x(0)=x$. *The invariant set is the set of all limit points of bounded trajectories (on* $[0,\infty)$*) of (2.1).*

and $X_n(\omega) \in$ compact $A \subset DA(X_0)$ infinitely often, then $X_n(\omega) \to x_0$. The proof follows from the techniques of the proof below and the proof of Theorem 2.3.1 of [6].

Proof. The proof uses an 'averaged' form of the Liapunov function $V(\cdot)$. We have

$$E_nV(X_{n+1})-V(X_n) = a_nV'_x(X_n)E_nh(X_n,\xi_n) + a_n^2\int_0^1 E_nh'(X_n,\xi_n)V_{xx}(X_n + s\delta X_n)h(X_n,\xi_n)(1-s)ds. \tag{2.2}$$

By (A2) and (A4), the last term is $\leq a_n\varepsilon_{1n}(1+|V'_x(X_n)\bar{h}(X_n)|)$, where $\varepsilon_{1n} \to 0$ w.p.1, as $n \to \infty$. We also have

$$E_nV_0(X_{n+1},n+1)-V_0(X_n,n) = E_n\sum_{n+1}^{\infty} a_jV'_x(X_{n+1})E_{n+1}[h(X_{n+1},\xi_j)-\bar{h}(X_{n+1})] - \sum_{n+1}^{\infty} a_jV'_x(X_n)E_n[h(X_n,\xi_j)-\bar{h}(X_n)] - a_nV'_x(X_n)[E_nh(X_n,\xi_n)-\bar{h}(X_n)], \tag{2.3}$$

which we rewrite as

$$-a_nV'_x(X_n)[E_nh(X_n,\xi_n)-\bar{h}(X_n)] + a_nE_nh'(X_n,\xi_n) \cdot \int_0^1 ds \sum_{j=n+1}^{\infty} a_j[E_{n+1}V'_x(X_n+s\delta X_n)(h(X_n+s\delta X_n,\xi_j)-\bar{h}(X_n+s\delta X_n))]_x. \tag{2.4}$$

By (A6) and (A7) and an application of Schwarz's inequality, the last term of (2.4) is bounded by

$$Ka_n^2(1+|V'_x(X_n)\bar{h}(X_n)|). \tag{2.5}$$

Define the 'averaged' Lipaunov function $\tilde{V}(n) = V(X_n) + V_0(X_n,n)$ and note that by (A3),

$$|V_0(X_n,n)| \le \alpha_{1n}(1 + |V_x'(X_n)\bar{h}(X_n)|) , \tag{2.6a}$$

$$\tilde{V}(n) \ge - O(\alpha_{1n}), \tag{2.6b}$$

where $\alpha_{1n} \to 0$ w.p.1 as $n \to \infty$. Combining (2.2) - (2.5),

$$E_n\tilde{V}(n+1) - \tilde{V}(n) = a_n(1+\delta_n)V_x'(X_n)\bar{h}(X_n) + \tilde{\delta}_n a_n , \tag{2.7}$$

where δ_n and $\tilde{\delta}_n$ go to zero w.p.1 as $n \to \infty$.

Define $\{m_n,M_n\}$ by

$$\tilde{V}(n) - \sum_{i=0}^{n-1} a_i(1+\delta_i)V_x'(X_i)\bar{h}(X_i) - \sum_{i=0}^{n-1} \tilde{\delta}_i a_i = \sum_{i=0}^{n-1} m_i = M_n. \tag{2.8}$$

By (2.7), $\{M_n\}$ is a martingale. By modifying $\{X_n,\delta_n,\tilde{\delta}_n,\xi_n\}$ on a set of arbitrarily small probability, we can suppose that there is an $N_0 < \infty$ such that $|\tilde{\delta}_i| \le \varepsilon_0/4$, $|\delta_i| \le 1/4$ for $i \ge N_0$. This modification will not alter the conclusions.

Let n_0 be a stopping time $\ge N_0$ and such that $X_{n_0} \notin Q_0$ (with n_0 equal to ∞ if $X_n \in Q_0$, all $n \ge N_0$). Define $n_1 = \min\{n:n>n_0, X_n \in Q_0\}$. Then by (2.6), (2.7), the sequence $\{\tilde{V}(n\cap n_1), n \ge n_0\}$ is a super martingale which is bounded below $-O(\alpha_{1n})$. The facts that $E_n\tilde{V}(n+1) - \tilde{V}(n) \le -\varepsilon_0 a_n/2$ for $X_n \notin Q_0$ and n large, and $\sum a_n = \infty$ imply that $X_n \in Q_0$ infinitely often w.p.1. Define $Q_1 = \{x:V(x) \le \lambda_1\}$ where $\lambda_1 > \lambda_0$. Let Q of (A1) be Q_1.

By a modification of the paths on a set of arbitrarily small measure, we can suppose that $|a_n h(x,\xi_n)| \le a_n\alpha_{0n} \le 1$ for large n and $x \in Q$ (say, for convenience, $n \ge N_0$). This modification will not affect the conclusions.

By (A1) and (A3), there are real $K_i(Q_1)$ such that if $X_n \in Q_1$,

$$|m_n|^2 \le (|V_0(X_n,n)|^2 + |V_0(X_{n+1},n+1)|^2 + K_1(Q_1)a_n^2 + |V(X_{n+1}) - V(X_n)|^2)K$$

$$\le K_2(Q_1)[a_n^2 + \alpha_{1n}^2 + \alpha_{1,n+1}^2 + a_n^2|h(X_n,\xi_n)|^2]$$

$$\le K_3(Q_1)[a_n^2 + \alpha_{1n}^2 + \alpha_{1,n+1}^2 + a_n^2\alpha_{0n}^2] \equiv K_3(Q_1)\beta_n. \tag{2.9}$$

Let n_2 be any stopping time such that $X_{n_2} \in Q_1$. Let $n_3 = \min\{n: X_n \notin Q_1, n \ge n_2\}$. Then by (2.9) (note that the right side is summable over all return periods in Q_1),

$$P\{\sup_{n_2 \le n < n_3} |\sum_{i=n_2}^{n} m_i| \ge \varepsilon\} \le K_3(Q_1)E\sum_{i=n_2}^{n_3-1} \beta_i/\varepsilon^2 . \tag{2.10}$$

If $\sum E(\alpha_{1n}^2 + a_n^2\alpha_{0n}^2) < \infty$, then we conclude from (2.8) and (2.10) and the recurrence of Q_0 and the facts that $a_n h(X_n,\xi_n) \to 0$ w.p.1 for $X_n \in Q_1$ and $V_x'(x)\bar{h}(x) \le -\varepsilon_0$ for $x \notin Q_0$ that eventually X_n remains in Q_1 for any $\lambda_1 > \lambda_0$. Hence M_n converges w.p.1. If only $\sum(\alpha_{1n}^2 + a_n^2\alpha_{0n}^2) < \infty$ w.p.1, then a stopping time argument yields the same result.

Furthermore, since M_n converges w.p.1 and $V_0(X_n,n) = O(\alpha_{1n}) \to 0$ w.p.1 (since X_n eventually remains in the bounded set Q_1),

$$\sup_{m \ge n} |V(X_m) - V(X_n) - \sum_{i=n}^{m-1} a_i V_x'(X_i)\bar{h}(X_i)| \to 0 \tag{2.11}$$

w.p.1 as $n \to \infty$. If $V_x'(x)\bar{h}(x) \le 0$ all x, then (2.11) implies that $X_n \to \{x: V_x'(x)\bar{h}(x) = 0\}$.

Next, fix $f(\cdot) \in \hat{C}_0^2$, and repeat the development with $f(\cdot)$ replacing $V(\cdot)$. This yields

$$\sup_{m\geq n}\left| f(X_m) - f(X_n) - \sum_{i=n}^{m-1} a_i f'_x(X_i)\bar{h}(X_i)\right| \to 0$$

w.p.1 as $n \to \infty$. In particular, since for large n, $X_n \in Q_1$, we have

$$\sup_{m\geq n} \left|X_m - X_n - \sum_{i=n}^{m-1} a_i \bar{h}(X_i)\right| \to 0 \quad w.p.1 \quad as\ n \to \infty. \tag{2.12}$$

Eqaution (2.12) implies that (w.p.1) the limit points are contained in the set of limit points of the bounded trajectories of (2.1). Q.E.D.

Remarks on the Unbounded Noise Case

(A6) and (A7), which were used only to get the bound (2.5), would not hold very often if $\{\xi_n\}$ were unbounded. If (2.5) were replaced by $a_n^2\alpha_{3n}(1+|V'_x(X_n)\bar{h}(X_n)|)$ where $\sup_n Ea_n^2\alpha_{3n}^2<\infty$, and if this inequality holds with an arbitrary continuously differentiable f(•) with compact support replacing V(•), then the proof goes through with only minor changes. The $\{\alpha_{3n}\}$ can depend on f(•).

An Extension for the Bounded Noise Case

Theorem 1 can sometimes be improved by use of a modified perturbed Liapunov function. We state the result in a form which is useful when the noise is bounded. Define $\delta h(x,\xi) = h(x,\xi) - \bar{h}(x)$, with $\bar{h}(\cdot)$ continuously differentiable. For $\mu \in (0,1)$, define

$$V_{1-\mu}(x,n) = \sum_{j=n}^{\infty} \mu^{n-j} a_j V'_x(x) E_n \delta h(x,\xi_j).$$

We will use

A3'. *For each* $\mu\varepsilon(0,1)$, *there is a* K_μ *such that* $(1-\mu)K_\mu \to 0$ *as* $\mu \to 1$ *and* $|V_{1-\mu}(x,n)| \leq K_\mu[1 + |V_x'(x)\bar{h}(x)|]O(a_n)$. *Similarly for* $f(\cdot)\varepsilon\hat{C}_0^2$ *replacing* $V(\cdot)$.

A6'. *For each* $\mu\varepsilon(0,1)$, *there is a* $\bar{K}_\mu < \infty$ *such that*

$$\left|\sum_{j=n+1}^{\infty} \mu^{j-n-1} a_j E_n[V_x'(X_{n+1})\delta h(X_{n+1},\xi_j) - V_x'(X_n)\delta h(X_n,\xi_j)]\right|$$

$$\leq O(a_n)\bar{K}_\mu(1 + |V_x'(X_n)\bar{h}(X_n)|),$$

and similarly for $f(\cdot)\varepsilon\hat{C}_0^2$ *replacing* $V(\cdot)$.

Theorem 1 *holds if* (A3', 6') *replace* (A3, 6, 7). Note that if $E_n\delta h(x,\xi_j) \to 0$ as $j-n\to\infty$ (uniformly in ω), then $\lim_{\mu\to 1}(1-\mu) \sum_{j=n}^{\infty} \mu^{j-n}E_n\delta h(x,\xi_j) = 0$. Thus, the extension is useful if $\{\xi_j\}$ satisfies a strong mixing condition with arbitrary mixing rate.

The proof of the extension is almost identical to that of Theorem 1, but the perturbed Liapunov function $V(X_n) + V_{1-\mu}(X_n,n) \equiv \tilde{V}_\mu(n)$ is used. Note that $E_n\tilde{V}_\mu(n+1)-\tilde{V}_\mu(n) = a_nV_x'(X_n)\bar{h}(X_n)$ + (term in (A6')) + (last term on right of (2.2)) $-(1-\mu) \sum_{n+1}^{\infty} \mu^{j-n-1}a_jE_n\delta h(X_n,\xi_j)$.

Now, choose μ such that $(1-\mu)K_\mu O(a_n)/a_n$ is small enough (see A3') and proceed as in the proof of Theorem 1.

Example. The algorithm $X_{n+1} = X_n-a_nY_n(Y_n'X_n-v_n)$ occurs in adaptive control and communications problems. Let $\{v_n, Y_n\}$ be bounded and second order stationary and let $E_nY_jY_j' \to M = EY_nY_n' > 0$, $E_nY_jv_j \to b = EY_nv_n$ (convergence uniform in ω, as $j-n\to\infty$). Then with $V(x) = x'M^{-1}x$, (A3', 6') hold.

III. THE PROJECTION METHOD

Recall the definition of G and π_G from Section 1. Let $\bar{\pi}(\bar{h}(\cdot))$ denote the (not necessarily unique) projection of the vector field $\bar{h}(\cdot)$ onto G; i.e.,

$$\bar{\pi}(\bar{h}(x) = \lim_{\Delta \downarrow 0} [\pi_G(x+\Delta\bar{h}(x)) - x]/\Delta .$$

We will use

A8. The $q_i(\cdot)$, $i = 1, \ldots, m$, *are continuously differentiable,* G *is bounded, and is the closure of its interior* $G^0 = G - \partial G = \{x: \ q_i(x) < 0, \ i = 1, \ldots, m\}$.

In lieu of (A6), we use the weaker ('unbounded' noise) condition:

A9. Let $\{\alpha_{2n}\}$ *be* (see (A4)) *a random sequence satisfying* $E_n|h(x,\xi_n)|^2 \leq \alpha_{2n}^2$ *for* $x\varepsilon G$. *Let* $\{\alpha_{4n}\}$ *be a random sequence satisfying (for* $x\varepsilon G$)

$$\left|\sum_{j=n}^{\infty} a_j[E_n h(x,\xi_j) - \bar{h}(x)]_x\right| \leq \alpha_{4n}$$

and let $\alpha_{2n}E_n^{\frac{1}{2}}\alpha_{4,n+1}^2 \to 0$ *and* $\alpha_{2n}^2 a_n \to 0$ w.p.1 *as* $n \to \infty$.

Define the cone

$$C(x) = \{y: \ y = \sum_{i\varepsilon A(x)} \lambda_i q_{i,x}(x), \ \lambda_i \geq 0\},$$

where A(x) is the set of constraints which are active at x.

Remark. The assumption that G is the closure of its interior is useful in visualizing constructions in the proof, and slightly simplifies the details. The theorem remains valid when there are only equality constraints. Then, of course, $\{X_n\}$ moves on the constraint surface. If $\{\xi_j\}$ is bounded and satisfies a sufficiently strong mixing condition, then the α_{4n} in (A9) is $O(a_n)$. Condition (A9) is used only to show that (3.4) is of the order given below (3.4).

Theorem 2. *Assume* (A1), (A3) *with the* V_x *term dropped and for* $x \in G$), (A8), (A9). *Then if there is only one constraint* (w.p.1) *the limit points of* $\{X_n\}$ *are those of the 'projected'* ODE

$$\dot{x} = \bar{\pi}(\bar{h}(x)). \tag{3.1a}$$

Let $H(\cdot) \geq 0$ *be a real valued function with continuous first and second partial derivatives and define* $\bar{h}(\cdot) = -H_x(\cdot)$. *Then, as* $n \to \infty$, $\{X_n\}$ *converges* w.p.1 *to the set* $KT = \{x: \bar{h}'(x)\bar{\pi}(\bar{h}(x)) = 0\}$. *In general, the limit points are those of*

$$\dot{x} = \bar{h}(x) + v, \quad \text{where} \quad v(t) \in C(x(t)). \tag{3.1b}$$

Remarks. The remarks after the statement of Theorem 1 also hold here. The form $\bar{h}(\cdot) = -H_x(\cdot)$ arises in the projected form of the Kiefer-Wolfowitz procedure, where we seek to minimize the regression $H(\cdot)$, subject to $x \in G$.

Proof. Except for the treatment of certain projection terms, the proof is quite similar to that of Theorem 1. Since G is bounded, the Liapunov function $V(\cdot)$ is not required and (A3, A6) and extensions need only be applied with $f(\cdot)$ an arbitrary real valued function with continuous second partial derivative, in order to characterize the limit points, similarly to what was done in Theorem 1.

We have

$$E_n f(X_{n+1}) - f(X_n) = a_n f_x'(X_n) E_n h(X_n, \xi_n) + a_n f_x'(X_n) E_n \tau_n$$

$$+ a_n^2 \int_0^1 E_n (\delta X_n / a_n)' \, f_{xx}(X_n + s\delta X_n)(\delta X_n / a_n)(1-s)ds \; ,$$

where τ_n is the 'projection error':

$$\tau_n = [\pi_G(X_n + a_n h(X_n, \xi_n)) - (X_n + a_n h(X_n, \xi_n))]/a_n \; . \tag{3.2}$$

If $X_n + a_n h(X_n, \xi_n) \notin G$, but there is a unique i such that $X_{n+1} \in \partial G_i = \{x: q_i(x) = 0\}$, then τ_n points 'inward' at X_{n+1} and, in fact, $\tau_n = -\lambda_i q_{i,x}(X_{n+1})$ for some $\lambda_i \geq 0$. In general, suppose that $X_n + a_n h(X_n,\xi_n) \notin G$, but (with a reordering of the indices, if necessary) X_{n+1} is in the intersection

$\prod_1^{\ell} \partial G_i$. Then for each y for which $q'_{i,x}(X_{n+1})y \leq 0$, each $i \leq \ell$,

we must have $\tau_n' y \geq 0$. (Otherwise τ_n would not be the 'projection error' or, equivalently, X_{n+1} would not be the *closest point* on ∂G to $X_n + a_n h(X_n,\xi_n) \notin G$.) Thus, by Farkas' Lemma, τ_n must lie in the cone $-C(X_{n+1})$.

Note also that since $\delta X_n \to 0$ w.p.1 (by (A1)), there is a real sequence $0 < \mu_n \to 0$ such that $\tau_n = 0$ for large n (w.p.1) if distance $(X_n, \partial G) \geq \mu_n$, and $a_n \tau_n$ and $a_n E_n \tau_n \to 0$ w.p.1. These facts will be used when characterizing the limit points below.

Now, following the argument in Theorem 1 but for smooth $f(\cdot)$ replacing $V(\cdot)$, define

$$f_0(x,n) = \sum_{j=n}^{\infty} a_j f_x'(x) E_n[h(x,\xi_j) - \bar{h}(x)]$$

and define $\tilde{f}(n) = f(X_n) + f_0(X_n,n)$. Analogously to the result in Theorem 1, there is a random sequence $\tilde{\delta}_n \to 0$ w.p.1 such that

$$E_n\tilde{f}(n+1) - \tilde{f}(n) - \tilde{\delta}_n a_n - a_n f_x'(X_n)\bar{h}(X_n) - a_n f_x'(X_n) E_n \tau_n = 0. \tag{3.3}$$

We do not need to average out the $E_n \tau_n$ term. In order to get (3.3), via the method of Theorem 1, we must show that the difference

$$E_n \sum_{j=n+1}^{\infty} f'_x(X_{n+1})E_{n+1}[h(X_{n+1},\xi_j)-\bar{h}(X_{n+1})]$$

$$- \sum_{j=n+1}^{\infty} f'_x(X_n)E_n[h(X_n,\xi_j)-\bar{h}(X_n)]$$

$$= E_n \; \delta X'_n \int_0^1 ds \sum_{j=n+1}^{\infty} a_j[f'_x(X_n+s\delta X_n)E_{n+1}\{h(X_n+s\delta X_n,\xi_j)$$

$$- \bar{h}(X_n+s\delta X_n)\}]_x \tag{3.4}$$

is of an order $a_n\alpha_{5n}$ where $\alpha_{5n} \to 0$ w.p.1 as $n \to \infty$. But by (A1), (A9), and an application of Schwarz's inequality, we have $\alpha_{5n} = \alpha_{2n}E_n^{\frac{1}{2}} \alpha_{4,n+1}^2$.]

We also have

$$\tilde{f}(n)-\tilde{f}(0)-\sum_{i=0}^{n-1} a_i f'_x(X_i)\bar{h}(X_i)-\sum_{i=0}^{n-1} a_i f'_x(X_i)\tau_i -$$

$$- \sum_{i=0}^{n-1} a_i\tilde{\delta}_i \equiv \sum_{i=0}^{n-1} m_i \equiv M_n \,, \tag{3.5}$$

where $\{M_n\}$ is a martingale and[3] $E\sum m_i^2<\infty$. Finally, letting $f(\cdot)$ equal an arbitrary coordinate variable in G, and using the above square integrability and the fact that $f_0(X_n,n) \to 0$ w.p.1, we get

$$\sup_{m\geq n} \left|X_m-X_n - \sum_{i=n}^{m-1} a_i\bar{h}(X_i) - \sum_{i=n}^{m-1} a_i\tau_i\right| \to 0 \text{ w.p.1 as } n \to \infty \,. \tag{3.6}$$

By the properties of the 'projection terms' $\{a_n\tau_n\}$, and the fact that the 'limit dynamics' implied by (3.6) is that of the 'projected' ODE (3.1), (3.6) implies that (w.p.1) all limit

[3] *To get the inequality, we might have to alter $\{X_n,\xi_n\}$ on a set of arbitrarily small probability, but as in Theorem 1, this does not alter the conclusions.*

points of $\{X_n\}$ must be limit points of (3.1). The $\sum a_i \tau_i$ term simply compensates for the part of $\sum a_i \bar{h}(X_i)$ which would take the trajectory out of G.

Now, let $\bar{h}(\cdot) = -H_x(\cdot)$, and use $H(\cdot)$ as a Liapunov function. Then

$$\dot{H}(x) = H_x(x)\bar{\pi}(-H_x(x)) \leq 0 \tag{3.7}$$

Equation (3.7) implies that the limit points of (3.1a) are contained in KT. Q.E.D.

IV. STATE DEPENDENT NOISE

It is often necessary to take explicit account of the way that the evolution of $\{\xi_j, j \geq n\}$, depends on $\{X_j, j \leq n\}$. We might use a parametrization of the type $\xi_n = g_n(\xi_{n-1}, X_n, X_{n-1}, \ldots, X_{n-k}, \psi_n)$, where $\{\psi_n\}$ is an "exogenous' sequence. Such a scheme was used in [5] and [6], where the g_n were assumed to be sufficiently smooth functions of the $X_n, X_{n-1}, \ldots$. In the development of this section, we suppose that $\{X_n, \xi_{n-1}, n \geq 1\}$ is a Markov process (not necessarily stationary). In fitting this format to particular applications, it might be required to 'Markovianize' the original (state, noise) process. Let E_n denote conditioning on $\{\xi_j, j < n, X_j, j \leq n\}$. Define the 'partial' transition function as follows. Define

$$P(\xi,n,\Gamma,n+1|x) = P\{\xi_{n+1} \in \Gamma | \xi_n = \xi, X_{n+1} = x\} .$$

In general, define $P(\xi,n,\Gamma,n+\alpha|x)$ by the convolution

$$P(\xi,n,\Gamma,n+\alpha+\beta|x) = \int P(\xi,n,dy,n+\alpha|x)P(y,n+\alpha,\Gamma,n+\alpha+\beta|x) .$$

Thus in calculating the above transition function, X_j is held fixed at x for $j \leq n + \alpha + \beta$. This partial transition function is useful because, loosely speaking, $\{\xi_n\}$ varies much faster than $\{X_n\}$ does. $V_0(x,n)$ is now written in the form

$$V_0(x,n) = \sum_{j=n}^{\infty} a_j V_x'(x)\{\int h(x,\xi)P(\xi_{n-1},n-1,d\xi,j|x)-\bar{h}(x)\}. \tag{4.1}$$

Define $\tilde{V}(n) = V_0(X_n,n) + V(X_n)$. Note the way the averaging is done in (4.1) compared to how it was done in the $V_0(X,n)$ of (A3). The integral in (4.1) could be written as $E[h(x,\xi_j(x))|\xi_{n-1}(x) = \xi_{n-1}]$ where for each x,n, $\{\xi_j(x), j\geq n-1\}$ is a process which evolves according to the law $P(\xi,\alpha,\Gamma,\beta|x)$, where $\beta \geq \alpha \geq n-1$ and $\xi_{n-1}(x)=\xi_{n-1}$. See [12] for other applications of this idea.

Suppose that the sum in (4.1) is continuously differentiable in x, and that the derivatives can be taken termwise. Then (4.2) replaces the sum in (A6).

$$\sum_{j=n}^{\infty} a_j[V_x'(x)\{\int h(x,\xi)P(\xi_{n-1},n-1,d\xi,j|x) - \bar{h}(x)\}]_x . \tag{4.2}$$

Theorem 3. Assume (A1) - (A7) *but with the above cited replacements* (4.1), (4.2). *Then the conclusions of Theorem* 1 *hold. The extensions to the unbounded noise case stated in the remark after Theorem* 1 *also hold here. Under the conditions of Theorem* 2, *subject to the above replacements, the conclusions of Theorem* 2 *hold.*

The proof is almost identical to that of Theorem 1 and (where appropriate) Theorem 2. We note only the following. By the Markov property,

$$E_n P(\xi_n,n,\Gamma,j|X_n) = P(\xi_{n-1},n-1,\Gamma,j|X_n), \quad j \geq n . \tag{4.3}$$

Note that the lowest term in (4.2) is (with $x = X_n$) $a_n V_x'(X_n)[E_n h(X_n,\xi_n)-\bar{h}(X_n)]$, exactly as in the sum in (A.3). In the proof we get (4.4), the analog of the first two terms on the right of (2.3).

$$\sum_{j=n+1}^{\infty} E_n a_j V_x'(X_{n+1})\{\int h(X_{n+1},\xi)P(\xi_n,n,d\xi,j|X_{n+1})-\bar{h}(X_{n+1})\}$$

$$-\sum_{j=n+1}^{\infty} E_n a_j V_x'(X_n)\{\int h(X_n,\xi)P(\xi_{n-1},n-1,d\xi,j|X_n)-\bar{h}(X_n)\} \tag{4.4}$$

The left side of (4.3) replaces the right side of (4.3) in the second sum of (4.4). Then the differentiality assumption, and the bounds in (A1) - (A7), yield a bound analogous to the one obtained for the sums on the right of (2.3).

Remark. All the foregoing results hold if $\{a_n\}$ is random, under the following additional conditions. a_n depends on $\{X_i, i \leq n\}$ only, $\sum_n a_n = \infty$, $\sum_n a_n^2 < \infty$, $\sum_n |a_{n+1}-a_n| < \infty$ w.p.1 and with

$$\hat{V}_0(x,n) = a_n \sum_{j=n}^{\infty} E_n V_x'(x)[h(x,\xi_j)-\bar{h}(x)] \tag{4.5}$$

replacing the $V_0(x,n)$ of (A3).

Remarks on Weak Convergence. Define $t_n = \sum_0^{n-1} a_i$, $m_n = \min\{n:t_n \geq t\}$ and $X^0(t) = X_n$ on $[t_n,t_{n+1})$ and $X^n(t) = X(t+t_n)$ for $t \geq -t_n$ and $X^n(t) = X_0$ for $t \leq -t_n$. Then the previous theorems imply various strong convergence properties for $\{X^n(\cdot)\}$. E.g., the proof of Theorem 1 implies that $\{X^n(\cdot)\}$ converges uniformly on finite time intervals to a solution of (2.1) and that the limit path is contained in the invariant set S cited there. Under weaker conditions, $\{X^n(\cdot)\}$ possesses various weak convergence (in $D^r[0,\infty)$) properties. Here we only cite a result; the proof is quite similar to that of Theorem 8 in [15].

Assume the conditions of Theorem 1, except replace $\sum(1 + \alpha_{0n}^2)a_n^2 < \infty$ by $a_n \to 0$ and $\alpha_{0n}a_n \to 0$ w.p.1. Weaken (A3) to require only $\alpha_{1n} \to 0$ w.p.1 and $E\alpha_{1n}^2 \to 0$. Then $\{X^n(\cdot)\}$ is tight, and all limit paths are in S. Also $X_n \to S$ in probability. More strongly, for each $T < \infty$, $\varepsilon > 0$,

$$\lim_{n\to\infty} P\{\sup_{|t|\leq T} \text{distance } (X^n(t), S) \geq \varepsilon\} = 0.$$

If the conditions $Ea_n^2\alpha_{3n}^2 \to 0$ and $a_n\alpha_{3n} \to 0$ w.p.1 replace $\sum a_n^2\alpha_{3n}^2 < \infty$, then the above result also holds for the 'unbounded noise' case. There are similar weak convergence versions of Theorems 2 and 3.

V. EXAMPLES

Va. Example 1. Convergence of an Adaptive Quantizer.

Frequently in telecommunications systems, the signal is quantized and only the quantized form is transmitted, in order to use the communications channel as efficiently as possible. It is desirable to adapt the quantizer to the particular signal [13, 14], in order to maximize the quality of the received signal. Here a stochastic approximation form of an adaptive quantizer will be studied. Let $\xi(\cdot)$ denote the original stationary signal process and Δ a sampling interval. Write $\xi(n\Delta) = \xi_n$. The signal $\xi(\cdot)$ is sampled at instants $\{n\Delta, n = 0, 1, \ldots,\}$, a quantization $Q(\xi_n)$ calculated, and only this quantization is transmitted.

The quantizer is defined as follows. Let L denote an interger, x a parameter, and $\{\rho_i, \eta_i\}$ real numbers such that $0 = \rho_0 < \rho_1 < \ldots < \rho_{L-1} < \rho_L = \infty$, $0 = \eta_1 < \eta_2 < \ldots < \eta_L$. If $\xi_n > 0$, define $Q(\xi_n) = x\eta_i$ if $\xi_n \in [x\rho_{i-1}, x\rho_i)$, and set

$Q(z) = Q(-z)$. In order to maintain the fidelity of the signal which is reconstrcuted from the sequence of received quantizations, the scaling parameter x should increase as the signal power increases.

Let $\beta \varepsilon (0,1]$ and let $0 < M_1 < \ldots < M_L < \infty$ and $M_1 < 1$, $M_L > 1$. A typical adaptive quantizer (adapting the scale x) is defined by (5.1), where X_n is the scale value in the *nth* sampling instant.

$$X_{n+1} = X_n^{\beta} B_n \, , \; B_n = M_i \qquad X_n \rho_{i-1} \leq |\xi_n| < X_n \rho_i \, . \tag{5.1}$$

We will analyze a stochastic approximation version of (5.1). Let $\alpha > 0$ be such that $a_n \alpha < 1$, and let $\{\ell_i\}$ be real numbers that $\ell_1 < 0$, $\ell_L > 0$, and $\ell_1 < \ell_2 < \ldots < \ell_L$. Let $0 < x_\ell < x_u < \infty$. Then we use

$$X_{n+1} = X_n^{(1-a_n\alpha)} (1 + a_n b_n) \Big|_{x_\ell}^{x_u} \tag{5.2}$$

where $b_n = \ell_i$ if $X_n \rho_{i-1} \leq |\xi_n| < X_n \rho_i$, and the bar $|$ denotes truncation. With $\alpha > 0$, the algorithm has some desirable robustness properties. The algorithm (5.2) can be rewritten as (use $y^{1-\varepsilon} = y(1-\varepsilon \log y) + 0(\varepsilon^2)$)

$$X_{n+1} = [X_n + a_n h(X_n, \xi_n) + 0(a_n^2)] \Big|_{x_\ell}^{x_u} \tag{5.3}$$

where

$$h(X_n, \xi_n) = X_n b_n - \alpha X_n \log X_n$$

or

$$h(x,\xi) = -\alpha x \log x + x \sum_{i=1}^{L} \ell_i I\{x\rho_{i-1} \le |\xi_n| < x\rho_i\} ,$$

$$\bar{h}(x) = -\alpha x \log x + x \sum_{i=1}^{L} \ell_i P\{x\rho_{i-1} \le |\xi_n| < x\rho_i\} . \tag{5.4}$$

For specificity, let $\xi(\cdot)$ be a *stationary Gaussian process*. in particular for a matrix M whose eigenvalues have negative real parts and a standard Wiener process $w(\cdot)$, define $v(\cdot)$, $\xi(\cdot)$ by $dv = Mv\,dt + Cdw$, $\xi = Dv$. Let $\sigma_0^2 = \text{var}\,\xi(t)$. Suppose that $\text{Cov}\,v(t) = \Sigma > 0$. We have

$$\frac{d}{dx}\left(\frac{\bar{h}(x)}{x}\right) = \frac{2}{\sqrt{2\pi}\,\sigma_0} \sum_{i=1}^{L} \ell_i \left[\rho_i \exp - \frac{\rho_i^2 x^2}{2\sigma_0^2} - \rho_{i-1} \exp - \frac{\rho_{i-1}^2 x^2}{2\sigma_0^2}\right] - \alpha/x$$

$$= \frac{2}{\sqrt{2\pi}\,\sigma_0} \sum_{i=1}^{L-1} (\ell_i - \ell_{i+1})\rho_i \exp - \rho_i^2 x^2/2\sigma_0^2 - \alpha/x. \tag{5.5}$$

Thus $\bar{h}(x)/x$ is the sum of two strictly convex functions, the first being bounded and having a negative slope and the second going to ∞ as $x \to 0$ and to $-\infty$ as $x \to \infty$. Thus there is a unique $\bar{x} \in (0,\infty)$ such that $\bar{h}(\bar{x}) = 0$. Also $\bar{h}(x) > 0$ for $x < \bar{x}$ and $\bar{h}(x) < 0$ for $x > \bar{x}$. We use the 'unbounded' noise version of Theorem 2. See the remark after that theorem statement.

Let $\sum a_n^2 < \infty$. Since $h(\cdot,\cdot)$ is bounded (for $x \in [x_\ell, x_u]$), we need only verify (A3) for $f(\cdot) \in \hat{C}_0^2$ and (as noted in the remark after the proof of Theorem 1) get the appropriate bound for the second term of (2.4), with $f(\cdot) \in \hat{C}_0^2$ replacing $V(\cdot)$. Let E_n denote conditioning on $v(i\Delta)$, $i < n$.

It can be verified that (the rate of convergence of the sum depends on $v(n\Delta-\Delta)$)

$$\sum_{j=n}^{\infty} E_n[h(x,\xi_j)-\bar{h}(x)] \leq a_n K(|v(n\Delta-\Delta)| + 1) . \tag{5.6}$$

The right-hand side of (5.6) goes to zero w.p.1 as $n \to \infty$, and (A3) holds. Next, using the fact that for some constant $c \geq 1$ and $j \geq n$, $P\{x\rho_{i-1} \leq |\xi_j| < x\rho_i | v(n\Delta-c\Delta)\}$ is a smooth and bounded function of x, and it and its x-derivatives converge (fast enough) to the *unconditional probability* and its x-derivatives as $j \to \infty$, we can get a bound of the form of the right-hand side of (5.6) on the last term of (2.4) (using $f(\cdot) \varepsilon \hat{C}_0^2$ in lieu of $V(\cdot)$). Thus all the conditions of the projection Theorem 2 hold. Hence if $\bar{x} \varepsilon [x_\ell, x_u]$, then $X_n \to \bar{x}$ w.p.1; otherwise X_n converges w.p.1 to the endpoint nearest to $\bar{x}$.

Vb. Example 2. A Kiefer-Wolfowitz Procedure with Observation Averaging

Let $f(\cdot)$ be a real valued function on R^1, whose first and second derivatives are bounded on R^1. Suppose that there is a unique θ (let $\theta = 0$) such that $f_x(\theta) = 0$ and let there be ε_1, ε_2 such that $f_x(x) \geq \varepsilon_1$ for $x \geq \varepsilon_2$, and $f_x(x) \leq -\varepsilon_1$, for $x \leq -\varepsilon_2$. Let $\{a_n, c_n\}$ be sequences of positive real numbers tending to zero and such that $\sum a_n = \infty$, $\sum a_n c_n < \infty$, $\sum a_n^2/c_n^4 < \infty$, $\limsup_n \sup_{j \geq n} a_j/a_n < \infty$, $\limsup_n \sup_{j \geq n} c_j/c_n < \infty$. Let $\{\psi_i\}$ be a sequence of mutually independent mean zero random variables with variance bounded by $\sigma^2 < \infty$, and *4th* moment by $m_4 < \infty$. Let $0 < \alpha < 1$, $\beta > 0$, and define $\{X_n, \xi_n\}$ by

$$X_{n+1} = X_n + a_n\xi_n \ , \ n \geq 1,$$

$$\xi_n = \alpha\xi_{n-1} - \beta\left[\frac{f(X_n+c_n) - f(X_n-c_n)}{2c_n} + \frac{\psi_n}{c_n}\right]$$

$$= \alpha\xi_{n-1} - \beta f_y(X_n) + O_n - \beta\psi_n/c_n \ , \tag{5.7}$$

where $|O_n| \leq Kc_n$.

With $\alpha = 0$, we have a form of the Kiefer-Wolfowitz (KW) process. Then $\sum a_n^2/c_n^2 < \infty$ can replace $\sum a_n^2/c_n^4 < \infty$. With $0 < \alpha < 1$, the observations are averaged with exponentially decreasing weights. The conditions on $\{\psi_n\}$ and on $f(\cdot)$ can be relaxed, but the technique will be well illustrated with the given conditions. Here $h(x,\xi) = \xi$, which is not *a priori* bounded, and in fact is 'state' dependent. We show that $X_n \to 0$ w.p.1, via Theorem 1 (extension for unbounded noise). Define $\bar{h}(x) = -\beta f_x(x)/(1-\alpha)$. For notational convenience, *we drop the* O_n in (5.7). It can readily be carried through with little additional difficulty. Define $V(x) = x^2$.

Conditions (A2) and (A5) obviously hold. Define $\{\bar{\xi}_n, \hat{\xi}_n\}$ by $\bar{\xi}_n = \alpha\bar{\xi}_{n-1} - \beta f_x(X_n)$, $\hat{\xi}_n = \alpha\hat{\xi}_{n-1} - \beta\psi_n/c_n$. Clearly $\{\bar{\xi}_n\}$ is uniformly bounded and $\hat{\xi}_n = \beta \sum_1^n \alpha^{n-i} \psi_i/c_i$. Now,

$$a_n^2 \, E \, \hat{\xi}_n^2 \leq \beta^2\sigma^2 \sum_{i=1}^{n} \alpha^{2n-2i} \, a_n^2/c_i^2$$

and

$$\sum a_n^2 \, E \, \hat{\xi}_n^2 < \infty \ .$$

Thus (A1) holds and $a_n\hat{\xi}_n \to 0$ w.p.1. Now check (A4). We have

$$Ea_n^2\Big(\sum_{i=1}^{n} \alpha^{n-i} \psi_i/c_i\Big)^4 \leq a_n^2 \sum_{i,j=1}^{n} \alpha^{2(n-i)}\alpha^{2(n-j)}\alpha^4/c_i^2c_j^2$$

$$+ a_n^2 \sum_{i=1}^{n} \alpha^{4(n-i)} m_4/c_i^4 \equiv \gamma_n . \qquad (5.8)$$

Since $\sum_n \gamma_n < \infty$, we have $a_n\hat{\xi}_n^2 \to 0$ w.p.1. and $a_nE_n\hat{\xi}_n^2 \to 0$ w.p.1. Thus (A4) holds.

Define $\{\bar{\xi}_j(x),\ j \geq n\}$, $\bar{V}_0(x,n)$, $\hat{V}_0(x,n)$, by

$$\bar{\xi}_j(x) = \alpha\bar{\xi}_{j-1}(x) - \beta f_x(x) , \quad \bar{\xi}_n(X_n) = \bar{\xi}_n .$$

$$\bar{V}_0(x,n) = \sum_{j=n}^{\infty} a_j V_x(x)[\bar{\xi}_j(x) - \bar{h}(x)] ,$$

$$\hat{V}(x,n) = \sum_{j=n}^{\infty} a_j V_x(x) E_n \hat{\xi}_j .$$

Then $|\bar{V}_0(x,n)| \leq Ka_n(1+|x|)$. Also

$$E_n\hat{\xi}_j = \alpha^{j+1-n}\hat{\xi}_{n-1}$$

and

$$\hat{V}(x,n) = \Big(\sum_{j=n}^{\infty} a_j\alpha^{j+1-n} V_x(x)\Big)\xi_{n-1} .$$

These representations can be used to readily show that both (A3) and the condition mentioned for the unbounded noise case after the proof of Theorem 1 hold. Thus, by (the unbounded noise extension of) Theorem 1, $X_n \to 0$ w.p.1.

REFERENCES

[1] Robbins, H. and Monro, S. (1951). "A Stochastic Approximation Method," *Ann. Math. Statist., 22*, 400-407.

[2] Kiefer, J. and Wolfowitz, J. (1952). "Stochastic Estimation of the Maximum of a Regression Function," *Ann. Math. Statist., 23*, 462-466.

[3] Wasan, M. T. (1969). *Stochastic Approximation.* Cambridge University Press, Cambridge.

[4] Nevelson, M. B. and Khazminskii, R. Z. (1972). *Stochastic Approximation and Recursive Estimation.* English Translation published by the American Mathematical Society, 1976.

[5] Ljung, L. (1977). "Analysis of Recursive Stochastic Algorithms," *IEEE Trans. on Automatic Control, AC-22*, 551-575.

[6] Kushner, H. J. and Clark, D. S. (1978). *Stochastic Approximation Methods for Constrained and Unconstrained Systems.* Applied Math Science Series # 26. Springer, Berlin.

[7] Farden, D. C. (1981). "Stochastic Approximation With Correlated Data," *IEEE Trans. on Information Theory, IT-27*, 105-113.

[8] Solo, V. (1979). "The Convergence of AML," *IEEE Trans. on Automatic Control, AC-24*, 958-962.

[9] Benveniste, A., Goursat, M. and Ruget, G. (1980). "Analysis of Stochastic Approximation Schemes with Discontinuous and Dependent Forcing Terms with Applications to Data Communication," *IEEE Trans. on Automatic Control, AC-25*, 1042-1057.

[10] Kushner, H. J. (1980). "A Projected Stochastic Approximation Method for Adaptive Filters and Identifiers," *IEEE Trans. on Automatic Control, AC-25*, 836-838.

[11] Kushner, H. J. and Huang, Hai (1981). "On the Weak Convergence of a Sequence of General Stochastic Difference Equations to a Diffusion," *SIAM J. on Applied Math., 40*, 528-541.

[12] Kushner, H. J. and Huang, Hai (1981). "Averaging Methods for the Asymptotic Analysis of Learning and Adaptive Systems, with Small Adjustment Rate," *SIAM J. on Control and Optimization, 19*, 635-650.

[13] Goodman, D. J. and Gersho, A. (1974). "Theory of an Adaptive Quantizer," *IEEE Trans. on Communication, COM-22*, 1037-1045.

[14] Mitra, D. (1979). "A Generalized Adaptive Quantization System with a New Reconstruction Method for Noisy Transmission," *IEEE Trans. on Communication, COM-27*, 1681-1689.

[15] Kushner, H. J. (1980). "Stochastic Approximation with Discontinuous Dynamics and State Dependent Noise; w.p.1 and Weak Convergence," *Proceedings of the Conference on Decision and Control*, Albuquerque.

ENLIGHTENED APPROXIMATIONS IN OPTIMAL STOPPING

P. Whittle

Statistical Laboratory
University of Cambridge
Cambridge, England

It seems quite improbable that Herman Chernoff has reached this valedictory stage - surely he is going as well as ever! For most of us the flame of creation is a somewhat fitful phenomenon. However, in Herman's case it seems always to have cast the same clear, relaxed and amiable glow. This is surely related to the fact that, of admittedly clever people, he is one of the kindest. He has written so well on design, but is himself notably undesigning. He has written so well on optimal stopping, but his own stopping is yet remote and unsignalled.

I. CHARACTERISATIONS

Consider a Markov stopping problem in discrete time with state variable x, stopping reward K(x), immediate continuation reward g(x) and constant discount factor β. Define the operator L by

$$(L\phi)(x) = g(x) + \beta E(\phi(x_{t+1}) | x_t = x) \tag{1}$$

where x_t is the value of state at time t. Under appropriate regularity conditions the maximal expected discounted reward

ISBN 0-12-589320-5

from state x is defined, F(x) say, and satisfies

$$F = \max(K, LF). \tag{2}$$

The set of x for which LF < K constitutes the optimal continuation set C; this is the set one wishes to determine in order to determine an optimal policy. We denote its complement, the optimal stopping set, by $\bar{C}$.

The optimal continuation set C and its boundary ∂C can be characterised in different ways. If the process is formulated in continuous time with the continuing x-process a diffusion process then F obeys a *tangency* condition at ∂C: not only is F itself continuous there, but also its vector of first derivatives F_x (see Chernoff, 1968, 1972). This tangency condition is what Shepp (1969) refers to as the "principle of smooth fit at the boundary". The condition has a version also for processes which, unlike the diffusion process, can show positive overshoot into the stopping set when they cross ∂C (see Whittle, 1983, Chapters 30 and 37).

Another characterisation of C has been given by Bather (1970). He gave this for the diffusion case, but it holds more generally, even for non-Markov processes (see Whittle, 1983, Chapter 30). Let A be an arbitrary continuation set and let $V^A(x)$ be the expected discounted reward from x if continuation set A is used. This will satisfy

$$V^A = \begin{cases} LV^A & (A) \\ K & (\bar{A}) \end{cases} \tag{3}$$

Let us say that A is *valid* if V^A is well-defined, and determined by (3). Let $\mathcal{A}$ be the set of A for which $V^A \leq K$. Then Bather shows that

$$C = \bigcup_{\mathcal{A}} A. \tag{4}$$

This suggests a technique for the exact or approximate determination of C. Rather than to determine V^A for a varying particular A and test it for the property $V^A \leq K$ one does better to attempt the reverse procedure: to determine possible cost functions first and their associated continuation regions later.

Theorem

Let ϕ *be a solution of*

$$\phi = L\phi \tag{5}$$

and A *the set of* x *for which* $\phi < K$. *Suppose* A *a valid continuation set. Then*

(i) $V^A \leq \phi$ *and* $A \in \mathcal{A}$, $A \subset C$. (ii). *To a no-overshoot approximation* $V^A \sim \phi$ *in* A.

So, one generates solutions of (5) in any convenient way, and the sets of x for which $\phi < K$ and their unions constitute lower bounds to C. Proof is direct, see Whittle (1983). The *validity condition*, that A be a valid continuation set, is necessary. Essentially, one must be able to guarantee that $\phi \geq F$, so that effective modification of K to ϕ cannot offer a smaller cost than could otherwise be obtained.

A systematic method of using the theorem to approximate C is the following. Assume the Markov case, which is the more likely in applications. One can often determine a whole family of solutions ϕ^θ of (5), θ being an indexing parameter. Assuming that the set $A^\theta = \{x : \phi^\theta < K\}$ obeys the validity condition one will choose θ so as to push the boundary of A^θ as far as possible in some direction. That is, until the validity condition fails, which will often be signalled at a boundary point of A^θ by imminent failure of the change of sign of ϕ-K at the boundary. That is, at such a boundary point one will have, in addition to the defining condition

$$\phi^\theta = K \tag{6}$$

also tangency in at least some directions

$$\frac{\partial \phi^{\theta}}{\partial x_j} = \frac{\partial K}{\partial x_j} . \tag{7}$$

The greater the number of disposable parameters one has in one's family of solutions, the greater the number of axes j along which can plausibly demand (7).

Relations (6) and (7) will determine a critical boundary point or points, and also the critical value of θ. The envelope of critical boundary points thus obtained will define the surface of a set Γ, which is the surface of a union of sets of A, and so an inner approximation to C. The set Γ is the largest lower bound to C obtainable from the given family of solutions ϕ^{θ}. The tangency condition (7), holding only at a critical boundary point for each critical value of θ, constitutes an approximation to the actual tangency condition.

II. THE MAXIMISATION OF A SAMPLE MEAN

This is the classic stopping problem considered by, among others, Taylor (1968), Shepp (1969), Walker (1969) and Thompson and Owen (1972). Let $S = \sum_1^n y_j$ be the sum of n independent observations y_j of known common distribution. The experimenter who collects these observations has an interest in establishing that the mean value $\mu = E(y)$ is large, and so (if he is unprincipled) wishes to publish his conclusions at a stage n when the sample value $\hat{\mu} = S/n$ is as large as it is likely to be.

We can take (S,n) as state variable with

$$K(S,n) = S/n$$

and

$$L\phi(S,n) = E\phi(S+y,n+1)$$

where the expectation is with respect to y. A solution of (5) is

$$\phi(S,n) = \alpha\psi(\lambda)^{-n} e^{\lambda S}$$

where α, λ are arbitrary constants and $\psi(\lambda) = E(e^{\lambda y})$. A more general solution is obtained by linear combination of these.

Consider the particular choice

$$\phi(S,n) = \alpha\psi(\lambda)^{-n} e^{\lambda S} + \alpha_2 . \tag{8}$$

If we choose $\alpha_2 = \mu$ then the decision rule generated is translation-invariant, i.e. depends upon μ and $\hat{\mu}$ only in the combination $\hat{\mu}-\mu$. We now apply the technique suggested in the last section to the trial solution (8), with $\alpha_2 = \mu$. That is, we write down equations expressing that ϕ-K and its derivatives with respect to S and n should all be zero at some boundary point. Eliminating α, λ from these relations we deduce an envelope which constitutes an inner approximation to the decision boundary

$$\log \psi((S-n\mu)^{-1}) = \frac{S}{n(S-n\mu)} .$$

In the case of normal observations $y \sim N(\mu,\sigma^2)$ this yields the continuation region

$$\sqrt{n}(\hat{\mu}-\mu) \leq \gamma\sigma \tag{9}$$

with

$$\gamma = \frac{1}{\sqrt{2}} \sim 0.7071 . \tag{10}$$

In fact, Shepp (1969) has shown that the exact optimal boundary in the diffusion limit is of the form (9) with

$$\gamma \sim 0.8399 . \tag{11}$$

So, relatively simple methods have yielded a qualitatively correct and quantitatively close approximation to the true optimal boundary. We note from (10), (11) that the continuation region deduced indeed lies inside the true optimal region.

It is shown in Whittle (1983) that the condition $\alpha_2 = \mu$ can indeed be deduced from consistent application of the technique itself rather than from the imposition of translation-invariance.

III. EVALUATION OF THE GITTINS INDEX

Suppose that the terminal reward K(x) is given a constant value M. Suppose that this constant value is however adjusted so that a given state value x falls exactly on the optimal decision boundary; let M(x) denote this value of M. Then the function M(x) is the *Gittins index* of the continuing process. It constitutes an equitable price to offer for the process if this is currently in state x, and if the offer is understood to remain fixed and open indefinitely. Gittins has shown that, with the determination of this index, one deduces an optimal policy for the general multi-armed bandit (see Gittins and Jones, 1974; Gittins and Nash, 1977; Gittins, 1979; Whittle, 1980).

However calculation of the index M(x) for cases of interest constitutes itself an outstanding problem. It is closely related to solution of the optimal stopping problem with constant terminal reward M, and so amenable to the technique suggested in Section I.

Consider first the case of what one might call a *Gaussian project*. One has a sequence of independent observations of unknown mean μ and known variance σ^2. The reward per unit time is $g = \mu$. On the basis of observations one deduces a posterior estimate $\hat{\mu}$ of μ with posterior variance v; these two quantities constitute the effective state variable of the decision process. Application of the technique suggested in Section I yields the approximate evaluation of the Gittins index

$$M(\hat{\mu},v) = \frac{1}{1-\beta}\left[\hat{\mu} + \frac{v}{\sqrt{2\alpha\sigma^2+v}}\right]$$

where β is the discount factor and $\alpha = -\log \beta$. Indeed, the expression reflects an expected total discounted reward $\hat{\mu}/(1-\beta)$ plus an allowance, favourable on balance, for the effect of uncertainty of estimate of μ. For details see Whittle (1983).

The other case of interest is that of a *Bernoulli project*, corresponding to the use of a single arm of the archetypal multi-armed bandit. Suppose g = p, where p is the unknown success probability of the arm, and that the distribution of p conditional upon current information is Beta with parameter r,s. That is, r and s can be regarded as the effective number of successes and failures observed with the arm. Define

$$n = r + s$$
$$\hat{p} = r/n$$
$$\hat{q} = s/n.$$

Then application of the technique suggested in Section I leads to the approximate evaluation

$$M(r,s) = \frac{1}{1-\beta}\left[\hat{p} + \frac{\hat{p}\hat{q}}{n\sqrt{(2\alpha+n^{-1})\hat{p}\hat{q}} + \frac{1}{2}(\hat{p}-\hat{q})}\right]$$

if $\alpha + \frac{1}{2n}$ is small.

IV. REFERENCES

Bather, J.A. (1970). "Optimal Stopping Problems for Brownian Motion." *Adv. Appl. Prob. 2,* 259-286.

Chernoff, H. (1968). "Optimal Stochastic Control." *Sankhya, Ser. A 30,* 221-252.

Chernoff, H. (1972). *Sequential Analysis and Optimal Design,* SIAM, Philadelphia.

Gittins, J.C. (1979). "Bandit Processes and Dynamic Allocation Indices." *J. Roy. Statist. Soc. B, 41,* 148-164.

Gittins, J.C. and Jones, D.M. (1974). A Dynamic Allocation Index for the Sequential Design of Experiments. In *Progress in Statistics* (J. Gani, ed.), Amsterdam: North Holland, 241-266.

Gittins, J.C. and Nash, P. Scheduling, Queues and Dynamic Allocation Indices. *Proc. EMS, Prague 1974.* Prague: Czechoslovak Academy of Sciences, 191-202.

Shepp, P.A. (1969). "Explicit Solutions to Some Problems of Optimal Stopping." *Ann. Math. Statist. 40,* 993-1010.

Taylor, H.M. (1968). "Optimal Stopping in a Markov Process." *Ann. Math. Statist. 39,* 1333-1344.

Thompson, M.F. and Owen, W.L. (1972). "The Structure of the Optimal Stopping Rule in the S_n/n Problem." *Ann. Math. Statist. 43,* 1110-1121.

Walker, L.H. (1969). "Regarding Stopping Rules for Brownian Motion and Random Walks." *Bull. Amer. Math. Soc. 75,* 46-50.

Whittle, P. (1980). "Multi-Armed Bandits and the Gittins Index." *J. Roy. Statist. Soc. B, 42,* 143-149.

Whittle, P. (1982). *Optimization Over Time, Vol. 1,* Wiley Interscience.

Whittle, P. (1983). *Optimization Over Time, Vol. 2,* Wiley Interscience.

SURVEY OF CLASSICAL AND BAYESIAN APPROACHES TO THE CHANGE-POINT PROBLEM: FIXED SAMPLE AND SEQUENTIAL PROCEDURES OF TESTING AND ESTIMATION[1]

S. Zacks

Department of Mathematical Sciences
State University of New York
Binghamton, New York

I. INTRODUCTION

The change-point problem can be considered one of the central problems of statistical inference, linking together statistical control theory, theory of estimation and testing hypotheses, classical and Bayesian approaches, fixed sample and sequential procedures. It is very often the case that observations are taken sequentially over time, or can be intrinsically ordered in some other fashion. The basic question is, therefore, whether the observations represent independent and identically distributed random variables, or whether at least one change in the distribution law has taken place.

This is the fundamental problem in the statistical control theory, testing the stationarity of stochastic processes, estimation of the current position of a time-series, etc. Accordingly, a survey of all the major developments in

[1] *Research supported in part by ONR Contracts N00014-75-0725 at The George Washington University and N00014-81-K-0407 at SUNY-Binghamton.*

ISBN 0-12-589320-5

statistical theory and methodology connected with the very general outlook of the change-point problem, would require review of the field of statistical quality control, the switching regression problems, inventory and queueing control, etc. The present review paper is, therefore, focused on methods developed during the last two decades for testing the null hypothesis of no change among given n observations, against the alternative of at most one change; the estimation of the location of the change-point(s) and some sequential detection procedures. The present paper is composed accordingly of four major sections. Section II is devoted to the testing problem in a fixed sample. More specifically, we consider a sample of n independent random variables. The null hypothesis is $H_0: F_1(x) = \ldots = F_n(x)$, against the alternative, H_1: $F_1(x) = \ldots = F_\tau(x)$; $F_{\tau+1}(x) = \ldots = F_n(x)$, where $\tau = 1, 2, \ldots, n-1$ designates a possible unknown change point. The studies of Chernoff and Zacks [14], Kander and Zacks [35], Gardner [20], Bhattacharya and Johnson [9], Sen and Srivastava [56] and others are discussed. These studies develop test statistics in parametric and non-parametric, classical and Bayesian frameworks. Section III presents Bayesian and maximum likelihood estimation of the location of the shift points. The Bayesian approach is based on modeling the prior distribution of the unknown parameters, adopting a loss function and deriving the estimator which minimizes the posterior risk. This approach is demonstrated with an example of a shift in the mean of a normal sequence. The estimators obtained are generally non-linear complicated functions of the random variables. From the Bayesian point of view these estimators are optimal. If we ask, however, classical questions concerning the asymptotic behavior of such estimators, or their sampling distributions under repetitive sampling, the analytical problems become very difficult and untractable. The classical efficiency of such estimators is often estimated in some special cases by extensive simulations. The maximum likelihood estimation of the location parameter of the change point is an attractive alternative to the Bayes estimators. The derivation of the asymptotic distributions of these estimators is very complicated. Section 4 is devoted to sequential detection procedures. We present the basic Bayesian and classical results in this area. The studies of Shiryaev [59, 60], Bather [7, 8], Lorden [42] and Zacks and Barzily [68] are discussed with some details. The study of Lorden [42] is especially significant in proving that Page's CUSUM procedures [46-48] are asymptotically minimax.

The important area of switching regressions have not been reviewed here in any details. The relevance of the switching regression studies to the change-point problem is obvious. Regression relationship may change at unknown epochs (change points), resulting in different regression regimes that should be detected and identified. The reader is referred to the important studies of Quandt [50, 51], Inselman and Arsenal [34], Ferreira [19], Marronna and Yohai [44] and others.

An annotated bibliography on the change-point problem was published recently by Shaban [58]. The reader can find there additional references to the ones given in the present paper.

II. TESTING HYPOTHESES CONCERNING CHANGE POINTS

The problem of testing hypotheses concerning the existence of shift points was posed by Chernoff and Zacks [14] in the following form.

Let $X_1, \ldots, X_n$ be a sequence of independent random variables having normal distributions $N(\theta_i, 1)$, $i = 1, \ldots, n$. The hypothesis of no shift in the means, versus the alternative of one shift in a positive direction is

$$H_0: \ \theta_1 = \ldots = \theta_n = \theta_0 \ .$$

vs.

$$H_1: \ \theta_1 = \ldots = \theta_\tau = \theta_0; \ \theta_{\tau+1} = \ldots = \theta_n = \theta_0 + \delta,$$

where $\tau = 1, \ldots, n-1$ is an unknown index of the shift point, $\delta > 0$ is unknown and the initial mean θ_0 may or may not be known.

Chernoff and Zacks showed in [14] that a Bayes test of H_0 versus H_1, for δ values close to zero, is given by the test statistic

$$T_n = \begin{cases} \sum_{i=1}^{n-1} (i+1) X_i & , \text{ if } \theta \text{ is known} \\ \sum_{i=1}^{n-1} (i+1) (X_i - \bar{X}_n) & , \text{ if } \theta \text{ is unknown,} \end{cases} \qquad (2.1)$$

where $\bar{X}_n$ is the average of all the n observations. It is interesting to see that this test statistic weighs the current observations (those with index close to n) more than the initial ones. Since the above test statistic is a linear function of normal random variables T_n is normally distributed and it is easy to obtain the critical value for size α test and the power function. These functions are given in the paper of Chernoff and Zacks [14] with some numerical illustrations.

The above results of Chernoff and Zacks were later generalized by Kander and Zacks [36] to the case of the one-parameter exponential family, in which the density functions are expressed, in the natural parameter form as $f(x;\theta) = h(x) \exp\{\theta U(x) + \psi(\theta)\}$ (see Zacks [69; pp. 95]). Again, Kander and Zacks established that the Bayes test of H_0, for small values of δ when θ_0 is known, is of the form (2.1), where X_i are replaced by $U(X_i)$ $(i=1, \ldots, n)$. The exact determination of the critical levels might require a numerical approach, since the exact distribution of T_n is not normal, if $U(X_i)$ are not normal. Kander and Zacks showed how the critical levels and the power functions can be determined exactly, in the binomial and the negative-exponential cases. If the samples are large, the null distribution of T_n converges to a normal one, according to the Liapunov version of the Central Limit Theorem. They provided numerical comparisons of the exact and asymptotic power functions of T_n, in the binomial and the negative exponential cases. Hsu [33] utilized the above test for testing whether a shift occurred in the variance of a normal distribution.

Gardner [20] considered the testing problem of H_0 versus H_1 for the normally distributed random variables, but with $\delta \neq 0$ unknown. He showed that the Bayes test statistics, with prior probabilities Π_t, t = 1, 2, ..., n-1, is

$$Q_n = \sum_{t=1}^{n-1} \Pi_t \left[\sum_{j=t}^{n-1} (X_{j+1} - \bar{X}_n) \right]^2$$

$$\sum_{t=1}^{n-1} \Pi_t (n-t)^2 (\bar{X}^*_{n-t} - \bar{X}_n)^2 \quad , \tag{2.2}$$

where $\bar{X}^*_{n-t}$ is the mean of the last n-t observations and $\bar{X}_n$ is the mean of all n observations. Gardner investigated the exact and the asymptotic distributions of Q_n, under the null hypothesis H_0 and under the alternative H_1, for the case of equal prior probabilities. Scaling Q_n, so that its expected value will be 1 for each n, by the transformation $Y_n = (6n/(n^2-1))Q_n$, n = 2, 3, ..., we obtain that, under H_0, Y_n is distributed like

$\sum_{k=1}^{n-1} \lambda_k U_k^2$, where $U_1, \ldots, U_{n-1}$ are i.i.d. standard normal r.v.'s

$$\lambda_k = \frac{6n^2}{\pi^2 (n^2-1) k^2} \left[\frac{2n}{k\pi} \cos (k\pi/2n)\right]^{-2}, \ k=1, \ldots, n-1 \tag{2.3}$$

Thus, as $n \to \infty$, the asymptotic distribution of Y_n, under H_0, is like that of

$$Y = \frac{6}{\pi^2} \sum_{k=1}^{\infty} \frac{1}{k^2} U_k^2 \tag{2.4}$$

The distribution of Y is that of the asymptotic distribution of Smirnov's statistic ω_n^2, normalized to have mean 1. Gardner refers the reader to Table VIII of VonMises [65] for the critical values of Y_n, for large n. Critical values $c_n(\alpha)$, for $\alpha = .10$, .05 and .01 and various values of n, can be obtained from Figure 1 of Gardner's paper. The power function of the test was determined by Garnder in some special cases by simulation.

Sen and Srivastava [55] discussed the statistic

$$U_n = \frac{1}{n^2} \sum_{i=1}^{n-1} \left(\sum_{j=i}^{n-1} X_{j+1} \right)^2$$

$$= \frac{1}{n^2} \sum_{i=1}^{n-1} (n-i)^2 \, (\bar{X}^*_{n-i})^2 \tag{2.5}$$

for testing H_0 versus H_1 with $\delta \neq 0$, when the initial mean, μ_0, is known. They derived the c.d.f. of U_n for finite values of n and its asymptotic distribution. A table in which these distributions are presented for n = 10, 20, 50 and ∞ is also provided. Moreover, Sen and Srivastava proposed test statistics which are based on the likelihood ratio test.

For testing H_0 versus H_1, with $\delta > 0$, when μ_0 is unknown, the likelihood function, when the shift is at a point t, is

$$L_t(\underset{\sim}{X}_n) = \frac{1}{(2\pi)^{n/2}} \exp\{-\tfrac{1}{2}[\sum_{i=1}^{t} (X_i - \bar{X}_t)^2 + \sum_{i=t+1}^{n} (X_i - \bar{X}^*_{n-t})^2]\} . \tag{2.6}$$

It follows that the likelihood ratio test statistic is then

$$\Lambda_n = \sup_{1 \leq t \leq n-1} (\bar{X}_t - \bar{X}^*_{n-t}) / (\frac{1}{t} + \frac{1}{n-t})^{\frac{1}{2}} \quad (2.7)$$

Power comparisons of the Chernoff and Zacks Bayesian statistic T_n and the likelihood ratio statistic Λ_n are given for some values of n and point of shift τ. These power comparisons are based on simulations, which indicate that the Chernoff-Zacks Bayesian statistic is generally more powerful than the Sen-Srivastava likelihood ratio statistic when $\tau \approx n/2$. On the other hand, when τ is close to 1 or to n, the likelihood ratio test statistic is more powerful.

Bhattacharya and Johnson [9] approached the testing problem in a non-parametric fashion. It is assumed that the random variables $X_1, X_2, \ldots, X_n$ are independent and have continuous distributions F_i, (i=1, ..., n). Two types of problems are discussed. One of *known* initial distribution, F_0, which is symmetric around the origin. The other one is that of *unknown* initial distribution, which is not necessarily symmetric. The hypotheses corresponding to the shift problem when F_0 is known are

$$H_0: \; F_0 = \ldots = F_n$$

for some *specified* F_0, symmetric about 0, versus

$$H_1: \; F_0 = \ldots = F_\tau > F_{\tau+1} = \ldots = F_n \; .$$

τ is an unknown shift parameter. $F_\tau > F_{\tau+1}$ indicates that the random variables after the point of shift are stochastically greater than the ones before it. For the case of known initial distribution $F_0(x)$, the test is constructed with respect to a translation alternative of the form $F_{\tau+1}(x) = F_0(x-\Delta)$, where $\Delta > 0$ is an unknown parameter. Bhattacharya and Johnson proved that, under some general smoothness conditions on the p.d.f.

$f_0(x)$, the form of the invariant test statistic, which maximizes the derivative of the average power

$\bar{\psi}(\Delta) = \sum_{i=1}^{n} q_i \psi(\Delta|i-1)$, at $\Delta = 0$, where $\psi(\Delta|t)$ is the power at Δ when the shift occurs at t, and $q_1, \ldots, q_n$ are given probability weights, is

$$T_n = \sum_{i=1}^{i} Q_i \operatorname{sgn}(X_i) E\{-f_0'(V^{(R_i)})/f_0(V^{(R_i)})\} . \qquad (2.8)$$

$V^{(1)} \leq \ldots \leq V^{(n)}$ is an ordered statistic of n i.i.d. random variables having a distribution $F_0(x)$, and $Q_i \sum_{j=1}^{n} q_j$. More specific formulae for the cases of double-exponential, logistics and normal distributions are given. The null hypothesis H_0 is rejected for large values of T_n. It is further proven that, any test of the form $T = \sum_{i=1}^{n} Q_i \operatorname{sgn}(X_i) U(R_i)$, where U is a strictly increasing function, is unbiased. Moreover, if the system of wieghts $\{q_{n,i};\ i = 1, \ldots, n\}$ satisfies the condition

$$\lim_{n\to\infty} \frac{1}{n} \sum_{i=1}^{n} Q_{n,i}^2 = b^2, \qquad 0<b^2<\infty \qquad , \qquad (2.9)$$

then, the distribution of $T_n/(nb^2(\int_0^1 \psi^2(u)\,du))^{\frac{1}{2}}$, as $n \to \infty$, converges to the standard normal distirbution, where

$$\psi(u) = -f_0'(F_0^{-1}((u+1)/2))/f_0(F_0^{-1}((u+1)/2)) \qquad (2.10)$$

Similar analysis is done for the case of unknown initial distribution F_0. In this case the test statistic is a function of the maximal invariant $(S_1, \ldots, S_n)$, which are the ranks of $(X_1, \ldots, X_n)$. The test statistic in this case is of the general form

$$T_n^* = \sum_{i=1}^{n} Q_i \, E\{-f'(V^{(S_i)})/f(V^{(S_i)})\} . \qquad (2.11)$$

In the normal case, for example, with equal weights for t = 2, ..., n and weight 0 for t = 1, the test statistic is $T_n^* \sum_{i=1}^{n} (i-1)S_i$. Notice the similarity in structure between the statistic T_n^* and that of Chernoff and Zacks, T_n. The difference is that the actual values of X_i are replaced by their ranks, S_i.

Hawkins [22] also considered the normal case, with two sided hypothesis, both θ_0 and δ unknown. Like Sen and Srivastava, he considered the test statistic $U_n = \max_{1 \le k \le n-1} |T_k|$, where

$$T_k = \left(\frac{n}{k(n-k)}\right)^{\frac{1}{2}} \sum_{i=1}^{n} (X_i - \bar{X}_n), \quad k = 1, ..., n - 1 \qquad (2.12)$$

The statistics $T_1, ..., T_{n-1}$ are normally distributed, having a correlation function

$$\rho(T_m, T_k) = \left(\frac{m(n-k)}{k(n-m)}\right)^{\frac{1}{2}}, \ m \le k \qquad (2.13)$$

Hawkins provided recursive formulae for the exact determination of the distribution of U. The critical level for a conservative testing can be determined by applying the Bonferroni inequality and using the appropriate fractile of the standard normal distribution. A numerical example is given to compare the exact and the Bonferroni approximation to the critical values of the test statistic U_n. In an attempt to understand the asymptotic properties of U_n, Hawkins considered the behavior of the maximum of a Gaussian process having the same covariance structure as that of $T_1, T_2, ...$. The asymptotic results are still not satisfactory.

Pettitt [49] discussed non-parametric tests different from those of Bhattacharya and Johnson. He defined for each $t=1, \ldots, n$, $U_{t,n} = \sum_{i=1}^{t} \sum_{j=t+1}^{n} \operatorname{sgn}(X_i - X_j)$ and studied the properties of the test statistic $K_n = \max_{1 \le t \le n} |U_{t,n}|$. The distribution of K_n was studied for Bernoulli random variables.

III. ESTIMATING THE LOCATION OF THE SHIFT POINT

Two types of estimators of the location of the shift point, τ, appear in the literature: Bayesian and maximum likelihood. El-Sayyad [17] , Smith [61], Broemeling [11], Zacks [69; pp. 311] and others, give the general Bayesian framework for inference concerning the location of the shift point, τ, in a model of at most one change. Hinkley [28] studies the maximum likelihood estimator. We start with an example concerning the Bayesian estimation and proceed then to present the maximum likelihood estimator.

III.I Bayesian Estimation of the Change Point

The Bayesian procedure is to derive the posterior distribution of the change point τ, and determine the estimator which minimizes the posterior risk, for a specified loss function. If the loss function is $L(\hat{\tau},\tau) = |\hat{\tau}-\tau|$, where $\hat{\tau}$ is an estimator of τ, then the Bayes estimator of the change point is the median of the posterior distribution of τ, given $\underset{\sim}{X}_n$. We illustrate this procedure with the following example. Let $X_1, \ldots, X_n$ be independent random variables having normal distributions $N(\theta_i,1)$, where

$$\theta_1 = \ldots = \theta_\tau = 0,\ \theta_{\tau+1} = \ldots = \theta_n = \delta.$$

Assume that the prior distribution of δ is normal, $N(0,\sigma^2)$, and that τ has prior probability function $\Pi(t)$, $t = 1, \ldots, n$. Here $\{\tau=n\}$ indicates the event of no change. The posterior probabilities of τ for this model are

$$\Pi(t|\underset{\sim}{X}_n) = \frac{\Pi(t)(1+(n-t)\sigma^2)^{\frac{1}{2}} \quad \exp \dfrac{(\bar{X}^*_{n-t})^2(n-t)^2\sigma^2}{2(1+(n-t)\sigma^2)}}{\sum\limits_{j=1}^{n} \Pi(j)(1+(n-j)\sigma^2)^{\frac{1}{2}} \exp \dfrac{(\bar{X}_{n-j})^2}{2} \dfrac{(n-j)^2\sigma^2}{(1+(n-j)\sigma^2)}} \tag{3.1}$$

where $\bar{X}^*_{n-t} = \frac{1}{n-t} \sum\limits_{i=t+1}^{n} X_i$ is the average of the last $(n-t)$ observations. The median of the posterior distribution is then the Bayes estimator of τ, namely

$$\hat{\tau}(\underset{\sim}{X}_n) = \textit{least positive integer } t\textit{, such that } \sum_{i=0}^{t} \Pi(i|\underset{\sim}{X}_n) \geq .5 \tag{3.2}$$

In the following table we present the posterior probabilities (3.1) computed for the values of four simulated samples. Each sample consists of $n = 20$ normal variates with means θ_i and variance 1. In all cases $\theta_0 = 0$. Case I consists of a sample with no change in the mean $\delta = 0$. Cases II-IV have a shift in the mean at $\tau = 10$, and $\delta = .5$, 1.0 and 2.0. Furthermore, the prior probabilities of τ are $\Pi(t) = p(1-p)^{t-1}$ for $t = 1, \ldots,$ $n - 1$ and $\Pi(n) = (1-p)^{n-1}$, with $p = .01$; and the prior variance of δ is $\sigma^2 = 3$.

TABLE III.I Posterior Probabilities of $\{\tau=t\}$

$t\backslash\delta$	0	.5	1.0	2.0
1	0.002252	0.012063	0.003005	0.000000
2	0.004284	0.016045	0.002885	0.000000
3	0.004923	0.016150	0.002075	0.000000
4	0.006869	0.022634	0.002193	0.000000
5	0.006079	0.008002	0.002202	0.000001
6	0.004210	0.006261	0.002291	0.000050
7	0.004020	0.006735	0.001954	0.000026
8	0.002867	0.015830	0.001789	0.000015
9	0.003534	0.015914	0.001959	0.001087
10	0.002972	0.011537	0.002228	0.068996
11	0.003033	0.019014	0.002708	0.908434
12	0.003070	0.010335	0.002661	0.016125
13	0.003395	0.006026	0.002996	0.005237
14	0.003087	0.003201	0.003017	0.000009
15	0.004064	0.003461	0.003096	0.000011
16	0.003355	0.002709	0.002820	0.000009
17	0.004991	0.002899	0.003078	0.000000
18	0.009664	0.003486	0.004004	0.000000
19	0.007255	0.006106	0.012432	0.000000
20	0.916077	0.811593	0.940607	0.000000

We see in Table III.I that the Bayes estimator for Cases I-III is $\hat{\tau} = 20$ (no change), while in Case IV it is $\hat{\tau} = 11$. That is, if the magnitude of change in the mean is about twice the standard deviation of the random variables, the posterior distribution is expected to have its median close to the true change point.

In many studies (for example, Smith [61]) the Bayesian model is based on the assumption of equal prior probabilities of $\{\tau=t\}$ which reflects a state of prior indifference concerning the location of the change parameter. The Bayesian approach illustrated above is a powerful one. It has been applied in various problems.

Smith [61] derived formulae of the Bayes estimators for cases of sequences of Bernoulli trials. Bayesian estimators for the location of the shift parameter for switching regression problems are given by Smith [62], by Ferriera [19], Holbert and Broemeling [31], Tsurumi [64] and others.

III.II Maximum Likelihood Estimators

Let $X_1, X_2, \ldots, X_n$ be a sequence of independent random variables. As before, assume that $X_1, X_2, \ldots, X_\tau$ have the c.d.f. $F_0(x)$ and $X_{\tau+1}, \ldots, X_n$ have the c.d.f. $F_1(x)$. τ is the unknown point of shift. The maximum likelihood estimator (MLE) of τ, $\hat{\tau}_n$, is the smallest positive integer t, t = 1, ..., n, maximizing $S_{n,t}$, where

$$S_{n,t} = \begin{cases} \sum_{i=1}^{t} \log f_0(X_i) + \sum_{i=t+1}^{n} \log f_1(X_i), & \text{if } t = 1, \ldots, n-1 \\ \sum_{i=1}^{n} \log f_0(X_i) & , \text{ if } t = n \end{cases} \tag{3.3}$$

$f_0(x)$ and $f_1(x)$ are the p.d.f.'s corresponding to $F_0(x)$ and $F_1(x)$. The above definition of the MLE, $\hat{\tau}_n$, is equivalent to the definition:

$\hat{\tau}_n$ = least positive integer t, t=1,...,n, maximizing $W_{n,t}$,

where

$$W_{n,t} = \begin{cases} \sum_{i=t+1}^{n} \log \dfrac{f_1(X_i)}{f_0(X_i)}, & \text{if } t=1, \ldots, n-1 \\ 0 & , \text{ if } t=n \end{cases} \tag{3.4}$$

In the case of normal distributions considered in the example of Section III.I, the MLE is the least integer t, maximizing

$$(n-t) \sum_{i=t+1}^{n} X_i, \quad t = 0, \ldots, n-1.$$ If the maximum occurs for t=0 the conclusion is that no change has occurred.

It is generally easy to determine the MLE of τ, but it is not a trivial matter to determine its exact distribution. Hinkley [27] applied random walk techniques to derive the asymptotic distribution of the MLE. Numerical approximations to the distribution of the MLE in large samples can be obtained by the method developed by Hinkley.

IV. DYNAMIC CONTROL PROCEDURES

There are numerous papers on dynamic control problems, all of which deal in one way or another with the problem of shift at unknown time points. In particular we mention here the papers of Girshick and Rubin [21], Bather [7, 8], Lorden [42], Yadin and Zacks [67], Shiryaev [59, 60], and Zacks and Barzily [68].

We present first the Bayesian theory, followed by discussion of the CUSUM procedure. Again, we consider a sequence of independent random variables $X_1, X_2, \ldots, X_{m-1}, X_m, \ldots$. Let τ be the point of shift, $\tau = 0, 1, \ldots$. If $\tau \leq 1$, all the observations are from $F_1(x)$. If $\tau = t$ (t=2, 3, ...) then the first t-1 observations are from $F_0(x)$ and $X_t, X_{\tau+1}, \ldots$ are from $F_1(x)$. The random variables $X_1, X_2, \ldots$ are observed sequentially and we wish to apply a stopping rule which will stop soon after the shift occurs, without too many "false alarms". The following objectives are considered in the selection of a stopping variable N:

1) If $\Pi(\tau)$ denotes the prior distribution of τ, then the prior risk

$$R(\Pi,N) = P_{\Pi}(N<\tau) + c\, P_{\Pi}(N \geq \tau)\, E_{\Pi}\{N-\tau \mid N\geq\tau\} \tag{4.1}$$

is minized, with respect to all stopping rules.

2) To minimize $E_{\Pi}\{N-\tau \mid N\geq\tau\}$ subject to the constraint $P_{\Pi}(N<\tau) \leq \alpha$, $0<\alpha<1$.

IV.I Bayesian Procedures

The shift index, τ, is considered a random variable, having a prior p.d.f. $\Pi(t)$, concentrated on the non-negative integers. Shiryaev [59] postulated the following prior distribution

$$\Pi(t) = \begin{cases} \Pi & , \text{ if } t=0 \\ (1-\Pi)p(1-p)^{t-1} & , \text{ if } t=1,2,\ldots \end{cases} \tag{4.2}$$

for $0<\Pi<1$, $0<p<1$. $(\Pi+(1-\Pi)p)$ is the prior probability that the shift has occurred before the first observation, and p is the prior probability of a shift occuring between two observations.

After observing $X_1, \ldots, X_n$, the prior probability function $\Pi(t)$ is converted to a posterior probability function on $n, n+1, \ldots$, namely,

$$\Pi_n(t) = \begin{cases} \Pi_n & , \quad t=n \\ (1-\Pi_n)p(1-p)^{t-1} & , \quad t=n+1, \ldots \end{cases} \tag{4.3}$$

where Π_n is the posterior probability that the shift took place before the n-th observation. This posterior probability is given by $\Pi_n = 1-q_n$, where

$$q_n = \frac{(1-\Pi)(1-p)^n}{D_n} \prod_{i=1}^{n} f_0(X_i) \quad , \tag{4.4}$$

and

$$D_n = (\Pi+(1-\Pi)p) \prod_{i=1}^{n} f_1(X_i) +$$

$$(1-\Pi)p \sum_{j=1}^{n-1} (1-p)^j \prod_{i=1}^{j} f_0(X_i) \prod_{i'=j+1}^{n} f_1(X_{i'}) + \tag{4.5}$$

$$(1-\Pi)(1-p)^n \prod_{i=1}^{n} f_0(X_i) \quad .$$

Let $R(X_i) = f_1(X_i)/f_0(X_i)$, $i=1, 2, \ldots$ then

$$q_{n+1} = \frac{(1-\Pi)(1-p)^{n+1}}{R(X_{n+1})[D_n-(1-\Pi)(1-p)^n] + B_{n+1}} \quad , \tag{4.6}$$

where

$$B_{n+1} = R(X_{n+1})(1-\Pi)(1-p)^n \, p + (1-\Pi)(1-p)^{n+1}$$

But $(1-\Pi)(1-p)^n = q_n D_n$. Hence,

$$q_{n+1} = \frac{q_n(1-p)}{R(X_{n+1})(1-q_n(1-p))+q_n(1-p)} \quad , \tag{4.7}$$

or

$$\Pi_{n+1} = \frac{(\Pi_n + (1-\Pi_n)p)\, R(X_{n+1})}{(\Pi_n+(1-\Pi_n)p)R(X_{n+1})+(1-\Pi_n)(1-p)} \quad , \tag{4.8}$$

$n = 0, 1, \ldots$ with $\Pi_0 = \Pi$ and $q_0 = 1-\Pi$. Accordingly, the sequence of posterior probabilities $\{\Pi_n; n \geq 0\}$ is Markovian. Shiryaev [59] has shown that when F_0 and F_1 are known, the optimal stopping variable, with respect to the above objectives, is to stop at the smallest n for which $\Pi_n \geq A^*$, for some $0 < A^* < 1$. Bather [7] proved that for the constraint of bounding the expected number of false alarms by N, $A^* = (N+1)^{-1}$ is the optimal stopping boundary.

When the distributions F_0 and F_1 are not completely specified, the above problem of finding optimal stopping variables becomes much more complicated. Zacks and Barzily [68] studied Bayes procedures for detecting shifts in the probability of success, θ, of Bernoulli trials, when the values θ_0, before the shift, and the value θ_1 after it, are unknown. The Bayesian model assumed that θ_0 and θ_1 have a uniform prior distribution over the simplex $\{(\theta_0, \theta_1);\ 0<\theta_0 \leq \theta_1 < 1\}$ and the point of shift, τ, has the prior distribution (4.2). In this case, the posterior probability Π_n depends on the whole vector of observations $\underset{\sim}{X}_n = (X_1, \ldots, X_n)'$, and not only on Π_{n-1} and the last observation X_n. It is shown that this posterior probability Π_n is given by the formula

$$\Pi_n = 1 - (1-\Pi)(1-p)^{n-1} B(T_n+1,\ n-T_n+2)/D_n(\underset{\sim}{X}_n) \tag{4.9}$$

where B(p,q) is the beta-function $T_j = \sum_{i=1}^{j} X_i$ and

$$D_n(\underset{\sim}{X}_n) = \Pi\, B(T_n + 2,\ n - T_n + 1) +$$

$$(1-\Pi)p \sum_{j=1}^{n-1} (1-p)^{j-1} B(T_{n-j}^{(n)} + 1,\ n-j-T_{n-j}^{(n)} + 1)$$

$$\sum_{i=0}^{T_{n-j}^{(n)}} \binom{n-j-1}{i} B(T_n + 1,\ n-T_n + 2)$$

$$+ (1-\Pi)(1-p)^{n-1} B(T_n + 1,\ n - T_n + 2)\ . \tag{4.10}$$

Here, $T_{n-j}^{(n)} = T_n - T_j$ (j=0, ..., n). The sequence $\{\Pi_n; n \geq 1\}$ is not Markovian, but is a submartingale. Zacks and Barzily considered the problem of determining the optimal stopping rule under the following cost conditions:

After each observation we have the option to stop observations and declare that a shift has occurred. The process is then inspected. If the shift has not yet occurred a penalty of 1 unit is imposed. If, on the other hand, the shift has already occurred, a penalty of C units per delayed observation (or time unit) is imposed. It is shown then that the optimal stopping variable is

$$N^0 = \text{least } n \geq 1, \text{ such that } \Pi_n \geq b_n(\underset{\sim}{X}_n) \, , \tag{4.11}$$

where the stopping boundary $b_n(\underset{\sim}{X}_n)$ is given implicitly, as the limit of the sequence (as $j \to \infty$)

$$b_n^{(j)}(\underset{\sim}{X}_n) = \min\left(\Pi^* - \frac{M_n^{(j-1)}(\underset{\sim}{X}_n)}{C+p}, \ 1\right) \, , \tag{4.12}$$

with $\Pi^* = p/(C+p)$. The functions $M_n^{(j)}(\underset{\sim}{X}_n)$ can be determined recursively, according to the formula

$$M_n^{(j)}(\underset{\sim}{X}_n) = E\{\min(0, C\,\Pi_{n+1}(\underset{\sim}{X}_n, X_{n+1}) - p(1-\Pi_{n+1}(\underset{\sim}{X}_n, X_{n+1})) + M_{n+1}^{(j-1)}(\underset{\sim}{X}_n, X_{n+1})) \mid \underset{\sim}{X}_n\}. \tag{4.13}$$

It is very difficult, if not impossible, to determine these functions explicitly, for large values of j. The authors therefore considered a suboptimal procedure based on $b_n^{(2)}(\underset{\sim}{X}_n)$ only. Numerical simulations illustrate the performance of the suboptimal procedure.

IV.II Asymptotically Minimax Rules and the CUSUM Control

Lorden [41, 42] considered the sequential detection procedure from a non-Bayesian point of view and proved that the well known CUSUM procedures of Page [46, 47, 48] are asymptotically *minimax*.

Let $X_1, X_2, \ldots$ be a sequence of *independent* random variables. The distributions of $X_1, \ldots, X_{m-1}$ is $F_0(x)$ and that of $X_m, X_{m+1}, \ldots$ is $F_1(x)$. The point of shift m is *unknown*, $F_0(x)$ and $F_1(x)$ are known. The family of probability measures is $\{P_m;\ m=1, 2, \ldots\}$, where $P_m(\underset{\sim}{X}_n)$ is the joint p.d.f. of $\underset{\sim}{X}_n = (X_1, \ldots, X_n)$, in which X_m is the first random variable with a c.d.f. $F_1(x)$. It is desired to devise a sequential procedure with a (possibly) extended stopping variable N, (i.e., $\lim_{n\to\infty} P_m[N>n] \geq d>0$, $m = 0, 1, \ldots$) which minimizes the largest possible expectation of delayed action, and does not lead to too many false alarms. More precisely, if $P_0(\underset{\sim}{X})$ denotes the c.d.f. under the assumption that all observations have $F_0(X)$ as a c.d.f.; and if $E_m\{\cdot\}$ denotes the expectation under $P_m(\cdot)$, the objective is to *minimize*

$$E_1\{N\} = \sup_{m\geq 1} \operatorname{ess\,sup} E_m\{(N-m-1)^+ | \mathcal{F}_{m-1}\} \tag{4.14}$$

subject to the constraint

$$E_0\{N\} \geq \gamma^* \quad , \quad 1<\gamma^*<\infty \quad . \tag{4.15}$$

$E_m\{\cdot|\mathcal{F}_{m-1}\}$ denotes the conditional expectation given the σ-field generated by $(X_1, \ldots, X_{m-1})$. It is proven by Lorden [42] that an asymptotically minimax procedure, as $\gamma^* \to \infty$, is provided by Page's procedure, which is described below. Let $R(X_i) = f_1(X_i)/f_0(X_i)$, $i = 1, 2, \ldots$ where $f_i(x)$ is the p.d.f. corresponding to $F_i(X)$, $i = 0, 1$. Let

$S_k = \sum_{i=1}^{k} \log R(X_i)$, k=1, 2, ... and $T_n = s_n - \min_{k\leq n} S_k$. Then for $\gamma = \log \gamma^*$

$$N^* = \text{least } n\geq 1 \text{ such that } T_n \geq \gamma \quad , \tag{4.16}$$

is Page's (extended) stopping variable. The statistic T_n can be computed recursively by the formula

$$T_{n+1} = (T_n + \log R(X_{n+1}))^+ \quad , \quad n = 0, 1, \ldots$$

$$T_0 = 0 \ . \tag{4.17}$$

The above detection procedure can be considered as a sequence of one-sided Wald's SPRT with boundaries $(0, \gamma)$. Whenever the T_n statistic hits the lower boundary, 0, the SPRT is recycled, and all the previous observations can be discarded. On the other hand, for the first time $T_n \geq \gamma$ the sampling process is stopped. The repeated cycles are *independent* and identically distributed. Thus, Wald's theory of SPRT can be used to obtain the main results of the present theory. Let α and β be the error probabilities in each such independent cycle of Wald's SPRT; i.e., $\alpha = P_0[T_n \geq \gamma]$ and $\beta = P_1[T_n = 0]$. Let N_1 be the length of a cycle.

$$E_0\{N^*\} = \frac{1}{\alpha} E_0\{N \} \tag{4.18}$$

and

$$E_1\{N^*\} = \frac{1}{1-\beta} E_1\{N_1\} \quad . \tag{4.19}$$

Set $\gamma^* = \frac{1}{\alpha}$, then the constraint (4.15) is satisfied, since $E_1\{N_1\} \geq 1$. Moreover, Lorden proved that $E_1\{N^*\} = \bar{E}_1\{N^*\}$. Finally, applying well known results on the expected sample size in Wald's SPRT, we obtain

$$E_1\{N^*\} \approx \frac{\log \alpha}{I_1} \quad , \qquad \alpha \to 0 \qquad , \tag{4.20}$$

where $I_1 = E_1 \{\log \frac{f_1(X)}{f_0(X)}\}$ is the Kullbakc-Leibler information for discriminating between F_0 and F_1. The right hand size of (4.20) was shown to be the asymptotically minimum expected sample size. Thus, Page's procedure is asymptotically minimax.

Lorden and Eisenberg applied in [41] the theory presented above to solve a problem of life testing for a reliability system. It is assumed that the life length of the system is distributed exponentially, with intensity (failure-rate) λ. At an unknown time point, θ, the failure rate shifts from λ to $\lambda(1+\eta)$, $0<\eta_1\leq\eta\leq\eta_2<\infty$. Approximations to the formulae of $E_0\{N^*\}$ and $E_\eta\{N^*\}$ are given, assuming that λ is *known*. By proper transformations of the statistics the detection procedure can be applied also to cases of unknown λ. It is interesting to present some of the numerical results of this study. For the case of $\lambda = 1$ and $\alpha = 1/\sigma$ the expected number of observations required is

η	γ	$E_0\{N\}$	$\bar{E}_\eta\{N\}$
.4	20	422	48
.6	50	676	36
.9	40	342	20

Page's CUSUM procedure is thus very conservative, relative to the Bayes procedures which detect the shifts fast, but have also small $E_0\{N\}$.

REFERENCES

[1] Bagshaw, M. and Johnson, R. A. (1975). "The Influence of Reference Values and Estimated Variance on the ARL of CUSUM Tests." *J.R.S.S., B, 37*, 413-420.

[2] Bagshaw, M. and Johnson, R. A. (1975). "Sequential Detection of a Drift Change in a Wiener Process." *Commun. Statist. 4*, 787-796.

[3] Bagshaw, M. and Johnson, R. A. (1975). "The Effect of Serial Correlation on the Performance of CUSUM Tests II." *Technometrics*, *17*, 73-80.
[4] Balmer, D. W. (1975). "On a Quickest Detection Problem with Costly Information." *J. Appl. Prob.*, *12*, 87-97.
[5] Balmer, D. W. (1981). "On Quickest Detection Problem with Variable Monitoring." *J. Appl. Prob.*, *18*, 760-767.
[6] Barnard, G. A. (1959). "Control Charts and Stochastic Processes." *J. Roy. Statist. Soc. B*, *21*, 239-271.
[7] Bather, J. A. (1967). "On a Quickest Detection Problem." *Ann. Math. Statist.*, *38*, 711-724.
[8] Bather, J. A. (1976). "A Control Chart Model and a Generalized Stopping Problem for Brownian Motion." *Math. Oper. Res.*, *1*, 209-224.
[9] Bhattacharya, G. K. and Johnson, R. A. (1968). "Non-parametric Tests for Shift at an Unknown Time Point." *Ann. Math. Statist.*, *39*, 1731-1743.
[10] Broemeling, L. (1977). "Estimating the Future Values of Changing Sequences." *Commun. Statist.*, *A*, *6*, 87-102.
[11] Broemeling, L. D. (1974). "Bayesian Inferences About a Changing Sequence of Random Variables." *Commun. Statist.*, *3*, 234-255.
[12] Brown, R. L., Durbin, J. and Evans, J. M. (1975). Techniques for Testing the Constancy of Regression Relationships Over Time." *J. Royal Statis. Soc. B.*, *37*, 149-192.
[13] Brown, R. L. and Durbin, J. (1968). "Methods of Investigation Whether a Regression Relationship is Constant Over Time." Selected Statistical Papers I. Amsterdam: Math Centrum, European Meeting, 37-45.
[14] Chernoff, H. and Zacks, S. (1964). "Estimating the Current Mean of a Normal Distribution Which is Subject to Changes in Time." *Ann. Math. Statist.*, *35*, 999-1028.
[15] Cobb, G. W. (1978). "The Problem of the Nile: Conditional Solution to a Change Point Problem." *Biometrika*, *62*, 243-251.
[16] Darkhovshi, B. S. (1976). "A Non-parametric Method for the A Posteriori Detection of the 'Disorder' Time of a Sequence of Independent Random Variables." *Theory of Prob. Appl.*, *21*, 178-183.
[17] El-Sayyad, G. M. (1975). "A Bayesian Analysis of the Change-Point Problem." *Egypt. Statist. J.*, *19*, 1-13.
[18] Farley, J. U. and Hinich, M. J. (1970). "Detecting "Small" Mean Shifts in Time Series." *Management Science*, *17*, 189-199.
[19] Ferreira, P. E. (1975). "A Bayesian Analysis of Switching Regression Model: Known Number of Regimes." *J. Amer. Statist. Assoc.*, *70*, 370-374.

[20] Gardner, L. A., Jr. (1969). "On Detecting Changes in the Mean of Normal Variates." *Ann. Math. Statist., 40,* 116-126.
[21] Girshick, M. A. and Rubin, H. (1952). "A Bayes Approach to a Quality Control Model." *Ann. Math. Statist. 23,* 114-115.
[22] Hawkins, D. M. (1977). "Testing a Sequence of Observations for a Shift in Location." *J. Amer. Statist. Assoc., 72,* 180-186.
[23] Hawkins, D. M. (1980). "A Note on Continuous and Discontinuous Segmented Regression." *Technometrics, 22,* 443-444.
[24] Hines, W. G. S. (1976). "A Simple Monitor of a System with Sudden Parameter Changes." *IEEE Trans. Inf. Theory, IT,* 210-216.
[25] Hinkley, D. V. and Hinkley, E. A. (1970). "Inference About the Change-Point in a Sequence of Binomial Random Variables." *Biometrika, 57,* 477-488.
[26] Hinkley, D. V. (1969). "Inference About the Intersection in Two-Phase Regression." *Biometrike, 56,* 495-504.
[27] Hinkley, D. V. (1970). "Inference About the Change-Point in a Sequence of Random Variables." *Biometrika, 57,* 1-16.
[28] Hinkley, D. V. (1971). "Inference in Two Phase Regression." *J. Amer. Statist. Assoc., 66,* 736-743.
[29] Hinkley, D. V. (1971). "Inference About the Change-Point from a Cumulative Sum Test." 509-523.
[30] Hinkley, D. V. (1972). "Time-Ordered Classification." *Biometrika, 59,* 509-523.
[31] Holbert, D. and Broemling, L. (1977). "Bayesian Inferences Related to Shifting Sequences and Two-Phase Regression." *Commun. Statist., A,* 6, 265-275.
[32] Hsu, D. A. (1977). "Tests for Variance Shift at an Unknown Time Point." *Applied Statist., 26,* 279.-284.
[33] Hsu, D. A. (1979). "Detecting Shifts of Parameter in Gamma Sequences with Applications to Stock Price and Air Traffic Flow Analysis." *J. Amer. Statist. Assoc., 74,* 31-40.
[34] Inselman, E. H. and Arsenal, F. (1968). "Tests for Several Regression Equations." *Ann. Math. Statist., 39,* 1362.
[35] Kander, A. and Zacks, S. (1966). "Test Procedures for Possible Changes in Parameters of Statistical Distributions Occurring at Unknown Time Points." *Ann. Math. Statist., 37,* 1196-1210.
[36] Khan, R. A. (1975). "A Sequential Detection Procedure." Tech. Rep. No. 17, Dept. of Statistics, CWRU.
[37] Khan, R. A. (1979). "Some First Passage Problems Related to CUSUM Procedures. *Stoch. Proc. Appl., 9,* 207-216.
[38] Lau, T. L. (1973). "Gaussian Processes, Moving Averages and Quickest Detection Problems." *Ann. Prob., 1,* 825-837.
[39] Lai, T. L. (1974). "Control Charts Based on Weighted Sums." *Ann. Statist., 2,* 134-147.

[40] Lee, A. F. S. and Heghinian, S. M. (1977). "A Shift of the Mean Level in a Sequence of Independent Normal Random Variables - a Baeysian Approach." *Technometrics, 19,* 503-506.
[41] Lorden, G. and Eisenberg, I. (1973). "Detection of Failure Rate Increases." *Technometrics, 15,* 167-175.
[42] Lorden, G. (1971). "Procedures for Reacting to a Change in Distribution." *Ann. Math. Statist. 42,* 1897-1908.
[43] Maronna, R. and Yohai, V. J. (1978). "A Bivariate Test for the Detection of a Systematic Change in the Mean." *J. Amer. Statist. Assoc., 73,* 640-645.
[44] Mustafi, C. K. (1968). "Inference Problems About Parameters Which Are Subjected to Changes Over Time." *Ann. Math. Statist., 39,* 840-854.
[45] Nadler, J. and Robbins, N. D. (1971). "Some Characteristics of Page's Two-Sided Procedure for Detecting a Change in the Location Parameter." *Ann. Math. Statist., 42,* 538-551.
[46] Page, E. S. (1954). "Continuous Inspection Schemes." *Biometrika, 41,* 100-115.
[47] Page, E. S. (1955). "A Test for a Change in a Parameter Occurring at an Unknown Point." *Biometrika, 42,* 523-526.
[48] Page, E. S. (1957). "On Problem in Which a Change in a Parameter Occurs at an Unknown Point." *Biometrika, 44,* 248-252.
[49] Pettitt, A. N. (1979). "A Non-Parametric Approach to the Change-Point Problem." *Appl. Statist., 28,* 126-135.
[50] Quandt, R. E. (1958). "The Estimation of the Parameters of a Linear Regression System Obeys Two Separate Regimes." *J. Amer. Statist. Assoc.*
[51] Quandt, R. E. (1960). "Tests of the Hypothesis That a Linear Regression System Obeys Two Separate Regimes." *J. Amer. Statist. Assoc., 55,* 324-330.
[52] Rao, P. S. R. S. (1972). "On Two Phase Regression Estimator." *Sankhya, 34,* 473-476.
[53] Schweder, T. (1976). "Some 'Optimal' Methods to Detect Structural Shift or Outliers in Regression." *J. Amer. Statist. Assoc., 71,* 491-501.
[54] Sen, A. and Srivastava, M. (1975). "On Tests for Detecting Change in the Mean When Variance is Unknown." *Ann. Inst. Statist. Math., 27,* 593-602.
[55] Sen, A. and Srivastava, M. S. (1975). "On Tests for for Detecting Changes in Means." *Annals of Statist., 3,* 98-108.
[56] Sen, A. and Srivastava, M. S. (1975). "On One-Sided Tests for Change in Level." *Technometrics, 17,* 61-64.

[57] Sen, A. K. and Srivastata, M. S. (1973). "On Multivariate Tests for Detecting Change in Mean." *Sankhya, 35,* 173-186.
[58] Shaban, S. A. (1980). "Change-Point Problem and Two-Phase Regression: An Annotated Bibliography." *Inst. Statist. Review, 48,* 83-93.
[59] Shiryaev, A. N. (1963). "On Optimum Methods in Quickest Detection Problems." *Theory Prob. Appl., 8,* 22-46.
[60] Shiryaev, A. N. (1973). "Statistical Sequential Analysis: *Optimal Stopping Rules*. Translations of Mathematical Monographs, Vol. 38, American Mathematical Society, Providence, RI.
[61] Smith, A. F. M. (1975). "A Bayesian Approach to Inference About Change-Point in Sequence of Random Variables." *Biometrika, 62,* 407-416.
[62] Smith, A. F. M. (1977). "A Bayesian Analysis of Some Time Varying Models." *Recent Developments in Statistics,* J. R. Barra, et. al., Editors, North-Holland, New York.
[63] Swamy, P. A. V. B. and Mehta, J. S. (1975). "Bayesian and Non-Bayesian Analysis of Switching Regressions and of Random Coefficient Regression." *J. Amer. Statist. Assoc., 70,* 593-602.
[64] Tsurumi, H. (1977). "A Bayesian Tests of a Parameter Shift and an Application." *Jour. of Econometrics, 6,* 371-380.
[65] VonMises, R. (1964). *Mathematical Theory of Probability and Statistics.* Academic Press, New York.
[66] Wichern, D. W., Miller, R. B. and Hsu, D. A. (1976). "Change of Variance in First-Order Autoregressive Time Series Models with An Application. *Appl. Statist., 25,* 25, 248-256.
[67] Zacks, S. and Yadin, M. (1978). "Adaptation of the Service Capacity in a Queueing System Which is Subjected to a Change in the Arrival Rate..." *Adv. Appl. Prob., 10,* 666-681.
[68] Zacks, S. and Barzily, Z. (1981). "Detecting a Shift in the Probability of Success in a Sequence of Bernoulli Trials." *J. Statist. Planning and Inf., 5,* 107-119.
[69] Zacks, S. (1981). *Parametric Statistical Inference: Basic Theory and Modern Approaches.* Pergamon Press, Oxford.
[70] Zacks, S. (1981). "The Probability Distribution and the Expected Value of a Stopping Variable Associated with One-Sided CUSUM Procedures..." *Commun. Statist., A, 10,* 2245-2258.

III. NONPARAMETRICS INCLUDING LARGE SAMPLE THEORY

LARGE DEVIATIONS OF THE MAXIMUM LIKELIHOOD ESTIMATE IN THE MARKOV CHAIN CASE

R. R. Bahadur[1]

Department of Statistics
The University of Chicago
Chicago, Illinois

I. INTRODUCTION

Let X be a finite set, say X = {1, 2, ..., k}, and let x_0, x_1, ... be a Markov chain with values in X. In order to effect certain simplications assume that the chain starts with a given point in X, say $x_0 \equiv 1$. Let $\theta = \{\theta_{ij}\}$ denote a one-step transition probability matrix for X, and let P_θ be the probability on X^∞ determined by the conditions $P_\theta(x_0 = 1) = 1$, $P_\theta(x_{n+1}=j \mid x_0 = i_0, x_1 = i_1, \ldots, x_n = i) = \theta_{ij}$. Let Θ be a given set of matrices θ and suppose that θ is restricted to Θ but otherwise unknown. It is assumed that, for each $\theta = \{\theta_{ij}\}$ in Θ, $\theta_{ij} > 0$ for all i and j.

Let Γ be a metric space of points γ, let $g\colon \Theta \to \Gamma$ be a continuous function, and suppose it is required to estimate g. Let $D(\gamma_1, \gamma_2 \mid \theta)$ be a real valued function defined for γ_1, γ_2 in Γ and θ in Θ such that, for fixed γ_2 and θ, D is a continuous function of γ_1. We may think of $D^2(\gamma_1, \gamma_2 \mid \theta)$ as the

[1] *Research supported by NSF grant MCS-8002217.*

ISBN 0-12-589320-5

loss when θ obtains, $g(\theta) = \gamma_2$, but g is estimated to be γ_1. For each positive integer n, let T_n be a function of $(x_0, \ldots, x_n)$ with values in Γ, to be thought of as an estimate of g, and let

$$\alpha_n(\varepsilon,\theta) = P_\theta(D(T_n, g(\theta)|\theta) > \varepsilon) \tag{1}$$

for $\varepsilon > 0$.

Let $\pi_i(\theta)$ denote the stationary probability of the state i when θ obtains. Some possible specifications of Γ, g, and D are the following. 1) $\Gamma = R^1$, $g(\theta) = \pi_i(\theta)$, or θ_{ij}, or θ_{ij}/θ_{im} for given i,j, and m, and $D(\gamma_1,\gamma_2|\theta) = \gamma_1-\gamma_2$, or $\gamma_2-\gamma_1$, or $|\gamma_1-\gamma_2|$, or $|\gamma_1-\gamma_2|w(\theta)$ where $w(\theta) > 0$. 2) $\Gamma = R^k$, $g(\theta) = (\pi_1(\theta),\ldots,\pi_k(\theta))$, and $D(\gamma_1,\gamma_2|\theta) = |\gamma_{1i}-\gamma_{2i}|$ for given i, or $D = \max_i|\gamma_{1i}-\gamma_{2i}|$. 3) $\Gamma = \Theta$, $g(\theta) = \theta$, and $D = |\gamma_{1ij}-\gamma_{2ij}|$ for given i and j, or $D = \max_{i,j}|\gamma_{1ij}-\gamma_{2ij}|$, or $D = d_\theta$ where d_θ is defined by (20) in the following section. For given Γ, g, and D, let $\Delta(\varepsilon,\theta) = \{\delta: \delta \text{ in } \Theta, D(g(\delta), g(\theta)|\theta) > \varepsilon\}$, and let

$$b(\varepsilon,\theta) = \inf\{K(\delta,\theta): \delta \text{ in } \Delta(\varepsilon,\theta)\} \tag{2}$$

if $\Delta(\varepsilon,\theta)$ is nonempty and $b = +\infty$ otherwise, where

$$K(\delta,\theta) = \sum_{i=1}^{k} \pi_i(\delta) \cdot \sum_{j=1}^{k} \delta_{ij} \log(\delta_{ij}/\theta_{ij}) \tag{3}$$

for δ and θ in Θ. For fixed θ in Θ, $0 \le K \le \max\{\log \theta_{ij}^{-1}: 1 \le i, j \le \infty\}$ for all δ, with $K = 0$ only for $\delta = \theta$. It is plain that $b \ge 0$, and that b is nonincreasing as ε decreases. In many cases

$$b(\varepsilon,\theta) = \frac{\varepsilon^2}{2} c(\theta) + O(\varepsilon^3) \text{ as } \varepsilon \to 0 \tag{4}$$

where $0 < c(\theta) < \infty$, but it will suffice to assume here that

$$0 < b(\varepsilon,\theta) \le \infty, \qquad \lim_{\varepsilon\to 0} b(\varepsilon,\theta) = 0. \tag{5}$$

Suppose now that T_n is a consistent estimate of g, i.e. $T_n \to g(\theta)$ in P_θ-probability for each θ in Θ. It is known that

$$\liminf_{n\to\infty} n^{-1} \log \alpha_n(\varepsilon,\theta) \ge -b(\varepsilon,\theta) \tag{6}$$

for all ε > 0 and θ in Θ. Since this holds for any choice of T_n, we shall say that a particular consistent T_n is optimal at θ if $n^{-1} \log \alpha_n(\varepsilon,\theta) \to -b(\varepsilon,\theta)$ for each sufficiently small ε; we shall say that T_n is locally optimal at θ if, with r_n defined for all sufficiently small ε by

$$n^{-1} \log \alpha_n(\varepsilon,\theta) = -b(\varepsilon,\theta) \cdot [1-r_n(\varepsilon,\theta)] , \tag{7}$$

we have

$$\lim_{\varepsilon\to 0} \limsup_{n\to\infty} |r_n(\varepsilon,\theta)| = 0. \tag{8}$$

Local optimality means, of course, that $\alpha_n \to 0$ exponentially fast as $n \to \infty$, and that the rate is nearly the optimal rate exp(-nb) if ε > 0 is very small.

The inequality (6) and consequent definitions of optimality are a specialization to the present case of some of the general considerations of Bahadur, Gupta, and Zabell (1980); see also Bahadur (1960, 1967, 1971), Fu (1973, 1975, 1982), Kester (1981), Rubin and Rukhin (1983), Rukhin (1983), Sievers (1978), and Wijsman (1971). In cases where g is real valued, $D = |\gamma_1-\gamma_2|$, and (4) holds, local optimality is equivalent to optimality in the sense of asymptotic effective variances as defined in Bahadur (1960).

It is shown in the following section, under certain general assumptions on Θ, that for any admissible specification of Γ, g, and D the maximum likelihood (ML) estimate of g is locally optimal at each θ. The proof is based on the facts that the ML estimate of g is a substitution estimate, i.e., it is of the form $g(U_n)$ where U_n is the ML estimate of θ itself, and that for substitution estimates the large-deviation probability α_n is bounded from above by a large-deviation probability which is often easier to estimate. This approach is available in general, and a version of it has been used previously by Kester (1981) in the case treated there. A different approach to proving locally optimality of ML estimates in the present Markov chain case is outlined in Bahadur (1980).

Local optimality is a rather weak property, and the methods of the following section can be used to show that various estimates other than the ML estimate are also locally optimal. Optimality is, however, a much stronger property; in particular, the ML estimate can be optimal only in certain special cases. To see this suppose henceforth in this paragraph that, for each θ in Θ, the rows of θ are identical so $\{x_n: n \geq 1\}$ is an i.i.d. sequence. Suppose first that Θ is a finite set, say $\Theta = \{\theta_1,\ldots,\theta_m\}$ with $m \geq 2$, that $g(\theta_r) = r$, and $D(r_1,r_2|\theta) = |r_1-r_2|$. In this case, according to Rukhin (1983), there exists no T_n which is optimal at each θ; here the second condition in (5) is not satisfied, but perhaps there are examples where even (4) holds but the present notion of optimality at each θ remains vacuous. Suppose next that Θ is one-dimensional, g is a 1-1 smooth function on Θ onto an open interval of the line, and $\alpha_n(\varepsilon,\theta) = P_\theta(|T_n-g(\theta)| > \varepsilon)$. Suppose further that, with T_n the ML estimate of g, $n^{-1} \log \alpha_n(\varepsilon,\theta)$ converges to a limit, say $\beta(\varepsilon,\theta)$, and that both β and b have expansions in powers of ε. In this case, results of Fu (1982) imply that although T_n enjoys a property stronger than local optimality, a necessary condition for optimality of T_n at θ is that the statistical curvature at θ be zero. Suppose finally that $\{x_n: n \geq 1\}$ is an i.i.d. sequence and that the common distribution of the x_n over $\{1, \ldots, k\}$ is positive but otherwise entirely unknown. It then follows from results of Kester (1981) that, for any admissible g, the ML estimate of g is optimal at each θ.

II. PROOF

We begin with some remarks about substitution estimates in general. For each n and $s_n = (x_0,\ldots,x_n)$ let U_n be a function of s_n with values in Θ, and suppose U_n is a consistent estimate of θ. Let

$$T_n = g(U_n). \tag{9}$$

Since $g: \Theta \to \Gamma$ is continuous, this T_n is a consistent estimate of g. It follows from (9) that $D(T_n, g(\theta)|\theta) > \varepsilon$ implies that U_n is in the set $\Delta(\varepsilon,\theta)$ and hence $K(U_n,\theta) \geq b(\varepsilon,\theta)$ by (2). Thus we have

$$\alpha_n(\varepsilon,\theta) \leq P_\theta(K(U_n,\theta) \geq t) \text{ with } t = b(\varepsilon,\theta) \tag{10}$$

by (1).

Suppose that

$$\limsup_{n \to \infty} n^{-1} \log P_\theta(K(U_n,\theta) \geq t) \leq -t \tag{11}$$

for all sufficiently small $t > 0$. It then follows from (5), (6), and (10) that $n^{-1} \log \alpha_n(\varepsilon,\theta) \to -b(\varepsilon,\theta)$ for each sufficiently small ε, so T_n is optimal at θ. The sufficient condition (11) is also necessary for optimality in the special case $g(\theta) \equiv \theta$, $D(\gamma_1,\gamma_2|\theta) \equiv K(\gamma_1,\gamma_2)$; the simplifying inequality (10) is therefore entirely satisfactory for present purposes.

Suppose for the moment that for given θ in Θ it is required to test the simple hypothesis that θ obtains against the composite alternative that some δ in $\Theta - \{\theta\}$ obtains. If U_n is strongly consistent, and if (11) holds for all t, then $K(U_n,\theta)$ is an optimal test statistic in the sense of exact slopes (see Bahadur and Raghavachari (1972)). Thus, as might be expected, optimality of the point estimate $T_n = g(U_n)$ when θ obtains is related to optimality of the test statistic $K(U_n, \theta)$ for testing θ; see also the proof of (6) in Bahadur, Gupta, and Zabell (1980).

Suppose now that, instead of (11), the weaker condition

$$\limsup_{t \downarrow 0} \limsup_{n \to \infty} (nt)^{-1} \log P_\theta(K(U_n,\theta) \geq t) \leq -1 \quad (12)$$

is satisfied. It follows easily from (5), (7), (10), and (12) that

$$\limsup_{\varepsilon \to 0} \limsup_{n \to \infty} \{r_n(\varepsilon,\theta)\} \leq 0. \quad (13)$$

On the other hand (5), (6), and (7) imply that

$$\liminf_{\varepsilon \to 0} \liminf_{n \to \infty} \{r_n(\varepsilon,\theta)\} \geq 0. \quad (14)$$

It follows from (13) and (14) and (8) holds, so T_n is locally optimal at θ.

It is thus seen that, whatever Γ, g, and D may be, to verify that a particular T_n of the form (9) is locally optimal it suffices to show that U_n is consistent and that U_n satisfies (12). The remainder of this section is devoted to these demonstrations for the case when $U_n = \hat{\theta}_n$ = the ML estimate of θ based on $s_n = (x_0, \ldots, x_n)$. It should perhaps be added here that proof of consistency is an intermediate step; if (12) holds for each θ in Θ then U_n is necessarily consistent in a very strong sense.

It is assumed henceforth that the following assumptions (i), (ii), and (iii) hold for the given set Θ. Let Λ denote the set of all $k \times k$ transition probability matrices, and Λ^+ the set of all λ in Λ with positive elements. Assumption (i): Θ is a nonempty subset of Λ^+. An important consequence of this assumption is that the stationary probabilities $\pi_i(\theta)$ are well defined, continuous and positive functions on Θ. Let $\overline{\Theta}$ be the closure of Θ. Assumption (ii): $\overline{\Theta}-\Theta$ is a (possibly empty) closed set. Assumptions (i) and (ii) suffice for the asymptotic existence and consistency of the ML estimate of θ. Let (λ,δ) be an inner product on the space of $k \times k$ matrices with real elements, and let $d(\lambda,\delta) = (\lambda-\delta,\lambda-\delta)^{\frac{1}{2}}$ be the corresponding distance function. For any λ in Λ, let $m(\lambda)$ be the distance from λ to Θ, i.e., $m(\lambda) = \inf\{d(\lambda,\delta): \delta \text{ in } \Theta\}$. Choose a θ in Θ. Assumption (iii): Given $\varepsilon_1 > 0$ there exist $r = r(\varepsilon_1,\theta) > 0$ and $\varepsilon_2 = \varepsilon_2(\varepsilon_1,\theta) > 0$ such that if λ in Λ and $\hat{\theta}$ in Θ satisfy

$$d(\lambda,\theta) \le r, \quad d(\lambda,\hat{\theta}) \le (1+\varepsilon_2) \cdot m(\lambda) \tag{15}$$

then $d(\hat{\theta},\theta) \le (1+\varepsilon_1) \cdot d(\lambda,\theta)$.

Assumption (iii) holds trivially if Θ is a finite set, or if $\Theta = \Lambda^+$. It is easily seen that the assumption holds also in the intermediate case when Θ is an open subset of an affine subspace of the space of $k \times k$ matrices; here it suffices if $(1+\varepsilon_2)^2 = 1+\varepsilon_1^2$, and $r = r(\theta)$ is sufficiently small. Presumably assumption (iii) holds in the general intermediate case provided only that Θ is a sufficiently smooth surface but this is not established here.

For given $n \ge 1$ and $s_n = (x_0,\ldots,x_n)$ let $\{f_{ij}^{(n)}\}$ be the transition count matrix, i.e., $f_{ij}^{(n)}$ = the number of indices m with $0 \le m \le n-1$ such that $x_m = i$ and $x_{m+1} = j$ for $i, j = 1, \ldots, k$. The probability of observing an s_n with $x_0 = 1$ when some δ in Λ obtains say $\ell_n(\delta)$, is

$$\ell_n(\delta) = \prod_{i,j} [\delta_{ij}]^{f_{ij}^{(n)}}, \qquad 0 \le \ell_n \le 1 . \tag{16}$$

Let $\ell_n(\Theta) = \sup\{\ell_n(\theta): \theta \text{ in } \Theta\}$, and $\Theta_n^* = \{\delta: \delta \text{ in } \bar{\Theta}, \ell_n(\delta) = \ell_n(\Theta)\}$. Since ℓ_n is continuous on Λ, and $\bar{\Theta}$ is compact, $\ell_n(\Theta) = \ell_n(\bar{\Theta})$ and Θ_n^* is nonempty; Θ_n^* is the set of all extended ML estimates based on s_n. Let $\hat{\Theta}_n = \Theta_n^* \cap \Theta$ be the (possibly empty) set of all ML estimates based on s_n. For each n choose and fix a function $\hat{\theta}_n$ of s_n with values in Θ such that $\hat{\theta}_n$ is in $\hat{\Theta}_n$ whenever the latter set is nonempty. Cf. Bahadur (1960).

Now choose and fix a θ in Θ and suppose henceforth that θ obtains. Given a sequence $\{A_n: n = 1, 2, ...\}$ of events A_n, let us say that A_n is negligible if there exists a $\rho = \rho(\theta)$, $0 < \rho < 1$, such that $P_\theta(A_n) < \rho^n$ for all sufficiently large n. Given a sequence $\{Y_n: n = 1, 2, ...\}$ of extended real valued random variables, we shall say that Y_n is negligible if the event $|Y_n| > \varepsilon$ is negligible for each $\varepsilon > 0$; we then write $Y_n \doteq 0[\theta]$. It is readily seen that finite unions of negligible events are negligible; sums and maxima of a finite number of negligible random variables are negligible; etc. As a consequence of assumption (i) we have

$$n^{-1} f_{ij}^{(n)} - \pi_i(\theta) \cdot \theta_{ij} \doteq 0[\theta] \tag{17}$$

for all i and j. Hence, with $\xi_i^{(n)} = n^{-1} \sum_{j=1}^{k} f_{ij}^{(n)}$,

$$\xi_i^{(n)} - \pi_i(\theta) \doteq 0[\theta] \tag{18}$$

for all i. One method of proof of these strong forms of the law of large numbers is to regard "$i \to j$" as a delayed recurrent event (see Feller (1968)), to express the event

$|n^{-1}f_{ij}^{(n)} - \pi_i\theta_{ij}| > \varepsilon$ in terms of waiting times for "i → j", and to use the Bernstein inequality; Chernoff's theorem yields the exact ρ. An alternative method for (17) and (18) is provided by results of Donsker and Varadhan (1975). The details of these suggested proofs are omitted. Let $\lambda_{ij}^{(n)} = n^{-1}f_{ij}^{(n)}/\xi_i^{(n)}$ if $\xi_i^{(n)} > 0$ and $\lambda_{ij}^{(n)} = 1/k$ if $\xi_i^{(n)} = 0$; then $\lambda^{(n)} = \{\lambda_{ij}^{(n)}\}$ is a version of the empirical transition matrix. It follows from (17) and (18) that

$$\lambda_{ij}^{(n)} - \theta_{ij} \doteq o[\theta] \tag{19}$$

for all i and j.

Let V denote the set of all $v = (v_1,\ldots,v_k)$ in R^k with $v_i \geq 0$ and $\sum_1^k v_i = 1$, and let V^+ be the set of all such v with $v_i > 0$ for each i. For u and v in V, let $||v-u|| = \max\{|v_i-u_i|: 1 \leq i \leq k\}$, and let $\phi(v,u) = \sum_1^k v_i \log (v_i/u_i)$; then $0 \leq \phi \leq \infty$. Here, and subsequently also, $0/0 = 1$ and $0 \log 0 = 0$. For u and v in V and y in V^+, let $\psi(v,u|y) = \frac{1}{2}\sum_1^k (v_i-u_i)^2/y_i$. It is easy to see that z if z is a point in V^+ then $\phi(v,u)/\psi(v,u|y)$ converges to 1 as u, v and y all converge to z. For any k×k matrix δ let δ_i denote the ith row of δ, and let the distance function $d_\theta \geq 0$ be defined by

$$d_\theta^2(\delta^{(1)},\delta^{(2)}) = \sum_{i=1}^{k} \pi_i(\theta) \cdot \psi(\delta_i^{(1)},\delta_i^{(2)}|\theta_i) . \tag{20}$$

On occasion we shall also use $||\delta^{(1)}-\delta^{(2)}|| = \max\{|\delta_{ij}^{(1)}-\delta_{ij}^{(2)}| : 1 \leq i, j \leq k\}$. Write the point $(\pi_1(\theta), \ldots, \pi_k(\theta))$ in V^+ as $\pi(\theta)$. It follows from (20) by the stated properties of ϕ and ψ that if u is a variable point in V, and $\delta^{(1)}$, $\delta^{(2)}$ are variable points in Λ, then

$$\frac{\sum_i u_i \cdot \phi(\delta_i^{(1)}, \delta_i^{(2)})}{d_\theta^2(\delta^{(1)}, \delta^{(2)})} \to 1 \tag{21}$$

as $||u-\pi(\theta)||$, $||\delta^{(1)}-\theta||$, and $||\delta^{(2)}-\theta||$ all tend to zero.

With the $\xi_i^{(n)}$ and $\lambda_{ij}^{(n)}$ defined as above, (16) can be written as

$$\ell_n(\delta) = h_n \cdot \exp[-n\, Y_n(\delta)] \tag{22}$$

where $h_n(s_n)$ is the probability of s_n when $\lambda^{(n)}$ obtains, $0 < h_n \leq 1$, and

$$Y_n(\delta) = \sum_{i=1}^{k} \xi_i^{(n)} \cdot \phi(\lambda_i^{(n)}, \delta_i), \quad 0 \leq Y_n \leq \infty. \tag{23}$$

It follows from (22) that Θ_n^* is the set of all δ in $\bar{\Theta}$ which minimize $Y_n(\delta)$ with δ restricted to $\bar{\Theta}$; let Y_n^* denote the minimum value. Since $Y_n(\theta)$ depends on s_n only through $\{f_{ij}^{(n)}\}$ it follows from (22) with δ replaced by θ, as in Section 6 of Bahadur and Raghavachari (1972), that

$$P_\theta(Y_n(\theta) \geq t) \leq (n+1)^{k^2} \exp(-nt) \tag{24}$$

for all n and $t > 0$. It is plain from (24) that if $K(\hat{\theta}_n, \theta)$ were identical with $Y_n(\theta)$ then (11) would hold for $\hat{\theta}_n$; we shall show in effect that $K(\hat{\theta}_n, \theta)$ and $Y_n(\theta)$ are asymptotically close in a sense which ensures that at least (12) is satisfied by $\hat{\theta}_n$.

It follows from (24) and $Y_n(\theta) \geq 0$ that $Y_n(\theta) \doteq 0[\theta]$. Choose i and choose q, $0 < q < \pi_i(\theta)$. Then for δ in Θ_n^*, $Y_n(\theta) \geq Y_n^* = Y_n(\delta) \geq \xi_i^{(n)} \phi(\lambda_i^{(n)}, \delta_i) \geq q\phi(\lambda_i^{(n)}, \delta_i) \geq 0$ if $\xi_i^{(n)} \geq q$, by (23). Since $\xi_i^{(n)} < q$ is negligible by (18), it follows that $q \sup\{\phi(\lambda_i^{(n)}, \delta_i) : \delta \text{ in } \Theta_n^*\} \doteq 0[\theta]$. Hence $\sup\{\phi(\lambda_i^{(n)}, \delta_i) : \delta \text{ in } \Theta_n^*\} \doteq 0[\theta]$; hence $\sup\{||\theta_i - \delta_i|| : \delta \text{ in } \Theta_n^*\} \doteq 0[\theta]$ by (19) and properties of the function ϕ. Since this holds for each i,

$$\sup\{||\delta-\theta|| \ : \ \delta \text{ in } \Theta_n^*\} \doteq o[\theta] \ . \tag{25}$$

It follows from assumption (ii) that δ in $\bar{\Theta}$ and $||\delta-\theta||$ sufficiently small implies that δ is in Θ. It follows hence from (25) that $\hat{\Theta}_n \neq \Theta_n^*$ is a negligible event; hence $\hat{\theta}_n$ not in Θ_n^* is negligible; hence

$$||\hat{\theta}_n-\theta|| \doteq o[\theta] \tag{26}$$

by (25). Thus $P_\theta(||\hat{\theta}_n-\theta|| > \varepsilon) \to 0$ exponentially fast as $n \to \infty$, for each $\varepsilon > 0$. Since θ in Θ is arbitrary, $\hat{\theta}_n$ is consistent.

In the following, $\eta^{(n)}$ denotes a well defined function of n, s_n, and θ such that $0 \leq \eta^{(n)} \leq \infty$ and such that $\eta^{(n)}-1 \doteq o[\theta]$. For example, if $\{\delta^{(n)}\}$ is a random sequence in Λ such that $||\delta^{(n)}-\theta|| \doteq o[\theta]$, it follows from (21) and (23) with $u = \xi^{(n)}$, $\delta^{(1)} = \lambda^{(n)}$ and $\delta^{(2)} = \delta^{(n)}$ by (18) and (19) that $Y_n(\delta^{(n)}) = \eta^{(n)} \cdot d_\theta^2(\lambda^{(n)},\delta^{(n)})$. In particular,

$$Y_n(\theta) = \eta_1^{(n)} \cdot \ d_\theta^2(\lambda^{(n)},\theta) \ , \tag{27}$$

and

$$Y_n(\hat{\theta}_n) = \eta_2^{(n)} \cdot d_\theta^2(\lambda^{(n)}, \hat{\theta}_n) \tag{28}$$

by (26). For each n, let $\mu^{(n)} = \mu^{(n)}(s_n,\theta)$ be a point in $\bar{\Theta}$ such that $d_\theta(\lambda^{(n)},\mu^{(n)})$ is the d_θ-distance from $\lambda^{(n)}$ to Θ. Since $d_\theta(\lambda^{(n)},\mu^{(n)}) \leq d_\theta(\lambda^{(n)},\theta)$ it follows from (19) and (20) that $||\mu^{(n)}-\theta|| \doteq o[\theta]$; hence, as above,

$$Y_n(\mu^{(n)}) = \eta_3^{(n)} \cdot d_\theta^2(\lambda^{(n)},\mu^{(n)}) \ . \tag{29}$$

We have seen in the proof of (26) that $\hat{\theta}_n$ not in Θ_n^* is negligible; hence, by the remark following (23), $Y_n(\mu^{(n)}) \geq Y_n(\hat{\theta}_n)$ except for a negligible event. On the other hand, $d_\theta(\lambda^{(n)},\mu^{(n)}) \leq d_\theta(\lambda^{(n)},\hat{\theta}_n)$ by the definition of $\mu^{(n)}$. It therefore follows from (28) and (29) that

$$d_\theta^2(\lambda^{(n)},\hat{\theta}_n) = \eta_4^{(n)} \cdot d_\theta^2(\lambda^{(n)}, \mu^{(n)}) . \tag{30}$$

It follows from (3) and (26) by letting $u = \pi(\hat{\theta}_n)$, $\delta^{(1)} = \hat{\theta}_n$, and $\delta^{(2)} = \theta$ in (21) that

$$K(\hat{\theta}_n,\theta) = \eta_5^{(n)} \cdot d_\theta^2(\hat{\theta}_n,\theta) . \tag{31}$$

Choose and fix $\varepsilon > 0$. Choose $\varepsilon_1 > 0$ so that $(1 + \varepsilon_1)^4 < 1 + \varepsilon$, and let r and ε_2 be positive constants such that the condition stated in assumption (iii) is satisfied for $d = d_\theta$. It follows from (27) that $d_\theta^2(\lambda^{(n)},\theta) > (1+\varepsilon_1) \cdot Y_n(\theta)$ is a negligible event. It follows from (19) and (30) by the definition of $\mu^{(n)}$ that, except for a negligible event, (15) holds with $d = d_\theta$, $\lambda = \lambda^{(n)}$, and $\hat{\theta} = \hat{\theta}_n$; hence $d_\theta^2(\hat{\theta}_n,\theta) > (1+\varepsilon_1)^2 \cdot d_\theta^2(\lambda^{(n)}, \theta)$ is negligible. Since $K(\hat{\theta}_n,\theta) > (1+\varepsilon_1) \cdot d_\theta^2(\hat{\theta}_n,\theta)$ is negligible by (31), it now follows from the choice of ε_1 that $K(\hat{\theta}_n,\theta) > (1+\varepsilon) \cdot Y_n(\theta)$ is a negligible event.

Let $\rho = \rho(\varepsilon,\theta)$ be such that $0 < \rho < 1$ and $P_\theta(K(\hat{\theta}_n,\theta) > (1+\varepsilon) \cdot Y_n(\theta)) < \rho^n$ for all sufficiently large n, say for $n \geq n_1$. With $U_n \equiv \hat{\theta}_n$ it then follows that, for $t > 0$,

$$P_\theta(K(U_n,\theta) \geq t) \leq P_\theta(Y_n(\theta) \geq (1+\varepsilon)^{-1}t) + \rho^n \tag{32}$$

for $n \geq n_1$. It follows from (24) and (32) that the left-hand side of (11) does not exceed $\max\{-(1+\varepsilon)^{-1}t, \log \rho\}$. Since this holds for any t, the left-hand side of (12) does not exceed $-(1+\varepsilon)^{-1}$. Since ε is arbitrary, (12) holds and this completes the proof.

The empirical transition matrix $\lambda^{(n)}$ is an ML estimate if the transition probabilities are entirely unknown, i.e., if Λ is the parameter space. The role of assumption (iii) in the preceding proof is to ensure that if the actual parameter space Θ is a proper subset of Λ, and if some θ in Θ obtains then $\hat{\theta}_n$ is at least as close to θ as $\lambda^{(n)}$.

REFERENCES

Bahadur, R. R. (1960). "Asymptotic Efficiency of Tests and Estimates," *Sankhyā, 22,* 229-252.

Bahadur, R. R. (1967). "Rates of Convergence of Estimates and Test Statistics," *Ann. Math. Statist., 38,* 303-324.

Bahadur, R. R. (1971). *Some Limit Theorems in Statistics.* SIAM, Philadelphia.

Bahadur, R. R., and Raghavachari, M. (1972). "Some Asymptotic Properties of Likelihood Ratios on General Sample Spaces," *Proc. Sixth Berkeley Symp. Math. Statist. Prob., I,* 129-152.

Bahadur, R. R., Gupta, J. C., and Zabell, S. L. (1980). "Large Deviations, Tests, and Estimates," *Asymptotic Theory of Tests and Estimation,* Hoeffding Festschrift (I. M. Chakravarti, ed.), Academic Press, 33-64.

Bahadur, R. R. (1980). "A Note on the Effective Variance of Randomly Stopped Means," *Statistics and Probability: Essays in Honor of C. R. Rao* (Kallianpur, Krishnaiah, and Ghosh, eds.), North-Holland, 39-43.

Donsker, M. D., and Varadhan, S. R. S. (1975). "Asymptotic Evaluation of Certain Markov Process Expectations for Large Time - I," *Comm. Pure Appl. Math., 27,* 1-47.

Feller, William (1968). *An Introduction to Probability and Its Applications, I,* 3rd Edition. Wiley, New York.

Fu, J. C. (1973). "On a Theorem of Bahadur on the Rate of Convergence of Point Estimators," *Ann. Statist., 1,* 747-749.

Fu, J. C. (1975). "The Rate of Convergence of Consistent Point Estimators," *Ann. Statist., 3,* 234-240.

Fu, J. C. (1982). "Large Sample Point Estimation: A Large Deviation Theory Approach," *Ann. Statist., 10,* 762-771.

Kester, A. D. M. (1981). "Large Deviation Optimality of MLE's in Exponential Families," Report 160, Wiskundig Seminarium, Vrije Universiteit, Amsterdam.

Rubin, H., and Rukhin, A. L. (1983). "Convergence Rates of Large Deviations Probabilities for Point Estimators," *Statistics and Probability Letters.* (To appear).

Rukhin, A. L. (1983). "Convergence Rates of Estimators of a Finite Parameter: How Small Can Error Probabilities Be?" *Ann. Statist.* (To appear).

Sievers, G. L. (1978). "Estimates of Location: A Large Deviation Comparison," *Ann. Statist.*, *6*, 610-618.

Wijsman, R. A. (1971). "Lecture Notes on Estimation," Department of Mathematics, University of Illinois, Urbana. (Mimeographed).

BAYESIAN DENSITY ESTIMATION BY MIXTURES OF NORMAL DISTRIBUTIONS[1]

Thomas S. Ferguson

Department of Mathematics
University of California
Los Angeles, California

I. INTRODUCTION

This paper is concerned with the estimation of an arbitrary density f(x) on the real line. We model this density as a mixture of a countable number of normal distributions in the form

$$f(x) = \Sigma_1^\infty \; p_i h(x|\mu_i,\sigma_i), \tag{1}$$

where $h(x|\mu,\sigma)$ is the density of $N(\mu,\sigma^2)$, the normal distribution with mean μ and variance σ^2. There are a countably infinite number of parameters of the model, $(p_1,p_2,\ldots,\mu_1,\mu_2,\ldots,\sigma_1,\sigma_2,\ldots)$. Using such mixtures, any distribution on the real line can be approximated to within any preassigned accuracy in the Lévy metric, and any density on the real line can be approximated similarly in the L_1 norm. Thus the problem may be considered nonparametric.

Let $x_1,\ldots,x_n$ represent a sample of size n from f(x). Consider the problem of estimating f(x) at some fixed x or of estimating some functional of f(x) such as the mean, $\int xf(x)dx$,

[1] *This research was partially supported by the National Science Foundation under Grant MCS77-2121.*

ISBN 0-12-589320-5

using squared error loss based on $x_1,\ldots,x_n$. We consider a Bayesian approach to this problem. This entails placing a joint distribution on the parameters of the model and attempting to evaluate the posterior expectation of f(x) or $\int xf(x)dx$ given the sample.

There are several advantages of such an approach. *1. Use of prior information.* It gives the statistician a formal method of combining some of his prior information with the data. *2. Consistency.* The argument of Doob (1948) shows that these estimates should be consistent for *almost all* f chosen by the prior. However, a direct consistency result seems difficult to obtain and rates of convergence look even more difficult, even though direct consistency and rates of convergence are easy to obtain for other methods of density estimation such as kernel estimation. *3. Automatic adaptation.* Asymptotic theory for kernel estimators involves problems of letting the window size tend to zero at some rate as the sample size tends to infinity. Such problems are automatically taken care of in the Bayesian framework. In particular, larger windows for more remote observations are seen to occur naturally. *4. Small sample optimality.* Classical methods have only a large sample justification and look rather ad hoc if the sample is small. However, Bayes estimates with squared error loss are generally admissible.

We are not interested in estimating the parameters of the model, $p_1,p_2,\ldots,\mu_1,\mu_2,\ldots,\sigma_1,\sigma_2,\ldots$. It would not make much sense to do so since the parameters are not identifiable. One may attempt to obtain identifiability by writing (1) in the form

$$f(x) = \int h(x \mid \mu,\sigma)dG(\mu,\sigma) \qquad (2)$$

where G is the probability measure on the half-plane $\{(\mu,\sigma):\sigma>0\}$ that gives mass p_i to the point (μ_i,σ_i), $i=1,2,\ldots$. It has been shown by Teicher (1960) that if G is restricted to the class of finite probability measures, then G is identifiable, but that if G is unrestricted then G is not identifiable. The identifiability of G when G is restricted to the class of countable probability measures is still an open question.

II. THE PRIOR

We describe the prior distribution of $(p_1,p_2,\ldots,\mu_1,\mu_2,\ldots,\sigma_1,\sigma_2,\ldots)$ as follows:

(a) $(p_1,p_2,\ldots)$ and $(\mu_1,\mu_2,\ldots,\sigma_1,\sigma_2,\ldots)$ are independent.

(b) Let $q_1,q_2,\ldots$ be i.i.d. q_i having the beta distribution, Be(M,1) (i.e. the common density of the q_i is $Mq^{M-1}I_{[0,1]}(q)$), and let $p_1 = 1-q_1$, $p_2 = q_1(1-q_2),\ldots,p_j = (\Pi_{i=1}^{j-1}q_i)(1-q_j),\ldots$.

(c) (μ_1,σ_1), $(\mu_2,\sigma_2),\ldots$ are i.i.d. with common distribution the usual gamma-normal conjugate prior for the two-parameter normal distribution, namely, the precision (reciprocal of variance) $\rho_i = 1/\sigma_i^2$ has the gamma distribution $G(\alpha,2/\beta)$ (i.e. the density of ρ_i is $[1/\Gamma(\alpha)](\beta/2)^\alpha e^{-\rho\beta/2}\rho^{\alpha-1}I_{(0,\infty)}(\rho)$) and given ρ_i,μ_i is distributed as the normal distribution with mean μ and precision $\rho_i\tau$.

There are five parameters of the prior, $M>0$, $\alpha>0$, $\beta>0$, μ, and $\tau>0$. Note that the distribution of $(p_1,p_2,\ldots)$ depends only on M, and that the distribution of $(\mu_1,\sigma_1,\mu_2,\sigma_2,\ldots)$ depends only on α, β, μ and τ.

The prior guess at f(x), denoted by $f_0(x)$, is the expectation of f(x) under the prior distribution.

$$f_0(x) = Ef(x) = \Sigma_1^\infty Ep_i\, Eh(x|\mu_i,\sigma_i) = Eh(x|\mu,\sigma)$$

$$= \frac{\Gamma(\alpha+\frac{1}{2})}{\Gamma(\alpha)\Gamma(\frac{1}{2})} \sqrt{\frac{\tau}{(\tau+1)\beta}}\left(1 + \frac{\tau}{(\tau+1)\beta}(x-\mu)^2\right)^{-(\alpha+\frac{1}{2})}$$

With the change of variable $y = \sqrt{\frac{2\alpha\tau}{(\tau+1)\beta}}(x-\mu)$, y has a t-distribution with 2α degrees of freedom (in the generalized sense since α need not be rational). The mean of $f_0(x)$ is μ and the variance is $(\tau+1)\beta/(2\tau(\alpha+1))$. Thus the prior distribution does not admit a prior guess that is not symmetric. A more general prior guess may be achieved through the use of mixtures, but the resulting formulas become more complicated and are not investigated in this paper.

Interpretation of M. There are two somewhat independent interpretations of M. The first concerns the *relative sizes of the probabilities*, p_i. A small value of M means there is a big difference in the p_i; generally, p_1 is large compared to p_2, p_2 is large compared to p_3,... etc. If M is large, there will be many small probabilities that tail off to zero slowly. As an aid to understanding this feature of M, Table I has been constructed which shows the expected values, standard deviations and the correlation of the two largest p_i.

The other interpretation of M is as *prior information*. A small M means that you don't trust your prior guess much (the estimate will be strongly influenced by the observations), and a large value of M means you do trust your prior guess (the estimate does not depend much on the observations.) In this regard, M is measured in units of sample size: M represents the number of observations for which you would be willing to trade your prior information.

Thus it appears that this prior cannot express the opinions of a persons who believes strongly that there is a big difference in the probabilities p_i. We will see below in (6) an estimate of f(x) in which the influence of M as prior information is at least partially removed.

Choice of prior parameters. The following considerations are an aid to choosing the parameters of the prior to express the opinions of the statistician.

Since $E\mu_i = \mu$, μ should be chosen as the statistician's prior guess at the center of mass. Since $\text{Var}\ \mu_i = E\sigma_i^2/\tau$, τ

should be chosen approximately as $E\sigma_i^2/\text{Var}\ \mu_i$. If the uncertainty in the values of the μ_i is greater than (equal to, less than) the average variance, then τ should be chosen less than (equal to, greater than) 1.

Since $E\rho_i = 2\alpha/\beta$ (or $E\sigma_i^2 = \beta/2(\alpha-1)$ for $\alpha>1$), choose $2\alpha/\beta \sim E\rho_i$, leaving one parameter, say β, to be chosen in a way that reflects how diffuse the ρ_i are thought to be. Since $\text{Var}\ \rho_i = 4\alpha/\beta^2$, choose β large if the ρ_i are expected to be close to $2\alpha/\beta$ and choose β small if the ρ_i, and hence the σ_i^2, are expected to be diffuse.

One should choose M to reflect the statistician's belief on the relative sizes of the probabilities. For this, moderate values of M, from .5 to 5 say, are appropriate; Table I will aid this choice.

TABLE I. Expectations, Standard Deviations and Correlation of the two Largest Probabilities for Various Values of M, based on 10,000 Monte Carlo Trials

M	EP_1	EP_2	$s.d.P_1$	$s.d.P_2$	Corr.
.1	.938	.058	.124	.115	-.984
.2	.881	.101	.158	.134	-.961
.5	.756	.172	.192	.137	-.890
1	.625	.210	.192	.112	-.758
2	.476	.213	.164	.080	-.487
5	.296	.170	.106	.049	-.031
10	.195	.125	.070	.032	.238
20	.122	.085	.040	.020	.383

III. CONNECTION WITH THE DIRICHLET PROCESS

The prior distribution of the parameters $(p_1, p_2, \ldots, \mu_1, \mu_2, \ldots, \sigma_1, \sigma_2, \ldots)$ has been chosen so that the distribution function, G, of (2) is a Dirichlet process with parameter $\alpha = MG_0$, where $G_0 = \mathcal{E}G$ is the conjugate prior for (μ, σ^2) for the normal distribution given in (c) of Section II. That this is so follows from the representation of the Dirichlet process given by Sethuraman and Tiwari (1981). Their representation of the Dirichlet process, G, with parameter MG_0 is of the form

$$G = \Sigma_1^\infty p_i \delta_{\theta_i}$$

where $(p_1, p_2, \ldots)$ and $(\theta_1, \theta_2, \ldots)$ are independent, the distribution of the p_i as in (b) of Section II, and $\theta_1, \theta_2, \ldots$ are i.i.d. G_0. (The following discussion is quite general; for the specific case introduced in Section II, θ_i represents (μ_i, σ_i), i=1,... .) This is similar to but simpler than the representation in Ferguson (1973) which describes the distribution of the p_i by their order statistics $p_{(1)} \geq p_{(2)} \geq \ldots$. The actual distribution of $p_{(1)}$ is difficult to exhibit and even $\mathcal{E}p_{(1)}$ seems difficult to obtain, as pointed out by J.F.C. Kingman (1975). However the representation of Sethuraman and Tiwari makes it easy to evaluate $\mathcal{E}p_{(1)}$, $\mathcal{E}p_{(2)}, \ldots$ etc. by Monte Carlo. The results of such a computation are found in Table I.

If $G \in \mathcal{D}(\alpha)$ with $\alpha = MG_0$, and if $X_1, \ldots, X_n$ is a sample from a distribution with density $f(x) = \int h(x|\theta)dG(\theta)$, then the posterior distribution of G given $X_1, \ldots, X_n$ has been found by Antoniak (1974) to be a mixture of Dirichlet processes

$$G|x_1, \ldots, x_n \in \int \ldots \int \mathcal{D}(\alpha + \Sigma_1^n \delta_{\theta_i}) dH(\theta_1, \ldots, \theta_n | x_1, \ldots, x_n). \tag{3}$$

One may consider the observations $X_1, \ldots, X_n$ to be chosen by first choosing $\theta_1, \ldots, \theta_n$ i.i.d. from $G(\theta)$, and then X_i from $h(x|\theta_i)$ i=1,...,n independently. With this interpretation

$H(\theta_1,\ldots,\theta_n|x_1,\ldots,x_n)$ is the posterior distribution of $\theta_1,\ldots,\theta_n$ given $x_1,\ldots,x_n$. Using (3), the posterior expectation of $G(\theta)$ given $x_1,\ldots,x_n$ may be written as follows. Let $\hat{G}_n$ denote the empirical distribution of $\theta_1,\ldots,\theta_n$, $\hat{G}_n = \frac{1}{n}\Sigma_1^n \delta_{\theta_i}$. Then, since the expectation of $\mathcal{D}(\alpha+\Sigma_1^n \delta_{\theta_i})$ is $(MG_0+n\,\hat{G}_n)/(M+n)$,

$$E(G\ \theta|x_1,\ldots,x_n) = \frac{M}{M+n} G_0(\theta) + \frac{n}{M+n} \int\ldots\int \hat{G}_n(\theta) dH(\theta_1,\ldots,\theta_n| x_1,\ldots,x_n) \tag{4}$$

and

$$f_n(x) = E(f(x)|x_1,\ldots,x_n) = \frac{M}{M+n} f_0(x) + \frac{n}{M+n} \hat{f}_n(x) \tag{5}$$

where

$$\hat{f}_n(x) = \frac{1}{n} \Sigma_1^n \int\ldots\int h(x\ |\theta_i) dH(\theta_1,\ldots,\theta_n|x_1,\ldots,x_n). \tag{6}$$

The estimate (6) may be considered as a partially Bayesian estimate of f(x) with the influence of the prior guess at f(x) removed.

This approach to density estimation including the special case of normal-gamma shape for G_0, has been treated by Lo (1978). Lo has found the following useful representation of the function H.

$$dH(\theta_1,\ldots,\theta_n|x_1,\ldots,x_n) = \frac{(\Pi_1^n h(x_i|\theta_i))\Pi_{i=1}^n d(MG_0+\Sigma_{j=1}^{i-1}\delta_{\theta_j})(\theta_i)}{M^{(n)}\ h(x_1,\ldots,x_n)} \tag{7}$$

where

$$h(x_1,\ldots,x_n) = \int\ldots\int (\Pi_{i=1}^n h(x_i|\theta_i))\Pi_{i=1}^n d(MG_0 + \Sigma_{j=1}^{i-1}\delta_{\theta_j})(\theta_i)/M^{(n)}. \tag{8}$$

With this notation we may write (5) as

$$E(f(x)|x_1,\ldots,x_n) = h(x,x_1,\ldots,x_n)/h(x_1,\ldots,x_n). \tag{9}$$

The two extreme cases of these estimates as $M \to 0$ and $M \to \infty$ are worth noting. As $M \to 0$, it becomes more and more likely that all the θ_i are equal, resulting in the estimate

$$\lim_{M\to 0} E(f(x)\,|\,x_1,\ldots,x_n) = \frac{\int h(x|\theta)\Pi_1^n h(x_i|\theta)\,dG_0(\theta)}{\int \Pi_1^n h(x_i|\theta)\,dG_0(\theta)} . \tag{10}$$

This is the parametric estimate of f(x) when θ is chosen from $G_0(\theta)$ and $x_1,\ldots,x_n$ are chosen i.i.d. from $h(x|\theta)$. As $M \to \infty$, the estimate (5) converges to $f_0(x)$ which is not very useful. However, the estimate (6) which is less dependent on M as a measure of prior information also converges. Since as $M \to \infty$ it becomes more and more likely that all θ_i are distinct,

$$\lim_{M\to\infty} \hat{f}_n(x) = \frac{1}{n}\,\Sigma_1^n f(x|x_i), \tag{11}$$

where

$$f(x|x_i) = \int h(x|\theta)h(x_i|\theta)\,dG_0(\theta)/\int h(x_i|\theta)\,dG_0(\theta)$$

is the Bayes estimate of the density $h(x|\theta)$ based on a single observation, x_i, from $h(x|\theta)$ when θ has prior distribution $G_0(\theta)$. This is a variable kernel estimate.

If the density $h(x|\theta)$ were $N(\theta,\tau^2)$ with known τ^2, and if θ were $N(\mu,\sigma^2)$ with known μ and σ^2, then $f(x|x_i)$ is

$$N\left(\frac{\tau^2\mu+\sigma^2 x_i}{\tau^2+\sigma^2}, \frac{\tau^2+2\sigma^2}{\tau^2+\sigma^2}\right).$$

This yields a variable kernel estimate with constant window size, but centered at a point between x_i and μ, as is typical of shrinkage estimates.

For the problem treated in this paper, θ represents (μ,σ^2), which has a normal-gamma prior, while $h(x|\theta)$ is $N(\mu,\sigma^2)$. In this case, $f(x|x_i)$ is the density of a $t_{2\alpha+1}$-distribution centered at

$\frac{\tau\mu+x_i}{\tau+1}$, with scale $((\tau+2)(\beta+\frac{\tau}{\tau+1}(x_i-\mu)^2)/((2\alpha+1)(\tau+1)))^{\frac{1}{2}}$.

In addition to the shrinkage phenomenon, the window size depends on the observations with larger windows for observations x_i farther from μ. Unfortunately, one has lost, through this interchange of limits as $M \to \infty$ and $n \to \infty$, the valuable Bayesian property of not having to worry about the window size as a function of n. However, one may hope to let α and τ depend on n, $\alpha_n \to \infty$ and $\tau_n \to 0$, to obtain good asymptotic properties of the estimate (11).

IV. COMPUTATIONS

Since computation of $E(f(x)|x_1,\ldots,x_n)$ depends on computing the ratio (9), let us concentrate on the denominator, $h(x_1,\ldots,x_n)$. For this we use the analogue of a Monte Carlo technique of Kuo (1980) developed for a Bayesian approach to the empirical Bayes decision problem.

First, we expand the product measure in Lo's representation.

$$\Pi_{i=1}^{n}(MG_0+\Sigma_{j=1}^{i-1}\delta_{\theta_j})(d\theta_i)/M^{(n)}$$

$$= G_0(d\theta_1)(\frac{M}{M+1}G_0(d\theta_2) + \frac{1}{M+1}\delta_{\theta_1}(d\theta_2))\ldots \tag{12}$$

$$(\frac{M}{M+N-1}G_0(d\theta_n) + \frac{1}{M+n-1}\delta_{\theta_1}(d\theta_n)+\ldots+\frac{1}{M+n-1}\delta_{\theta_{n-1}}(d\theta_n)).$$

When the product is expanded, there are n! terms, but some of them are equal. For example, a term that begins $G_0(d\theta_1)\delta_{\theta_1}(d\theta_2)\delta_{\theta_2}(d\theta_3)\ldots$ is the same as a term that begins $G_0(d\theta_1)\delta_{\theta_1}(d\theta_2)\delta_{\theta_1}(d\theta_3)\ldots$; in both, $\theta_1 = \theta_2 = \theta_3$. Each term of the expansion determines a partition, $Q = \{K_1,\ldots,K_m\}$, of $\{x_1,\ldots,x_n\}$ with the property that $\theta_1 = \theta_j$ in the term if and only if x_i and x_j are in the same set $K \in Q$. Thus,

$$h(x_1,\ldots,x_n) = \sum_Q P_M(Q)Z(Q) \tag{13}$$

where $P_M(Q)$ represents the probability that partition Q is selected, and

$$Z(Q) = \prod_{K\in Q} \int \prod_{x_i\in K} h(x_i|\theta)dG_0(\theta). \tag{14}$$

The Monte Carlo technique of Kuo entails sampling partitions, Q, at random according to $P_M(Q)$, and evaluating Z(Q). This is repeated N times with partitions $Q_1,\ldots,Q_N$, and the average $\overline{Z(Q)} = \Sigma_1^N Z(Q_i)/N$ is taken as the Monte Carlo estimate of $h(x_1,\ldots,x_n)$.

As the computations are being carried out, one may obtain an estimate of the standard error of the Monte Carlo estimate, namely

$$(\Sigma_1^N (Z(Q_i) - \overline{Z(Q)}\ )^2)\ /\{N(N-1)\}^{\frac{1}{2}}.$$

Though useful in most situations, it should be realized that this can be very misleading as an estimate of error, since typical situations arise in which the true variance is very large due to a few values of Z(Q) that are very large and have small probabilities of appearing in the sample. Error reduction techniques are discussed in the next section.

Kuo's method of choosing Q is as follows. Take any ordering of the x_i. Start a set of the partition with x_1. For k=1,... n-1, repeat the following operations: *Let* x_{k+1} *start a new set of the partition with probability* M/(M+k); *otherwise, with probability* 1/(M+k) *each, put* x_{k+1} *into the set containing* x_i *for* i=1,...k.

To adapt this method to compute the estimate, (6), we randomly choose a partition Q in this manner, and in addition to computing the value of the denominator, Z(Q), we also compute the value of the numerator, call if Y(Q), namely,

$$Y(Q) = Z(Q) \sum_{K \in Q} \frac{|K|}{n} \frac{\int f(x|\theta) \prod_k f(x_i|\theta)\, dG_0(\theta)}{\int \prod_K f(x_i|\theta)\, dG_0(\theta)} .$$

The Monte Carlo estimate of $\hat{f}(x)$ is then

$$\Sigma Y(Q_i)/\Sigma Z(Q_i)$$

and its standard error can be estimated using the usual asymp- formula for the variance of a ratio of means,

$$\mathrm{Var}(\overline{Y}/\overline{Z}) \sim \frac{1}{n\mu_z^2} \left(\sigma_y^2 - 2\sigma_{yz} \frac{\mu_y}{\mu_z} + \sigma_z^2 \frac{\mu_y^2}{\mu_z^2}\right).$$

To test the computational method, a specific case with n = 5 was chosen so that exact expectations could be made for comparison. The observed values are taken to be $x_1 = 1.0$, $x_2 = 1.1$, $x_3 = 1.9$, $x_4 = 2.3$, and $x_5 = 2.6$. Simple values of the parameters were chosen and a small Monte Carlo of size N = 10 was carried out. The results are found in Table II. The estimates seem very good for so small a value of N, but with a larger value of n it is expected that a larger value of N is also needed.

TABLE II. Density Estimate with Prior Parameters M = 1, α = 1, β = 1, μ = 2, and τ = .5, and with Observations 1, 1.1, 1.9, 2.3, and 2.6. Based on a Monte Carlo of size N = 10

x	Prior guess $f_0(x)$	Estimated $\hat{f}_n(x)$	Estimated $f_n(x)$	Estimated st. err. $\hat{f}_n(x)$	Exact $\hat{f}_n(x)$
0	.081	.055	.059	.002	.053
.5	.125	.128	.128	.005	.127
1.0	.188	.272	.258	.011	.270
1.5	.256	.434	.404	.009	.432
2.0	.289	.457	.429	.015	.467
2.5	.256	.315	.306	.009	.329
3.0	.188	.159	.163	.003	.159
3.5	.125	.069	.078	.003	.065
4.0	.081	.030	.039	.002	.027

Consider now an example that illustrates the way one goes about choosing the parameters to express prior beliefs or fit data. Taking the five data points of the previous example, suppose we expect one main peak and one smaller local maximum, the rest being quite small in comparison; then we choose $M = 1$ or $M = 2$ for definiteness. We expect the center of mass to be around 2 so we choose $\mu = 2$. (If we use data-aided choice we might choose $\mu = \bar{x} = 1.8$). We might set the standard deviation of the μ_i at $(2.3 - 1.1)/2 = .6$, and if we expect the σ_i to range from near 0 to .5 we might set $E\sigma_i^2 = .06$ giving a ratio of $\tau = .12/.36 = 1/3$, so $\tau = .1$ or $\tau = .5$ might do. We choose a small value of β, say $\beta = .5$, and solve $E\sigma_i^2 = \beta/(2\alpha-1)$ giving an α close to 5. Table III contains the exact values of the estimate (6) of the density for certain parameter values close to these, the first column being the one of choice. The density estimate is seen to have one large peak close to 2.25 and a smaller one close to 1.25.

TABLE III. Exact Values of the Estimate (6) for the Data Points 1, 1.1, 1.9, 2.3, 2.6. Parameter values $M = 1$, $\mu = 2$

	$\alpha = 5$, $\beta = .5$		$\alpha = 5$, $\beta = .1$		$\alpha = 1$, $\beta = .1$	
x	$\tau=.5$	$\tau=.1$	$\tau=.5$	$\tau=.1$	$\tau=.5$	$\tau=.1$
0	.002	.003	0	0	.022	.015
.25	.010	.013	.002	0	.042	.032
.50	.039	.063	.013	.005	.082	.082
.75	.136	.238	.085	.103	.160	.220
1.00	.337	.512	.374	.824	.282	.460
1.25	.509	.511	.663	.604	.400	.477
1.50	.488	.316	.394	.069	.452	.342
1.75	.464	.309	.307	.272	.511	.362
2.00	.607	.498	.671	.537	.619	.496
2.25	.694	.659	.831	.623	.614	.597
2.50	.475	.538	.575	.776	.418	.504
2.75	.181	.247	.074	.163	.198	.241
3.00	.046	.072	.005	.010	.085	.088
3.25	.010	.016	.001	.001	.039	.034
3.50	.002	.003	0	0	.019	.015
3.75	0	.001	0	0	.010	.007

V. IMPROVING THE MONTE CARLO

In some situations, it is possible to change the Monte Carlo sampling method to reduce the variance of the estimate. See Rubenstein (1981) for a review of such methods. In the problems treated here, there may be a partition Q with a large value of Z(Q) and a small value of $P_M(Q)$ that is rarely chosen in the Monte Carlo sampling even though it contributes significantly to the expectation. As an example, suppose there are 3 Q's with probabilities .01, .50, and .49 and values of Z(Q) 1000, 2, 1, respectively. The true mean is h = .01 (1000) + .50 (2) + .49 (1) = 11.49. With a Monte Carlo of size N = 10, there is a large variance; if the sample contains the value 1000, the estimate is greater than 100, and if it doesn't, the estimate is at most 2. If it is known which Z(Q) are expected to be large, one may change the probabilities and values; for example, we may write

$$h = .85\ (\frac{.01}{.85}\ 1000) + .10\ (\frac{.5}{.1}\ 2) + .05\ (\frac{.49}{.05}\ 1)$$

$$= .85\ (11.76) + .10\ (10) + .05\ (9.8) = 11.49\ .$$

The Monte Carlo of size N = 10, on values 11.76, 10, 9.8 with probabilities .85, .10, .05 respectively has a small variance.

The problem is to tell which Z(Q) are likely to be large and by how much. It is useful to consider two separate sources of variation in the Z(Q): the number of sets in the partition, and their within set variation.

Often the number of sets in the partition significantly affects Z(Q), the larger the number of sets the smaller the Z(Q). In the example with n = 5 observations, 1, 1.1, 1.9, 2,3, and 2.6, and paramter values M = 1, $\alpha = 1$, $\beta = 1$, $\mu = 0$, and $\tau = 1$, Z(Q) goes down roughly as $.5^m$ where $m = |Q|$. In the more reasonable case that $\mu = 2$, it goes down roughly as $.8^m$. This being so, we can improve the Monte Carlo as follows. *Choose the partition using Kuo's method but using a different value of* M *say* M'; *after* Z(Q) *has been evaluated multiply it by* $(M/M')^{m}M'^{(n)}/M^{(n)}$ *where* $m = |Q|$. In the cases mentioned M'/M should be chosen roughly as .5 or .8.

For a fixed number of sets in the partition, the value of Z(Q) is largest if the sets contain contiguous order statistics so that the within set variation is small. In the method of Kuo, the distribution of the assignment of the x_i to the sets of the partition is invariant under interchange of subscripts. Needed is a method of choosing "clumpy" partitions in which x_i's that are close in value have a higher probability of being in the same set. Here is one possibility involving rank-dependent grouping.

Order the x_i's, $x_1 < x_2 < \ldots < x_n$, choose a value of $t \geq 1$, and let x_1 start a set of the partition. For k=1,2,...,n-1, repeat the following operation. *Let* x_{k+1} *start a new set of the partition with probability* M/(M+k). *Otherwise, put* x_{k+1} *into the already created set* K *with probability proportional to* $\Sigma_{i\in K}\, t^i$. *The ratio of the old probability of* K *to the new probability is*

$$ratio = \frac{|K|}{k}\,\frac{\Sigma_1^k\, t^i}{\Sigma_{i\in K}\, t^i}\,.$$

Keep a running product of the ratios as you go along and multiply Z(Q) *by this product when finished.*

The old probabilities are those given by the method of Kuo which uses t = 1. One drawback of this method is that $Q_1 = \{\{1,2,3\}, \{4,5\}\}$ and $Q_2 = \{\{1,2\}, \{3,4,5\}\}$ have different probabilities even though they are symmetric. One can avoid this by randomizing with probability ½ ordering the x_i in ascending or descending order.

In some cases it may be preferable to use value-dependent grouping, in which x_{k+1} is assigned to a set K with probability proportional to some function of the values of the observations in K such as $\Sigma_{i\epsilon K} \exp\{-t|x_i - x_{k+1}|\}$ or $|K| \exp\{-t\Sigma_{i\epsilon K}(x_i - x_{k+1})^2/|K|\}$ for some $t \geq 0$.

It is easy to combine these two methods simultaneously changing M to M' and changing the probability of assignment to the already created sets K. Both methods have been tried on the 5-point data set used in the example with parameters similar to those mentioned. In both cases, a small reduction in the variance of the estimate (about 20%) was noted.

VI. SUMMARY

The problem of nonparametric density estimation is considered from a Bayesian viewpoint. The density is assumed to be a countable mixture of normal distributions with arbitrary means and variances, and a prior joint distribution is placed on the mixing probabilities and the means and variances. The resulting model is seen to be equivalent to Lo's model, and a method of Kuo is adapted to carry out the computations. A simple example is investigated to show the feasibility of the method.

REFERENCES

Antoniak, Charles E. (1974). "Mixtures of Dirichlet Processes with Application to Bayesian Nonparametric Problems." *Ann. Statist. 2*, 1152-1174.

Doob, J. (1948). "Application of the Theory of Martingales." *Le Calcul des Probabilités et ses Applications. Colloques Internationaux du Centre National de la Recherche Scientifique*, Paris, 23-28.

Ferguson, Thomas S. (1973). "A Bayesian Analysis of Some Non-parametric Problems." *Ann. Statist. 1*, 209-230.

Kingman, J.F.C. (1975). "Random Discrete Distributions." *J.R.S.S. Series B 37*, 1-15.

Kuo, Lynn (1980). Computations and Applications of Mixtures of Dirichlet Processes, Ph.D. thesis, UCLA.

Lo, Albert Y. (1978). On a Class of Bayesian Nonparametric Estimates: I. Density Estimates, Mimeographed preprint, Rutgers, University, New Brunswick, N.J.

Rubenstein, Reuven Y. (1981). *Simulation and the Monte Carlo Method,* John Wiley and Sons, New York.

Sethuraman J. and Tiwari, Ram C. (1982). Convergence of Dirichlet Measures and the Interpretations of their Parameter, *Statistical Decision Theory and Related Topics, III,* Editors, S.S. Gupta and J.O. Berger, Academic Press, New York.

Teicher, Henry (1960). "On the Mixture of Distributions." *Ann. Math. Statist. 31,* 55-73.

AN EXTENSION OF A THEOREM OF H. CHERNOFF AND E. L. LEHMANN

Lucien Le Cam[1]

Department of Statistics
University of California
Berkeley, California

Clare Mahan

Public Health Program
Tufts University
Medford, Massachusetts

Avinash Singh[2]

Department of Mathematics and Statistics
Memorial University of Newfoundland
St. John's, Newfoundland, Canada

I. INTRODUCTION

This is about certain quadratic forms, or related objects, that are often used for tests of goodness of fit. The most familiar of them is the usual K. Pearson's chi-square test of fit for distributions of identically distributed independent observations. We shall use it throughout as basic example, but the arguments given here apply to a large variety of other situations. For some of them, see [14], [15] and [16].

[1] *Research supported in part under grant ARO-DAAG29-C-0093.*
[2] *Research supported in part under grant NIEHS-2R01ES01299.*

ISBN 0-12-589320-5

The present paper gives only a general theory. We met the need for it in a very practical context that will be described elsewhere.

The usual Pearson chi-square evaluates the discrepancy between an observed multinomial vector O and its expectation E under a model θ. When the model specifies E entirely, the asymptotic distribution is that of a Gaussian χ^2 with a number of degrees of freedom k equal to the effective dimension of O. If E depends on parameters θ that need to be estimated, the usual recipe is to decrease the degrees of freedom k by a number r equal to the dimension of θ.

This was the recipe introduced by R. A. Fisher. It presumes that θ is estimated by a procedure equivalent to the minimization of the chi-square.

In a paper [2] published in 1954 Chernoff and Lehmann showed that, for the usual test of goodness of fit, if one estimates θ by a maximum likelihood (m.ℓ) procedure applied to the entire set of observations, instead of the grouped ones, the distributions need no longer be χ^2. Instead the resulting quadratic behaves as $\chi^2_{k-r} + \Sigma^r_{i=1} c_i u_i^2$ for independent terms with $\mathcal{L}(u_i) = \mathcal{N}(0,1)$ and $c_i \in [0,1]$.

It has been observed (see for instance Molinari [13]) that the Chernoff-Lehmann result does not remain valid for certain estimates other than m.ℓ.e.

Some authors, e.g., Djaparidze and Nikulin [5], R. M. Dudley [7], have proposed procedures that remove the extra terms $c_i u_i^2$. Other authors, and most particularly Rao and Robson [17] and D. S. Moore have proposed procedures which amount to computing the covariance matrix of the vector $O - E(\hat{\theta})$ and using its (estimated) inverse to produce a quadratic with asymptotic χ^2_k behavior.

In the process Rao and Robson show that, under their conditions, if $\hat{\theta}$ is the m.ℓ.e. the χ^2_k behavior of $\sum_i [O_i - E_i(\hat{\theta})]^2/E_i(\hat{\theta})$ can be restored by adding to it another quadratic expression.

In the present paper, we consider a general asymptotically Gaussian situation and exhibit conditions under which the behavior described by Chernoff and Lehmann can occur. Stronger conditions, involving an asymptotic sufficiency property of $\hat{\theta}$, yield formulas of the special Rao-Robson type. The particular (logistic) case, [12], that spurred our investigation presented some problems involving nearly singular covariance matrices. We describe families of test criteria that can be used under such circumstances.

The choice of the test criteria needs to be made keeping in mind the power of the tests. We show that any two distinct members of a family of criteria lead to risk functions that are eventually not comparable, but that, under suitable restrictions, all these tests are asymptotically admissible.

The pattern of the proofs is to use a reduction to the case where the observable vectors are exactly Gaussian, with specified covariance structure. Sections II and III deal with this Gaussian case. First we give conditions for the validity of the special Rao-Robson formula and weaker conditions for the validity of the Chernoff-Lehmann type of result. The strong conditions involve a sufficiency property for the estimates. Consequences of this sufficiency are described in Section III for the Gaussian case in Section IV for the asymptotically Gaussian situation. Section V deals with the asymptotic optimality properties of the tests. It relies on an assumption that implies convergence of experiments in the sense of [8] and contains an example showing that if such assumptions are not satisfied the optimality properties are valid only if one restricts the possible families of tests by regularity conditions.

II. TEST CRITERIA FOR THE GAUSSIAN CASE

The standard χ^2 test of goodness of fit formula $\Sigma[n_j - Np_j(\hat{\theta})]^2/Np_j(\hat{\theta})$ involves two vectors: a vector X^* with coordinates $\frac{1}{\sqrt{N}}[n_j - Np_j(\theta_0)]$ and a vector Z^* with coordinates $\sqrt{N}[p_j(\hat{\theta}) - p_j(\theta_0)]$, where θ_0 is the true value of the parameter. In this section and the next we shall consider an analogue Gaussian situation involving a pair vectors (X,Z). It will be assumed throughout that X takes its values in a finite dimensional vector space $\mathcal{X}$ and that Z takes its values in a finite dimensional vector space $\mathcal{Z}$. The expectation of X will be denoted ξ and that of Z will be ζ. Under the general hypothesis the joint distribution P of (X,Z) belongs to a class $\mathcal{D}_1$. The hypothesis to be tested is that P belongs to a subset $\mathcal{D}_0 \subset \mathcal{D}_1$. It will be assumed here that the following assumptions are satisfied.

(A1). For $P \in \mathcal{D}_0$ *the distribution of Z is Gaussian* $N(\zeta, B)$ *with an expectation* $\zeta = \zeta(P)$ *depending on P but a covariance matrix B independent of P. As P varies through* $\mathcal{D}_0$ *the expectation* $\zeta(P)$ *ranges through the entire space* $\mathcal{Z}$. *There is a* $P_0 \in \mathcal{D}_0$ *such that* $E_{P_0} X = E_{P_0} Z = 0$.

This assumption will occur again in Section III. In the present section we shall also make the following assumptions.

(B1). For every $P \in \mathcal{D}_1$ *the joint distribution of (X,Z) is Gaussian. Its covariance matrix* $\begin{pmatrix} A & C \\ C' & B \end{pmatrix}$ *is independent of P.*

(B2). As P varies through $\mathcal{D}_1$, *the expectations* $\begin{pmatrix} \xi \\ \zeta \end{pmatrix}$ *range through the entire space* $\mathcal{X} \times \mathcal{Z}$.

(B3). There is a linear map S from $\mathcal{Z}$ *to* $\mathcal{X}$ *such that if* $P \in \mathcal{D}_0$ *then* $\xi = S\zeta$.

(B4). The covariance matrix $\begin{pmatrix} A & C \\ C' & B \end{pmatrix}$ *is nonsingular.*

Condition (B4) calls for some comments. In practical situations, it is usually very easy to insure that both A and B are nonsingular. The same need not be true of $\begin{pmatrix} A & C \\ C' & B \end{pmatrix}$. Under (B1), the vector $\begin{pmatrix} X-\xi(P) \\ Z-\zeta(P) \end{pmatrix}$ is always in the range of that covariance matrix. This need not be true of $\begin{pmatrix} X-S\xi(P) \\ Z-\zeta(P) \end{pmatrix}$. We shall mention this fact occasionally, indicating precautions to be taken in such a case.

Introduce norms by $||x||^2 = x'A^{-1}x$ and $|z| = z'B^{-1}z$. Then some of the statistics used in connection with the Pearson Chi-square formula (and the more general situations of [15]) are as follows:

a) The K-Pearson formula: $Q_p = ||X - SZ||^2$.

b) The Pearson-Fisher formula: $Q_F = \inf\{||X-S\zeta||^2;\ \zeta \in Z\}$.

c) The general "chi-square" formula, (for which see Chiang [3], Rao and Robson [17] and Moore [15])

$$Q_M = (X - SZ)'\, M^{-1}(X - SZ)$$

where M is the covariance matrix of $X - SZ$.

(d) The quadratic in the Neyman-Pearson likelihood ratio test

$$Q_{LR} = \inf_{\zeta} \begin{pmatrix} X-S\zeta \\ Z-\zeta \end{pmatrix}' \begin{pmatrix} A & C \\ C' & B \end{pmatrix}^{-1} \begin{pmatrix} X-S\zeta \\ Z-\zeta \end{pmatrix}.$$

e) A special formula of Rao and Robson

$$Q_{RR} = Q_p + (X - SZ)'\, K(X - SZ),$$

where

$$K = R'[B - C'A^{-1}C]^{-1} R \quad \text{for} \quad R = C'A^{-1}.$$

The value $S\hat{\zeta}$ that achieves the minimum in the Pearson-Fisher formula (b) is the orthogonal projection ΠX of X on the range $\{S\zeta;\ \zeta \in Z\}$ using the inner product defined by the norm $||\cdot||$. Thus, by Pythagoras' theorem one may write $Q_p = Q_F + ||\Pi X - SZ||^2$. There is a formula of Djaparidze and Nikulin [5] in which one evaluates Q_p and $||\Pi X - SZ||^2$ separately and then recover Q_F by difference.

The formula (d) given here for Q_{LR} appears complex. It can be easily simplified. Let U = X - SZ and V = Z - TU for a linear map T computed so that U and V are uncorrelated. This map T is given by the relation C' - BS' = TM where M is the covariance matrix of U, as in (c).

In terms of U and V, the hypothesis $\mathcal{D}_0$ becomes $E_p U = 0$ and $E_p V = \zeta$. The quadratic form in the definition of Q_{LR} becomes $U'M^{-1}U + (V-\zeta)' M_2^{-1}(V-\zeta)$ for the covariance matrix M_2 of V. Thus $Q_{LR} = U'M^{-1}U$ is identical to the quadratic Q_M of the "general chi-square" in (c).

The minimum value Q_{LR} is achieved at a well determined value ζ^* such that

f) $\zeta^* = V = Z - T(X - SZ)$.

(Note that when M is singular but X - SZ is in the range of M, there is no difficulty in defining Q_M. If on the contrary X - SZ is *not* in the range of M one should take $Q_{LR} = \infty$. This is not the same thing as replacing M^{-1} by a generalized inverse.)

Note that, by construction, the inequalities $Q_F \le Q_p \le Q_{RR}$ always hold, and that, under assumptions (A) (B), Q_F and Q_M have χ^2-distributions with respective number of degrees of freedom rank (I-Π) and rank M.

All these formulas appear to be intended for the case where under $\mathcal{D}_1$ the expectations range through the entire product space $\mathcal{X} \times \mathcal{Z}$. In such a case Wald [20] showed that $Q_{LR} = Q_M$ possesses a number of optimality properties: It gives tests with best constant and best average power over certain cylinders around $H = S\mathcal{Z}$. It yields tests that are "most stringent". Of course if the possible range of $\binom{\xi}{\zeta}$ under $\mathcal{D}_1$ is not the entire $\mathcal{X} \times \mathcal{Z}$, the quadratic of the relevant likelihood ratio is not Q_{LR} but

$$Q_{LR} - \inf \begin{pmatrix} X-\xi \\ Z-\zeta \end{pmatrix}' \begin{pmatrix} A & C \\ C' & B \end{pmatrix}^{-1} \begin{pmatrix} X-\xi \\ Z-\zeta \end{pmatrix}$$

for an infimum taken over the range of $\binom{\xi}{\zeta}$. This explains some of the remarks about non-comparability of tests made in [15]. Further results on non-comparability will be found in Section 5.

In cases of the type described by Rao and Robson, one can obtain their special formula by straightforward matrix algebra. It is more instructive to proceed as follows.

Introduce a variable $W = Z - RX$ with $C' = RA$, so that X and W are uncorrelated.

Then, under $\mathcal{D}_0$ the quadratic form appearing in the normal densities, and used to define Q_{LR}, takes the form

$$(X - S\zeta)'A^{-1}(X - S\zeta) + (W - \phi)'[B - C A^{-1}C]^{-1}(W - \phi)$$

with $\phi = E_p W = (I - RS)\zeta$. If in this formula one replaces ζ by the estimate Z, one obtains

$$(X - SZ)'[A^{-1} + R'[B - C'A^{-1}C]^{-1}R](X - SZ),$$

which is the Rao-Robson statistic Q_{RR}.

This leads to the following observation.

Lemma 1

Let the conditions (A) *and* (B) *hold.* Then Q_{LR} *is identically equal to* Q_{RR} *if and only if* $C = SB$. *This is equivalent to the fact that the vector* Z *is sufficient for the family* $\{L[\binom{X}{Z}|P];\ P \in \mathcal{D}_0\}$.

Note. We shall prove only that $Q_{LR} = Q_{RR}$ is equivalent to $C = SB$ and that this implies the stated sufficiency property. The reverse implication will be proved under more general conditions in Section III.

Proof. The value ζ^* that yields the minimum Q_{LR} is well determined and equal to $Z - T(X - SZ)$. The difference $Z - \zeta^*$ has a covariance matrix $(C - SB)'M^{-1}(C - SB)$. This vanishes identically only if $C = SB$. For the sufficiency, write the exponent of the normal distribution in the alternate form $U'M^{-1}U + (V - \zeta)'M_2^{-1}(V - \zeta)$ used to compute Q_{LR} above. Take the logarithm of likelihood ratio $\log dP/dP_0$ for a measure $P \in \mathcal{D}_0$ with expectations $E_pX = S\zeta$ and $E_pZ = \zeta$. This has the form

$$\zeta'M_2^{-1}[Z - T(X - SZ)] - \frac{1}{2}\zeta'M_2^{-1}\zeta.$$

Here $T = (C - SB)'M^{-1} = 0$ and the logarithm of likelihood ratio is $\zeta'B^{-1}Z - \frac{1}{2}\zeta'B^{-1}\zeta$. The sufficiency follows.

The Chernoff-Lehmann paper of 1954 contains several observations. It starts with the Pythagorean formula $Q_p = Q_F + ||\Pi X - SZ||^2$ with $Q_F = ||X - \Pi X||^2$ and, although perhaps not quite in those terms, notes that, under the conditions used there:

a) the vectors $X - \Pi X$ and $\Pi X - SZ$ are independent,

b) the difference between $\Pi A \Pi'$ and the covariance matrix of $\Pi X - SZ$ is positive semidefinite.

In the present context, the natural definition of the matrix A is as a linear map from the dual $\mathcal{X}'$ of $\mathcal{X}$ to $\mathcal{X}$. Similarly C is a map from the dual $\mathcal{Z}'$ of $\mathcal{Z}$ to $\mathcal{X}$. With this understanding, the Chernoff-Lehmann observation leads to the following.

Lemma 2

Let the conditions (A1) and (B1) to (B4) *be satisfied. Then* $X - \Pi X$ *and* $\Pi X - SZ$ *are independent if and only if the range of* CS' *is contained in that of* S. *The condition is satisfied whenever* $Q_{LR} \equiv Q_{RR}$ *and in that case the difference* SBS' *between the covariance of* ΠX *and that of* $\Pi X - SZ$ *is positive semidefinite.*

Remark. When S is injective, the vectors $X - \Pi X$ and $\Pi X - SZ$ are independent if and only if the range of C is contained in that of S. Indeed in the injective case, the map S' is a map of $\mathcal{X}'$ *onto* $\mathcal{Z}'$.

Proof. By construction $(X - \Pi X)$ and ΠX are independent. Thus $(X - \Pi X)$ and $\Pi X - SZ$ are independent only if the covariance of $X - \Pi X$ and SZ vanishes. This means that $CS' = \Pi CS'$ and yields the first assertion. For the second, Lemma 1 says that Q_{LR} and Q_{RR} coincide if and only if $C = SB$. This clearly implies the condition on ranges. Also $E_0(\Pi X - SZ)(\Pi X - SZ)' = \Pi A \Pi' - SBS'$ and therefore $E_0(\Pi X)(\Pi X)' - E_0(\Pi X - SZ)(\Pi X - SZ)' = SBS'$, a positive semi-definite matrix.

It is perhaps worth noting that the independence of $(X - \Pi X)$ and $(\Pi X - SZ)$, even coupled with the condition that $\Pi A \Pi' - E_0(\Pi X - SZ)(\Pi X - SZ)'$ be positive semi-definite does not imply that $C = SB$. If indeed $C = SB$, the conditions remain satisfied if one replaces S by $(1 + \alpha)S$ for any α such that $|\alpha| \leq 1$.

The independence of $(X - \Pi X)$ and $(\Pi X - SZ)$ implies that the quadratic form Q_{LR} can also be written

$$Q_{LR} = ||X - \Pi X||^2 + (\Pi X - SZ)' M_1^{-1} (\Pi X - SZ)$$

where M_1 is the covariance matrix of $\Pi X - SZ$. Thus the likelihood ratio test differs from the Pearson-Fisher test only in that it takes into account the difference between the two estimates ΠX and SZ of $S\zeta$. This is also true of Q_p and of the various other statistics that will be described later in Section IV.

It turns out that the sufficiency of Z under $\mathcal{D}_0$ is rich in implications that can easily be translated into an asymptotic framework. Thus, we shall now devote some space to that situation.

III. GAUSSIAN THEORY UNDER THE NULL HYPOTHESIS

In this section we retain the assumption (A1) of Section II and add the following:

(A2). If $P \in \mathcal{D}_0$ *then* $X - \xi$ *is Gaussian,* $N(0,A)$ with A independent of P.

(A3). For the family $L[\binom{X}{Z}|P]$; $P \in \mathcal{D}_0$, *the vector* Z *is a sufficient statistic.*

Consider also the following properties.

(P1). For the particular measure P_0, the joint distribution of (X,Z) is Gaussian.

(P2). For all $P \in \mathcal{D}_0$, the joint distribution of (X,Z) is Gaussian, with a covariance independent of $P \in \mathcal{D}_0$.

(P3). There is a linear map S from $\mathcal{Z}$ to $\mathcal{X}$ such that $P \in \mathcal{D}_0$ implies $E_p X = S(E_p Z)$.

(P4). As P ranges through $\mathcal{D}_0$, the expectations $\binom{\xi}{\zeta}$ range through a linear subspace of $\mathcal{X} \times \mathcal{Z}$.

Lemma 3

Let the conditions (A1) *to* (A3) *hold. Then the four properties* (P1) to (P4) *are equivalent. If one of them holds the vectors* Z *and* X - SZ *are independent.*

Proof. According to the sufficiency condition (A3), for every $P \in \mathcal{D}_0$ the density dP/dP_0 is the same as that of $L(Z|P)$ with respect to $L(Z|P_0)$. According to (A1), this density is

$$\frac{dP}{dP_0} = \exp\{Z'B^{-1}\zeta - \frac{1}{2}\zeta'B^{-1}\zeta\}.$$

Now take a v in the dual of $\mathcal{X}$ and compute $E_p \exp\{vX\}$. The condition (A2) yields $E_p \exp\{vX\} = \exp\{v\xi + \frac{1}{2}||v||^2\}$ for a norm defined by $||v||^2 = vAv'$.

One can also write $E_p \exp\{vX\} = \int \exp\{vX\} \frac{dP}{dP_0} dP_0$. This yields

$$E_{P_0} \exp\{vX + Z'B^{-1}\zeta\} = \exp\{\frac{1}{2}[\{|\zeta|^2 + ||v||^2 + 2v\xi]\}$$

for $|\zeta|^2 = \zeta' B^{-1}\zeta$. By sufficiency again, one must have $\xi = E_p X = E_p\, \phi_1(Z)$ for some function of Z. Thus ξ is an analytic function $\xi = \phi(\zeta)$ of ζ.

If (P4) holds, then the graph of the function ϕ so obtained is linear. In other words $\xi = E_p X = S\zeta$. According to the above expression for the Laplace transforms, this means that, under P_0, the vectors X and Z are jointly Gaussian with $E_0(vX)(Z'B^{-1}\zeta) = vS\zeta$ and therefore $C = SB$. Thus (P4) implies both (P3) and (P1).

Now suppose instead that (P1) is satisfied and take a point $w \in \mathcal{Z}$. Then, by the same kind of argument

$$E_p \exp\{vX + w'Z\} = E_{P_0} \exp\{vX + Z'(w + B^{-1}\zeta) - \frac{1}{2}|\zeta|^2\}$$

$$= \exp\{\frac{1}{2}\,||v||^2 + \frac{1}{2}[(w + B^{-1}\zeta)'B(w + B^{-1}\zeta)$$

$$- \zeta' B^{-1}\zeta] + v\,C(w + B^{-1}\zeta)\}.$$

It follows that X and Z are jointly Gaussian for all $P \in \mathcal{D}_0$. Also, putting $w = 0$ in the preceding formula one obtains

$$E_p \exp\{vX\} = \exp \frac{1}{2}[||v||^2 + 2\,vCB^{-1}\zeta]\}.$$

Thus $E_p X = CB^{-1}\zeta$ and therefore $\xi = S\zeta$ for $C = SB$. This completes the proof of the lemma.

The foregoing Lemma 3 has several consequences. One of them is that, under the null hypothesis $P \in \mathcal{D}_0$, one has $Q_{LR} = Q_{RR}$ almost surely. This can be stated as follows:

Corollary 1

Let the conditions (A1) *to* (A3) *and one of the* (Pi) *be satisfied. Assume also that the covariance matrix* M *of* (X - SZ) *is non singular. Then* Q_{LR} *and* Q_{RR} *are identical. The matrix* M^{-1} *can be written in the form* $M^{-1} = A^{-1} + K$ *where* K *is the positive definite matrix defined by*

$$K = A^{-1} C[B - C'A^{-1}C]^{-1} C'A^{-1}$$

$$= A^{-1} S[B^{-1} - S'A^{-1}S]^{-1} S'A^{-1}$$

Proof. This follows immediately from Lemma 1, Section 2 and the identity C = SB given by Lemma 3. Note that even if M is singular the foregoing identities remain meaningful and valid, since, under $\mathcal{D}_0$ and our assumptions, the vector $\binom{X}{Z}$ is always in the range of its covariance matrix.

We have given two alternative forms for the matrix K. They may be used according to convenience. For instance, standard arguments often yield the matrices A and B. The relation $E_p X = S E_p Z$ may be used to determine S. Then the form

$$K = A^{-1} S[B^{-1} - S'A^{-1}S]^{-1} S'A^{-1}$$

may be convenient.

Another immediate corollary is as follows.

Corollary 2

Let the conditions (A1) *to* (A3) *and one of the* (Pi) *be satisfied. Then, for* $P \in \mathcal{D}_0$, *the expression* Q_{RR} *has a* χ^2 *distribution with a number of degrees of freedom equal to the rank* of M. *The Pearson statistic* Q_p *may be written* $Q_p = ||X - \Pi X||^2 + \sum_i c_i U_i^2$ *where* $0 \leq c_i \leq 1$, *where the* U_i *are independent* $N(0,1)$, *independent* of $||X - \Pi X||^2$, *and where the number of terms in* $\sum c_i U_i^2$ *is the dimension of the range of* S.

This is immediate. One can also note that the sufficiency condition insures that A cannot be singular on S $\mathcal{Z}$. Indeed if $\xi' A \xi = 0$ for some $\xi \neq 0$, $\xi \in S\mathcal{Z}$ then $L(X;0)$ and $L(X;\xi)$ are disjoint while $L(Z;0)$ and $L(Z;\zeta)$ are mutually absolutely continuous. This gives the evaluation rank A - dim S $\mathcal{Z}$ for the number of degrees of freedom of $Q_F = ||X - \Pi X||^2$.

Remark 1. The procedure by which we obtained the identity $M^{-1} = A^{-1} + K$ in Section II shows also that

$$Q_{LR} = ||X - SZ||^2 + \inf_{\phi} (W - \phi)'D^{-1}(W - \phi)$$

where $W = Z - RX$, where D is the covariance matrix of W and where $\phi = (I-RS)\zeta$ for $\zeta \in \mathcal{Z}$. This may be another convenient way of determining Q_{LR}.

Remark 2. In this section we considered only the situation where $P \in \mathcal{D}_0$. If one returns to the class $\mathcal{D}_1$ and the additional conditions of Section II one can obtain similar results for the distribution of Q_{RR} or Q_p. Now they will involve non central chi-squares.

IV. ASYMPTOTICS UNDER THE NULL HYPOTHESIS

The aim of the present section is to show that if the assumptions called (A) and (P) in Section III are "approximately" satisfied, then the conclusions of Section III also hold "approximately". Traditionally, such arguments are stated as limit theorems as a certain integer n tends to infinity. In such a case, the framework of Section III becomes one involving sequences $\{\mathcal{X}_n, \mathcal{Z}_n, \mathcal{D}_{o,n}\}$ where $\mathcal{X}_n$ and $\mathcal{Z}_n$ are finite dimensional spaces and where $\mathcal{D}_{o,n}$ is a class of probability measures on a suitable measurable space. One needs also random vectors X_n and Z_n taking values respectively in $\mathcal{X}_n$ and $\mathcal{Z}_n$. The role of the Gaussian expectations ξ and ζ of Section III is now played by "centering constants" $\xi_n(P_n)$ and $\zeta_n(P_n)$ defined as functions on $\mathcal{D}_{o,n}$.

We shall consider first a direct translation of Section III. To keep the roles of the various objects in mind, the reader may refer to a standard χ^2 test of goodness of fit situation where the role of the vector X_n is played by the vector with coordinates $\frac{1}{\sqrt{N}}[n_j - NP_j(\theta_0)]$ where the role of Z_n is played by a vector $\sqrt{n}(\hat{\theta}_n - \theta_0)$ for some estimate of $\hat{\theta}_n$ of θ_0. However, it should be clear that, in the general framework, the index n plays no essential role except that of tending to infinity. *Thus we have dropped it from the notation* and will talk about classes $\mathcal{D}_0$ with various properties as $n \to \infty$. Certain objects will not depend on n. If so they will be labeled as such in any one of many ways and in particular by calling them "*fixed*".

A first assumption is as follows:

(C1). The vector spaces $\mathcal{X}$ *and* $\mathcal{Z}$ *are finite dimensional and their dimension is fixed. The random vector* X *takes value in* $\mathcal{X}$. *The random vector* Z *takes values in* $\mathcal{Z}$. *There are maps* ξ *and* ζ *from* $\mathcal{D}_0$ *to* $\mathcal{X}$ *and* $\mathcal{Z}$ *respectively.*

The next assumptions will involve convergence properties for certain distributions. For this one needs appropriate metrics. It will be assumed that $\mathcal{Z}$ comes equiped with a certain Euclidean norm, denoted $|\cdot|$.

For convergences of measures, the norm $|\cdot|$ on $\mathcal{Z}$ can be used to define a corresponding dual-Lipschitz norm by the relation $||\mu|| = \sup_f\{|\int f d\mu|\}$ where the supremum is taken over all functions f such that $|f| \leq 1$ and $|f(z_1) - f(z_2)| \leq |z_1 - z_2|$.

Another item used below is a function q that may or may not be a metric. To define it, for each $P \in \mathcal{D}_0$, let P' be the corresponding joint distribution of (X,Z) on $\mathcal{X} \times \mathcal{Z}$. Let $\rho(P_1', P_2')$ be the Hellinger affinity $\int \sqrt{dP_1' dP_2'}$ and let q be the positive square root of $q^2(P_1', P_2') = -8 \log \rho(P_1', P_2')$.

An analogue of condition (A1) of Sections II and III can be stated in the following manner. Let B denote a covariance matrix such that $|z|^2 = z'B^{-1}z$.

(C2). There is a $P_0 \in \mathcal{D}_0$ *such that* $\xi(P_0)$ *is the origin of* $\mathcal{X}$ *and such that* $\zeta(P_0)$ *is the origin of* $\mathcal{Z}$. *For any* $\zeta \in \mathcal{Z}$ *with* $|\zeta|$ *bounded* (independently of n) *there are measures* $P \in \mathcal{D}_0$

such that $q(P'_0,P')$ *remains bounded and such that* $|\zeta(P) - \zeta| \to 0$. *If* $q(P'_0,P')$ *remains bounded, the dual Lipschitz distance between* $L(Z|P)$ *and the Gaussian* $N[\zeta(P),B]$ *tends to zero.*

An analogue of condition (A3), Section III, given by the following condition (C3) or any equivalent form described in Lemma 4.

(C3). Let $P \in \mathcal{D}_0$ *be such that* $q(P'_0,P')$ *remains bounded. Then* $q^2(P'_0,P') - |\zeta(P)|^2$ *tends to zero.*

This condition is intended to imply that the statistics Z are asymptotically sufficient and "distinguished" in the sense of [9]. The adjective "distinguished" means that if $|Z^* - Z| \to 0$ then the Z^* are also asymptotically sufficient. The condition (C3) may look peculiar. Some of the equivalent forms given in Lemma 4 are more familiar.

Lemma 4

Let conditions (C1) *and* (C2) *be satisfied. Then for bounded* $q(P'_0,P')$, *the following two conditions are equivalent and equiv- to the condition given in* (C3).

a) *The distance between the experiments* $\{P'_0,P'\}$ *and* $\{N(0,B),\ N[\zeta(P),B]\}$ *tends to zero.*

b) *The difference*

$$\log \frac{dP'}{dP'_0} + \frac{1}{2}\{|Z - \zeta(P)|^2 - |Z|^2\}$$

tends to zero in probability.

Proof. The spaces $(Z,|\cdot|)$ are Euclidean of fixed dimension and can therefore be treated as if independent of n.

Assume that $q(P'_0,P'_1)$ remains bounded and take a subsequence along which the experiments $\{P'_0,P'_1\}$ tends to a limit, say $\{P^*_0,P^*_1\}$. Suppose also that along the subsequence, the $\zeta(P_1)$ tend to a limit ζ^* (which may be a point at infinity). Then by (C2) and a result of [9], the experiment formed by the pair $\{N(0,B),\ N(\zeta^*,B)\}$ is weaker than $\{P^*_0,P^*_1\}$. This implies in particular that ζ^* cannot be a point at infinity. Under (C3), the Hellinger affinity between P_0^* and P_1^* must be the same as that between $N(0,B)$ and $N(\zeta^*,B)$. From this it follows by a standard convexity argument that the experiments $\{P_0^*, P_1^*\}$ and $\{N(0,B),\ N(\zeta^*,B)\}$ must be equivalent. This yields the assertion (a). That (a) implies (b) is a consequence of Theorem 1 of [10]. The implications (b) $\Rightarrow$ (a) $\Rightarrow$ (C3) are easy and well known. Hence the result.

Remark 1. The condition (C3) does not imply that if $|\zeta(P)|$ remains bounded so does $q(P_0', P')$. Note also that, in most papers, there is a parameter θ that is essentially another name for the measure P. (The measure in question is then written $P_{n,\theta}$). Here ζ plays the role of a "renormalized" θ, for instance $\sqrt{n}(\theta - \theta_0)$, but P is not a function of θ. It is ζ that is a function of P. There are many situations where the framework used here is naturally applicable while the usual one is not. This happens for instance if the measures P are possible distributions for the trajectories over n generations of k_n independent supercritical branching processes. In such a case P depends on the distribution, say F, of the progeny of any given individual. However, as shown by C. Duby [6] and R. Davies [4], under very weak conditions the asymptotic behavior of P is determined by $\mu = \int x dF(x)$ and $\sigma^2 = \int (x - \mu)^2 dF(x)$, thus reducing the situation to one in which our Z is two dimensional.

Remark 2. We have used the same Euclidean norms $|\cdot|$ to define both the approximating Gaussian distributions and the dual Lipschitz norms that measure the degree of approximation. In many cases the problems are given with two separate norms for these different purposes. There must then be adequate connections between them. We shall not enter into discussion of this point here to save space.

A remark similar to the one made above for Z applies to the translation of condition (A2), Section III about the behavior of the vectors X. Although one can use apparently more general conditions, we shall restrict our attention to the simplest one.

We shall assume given, for each n, a Euclidean norm $||\cdot||$ on $\mathcal{X}$. It will be used to define a dual Lipschitz distance on measures carried by $\mathcal{X}$. It will also be denoted $||x||^2 = x'A^{-1}x$.

(C4). If $q(P_0', P')$ *remains bounded, then the dual Lipschitz distance between* $\mathcal{L}[X - \xi(P)|P]$ *and* $\mathcal{N}(0,A)$ *tends to zero.*

Before passing to analogues of the conditions (Pi) of Section III, let us describe i) a construction that will be used in several proofs below and, ii) some implications of the conditions (C1) - (C4).

For some proofs it is convenient to select particular measures P_t among those that are such that $\zeta(P) = t$. We shall do it as follows:

For each integer n, select an $\varepsilon_n > 0$ so that if $q(P_1', P_2') < \varepsilon_n$ the difference between P_1' and P_2' is considered of little practical relevance. For theoretical purposes, we shall assume that $\varepsilon_n \to 0$. An appropriate selection P_t will be one such that, if $\mathcal{D}_0(b) = \{P : q(P_0', P') \leq b\}$ then

$$|t - \zeta(P_t)| \leq \varepsilon_n + \inf_P\{|t - \zeta(P)|;\ P \in \mathcal{D}_0[1 + |t|]\}.$$

If the infimum is achieved, one can dispence with the ε_n and take for P_t one element of $\mathcal{D}_0[1 + |t|]$ that achieves the infimum.

We shall also define a function ϕ from $\mathcal{Z}$ to $\mathcal{X}$ by letting $\phi(t) = \zeta(P_t)$.

In the course of the proof of Lemma 3, Section III, it was noted that the expectations ξ were functions of the expectations ζ. The analogue for the present is as follows:

Lemma 5

Let the conditions (C1) *to* (C4) *be satisfied. Let* $\mathcal{D}_0(b) = \{P : q(P_0', P') \leq b\}$. *Then for fixed* ε *and* b *one will eventually have*

$$\|\xi(P_1) - \xi(P_2)\| \leq |\zeta(P_1) - \zeta(P_2)| + \varepsilon$$

for all pairs (P_1, P_2) *of elements of* $\mathcal{D}_0(b)$.

The maps ϕ *defined above are such that*

$$\sup\{|\xi(P) - \phi[\zeta(P)]| \; ; \; P \in \mathcal{D}_0(b)\}$$

tends to zero as $n \to \infty$.

Proof. The spaces $(\mathcal{X}, \|\cdot\|)$ and $(\mathcal{Z}, |\cdot|)$ are Euclidean spaces of fixed dimension and can be treated as if independent of n. Take a pair (P_1, P_2) in $\mathcal{D}_0(b)$. Let $F_i = \mathcal{L}(X|P_i)$. The Hellinger affinity $\rho(F_1, F_2)$ satisfies the inequality $-8 \log \rho(F_1, F_2) \leq q^2(P_1', P_2')$. Also, by the same argument as in Lemma 4,

$$\lim_n \sup\{\|\xi(P_1) - \xi(P_2)\| + 8 \log \rho(F_1, F_2)\} \leq 0.$$

Since $q^2(P_1', P_2') - |\zeta(P_1) - \zeta(P_2)|^2 \to 0$ by (C3) this gives the first assertion.

To prove the second, note that if b remains fixed, the representative P_t selected for a $P \in \mathcal{D}_0(b)$ and $t = \zeta(P)$ will eventually be an element of $\mathcal{D}_0(b + 2)$. Thus $|\zeta(P_t) - \zeta(P)| \to 0$ and so does $\|\xi(P) - \xi(P_t)\|$. Hence the result.

To translate the properties (Pi) of Section III, metrize $\mathcal{X} \times \mathcal{Z}$ by the norm defined by $\|x, z\|^2 = \|x\|^2 + |z|^2$. The analogues of the (Pi) are as follows.

(P'1). There are Gaussian measures G_0 *on* $\mathcal{X} \times \mathcal{Z}$ *such that the dual Lipschitz distance between* $\mathcal{L}\{(X,Z)|P_0\}$ *and* G_0 *tends to zero.* (Recall that everything here depends on n.)

(P'2). There are Gaussian measures G_0 *on* $X \times Z$ *such that if* $q(P'_0, P')$ *remains bounded then the distance between* G_0 *and* $L[(X - \xi(P), Z - \zeta(P))|P]$ *tends to zero.*

(P'3). There are linear maps S *from* Z *to* X *such that if* $q(P'_0, P')$ *remains bounded then* $||\xi(P) - S\zeta(P)|| \to 0$.

(P'4). There are linear subspaces L *of* $X \times Z$ *such that* 1) *if* $q(P'_0, P')$ *remains bounded, the distance from* $(\xi(P), \zeta(P))$ *to* L *tends to zero, and* 2) *if* $(\xi, \zeta) \in L$ *and if* $||\xi, \zeta||$ *remains bounded, there are* $P \in \mathcal{D}_0$, *with* $q(P'_0, P')$ *bounded, such that* $||\xi - \xi(P), \zeta - \zeta(P)|| \to 0$.

The asymptotic equivalent of Lemma 3, Section III can be stated in the following manner.

Proposition 1. Let the conditions (C1) *to* (C4) *be satisfied. Then the properties* (P'i), i = 1,...,4, *are all equivalent. If one of them holds, then for* $q(P'_0, P')$ *bounded, the dual Lipschitz distance between the joint distributions* $L[(X - SZ, Z)|P]$ *and the products* $L[(X - SZ)|P] \otimes L[Z|P]$ *tends to zero as* $n \to \infty$.

Proof. As before one may argue as if the spaces $(X, ||\cdot||)$ and $(Z, |\cdot|)$ were fixed. Consider the selection map $t \to P_t$ described before Lemma 4. Let $E = \{P': \zeta \in Z\}$. This is a certain sequence of experiments. The condition (C3) implies that the maps $\zeta \to P'_\zeta$ satisfy a tail equicontinuity condition for the Hellinger distance on the space M of measures on $X \times Z$.

Thus, one can extract subsequences such that for each $t \in Z$ the P'_t converge to a limit Q_t in M metrized by the weaker dual Lipschitz norm. The convergence will be uniform on the compacts of Z. The sequences can be selected so that the maps ϕ, chosen before Lemma 4, also converge uniformly on compacts to a certain limit, say ψ.

Let $F = \{Q_\zeta : \zeta \in Z\}$ be the experiment obtained in this manner. The arguments of Lemma 4, or the form of the likelihood ratios exhibited in that lemma, imply that for Q_ζ the distribution of Z is $\mathcal{N}(\zeta,B)$ and that Z is sufficient for the family $\{Q_\zeta: \zeta \in \mathcal{Z}\}$.

In summary, the limit $\{Q_\zeta: \zeta \in \mathcal{Z}\}$ satisfies the conditions (A1) (A2) (A3) of Section III.

By passage to the limit each condition (P'i) becomes the corresponding (Pi) of Section III.

It follows immediately that (P'1) implies (P'2) and therefore (P'4).

To show that (P'4) implies (P'3) and thus (P'1) one can proceed as follows. Take n so large that every point of the unit ball of $\mathcal{Z}$ is within 10^{-r} of some $\zeta(P)$, $P \in \mathcal{D}_0[1 + 2\ 10^{-r}]$. Let u_j; $j = 1,\ldots,\dim \mathcal{Z} = r$, be an orthonormal basis for $\mathcal{Z}$. Let $v_j = \phi(u_j)$. Extend this to a linear map of S of $\mathcal{Z}$ into $\mathcal{X}$ by writing $S\,u = \Sigma\, \alpha_j v_j$ if $u = \Sigma\, \alpha_j u_j$. We claim that whenever $|u|$ remains bounded then $||Su - \phi(u)|| \to 0$. To show this, pass to a subsequence as before, but in such a way that the maps S also possess a limit, say S_0. The limit of ϕ is a certain map ψ. By (P'4), this map is linear. It coincides with S_0 on the basis $\{u_j\}$ and therefore on the entire space $\mathcal{Z}$. Hence the result.

The asymptotic independence of X - SZ and Z follows from the independence of these variables for the measures Q_ζ. This completes the proof of the proposition.

Remark 1. The asymptotic independence of X - SZ and Z stated in Proposition 1 can also be restated in a different manner. Let C be the covariance of X and Z in the approximating joint normal distribution of (X,Z). Then, if S is as described in (P'3), the difference C - SB will tend to zero. This implies in particular that the maps S can be evaluated in different ways. One can evaluate them as in the proof of Lemma 5. If so, the resulting SB is an approximation of C. On the opposite side, one can evaluate C and determine S by the equality C = SB.

Remark 2. The assumptions (P'1) and (P'4) are very different in appearance. They may have different ranges of convenience in different contents. For instance (P'4) is often an immediate consequence of so called "regularity conditions" that imply a linear relation between the $\xi(P)$ and $\zeta(P)$ through differentiability assumptions. Condition (P'1) is often automatically fulfilled because the vectors X and Z are approximated by sums of independent uniformly negligible summands. Indeed suppose that the conditions (C1) to (C4) hold and that, under P_0, every cluster point of the sequence of distributions $P'_0 = \mathcal{L}\{(X,Z)|P_0\}$ is infinitely divisible. Then (P'1) holds.

This being acquired, let us pass to the asymptotic behavior, under the null hypothesis, of the analogues of the test criteria of Section II.

The simplest are the analogues of Q_p. There are several of them. One possibility is $Q_p^{(1)} = ||X - SZ||^2$ for a map S that satisfies (P'3). Another possibility is $Q_p^{(2)} = ||X - \phi(Z)||^2$ where ϕ is a map as described for Proposition 1. In all of these, one may replace the square norms $||x||^2 = x' A^{-1} x$ by $||x||^2_{\hat{}} = x' \hat{A} x$ where $\hat{A}$ is any estimate of A^{-1} such that, whenever $||x||$ remains bounded, the difference $||x||^2_{\hat{}} - ||x||^2 \to 0$ in P_0 probability. This last modification does not make any difference in the asymptotic behavior of the criteria. Thus we shall not discuss it in detail even though it is absolutely essential for applications.

The asymptotic behavior of $Q_p^{(1)}$ is easily describable: If $q(P'_0,P')$ remains bounded, then $Q_p^{(1)}$ and also $Q_p^{(2)}$ behave asymptotically as $\chi^2 + \Sigma c_i U_i^2$ where χ^2 has dim $\mathcal{X}$ - dim S $\mathcal{Z}$ degrees of freedom, where the sum $\Sigma c_i U_i^2$ contains dim S $\mathcal{Z}$ terms, with $c_i \in [0,1]$ and where the U_i are $\mathcal{N}(0,1)$ independent and independent of the χ^2.

The trouble with $Q_p^{(1)}$ or $Q_p^{(2)}$ is that they involve either maps S or ϕ that are defined only very locally. For applications one needs to look at expressions such as

$$Q_p^{(3)} = ||X - \xi(\hat{P})||^2$$

where $\hat{P}$ is some global estimate of the underlying measure P. Thus we shall assume that one has such estimates and subject them to the condition

(C5). The estimates $\hat{P}$ *are such that, if* $q(P_0', P')$ *remains bounded then* $|Z - \zeta(\hat{P})|$ *tends to zero in probability.*

As will be seen by reference to ordinary chi-square tests and many other situations, condition (C5) is satisfied often enough to be of interest.

Under these circumstances, the behavior of $Q_P^{(3)}$, or appropriate analogues, is that described by Chernoff and Lehmann.

For the Pearson-Fisher criterion Q_F the situation is analogous. Let Π be the orthogonal projection on the range of S for the matrices A^{-1} or their estimates $\hat{A}$. Then $Q_F^{(1)} = ||X - \Pi X||^2$ is asymptotically χ^2 with dim $\mathcal{X}$ - dim S $\mathcal{Z}$ degrees of freedom. However a more usable criterion would be $Q_F^{(2)} = \inf\{||X - \xi(P)||^2; P \in \mathcal{D}_0\}$. Under our assumptions this need not behave at all like $Q_P^{(1)}$. Indeed the assumptions in question do not preclude the possibility that there may be measures $P \in \mathcal{D}_0$ with $q(P_0', P')$ large and $||\xi(P), \zeta(P)||$ small. To exclude this one may add the assumption.

(C6). If $||\xi(P), \zeta(P)||$ *remains bounded so does* $q(P_0', P')$.

This is not enough to insure that $Q_F^{(1)} - Q_F^{(2)} \to 0$ in P_0 probability, but one can remedy the situation. One possibility, for practical applications, is to select some sufficiently large b and replace $Q_F^{(2)}$ by an analogous quantity $Q_F^{(3)} = \inf\{||X - \xi(P)||^2; P \in \mathcal{D}_0, |\zeta(P) - \zeta(\hat{P})| \leq b\}$, where $\hat{P}$ is as in (C5).

The situation for Q_{LR} or Q_{RR} is more complex. Indeed, let $\begin{pmatrix} A & C \\ C' & B \end{pmatrix}$ be a covariance matrix for a Gaussian approximation to the joint distribution of $\binom{X}{Z}$ under P_0. This gives a matrix $M = A - SC' - CS' + SBS'$ for a covariance of a Gaussian approximation of the distribution of X - S Z. Even if M is not singular for any particular n, there is nothing to prevent it from degenerating compared to A, as $n \to \infty$.

Since C is only approximately equal to SB, matrices such as A - SBS' or their estimates may not even be covariance matrices.

A possible analogue of the criterion Q_{LR} is given by

$$Q_{RR}^{(1)} = Q_p^{(3)} + (X - SZ)'K(X - SZ)$$

where $K = A^{-1}S[B^{-1} - S'A^{-1}S]^{-1} S'A^{-1}$ or any consistent estimate thereof. For these statistics one can assert the following:

Proposition 2. Assume that the conditions (C1) *to* (C5) *and* (P'i) *are satisfied. Assume also that if* $||v|| = 1$ *then* $v'Mv$ *stays bounded away from zero. Then the distribution of* $Q_{RR}^{(1)}$ *tends to that of a* χ^2 *with number of degrees of freedom equal to the dimension of* $\mathcal{X}$.

This is quite obvious, and stated only for reference purposes.

In the event that degeneracy would occur, which means in practice that if A is taken as the identity, some eigenvalues of M are exceedingly small, it may be better to use statistics other than $Q_{RR}^{(1)}$. The problem is that even if, for the dual Lipschitz distance defined by $||x||^2 = x'A^{-1}x$, the vector (X - SZ) is asymptotically Gaussian $\mathcal{N}(0,M)$, there is no assurance that such a statement will hold for a dual Lipschitz norm defined by M^{-1} or $A^{-1} + K$.

In some cases a cure is fairly obvious. This happens for instance in the standard treatment of tests for independence in contingency tables. There degeneracy does occur and M is actually singular. However the form of degeneracy is exactly the same under the null hypothesis and the alternatives. All one needs to do is reduce the dimension of $\mathcal{X}$.

In cases where possible near singularity of M is suspected, one may be tempted to remedy the situation by boosting up the matrices M or driving down the matrices K. An argument is as follows:

Let Y* be a random vector with values in $\mathcal{X}$. Assume that Y* is Gaussian $\mathcal{N}(0,D)$ independently of the other variables involved in the problem. Then if the conditions (C1) - (C5) and (P'i) are satisfied for (X,Z) they are also satisfied for (X + Y*, Z). One can take D in such a way that the matrices $\begin{pmatrix} A+D & C \\ C' & B \end{pmatrix}$ no longer degenerate. To achieve this it is sufficient to select some fixed $a > 0$ and take $D = a A$.

If so, let $\bar{K}$ and K_2 be the matrices defined by

$$\bar{K} = (A + D)^{-1}S[B^{-1} - S'(A + D)^{-1}S]^{-1}S'(A + D)^{-1}$$

and

$$K_2 = A^{-1}S[B^{-1} - S'(A + D)^{-1}S]^{-1}S'A^{-1}.$$

Consider the expression

$$Q^*_{RR} = [X + Y^* - \xi(\hat{P})]'[(A + D)^{-1} + \bar{K}][X + Y^* - \xi(\hat{P})]$$

This will be asymptotically chi-square with a number of degrees of freedom equal to the dimension of X. Here Y* is just extra noise. This suggests the use of expressions such as

$$Q^{(2)}_{RR} = ||X - \xi(\hat{P})||^2 + [X - \xi(\hat{P})]'K_2[X - \xi(\hat{P})].$$

$$Q^{(3)}_{RR} = ||X - SZ||^2 + (X - SZ)'K_2(X - SZ)$$

One could also use analogous expressions where $\bar{K}$ is used instead of K_2, but K_2 is easier to discuss here, except in the special case $D = a A$ where $K_2 = (1 + a)^2 \bar{K}$.

The distributions of $Q^{(2)}_{RR}$ or $Q^{(3)}_{RR}$ are no longer approximable by χ^2 distributions. However these quadratic expressions still behave in the manner described by Chernoff and Lehmann.

Proposition 3. Let the conditions (C1) *to* (C5) *and* (P'i) *be satisfied. Let the matrices* D *be selected so that the norm of* K_2 (or $[B^{-1} - S'(A + D)^{-1}S]^{-1}$) *remains bounded. Then if* $q(P'_0,P')$ *remains bounded, the quadratics* $Q^{(2)}_{RR}$ *and* $Q^{(3)}_{RR}$ *are asymptotically equivalent to* $||X - \Pi X||^2 + (\Pi X - SZ)'\, K^*(\Pi X - SZ)$

where K^* *is a positive semi-definite matrix whose eigenvalues on the space* $\mathcal{X}$ *metrized by* $(x'M^{-1}x)^{1/2}$ *are all between zero and unity. The terms* $X - \Pi X$ *and* $\Pi X - SZ$ *are asymptotically independent.*

Proof. The last statement about independence is already a part of Proposition 1. For the other statements, note that the matrix K_2 has the form $F\ S'\ A^{-1}$ for a certain matrix F. Now $S'\ A^{-1}\ (I - \Pi) = 0$. Thus $(X - SZ)'\ K_2\ (X - SZ) = (\Pi X - SZ)'\ K_2\ (\Pi X - SZ)$. The formula given in the proposition is therefore valid for a matrix $K^* = K_2 + \Pi'\ A^{-1}\ \Pi$. Since $K^* \le K$, the assertion concerning eigenvalues follows.

For the particular choice $D = a\ A$, a similar result holds if one replaces K_2 by D.

There are many other variants of the Rao-Robson special formula. The asymptotic independence of Z and $X - SZ$ implies that of Z and $\pi(X - SZ)$ for any bounded projection π. If H is the range of S, one can take for π the orthogonal projection Π followed by a projection on a subspace L of H. One can then use quadratic expressions of the form

$$Q = ||X - SZ||^2 + [\pi(X - SZ)]'K^{\pi}[\pi(X - SZ)]$$

where K^{π} is the analogue of K defined on L by

$$(\pi A\pi')^{-1}\pi S[B^{-1} - S'\pi'(\pi A\pi)^{-1}\pi S]^{-1}S'\pi'(\pi A\pi')^{-1}.$$

If π is an orthogonal projection for the norm $||\cdot||$ defined by A^{-1}, the expression Q_{π} can also be written

$$Q_{\pi} = ||X - \Pi X||^2 + ||\pi(X - SZ)||^2 + [\pi(X - SZ)\ 'K^*\ \ (X - SZ)]$$
$$+ ||(I - \pi)(X - SZ)||^2.$$

Here the sum of the first three terms would be asymptotically χ^2. The last term behaves according to the Chernoff-Lehmann assertions.

Now consider the case where, for some $v \in H$ such that $||v|| = 1$, the value $v'\ M_1\ v$ is very small. This means that in the direction v the two estimates ΠX and SZ of $S\zeta$ are very close to one another. They could be so close that their difference is within the range of accuracy of the arithmetical operations yielding them. In such a case it would not be wise to boost up essentially meaningless differences by multiplying them by the matrix K_1. It may be more sensible to use a matrix such as K^{π} for a projection π that deletes the unwanted direction v.

This suggests the use of a certain family of matrices and forms $Q_{RR}^{(\tau)}$ that will be described now in special coordinate systems. In the space X, take an orthonormal basis (for the norm $||\cdot||$) in such a way that the first dim H coordinates vectors are in the range H of S. Place them in such a manner that the matrix M_1, approximate covariance of $\Pi X - SZ$, becomes diagonal, with eigenvalues listed in decreasing order. In such a coordinate system, the first dim H coordinates yield a vector $Y_1 = \Pi X - SZ$. The remaining ones yield $Y_2 = X - \Pi X$ and the covariance matrix of the entire vector (Y_1, Y_2, Z) is formed of diagonal blocks.

Now take a value $\tau \in [0,1]$ and let L_τ be the subspace of H spanned by the coordinate vectors corresponding to eigenvalues of M_1 at least equal to $1 - \tau$. Let π_τ be the orthogonal projection of H onto L_τ. (We shall also use the notation π_τ for the projection $\pi_\tau \Pi$ of X onto L_τ). Finally, let $Q_{RR}^{(\tau)}$ be equal to Q_{π_τ} for the expression defined previously. One can rewrite $Q_{RR}^{(\tau)}$ in the form

$$Q_{RR}^{(\tau)} = ||Y_2||^2 + ||Y_1||^2 + Y_1' K_\tau Y_1$$

with K_τ equal to the previously defined K^τ with π replaced by π_τ.

Note that $Q_{RR}^{(0)}$ is the Pearson Statistic $Q_P^{(1)}$. Also for $\tau = 1$, the new $Q_{RR}^{(1)}$ is precisely the Rao-Robson statistic.

One can write

$$Q_{RR}^{(\tau)} = \{||Y_2||^2 + ||\pi_\tau Y_1||^2 + Y_1' K_\tau Y_1\}$$
$$+ ||(I - \pi_\tau)Y_1||^2.$$

For a fixed $\tau \in (0,1)$ the sum of the three terms in the curly brackets will be asymptotically χ^2 with a number of degrees of freedom dim X - dim $L_\tau^\perp$ where $L_\tau^\perp$ is the orthogonal complement of L_τ. The extra term $||(I - \pi_\tau) Y_1||^2$ is a sum of the Chernoff-Lehmann type $\Sigma c_i U_i^2$ with all the coefficients c_i such that $0 \leq c_i \leq 1 - \tau$.

V. ASYMPTOTIC BEHAVIOR UNDER ALTERNATIVES

In this section we shall retain the assumptions (C1) - (C5) and (P'i) of Section IV. The index n will be suppressed from the notation. The space $(X \times Z)$ will be metrized by the square norm $||x,z||^2 = ||x||^2 + |z|^2$. The class $\mathcal{D}_0$ will be a subclass of a larger class $\mathcal{D}_1$ of probability measures.

The condition (B1) of Section II will be replaced by the following:

(D1). The maps ξ and ζ are defined on $\mathcal{D}_1$. There are centered Gaussian distributions G_0 *such that, if* $q(P_0',P')$ *remains bounded, then the distance between* $L\{[X - \xi(P), Z - \zeta(P)]|P\}$ *and* G_0 *tends to zero.*

The condition called (B2) in Section 2 becomes:

(D2). If $(x,z) \in X \times Z$ *is such that* $||x||^2 + |z|^2$ *remains bounded, then there are* $P \in \mathcal{D}_1$ *such that* $||x - \xi(P)||^2 + |z - \zeta(P)|^2 \to 0$ *and such that* $q(P_0',P')$ *remains bounded.*

The non degeneracy stated in (B3, Section II) could be translated as follows:

(D3). Let $(x,z) \in X \times Z$ *be such that* $||x,z||$ *remains bounded. Let* $G_{x,z}$ *be* G_0 *shifted by* (x,z). *Then the sequences* $\{G_{x,z}\}$ *and* $\{G_0\}$ *are contiguous.*

(Equivalently $\int \sqrt{dG_{x,z}\, dG_0}$ stays bounded away from zero.)

A stronger assumption would be

(D'3). If $P_1 \in \mathcal{D}_1$ *is such that* $q(P_0', P_1')$ *remains bounded then* $\{P_0'\}$ *and* $\{P_1'\}$ *are contiguous sequences.*

Finally, we shall need to consider an even stronger assumption. To describe it, let $\omega(P)$ be the point $\omega(P) = \begin{bmatrix} \xi(P) \\ \zeta(P) \end{bmatrix}$ of $X \times Z$. Let Γ be the covariance matrix of the approximating Gaussian measure G_0 and let $||\omega||_\Gamma^2 = \omega' \, \Gamma^{-1} \omega$.

(D4). If $P_1 \in \mathcal{D}_1$ *is such that* $q(P_0', P_1')$ *remains bounded, then* $q^2(P_0', P_1') - ||\omega(P_1)||_\Gamma^2$ *tends to zero as* $n \to \infty$.

This (D4) is similar to (C3) of Section IV. Let $\mathcal{D}_1(b) = \{P \in \mathcal{D}_1;\ q(P_0', P') \leq b\}$. The combination (D1) - D4) implies that for every fixed b the distance between the experiments $\{P';\ P \in \mathcal{D}_1(b)\}$ and the corresponding Gaussian experiment $\{G_{\omega(P)};\ P \in \mathcal{D}_1(b)\}$ tends to zero. This can be seen exactly as in Lemma 4 of Section IV.

When such a convergence occurs the optimality properties of $Q_{LR} = Q_{RR}$ recalled in Section II admit immediate translations to the asymptotic situation. Let $\Omega = X \times Z$ and let S be one of the linear maps whose existence is proved in Section IV. For each fixed $a > 0$ let Σ_a be the surface

$$\Sigma_a = \{\omega : \omega \in \Omega,\ \inf[||\omega - S\zeta||_\Gamma;\ \zeta \in Z] = a\}.$$

It can be shown that, in a certain complicated but natural sense, the tests defined by rejecting $\mathcal{D}_0$ if $Q_{RR}^{(1)}$ exceeds a certain fixed number have "best asymptotically constant power over the surfaces Σ_a." A similar statement can be made for "best asymptotic average power over sections of Σ_a at fixed ζ". Finally, the tests in question can also be called "asymptotically most stringent." The precise description of these properties is complex. One can adapt it from similar definitions given by A. Wald in [20], taking into account that our convergences are not uniform on $\mathcal{D}_1$ but only on the bounded subsets of the type

$\mathcal{D}_1(b)$. For instance, for the "most stringent" definition, one can proceed as follows. Take a number b and a $P_1 \in \mathcal{D}_1(b)$. Let $\Phi(\alpha,b)$ be the family of tests ϕ such that $E[\phi|P] \leq \alpha$ for all $P \in \mathcal{D}_0(b)$. Let $\beta[P_1; \alpha,b] = \sup\{E(\phi|P_1); \phi \in \Phi(\alpha,b)\}$. Let $r[\phi,P_1; \alpha,b] = \beta(P_1; \alpha,b) - E[\phi|P_1]$. Let a be the number such that $\text{Prob}[\chi^2 > a] = \alpha$ for a χ^2 with dim X degrees of freedom and let ϕ_0 be the indicator of $\{Q_{RR}^{(1)} > a\}$.

Proposition 4. Let the conditions (C1) - (C5), (P'i) *and* (D1), (D2), (D3), (D4) *be satisfied. Let* $\alpha \in (0,1)$ *and* $\varepsilon > 0$ *be fixed. Then there is a fixed b such that, for* n *sufficiently large,*

$$\sup\{r[\phi_0,P; \alpha + b^{-1},b]; P \in \mathcal{D}_1(b)\}$$
$$\leq \varepsilon + \sup\{r[\phi_1,P; \alpha + b^{-1},b]; P \in \mathcal{D}_1(b)\}$$

for every $\phi \in \Phi(\alpha + b^{-1},b)$. *Also, for any fixed* b *the test* ϕ_0 *will eventually belong to* $\Phi[\alpha + b^{-1},b]$.

Proof. The proof can be carried out easily using the general principle of [9], [11] according to which a risk function that is not achievable on a limit experiment F is eventually not achievable either along a sequence of experiments that tend to F. This reduces the problem to that of proving that the likelihood ratio test is most stringent in the Gaussian case. This was proved by A. Wald in [20].

This most stringency property of $Q_{RR}^{(1)}$ could be viewed as an argument in favor of the use of $Q_{RR}^{(1)}$ instead of for instance the Pearson-Fisher criterion or the Pearson $Q_P^{(1)}$. However the situation is not that simple.

Let ψ be a test function that is the indicator of a set of the type $\{||X - \Pi X||^2 \geq c\}$, or $\{Q_{RR}^{(i)} \geq c_i\}$ for any one of the statistics considered at the end of Section IV.

Proposition 5. Let the conditions (C1) - (C5), (P'i) *and* (D1) - (D4) *be satisfied. Then, for fixed constants* c *or* c_i, *the tests* ψ *just described are all asymptotically admissible among tests based on the vectors* (X,Z) *only.*

Here again one should define "asymptotically admissible". One possibility is as follows: Let $\delta(\phi,P) = E(\phi|P)$ if $P \in \mathcal{D}_0$ and $\delta(\phi,P) = 1 - E(\phi|P)$ if $P \in \mathcal{D}_1 \setminus \mathcal{D}_0$.

Let $\gamma \in (0,1)$ and $b \geq 1$ be fixed numbers. Then there is a fixed $\gamma_1 > 0$ and a fixed $b_1 \geq b$ with the following property.

Suppose that ψ is one of the tests described in Proposition 5 and that ϕ is another test. For sufficiently large n, if $\delta(\phi,P) \leq \delta(\psi,P) - \gamma$ for some $P \in \mathcal{D}_1(b)$, then there is some $Q \in \mathcal{D}_1(b_1)$ such that $\delta(\phi,Q) \geq \gamma_1 + \delta(\psi,Q)$.

The proof of Proposition 5 can be carried out as that of Proposition 4 by reducing the problem to the case of Gaussian distributions. The corresponding tests in the Gaussian case have convex acceptance regions. Their admissibility follows from [1], [18], [19].

To describe certain other features of these tests, we shall use the special coordinate systems introduced at the end of Section IV.

The local equivalent of the test criteria called $Q_{RR}^{(i)}$ at the end of Section IV are all of the type $||Y_2||^2 + ||Y_1||^2 + Y_1' K_i Y_1$ for certain matrices K_i. The Pearson-Fisher $Q_F^{(1)}$ is $||Y_2||^2$ and the inequalities

$$Q_R^{(1)} \leq Q_{RR}^{(i)} \leq Q_{RR}^{(1)}$$

It can be shown that, if $K_i - K_j$ stays away from zero, the tests based on the two different K_i and K_j *have risk functions that are asymptotically incomparable.*

To show this consider first the situation where the distributions of (Y_1,Y_2,Z) are exactly Gaussian, with covariance matrix non singular and independent of $P \in \mathcal{D}_1$. Consider two

K_i and K_j and sets $\{||Y_2||^2 + ||Y_1||^2 + Y_1' K_i Y_1 \le c_i\}$ with indicator ϕ_i. Define ϕ_j similarly replacing K_i by K_j and c_i by c_j. If $K_i \ne K_j$ then $\phi_i \ne \phi_j$ and the risk functions $\delta(\phi_i, P)$ and $\delta(\phi_j; P)$ are also different. If one lets $\eta_1 = (\Pi - S\zeta)$ and $\eta_2 = (I - \Pi)\xi$, the set of pairs (η_1, η_2) where these functions coincide is a closed subset of X without interior points. Since both ϕ_i and ϕ_j are admissible, the risk functions cannot be comparable. A similar result applies also to Q_F.

One can conclude from this that under the conditions of Propositions 4 and 5 the tests based on the various $Q_{RR}^{(i)}$ or $Q_F^{(1)}$ with also eventually have incomparable risk functions unless they are asymptotically equivalent tests.

Now note for instance that, in the Gaussian situation, the non-centrality parameter applicable to Q_F is $||\eta_2||^2$. That applicable to $Q_{RR}^{(1)}$ is $||\eta_2||^2 + ||\eta_1||^2 + \eta_1' K_1 \eta_1$. Thus $Q_{RR}^{(1)}$ can yield better power than Q_F only in directions where $||\eta_1||^2 + \eta_1' K_1 \eta_1$ is substantial enough to overcome the differences due to the fact that the number of degrees of freedom of $Q_{RR}^{(1)}$ exceeds that of Q_F by dim H.

For the intermediate statistics $Q_{RR}^{(i)}$ (including $Q_{RR}^{(0)} = Q_P^1$), the situation is more complex since they are no longer asymptotically central or non-central chi-squares. However, for the Gaussian case, the Laplace transform $E \exp\{-\frac{t^2}{2} Q\}$ of a quadratic $Q = ||Y_1||^2 + Y_1' K Y_1$ is a multiple of $\exp\{-\frac{t^2}{2} \eta_1' K[I + t^2 K]^{-1} \eta_1\}$. This suggests that the test based on the various $Q_{RR}^{(i)}$ will be better than Q_F, or $Q_{RR}^{(0)}$, only for those measures for which η_1 is already substantial.

In other words, unless the difference between the two estimates ΠX and SZ is of importance, the various $Q_{RR}^{(i)}$ have no special advantage over Q_F. Note also that if k is large and r is small, $k^{-1} \chi_k^2$ and $(k - r)^{-1} \chi_{k-r}^2$ are very much alike. Thus, if dim X is large but rank S is small, treating the Pearson Q_P as if it was a chi-square will usually yield practically acceptable results.

The foregoing assertions depend strongly on the validity of condition (D4). This condition is equivalent to something that can be stated informally as follows: Let X^* and Z^* be other random vectors such that $||X - X^*||^2 + |Z - Z^*|^2$ tends to zero in probability. Then (X,Z) and (X^*,Z^*) are about "equally informative". This is part of Lemma 4, Section IV or of Theorem 1, [10]. It implies that one may restrict oneself to tests ϕ that depend "smoothly" on (X,Z). When (D4) does not hold counterexamples to Proposition 4 or analogous statements can be constructed. This is even so for the standard chi-square test of goodness of fit. To construct such an example, let λ be the Lebesgue measure on $[0,1]$. Take independent identically distributed observations $\{U_j;\ j = 1,\dots,n\}$ and a partition of $[0,1]$ into m equal intervals.

Let H be the space of all functions defined on $[0,1]$ such that $||h||^2 = \int h^2 d\lambda < \infty$ and $\int h d\lambda = 0$. Define $\gamma\,[||h||] \in [0,1]$ by $\{1 - \gamma[||h||]\}^2 = 1 - ||h||^2$. For any h such that $1 - \gamma_h[\frac{||h||}{\sqrt{n}}] + \frac{h}{\sqrt{n}} \geq 0$ let f_n be the density

$$f_{n,h} = \left[1 - \gamma(\frac{||h||}{\sqrt{n}}) + \gamma\,\frac{h}{\sqrt{n}}\right]^2.$$

Now take an $h_0 \in H$ that takes only a finite or countable set of values and assume that these values are *rationally independent*. Let h_k: $k = 1,2,\dots,r$ be other elements of H such that h_k is constant over any set where h_0 is constant. In such a case one can recompute the values of the $\sum_j h_k(U_j)$ from the values taken by $\sum_j h_0(U_j)$.

Consider a class $\mathcal{D}_0$ where the densities of the U_j are of the form $\left[1 - \gamma(\frac{||\zeta h||}{\sqrt{n}}) + \zeta\,\frac{h}{\sqrt{n}}\right]^2$ with $h = h_0$ and ζ real. Let $\mathcal{D}_1$ be the class of measures with densities of the form $f_{n,h}$, $h = \zeta h_0 + \sum_{k=1}^{r} \xi_k h_k$. Let $Z_n = \frac{1}{\sqrt{n}} \sum_{j=1}^{n} h_0(U_j)$. Let X_n be the normalized vector of counts $\frac{1}{\sqrt{n}}(n_i - np_i)$ $i = 1,\dots,m$ for the intervals used for the χ^2 test.

One can take h_0 so that it sets of constancy do not contain any of these intervals. Then it is clearly possible to select the h_k so that, under $\mathcal{D}_1$, the probabilities of the m intervals range over the entire simplex of possibilities.

Under these circumstances, it is easily verifiable that all the conditions of Proposition 4 except (D4) are satisfied. However here the logarithm of likelihood ratio $\sum_{j=1}^{n} \log f_{n,h}$ have for main random term the sum $2 \frac{1}{\sqrt{n}} \sum_{j=1}^{n} h(U_j)$. Thus, for an $f_{n,k}$ from $\mathcal{D}_1$, the likelihood ratios are closely approximated by functions of Z_n alone. On the contrary, the vector of counts X_n may contain exceedingly little information since the h_k may vary rapidly in each interval of the division. Note also that in the present case the condition (D'3) is also satisfied.

(One could still claim an asymptotic minimax property for the χ^2 test, allowing *all* alternatives of the form $f_{n,h}$, $h \in H$. However this is a rather weak property.)

In the preceding example, Z_n plays the role of an asymptotically sufficient statistic for the class $\mathcal{D}_1$. It is not "distinguished" on $\mathcal{D}_1$, but only on $\mathcal{D}_0$. To be able to extract information from it, one needs to look at it with a strong magnifying glass.

Without condition (D4) one can still obtain analogues of Proposition 4, and of the other optimality properties mentioned earlier, *provided one limits the class of possible tests*. Let $\{\phi_n\}$ be a sequence of test functions such that, for $q(P_0,P)$ bounded, the distances between $L[\phi_n(X,Z)|P]$ and $L[\phi_n(U,V)|P]$ for (U,V) distributed according to $G_{\omega(P)}$ tend to zero. For any such sequence, an analogue of Proposition 4 will hold. However, this may be of little comfort to a statistician working for a specific value of n, since it is not clear how smooth each individual ϕ_n needs to be.

There are other aspects of the optimality properties. Here we have assumed that the vectors (X,Z) have already been selected somehow. The *choice* of such vectors in a practical problem is a matter of major importance. For instance, in the standard goodness of fit situation, Neyman's smooth tests, or variations, may be much "better" than a standard χ^2. However this is another matter altogether.

VI. REFERENCES

[1] Birnbaum, A. (1950). "Characterization of Complete Classes of Tests of Some Multiparametric Hypotheses, with Applications to Likelihood Ratio Tests." *Ann. Math. Statist. 26,* 21-36.

[2] Chernoff, H. and Lehmann, E.L. (1954). "The Use of Maximum Likelihood Estimates in χ^2-tests of Goodness of Fit." *Ann. Math. Statist. 25,* 579-586.

[3] Chiang, C.L. (1956). "On Regular Best Asymptotically Normal Estimates." *Ann. Math. Statist. 27,* 336-351.

[4] Davies, R. (1980). Preprint.

[5] Djaparidze, K.O. and Nikulin, M.S. (1974). "On a Modification of the Standard Statistics of Pearson." *Theory Probab. Appl. 19,* 851-853.

[6] Duby, C. (1980). Personnal Communication.

[7] Dudley, R.M. (1976). "Probabilities and Metrics - Convergence of Laws on Metric Spaces with a View to Statistical Testing." Lecture Notes, Aarhus Universitet, *45*.

[8] Le Cam, L. (1964). "Sufficiency and Approximate Sufficiency." *Ann. Math. Statist. 35,* 1419-1455.

[9] Le Cam, L. (1972). "Limits of Experiments." *Proc. Sixth Berkeley Symp. Math. Statist. Prob. 1,* 245-262.

[10] Le Cam, L. (1977). "On the Asymptotic Normality of Estimates." *Proceedings of the Symposium to honor Jerzy Neyman,* Warsaw, 203-217.

[11] Le Cam, L. (1979). "On a Theorem of J. Hájek." *Contributions to Statistics The Hajek Memorial Volume,* Prague, 119-136.

[12] Mahan, C.M. (1979). "Diabetes Mellitus: Biostatistical Methodology and Clinical Implicatons." Thesis, Univ. of California, Berkeley.

[13] Molinari, L. (1977). "Distribution of the chi-square Test in Nonstandard Situations." *Biometrika 64,* 119-122.

[14] Moore, D.S. (1971). "A chi-square Statistic with Random Cell Boundaries." *Ann. Math. Statist. 42,* 147-156.

[15] Moore, D.S. (1977). "Generalized Inverses, Wald's Method, and the Construction of chi-square Tests of Fit." *J. Amer. Statist. Assoc. 72,* 131-137.

[16] Pollard, D. (1981). "Limit Thoerems for Empirical Processes." *Z. Wahrsch. Verw. Gebiete 55,* 91-108.

[17] Rao, K.C. and Robson, D.S. (1974). "A chi-square Statistic for Goodness of Fit Tests within the Exponential Family." *Comm. Statist. 3,* 1139-1153.

[18] Schwartz, R.E. (1967). "Admissible Tests in Multivariate Analysis of Variance." *Ann. Math. Statist. 38,* 698-710.

[19] Stein, C.M. (1956). "The Admissibility of Hotelling's T^2-test." *Ann. Math. Statist. 27,* 616-623.

[20] Wald, A. (1943). "Tests of Statistical Hypotheses Concerning Several Parameters when the Number of Observations is Large." *Trans. Amer. Math. Soc. 54,* 426-482.

THE LIMITING BEHAVIOR OF MULTIPLE ROOTS OF THE LIKELIHOOD EQUATION[1]

Michael D. Perlman

Department of Statistics
University of Washington
Seattle, Washington

I. INTRODUCTION

Let $X_1, X_2, \ldots$ be an infinite sequence of independent, identically distributed (i.i.d.) random variables (possibly vector-valued) with common density $p(x, \theta_0)$ with respect to some measure μ. This includes both absolutely continuous and discrete distributions. Here θ is a *real-valued* unknown parameter whose range we denote by Ω, a subset of the real line. It is assumed throughout that Ω is an open *interval* (a, b), with $-\infty \leq a < b \leq +\infty$.

For a given sample outcome $(x_1, \ldots, x_n)$, the likelihood function (abbreviated LF) is

$$L_n(\theta) = L_n(\theta;\ x_1, \ldots, x_n) = \prod_{i=1}^{n} p(x_i, \theta)$$

[1] *This research was supported in part by National Science Foundation Grants MPS 72-04364 A03 at the University of Chicago and MCS 80-02167 at the University of Washington.*

ISBN 0-12-589320-5

and the likelihood equation (LEQ) is

$$(\partial/\partial\theta) \log L_n(\theta;\ x_1, \ldots, x_n) = 0. \tag{1.1}$$

Let

$$S_n \equiv S(x_1, \ldots, x_n)$$

denote the (possibly empty) set of all solutions of the LEQ. In this paper we study the behavior of the set S_n as $n \to \infty$, with emphasis on the case of multiple roots where S_n may contain more than one solution for some sample sequences.

Huzurbazar (1948) proved a result which is commonly described by the phrase "a consistent root of the LEQ is unique" (see Proposition 4.2). There are, however, ambiguities in the notions "consistent" and "root of the LEQ" which make this statement unclear. First, consistency is a limiting property of a *sequence* of estimators, say $\{T_n\}$, so for any $N < \infty$, T_n can be altered arbitrarily for all $n \leq N$ without destroying consistency. In fact, if $\{T_n\}$ is a *strongly consistent* sequence of estimators of θ, (i.e., $T_n \to \theta_0$ with probability one) and if $\{T_n^*\}$ is another sequence such that $T_n^* = T_n$ for all sufficiently large n, then $\{T_n^*\}$ is also strongly consistent for θ. Next, by a "root of the LEQ" is meant any sequence of measurable functions $\{\hat{\theta}_n\} = \{\hat{\theta}_n(x_1, \ldots, x_n)\}$ such that for each n and each sample outcome $(x_1, \ldots, x_n)$, $\hat{\theta}_n(x_1, \ldots, x_n)$ is a solution of (1.1), i.e., $\hat{\theta}_n \in S_n$. If, in general, S_n may contain more than one point, then there are infinitely many "roots of the LEQ" as here defined. (The Cauchy family of densities with an unknown location parameter illustrates this situation - see Sections 6 and 7, and also Barnett (1966).) Thus if $\{\hat{\theta}_n\}$ is a strongly consistent root of the LEQ and if $\{\hat{\theta}_n^*\}$ is another root such that $\hat{\theta}_n^* = \hat{\theta}_n$ for all sufficiently large n with probability one, then $\{\theta_n^*\}$ is also a strongly consistent root. Therefore there may be many consistent roots of the LEQ, despite Huzurbazar's result.

In this paper these ambiguities are clarified by studying the limiting behavior of the *set* S_n as $n \to \infty$. From (1.1) it is seen that this limiting behavior is determined by that of $n^{-1}(\partial/\partial\theta)\log L_n(\theta)$ which, by the Strong Law of Large Numbers, converges to $g_1(\theta, \theta_0)$ (see (2.1)) with probability one pointwise in θ. In order to specify the limiting behavior of S_n, therefore, conditions guaranteeing that this convergence be uniform in θ are required, together with conditions guaranteeing that $g_1(\theta, \theta_0)$ has no zeroes other than at $\theta = \theta_0$. Because the standard conditions for uniformity (see G2a in Section III) also imply that $g_1(\theta, \theta_0) = I_1(\theta, \theta_0)$ [the derivative of the Kullback-Leibler function $I(\theta, \theta_0)$ - see (2.1)], this requirement on $g_1(\theta, \theta_0)$ reduces to the assumption that $I(\theta, \theta_0)$ decreases in a strong sense as θ moves away from θ_0, i.e., that the family of distributions $\{p(x,\theta): \theta \text{ in } \Omega\}$ is *strongly Kullback-Leibler* (K-L) *ordered* (see Definition 3.2).

After presenting notation and preliminary lemmas regarding uniform convergence of random functions in Section II, the uniform convergence $n^{-1}(\partial/\partial\theta)\log L_n(\theta)$ over compact subsets of Ω is discussed in Section III. If the family $\{p(x,\theta): \theta \text{ in } \Omega\}$ is strongly K-L ordered, this uniform convergence implies that all members of S_n approach either θ_0 or $\partial\Omega$ as $n \to \infty$. If in addition the convergence can be shown to be uniform over Ω itself and if $I_1(\theta, \theta_0)$ is bounded away from 0 as $\theta \to \partial\Omega$, then the set S_n converges to θ_0. In Section IV we clarify Huzurbazar's result by showing that under suitable assumptions on $(\partial^2/\partial\theta^2)\log L_n(\theta)$, exactly one member of S_n converges to θ_0, and this exceptional solution is a relative maximum of the LF.

These results are applied to monotone likelihood ratio families and location parameter families in Section V. Some special cases are discussed in Section VI, including the Cauchy distribution and the bivariate normal distribution with known variances and unknown correlation. Section VII contains some remarks concerning maximum likelihood estimation which were suggested by Barnett's (1966) study of the Cauchy distribution with unknown location parameter.

It is assumed that Ω is an open interval in order to avoid difficulties arising from the non-existence of solutions of the LEQ that can occur when the true parameter value lies on the boundary of the parameter space. The latter case was carefully treated by Chernoff (1954) in the context of hypothesis testing.

II. NOTATION AND PRELIMINARY LEMMAS

Throughout this paper, θ_0 denotes the true (but unknown) parameter value, and the probability and expectation symbols P_0 and E_0 refer to the probability distribution determined by $p(x,\theta_0)$. We assume for simplicity that the support of P_0 does not vary with θ_0. Let X denote a random variable with density $p(\cdot, \theta_0)$ and introduce the following notation:

$$\begin{aligned} I(\theta,\theta_0) &= E_0 \log [p(X,\theta)/p(X,\theta_0)] \\ I_1(\theta,\theta_0) &= (\partial/\partial\theta)\, I(\theta,\theta_0) \\ I_2(\theta,\theta_0) &= (\partial^2/\partial\theta^2)\, I(\theta,\theta_0) \\ g_1(\theta,\theta_0) &= E_0[(\partial/\partial\theta) \log p(X,\theta)] \\ g_2(\theta,\theta_0) &= E_0[(\partial^2/\partial\theta^2) \log p(X,\theta)] \end{aligned} \tag{2.1}$$

whenever these quantities exist (possibly infinite). Recall that $I(\theta, \theta_0)$ always exists, with $-\infty \leq I(\theta, \theta_0) \leq 0$, and that $I(\theta, \theta_0) < 0$ provided that $p(\cdot,\theta)$ and $p(\cdot, \theta_0)$ determine distinct distributions. Furthermore,

$$\begin{aligned} I(\theta_0,\theta_0) &= 0 \\ I_1(\theta_0,\theta_0) &= 0 \\ I_2(\theta_0,\theta_0) &\leq 0 \end{aligned} \tag{2.2}$$

provided the latter two exist, while

$$g_1(\theta,\theta_0) = I_1(\theta,\theta_0)$$
$$g_2(\theta,\theta_0) = I_2(\theta,\theta_0) \tag{2.3}$$

provided that $I(\theta, \theta_0)$ can be differentiated under the integral sign once or twice, respectively, at θ. [See the paragraphs following conditions G2a and L5a in Sections III and IV.]

For all sufficiently small $\delta > 0$ define the compact intervals J_δ, K_δ^-, $K_\delta^+ \subseteq \Omega$ as follows:

$$J_\delta = [\theta_0-\delta, \theta_0+\delta],$$

$$K_\delta^- = \begin{cases} [a+\delta, \theta_0-\delta] & \text{if } a > -\infty \\ [-\delta^{-1}, \theta_0-\delta] & \text{if } a = -\infty, \end{cases}$$

$$K_\delta^+ = \begin{cases} [\theta_0+\delta, b-\delta] & \text{if } b < +\infty \\ [\theta_0+\delta, \delta^{-1}] & \text{if } b = +\infty. \end{cases}$$

Also, define the neighborhoods U_δ^-, $U_\delta^+ \subseteq \Omega$ of a, b, respectively, by

$$U_\delta^- = \begin{cases} (a, a+\delta] & \text{if } a > -\infty \\ (-\infty, -\delta^{-1}] & \text{if } a = -\infty \end{cases}$$

$$U_\delta^+ = \begin{cases} [b-\delta, b) & \text{if } b < +\infty \\ [\delta^{-1}, \infty) & \text{if } b = +\infty. \end{cases}$$

Because we are concerned with the behavior of $S_n \equiv S(x_1, \ldots, x_n)$ as $n \to \infty$, we must consider the space $\mathcal{X}$ of all possible infinite sample sequences $(x_1, x_2, \ldots)$. For each n let A_n denote a subset of $\mathcal{X}$ which depends only on $(x_1, \ldots, x_n)$, and define the event $\{A_n \text{ a.e.n.}\}$ by

$$\{A_n \text{ a.e.n.}\} = \liminf A_n \equiv \bigcup_{n=1}^{\infty} \bigcap_{r=n}^{\infty} A_r,$$

where "a.e.n." abbreviates "for almost every n". Thus for each $(x_1, x_2, \ldots)$ in $\{A_n \text{ a.e.n.}\}$ there exists an integer $N = N(x_1, x_2, \ldots)$ such that $(x_1, \ldots, x_n) \in A_n$ for every $n \geq N$. It is clear that

$$\{A_n \text{ a.e.n.}\} \cap \{B_n \text{ a.e.n.}\} = \{A_n \cap B_n \text{ a.e.n.}\} .$$

We now recall several facts from Perlman (1972, Sections II and V) concerning the a.s. convergence of quantities of the form

$$\sup_{\theta\varepsilon\Gamma} y_n(\theta) \quad \text{and} \quad \inf_{\theta\varepsilon\Gamma} y_n(\theta) , \tag{2.4}$$

where

$$y_n(\theta) = \frac{1}{n} \sum_{i=1}^{n} y(X_i,\theta) , \tag{2.5}$$

$y(x,\theta)$ is a real-valued function, and Γ is a subset of Ω such that the quantities in (2.4) are measurable. These facts will be applied in Sections III and IV with $y(x,\theta)$ given by $(\partial/\partial\theta) \log p(x,\theta)$ and $(\partial^2/\partial\theta^2) \log p(x,\theta)$, respectively. We remark that the need to allow $k \geq 2$ (see Definition 2.1) in the dominance assumptions imposed on these functions arises mainly in the case where θ is a vector-valued parameter, rather than real-valued as assumed in this paper - cf. Perlman (1972, page 269).

Definition 2.1. The function $y = y(x,\theta)$ is *dominated (dominated by* 0) on Γ with respect to P_0 if

$$E_0 \sup_{\Gamma} y_k(\theta) < \infty \; (<0)$$

for some integer $k \geq 1$. The function y is *locally dominated (locally dominated by* 0) on Γ with respect to P_0 if for each θ' in Γ there exists a neighborhood Γ' of θ' ($\theta'\varepsilon\Gamma' \subseteq \Gamma$) such that y is dominated (dominated by 0) on Γ'.

Lemma 2.2. If y is dominated (dominated by 0) on Γ_i, $1 \leq i \leq r$, then y is dominated (dominated by 0) on $\cup\{\Gamma_i : 1 \leq i \leq r\}$.

Lemma 2.3. If $E_0 y(X,\theta)$ is well-defined (possibly infinite) for every θ in Γ, then

$$P_0[\sup_{\Gamma} E_0 y(X,\theta) \leq \liminf_{n\to\infty} \sup_{\Gamma} y_n(\theta)] = 1.$$

Lemma 2.4. If $E_0 y(X,\theta)$ and $E_0 \sup_{\Gamma} y_n(\theta)$ are well-defined for every θ in Γ and almost every n, respectively, then

$$\sup_{\Gamma} E_0 y(X,\theta) \leq \downarrow \lim_{n\to\infty} E_0 \sup_{\Gamma} y_n(\theta) .$$

Lemma 2.5. If y is dominated on Γ, then

$$P_0[\limsup_{n\to\infty} \sup_{\Gamma} y_n(\theta) = \downarrow \lim_{n\to\infty} E_0 \sup_{\Gamma} y_n(\theta)] = 1 .$$

Lemma 2.6. Suppose that Γ is a compact subset of Ω such that y is locally dominated on Γ and $y(x,\cdot)$ is upper semicontinuous on Γ for almost every x. Then y is dominated on Γ,

$$P_0[\limsup_{n\to\infty} \sup_{\Gamma} y_n(\theta) = \sup_{\Gamma} E_0 y(X,\theta)] = 1 ,$$

and $E_0 y(X,\cdot)$ is upper semicontinuous on Γ.

For any compact subset $\Gamma \subseteq \Omega$, let $C(\Gamma)$ denote the separable Banach space of all continuous real-valued functions on Γ endowed with the sup norm.

Lemma 2.7. (Strong Law of Large Numbers for $C(\Gamma)$-valued random variables). Suppose that Γ is a compact subset of Ω such that $|y|$ is locally dominated on Γ and $y(x,\cdot)$ is continuous on Γ for almost every x. Then $|y|$ is dominated on Γ,

$$P_0[\sup_{\Gamma} |y_n(\theta) - E_0 y(X,\theta)| \to 0] = 1, \tag{2.6}$$

and $E_0 y(X,\cdot)$ is continuous on Γ.

The preceding result can be extended to the case where Γ is a separable subset of Ω and $C(\Gamma)$ denotes the separable Banach space of all bounded continuous real-valued functions on Γ endowed with the sup norm.

Lemma 2.8. Suppose that Γ is a separable subset of Ω such that $|y|$ is dominated on Γ and $y(x,\cdot)$ is continuous on Γ for almost every x. Then (2.6) holds, and $E_0 y(X,\cdot)$ is continuous on Γ.

III. GLOBAL LIMITING BEHAVIOR OF S_n

The global first-order conditions G1-G3 presented in this section are minimal assumptions that determine the limiting behavior of S_n away from θ_0. More restrictive but simpler conditions G2a, b and G3a, b also will be given.

G1. For almost every x, $(\partial/\partial\theta) \log p(x,\theta)$ exists for all θ in Ω.

G2. The functions $y^-(x,\theta) \equiv -(\partial/\partial\theta) \log p(x,\theta)$ and $y^+(x,\theta) \equiv (\partial/\partial\theta) \log p(x,\theta)$ are locally dominated by 0 on (a, θ_0) and (θ_0, b), respectively.

G3. There exists $\delta_0 \equiv \delta(\theta_0) > 0$ such that y^- and y^+ are dominated by 0 on $U^-_{\delta_0}$ and $U^+_{\delta_0}$, respectively.

Our first result follows immediately upon applying Lemma 2.5 with $(y, \Gamma) - (y^{\pm}, K^{\pm}_{\delta})$ and $(y^{\pm}, U^{\pm}_{\delta})$.

Theorem 3.1. (i) Suppose that conditions G1 and G2 hold. Then for all sufficiently small $\delta > 0$,

$$P_0[\exists \text{ no solutions of the LEQ in } K_\delta^- \cup K_\delta^+ \text{ a.e.}n\,] = 1\ . \qquad (3.1)$$

(ii) If in addition G3 holds, then for all sufficiently small $\delta > 0$

$$P_0[\exists \text{ no solutions of the LEQ in } \Omega - J_\delta \text{ a.e.}n\,] = 1. \qquad (3.2)$$

More precisely, if G1 and G2 are satisfied then for almost all sample sequences $(x_1, x_2, \ldots)$ and any $\delta > 0$, there exists an integer N (depending on δ, θ_0, and the sample sequence) such that

$$S_n \subset U_\delta^- \cup J_\delta \cup U_\delta^+ \qquad (3.3)$$

for all $n \geq N$. If G3 is satisfied as well, then (3.3) can be strengthened to

$$S_n \subset J_\delta\ . \qquad (3.4)$$

By Fatou's Lemma, conditions G1 and G2 together imply that if $I(\cdot,\theta_0)$ is finite in a neighborhood of θ, then

$$\begin{cases} I_1^-(\theta,\theta_0) \geq g_1(\theta,\theta_0) > 0 & \text{for} \quad \theta < \theta_0 \\ I_1^+(\theta,\theta_0) \leq g_1(\theta,\theta_0) < 0 & \text{for} \quad \theta > \theta_0, \end{cases} \qquad (3.5)$$

where $I_1^-(\theta,\theta_0)$ and $I_1^+(\theta,\theta_0)$ denote the lower and upper derivatives of $I(\cdot,\theta_0)$ at θ. Thus, G1-G2 imply that $\{p(x,\theta):\ \theta \text{ in } \Omega\}$ is strongly K-L ordered at θ_0, according to the following definition:

Definition 3.2. The family $\{p(x,\theta):\ \theta \text{ in } \Omega\}$ is *Kullback-Leibler ordered* (briefly, *K-L ordered*) at θ_0 if $0 > I(\theta_1, \theta_0) > I(\theta_2,\theta_0)$ whenever $\theta_0 < \theta_1 < \theta_2$ or $\theta_2 < \theta_1 < \theta_0$. The family is *strongly K-L ordered* at θ_0 if

$$\begin{cases} I_1^-(\theta,\theta_0) > 0 & \text{for} \quad \theta < \theta_0 \\ I_1^+(\theta,\theta_0) < 0 & \text{for} \quad \theta > \theta_0\ . \end{cases} \qquad (3.6)$$

This suggests the introduction of two conditions stronger than G1-G2 but which may be easier to verify:

G1a. For almost every x, $(\partial/\partial\theta) \log p(x,\theta)$ exists and is continuous at each θ in Ω.

G2a. The function $|y|$ is locally dominated on $\Omega-\{\theta_0\}$, where $y(x,\theta) = (\partial/\partial\theta) \log p(x,\theta)$, and the family $\{p(x,\theta): \theta \text{ in } \Omega\}$ is strongly K-L ordered at θ_0.

By the Dominated Convergence Theorem, G1a-G2a imply that $I_1(\theta, \theta_0) = g_1(\theta,\theta_0)$ for θ in $\Omega-\{\theta_0\}$, and that both are continuous in θ. It is readily verified that G1a-G2a imply G1-G2. Alternatively, it follows directly from Lemma 2.7 that G1a-G2a imply the conclusion of Theorem 3.1(i).

Again by Fatou's Lemma, G3 implies that $g_1(\theta, \theta_0)$ is bounded away from 0 as $\theta \to \partial\Omega \equiv \{a,b\}$, strengthening (3.5). If $I_1(\theta,\theta_0)$ exists, this implies that $I_1(\theta, \theta_0)$ also is bounded away from 0 as $\theta \to \partial\Omega$. This suggests the following condition, stronger than G3.

G3a. There exists $\delta_0 \equiv \delta(\theta_0) > 0$ such that $|y|$ in G2a is dominated on $U^-_{\delta_0}$ and $U^+_{\delta_0}$, and $I_1(\theta,\theta_0)$ is bounded away from 0 as $\theta \to \partial\Omega$.

It again can be verified that G3a implies G3, while the fact that G1a-G3a imply the conclusion of Theorem 3.1(ii) follows alternatively from Lemma 2.8.

The following two conditions, based on continuity, also imply G2 and G3, respectively.

G2b. For some $k \geq 1$, $(\partial/\partial\theta) \log L_k(\theta)$ is continuous at each θ in $\Omega-\{\theta_0\}$ uniformly in $(x_1, \ldots, x_k)$, and $g_1(\theta, \theta_0) > 0(<0)$ if $\theta < \theta_0(>\theta_0)$.

G3b. For some $k \geq 1$, the limits $\lim_{\theta \to a,b} (\partial/\partial\theta) \log L_k(\theta)$ exist uniformly in $(x_1, \ldots, x_k)$, and

$$E_0[\lim_{\theta \downarrow a} (\partial/\partial\theta) \log L_k(\theta)] > 0 > E_0[\lim_{\theta \uparrow b} (\partial/\partial\theta) \log L_k(\theta)].$$

Finally, we remark that G1 and G2-2a-2b imply that $0 > I(\theta,\theta_0)$ $(\geq -\infty)$ if $\theta \neq \theta_0$, so that θ is an *identifiable* parameter, i.e., if $\theta \neq \theta_0$ then $p(\cdot,\theta)$ and $p(\cdot, \theta_0)$ determine distinct distributions. The latter obviously is a necessary condition for the existence of a unique consistent root of the likelihood equation.

IV. LOCAL LIMITING BEHAVIOR OF S_n

We now consider the limiting behavior of S_n, the set of solutions of the LEQ, in a neighborhood of θ_0. The following set of minimal local regularity conditions will be used: There exists $\delta_0 \equiv \delta(\theta_0) > 0$ such that:

L1. $\theta \in J_{\delta_0}$, $\theta \neq \theta_0 \Rightarrow p(\cdot, \theta)$ and $p(\cdot, \theta_0)$ determine distinct probability distributions;

L2. For almost every x, $p(x,\theta)$ is an upper semicontinuous function of θ on J_{δ_0};

L3. For almost every x, $(\partial/\partial\theta) \log p(x,\theta)$ exists for all θ in J_{δ_0};

L4. For almost every x, $(\partial^2/\partial\theta^2) \log p(x,\theta)$ exists for all θ in J_{δ_0};

L5. $y(x,\theta) \equiv (\partial^2/\partial\theta^2) \log p(x,\theta)$ is dominated by 0 on J_{δ_0}.

Clearly, L4 $\Rightarrow$ L3 $\Rightarrow$ L2, while L5 implies the local identifiability condition L1.

Proposition 4.1 below requires conditions L1-L3 and is well-known (cf. Cramer (1946) or Rao (1972)). Proposition 4.2, a traditional basis for the statement "there exists a unique consistent root of the LEQ," appears in Huzurbazar (1948) under slightly stronger conditions than L4-L5 assumed here. Our conditions L4-L5 imply the conclusion of Theorem 4.3, which is stronger than Proposition 4.2. A version of Theorem 4.3 under slightly stronger assumptions was stated by Le Cam (1953, p. 308).

Proposition 4.1. If L1-L2 are satisfied, then for all $\delta > 0$,

$$P_0[\exists \text{ at least one relative maximum of } L_n(\theta) \text{ in } J_\delta \; a.e.n.] = 1.$$

If L3 also holds, then for all $\delta > 0$,

$$P_0[\exists \text{ at least one solution of the LEQ in } J_\delta \; a.e.n.] = 1,$$

and this solution gives a relative maximum of the LF. Hence, there exists *at least one* strongly consistent root of the LEQ.

Proposition 4.2. (Huzurbazar). Suppose that L4-L5 are satisfied. Let $\{\bar{\theta}_n\} \equiv \{\bar{\theta}_n(X_1,\ldots,X_n)\}$ be any strongly consistent sequence of estimators for θ_0 (not necessarily roots of the LEQ). Then

$$P_0[(\partial^2/\partial\theta^2) \log L_n(\bar{\theta}_n) < 0 \; a.e.n.] = 1.$$

If $\{\hat{\theta}_n\}$ and $\{\hat{\theta}_n^*\}$ are strongly consistent estimating sequences for θ_0 such that both are roots of the LEQ, then

$$P_0[\hat{\theta}_n = \hat{\theta}_n^* \; a.e.n.] = 1 .$$

Theorem 4.3. Suppose that L4-L5 are satisfied. Then

$$P_0[\sup_{J_{\delta_0}} \frac{1}{n} (\partial^2/\partial\theta^2) \log L_n(\theta) < 0 \ a.e.n.] = 1 .$$

Thus (by Proposition 4.1) for all sufficiently small $\delta > 0$,

$$P_0[\exists \text{ exactly one solution of the LEQ in } J_\delta \ a.e.n.] = 1,$$

and this solution gives a relative maximum of the LF.

Proof. Apply Lemma 2.5 with $y(x,\theta) = (\partial^2/\partial\theta^2) \log p(x,\theta)$ and $\Gamma = J_{\delta_0}$.

Together, Theorems 3.1 and 4.3 imply that if L4-L5 and G1-G2 hold, then with probability one all members of S_n except one must approach $\partial\Omega \equiv \{a, b\}$ as $n \to \infty$. The exceptional solution is a relative maximum of the LF and converges to θ_0 as $n \to \infty$. If in addition G3 is satisfied, then with probability one there exists a unique solution of the LEQ for a.e.n. This solution gives the unique *absolute* maximum of the LF, hence is the maximum likelihood estimate (MLE), and converges to θ_0. [Hence with probability one, the MLE exists and is unique for a.e.n, and is strongly consistent. Of course, standard conditions for the existence, uniqueness, and strong consistency of the MLE do not require differentiability of $p(x,\theta)$ - cf. Wald (1949), Le Cam (1953), or Perlman (1972).]

By Fatou's Lemma, conditions L4-L5 imply that if $I(\cdot, \theta_0)$ is finite in a neighborhood of θ, then

$$I_2^+(\theta,\theta_0) \leq g_2(\theta,\theta_0) < 0 \qquad \text{for } |\theta-\theta_0| < \delta_0 \tag{4.1}$$

where $I_2^+(\theta, \theta_0)$ denotes the upper symmetric second derivative of $I(\theta, \theta_0)$, i.e.,

$$I_2^+(\theta,\theta_0) = \limsup_{\varepsilon \to 0} \varepsilon^{-2}[I(\theta+\varepsilon,\theta_0)-2I(\theta,\theta_0)+I(\theta-\varepsilon,\theta_0)]. \quad (4.2)$$

If $I_2(\theta_0,\theta_0)$ exists, (4.1) implies that $I_2(\theta_0, \theta_0) < 0$. This suggests the following two conditions, which are stronger than L4-L5 but possibly easier to verify:

There exists $\delta_0 \equiv \delta(\theta_0) > 0$ such that:

L4a. For almost every x, $(\partial^2/\partial\theta^2) \log p(x,\theta)$ exists and is continuous at each θ in J_{δ_0};

L5a. The function $|y|$ is dominated on J_{δ_0}, where $y(x,\theta) = (\partial^2/\partial\theta^2) \log p(x,\theta)$, and $I_2(\theta_0,\theta_0) < 0$.

By the Dominated Convergence Theorem and (4.2), L4a-L5a imply that $I_2(\theta, \theta_0) = g_2(\theta, \theta_0)$ for θ in J_{δ_1}, $\delta_1 < \delta_0$, and that both are continuous in θ on J_{δ_1}. By Lemmas 2.5 and 2.6 it can be seen that L4a-L5a imply L4-L5 (with δ_0 replaced by δ_1). Alternatively, Lemma 2.7 can be applied to show that L4a-L5a also imply the conclusion of Theorem 4.3.

The following condition also is stronger than L5:

L5b. For some $k \geq 1$, $(\partial^2/\partial\theta^2) \log L_k(\theta)$ is continuous at θ_0 uniformly in $(x_1, \ldots, x_k)$, and $g_2(\theta_0,\theta_0) < 0$.

Finally note that if the equation $\int p(x,\theta)\, d\mu(x) = 1$ can be differentiated twice under the integral sign then

$$g_2(\theta_0,\theta_0) = -I(\theta_0) \equiv -E_0[(\partial/\partial\theta) \log p(X,\theta_0)]^2 , \quad (4.3)$$

where $I(\theta_0)$ denotes the Fisher Information number. The assumption $I(\theta_0) > 0$ commonly occurs in the statements of standard results about likelihood estimation, but our treatment suggests that $g_2(\theta_0,\theta_0) < 0$ may be the more natural condition.

V. SOME EXAMPLES OF K-L ORDERED FAMILIES

First, consider a family of probability densities $\{p(x,\theta)\colon\ \theta \text{ in } \Omega\}$ such that $\log p(x,\theta)$ is concave in θ on Ω, so that $I(\theta,\theta_0)$ also is concave in θ. If the family satisfies a weak identifiability condition such as L1 (see Section IV), then $I(\theta, \theta_0)$ has a unique maximum at $\theta = \theta_0$, so by concavity must strictly decrease as θ moves away from θ_0 as long as it $(I(\theta, \theta_0))$ remains finite. Even more, (3.6) holds provided that $I(\cdot, \theta_0)$ is finite in a neighborhood of θ. Thus, if $I(\theta, \theta_0)$ is finite for all θ in Ω, log concavity of $p(x,\cdot)$ plus identifiability implies that the family of densities must be strongly K-L ordered at θ_0. Furthermore, under these assumptions $I_1^-(\theta, \theta_0)$ and $I_1^+(\theta, \theta_0)$ are decreasing in θ, hence are bounded away from 0 as $\theta \to \partial\Omega$ (see conditions G3a in Section III). Thus under the necessary dominance and smoothness assumptions, Theorems 3.1 and 4.3 hold in this case.

The assumptions on $p(x,\theta)$ in the preceding paragraph are satisfied in the well-known case of an exponential family, i.e., where $p(x,\theta)$ is of the form

$$p(x,\theta) = \beta(\theta) \exp\{\theta T(x)\}$$

for some real-valued statistic $T(x)$. Here, of course, under slight additional assumptions the conclusions of Theorems 3.1 and 4.3 hold not merely for a.e.n but in fact for every n - see Berk (1972). Haberman (1979), Mäkeläinen, Schmidt, and Styan (1981), Barndorff-Nielsen and Blaesild (1980), and Scholz (1981) also are relevant references.

Next, we consider the case where $\{p(x,\theta)\colon\ \theta \text{ in } \Omega\}$ is a (strict) monotone likelihood ratio family, i.e., where $p(x,\theta_1)/p(x,\theta_2)$ is a (strictly) increasing function of x whenever $\theta_1 > \theta_2$ (or, more generally, a (strictly) increasing function of some real-valued statistic $T(x)$). Throughout this discussion it is assumed that $0 > I(\theta_1, \theta_2) > -\infty$ for all $\theta_1 \neq \theta_2$ in Ω, so that $p(x,\theta_1)/p(x,\theta_2)$ is not a degenerate function of x.

Proposition 5.1. (i) A monotone likelihood ratio (MLR) family is K-L ordered.

(ii) Suppose that (a) the differentiability condition G1 holds; (b) the Fisher Information number $I(\theta) > 0$ for all θ in Ω (see (4.3)); (c) the equation $\int p(x,\theta)\, d\mu(x) = 1$ can be differentiated once under the integral sign at each θ in Ω. If $\{p(x,\theta)\colon \theta \text{ in } \Omega\}$ is a strict MLR family, then $g_1(\theta,\theta_0)>0(<0)$ for $\theta < \theta_0 (>\theta_0)$. If in addition the functions y^- and y^+ in G2 are locally dominated on (a,θ_0) and (θ_0, b) respectively, then (3.5) holds, so $\{p(x,\theta)\}$ is strongly K-L ordered.

Proof: (i) Suppose that $\theta_2 < \theta_1 < \theta_0$ (the proof for the case $\theta_0 < \theta_1 < \theta_2$ is similar). Then

$$\begin{aligned} I(\theta_1,\theta_0)-I(\theta_2,\theta_0) &= \int \left\{ \log\left[\frac{p(x,\theta_1)}{p(x,\theta_2)}\right] \right\} p(x,\theta_0)\, d\mu(x) \\ &= \int \left\{ \log\left[\frac{p(x,\theta_1)}{p(x,\theta_2)}\right] \right\} \left\{ \frac{p(x,\theta_0)}{p(x,\theta_1)} \right\} p(x,\theta_1)\, d\mu(x) \\ &\geq \{-I(\theta_2,\theta_1)\}\{\textstyle\int p(x,\theta_0)\, d\mu(x)\} \\ &> 0, \end{aligned}$$

where the first inequality follows from the fact that the covariance of two increasing functions of x is nonnegative.

(ii) By (i), (3.6) holds with $>0(<0)$ replaced by $\geq 0(\leq 0)$; it must be shown that in fact (3.5) holds. For $\theta < \theta_0$,

$$
\begin{aligned}
g_1(\theta,\theta_0) &= \int\{(\partial/\partial\theta)\log p(x,\theta)\}\, p(x,\theta_0)\, d\mu(x) \\
&= \int\{(\partial/\partial\theta)\log p(x,\theta)\}\left\{\frac{p(x,\theta_0)}{p(x,\theta)}\right\} p(x,\theta)\, d\mu(x) \\
&> g_1(\theta,\theta)\int p(x,\theta_0)\, d\mu(x) \\
&= 0 .
\end{aligned}
$$

The inequality follows since $p(x,\theta_0)/p(x,\theta)$ is strictly increasing in x, while $(\partial/\partial\theta)\log p(x,\theta)$ is increasing and nondegenerate (by (b)). The final equality follows from (c). If y^- is locally dominated at θ, then $I_1^-(\theta,\theta_0) \geq g_1(\theta,\theta_0)$ by Fatou's Lemma, so (3.5) is valid.

As a third example, consider the case where $\{p(x,\theta)\}$ is a location-parameter family, i.e., $p(x,\theta) = f(x-\theta)$. Here, x is a real variable, $\Omega = (-\infty,\infty)$, and f is a probability density (with respect to Lebesgue measure) such that $f(x) > 0$ for $-\infty < x < \infty$. Clearly, $I(\theta_1,\theta_2) < 0$ whenever $\theta_1 \neq \theta_2$; we assume further that $I(\theta_1, \theta_2) > -\infty$.

If $p(x,\theta) \equiv f(x-\theta)$ has a MLR then f must be log concave, so the strong conclusions in the first paragraph of this section obtain. Rather than assuming the MLR property, however, we shall show that the conclusions of Proposition 5.1 remain valid if, instead, it is assumed that f is symmetric and unimodal: $f(x) = f(-x)$ for all x, and f is decreasing on $[0,\infty)$ (strict monotonicity is not assumed).

Proposition 5.2. (i) If f is a symmetric and unimodal density on $(-\infty,\infty)$ with $f(x) > 0$ for all x, then the family $\{f(x-\theta)\colon\ -\infty < \theta < \infty\}$ is K-L ordered.

(ii) Suppose in addition that f is continuously differentiable on $(-\infty,\infty)$. Then $g_1(\theta,\theta_0) > 0(<0)$ for $\theta < \theta_0(>\theta_0)$. If also y^- and y^+ in G2 are locally dominated on

$(-\infty,\theta_0)$ and (θ_0,∞), respectively, then (3.5) holds, so $\{f(x-\theta)\}$ is strongly K-L ordered.

Proof. (i) We may take $\theta_0 = 0$ and choose $\theta_2 > \theta_1 > 0$. We must show that

$$J \equiv \int_{-\infty}^{\infty} \log\left[\frac{f(x-\theta_1)}{f(x-\theta_2)}\right] f(x)\,dx > 0.$$

Defining $\nu = \frac{1}{2}(\theta_1+\theta_2)$, $\delta = \frac{1}{2}(\theta_2-\theta_1)$, and $t(y) = \log[f(y+\delta)/f(y-\delta)]$, we find that

$$J = \int_{-\infty}^{\infty} t(x-\nu)f(x)\,dx = \int_{-\infty}^{\infty} t(x)f(x+\nu)\,dx .$$

Since $f(x)$ is symmetric, $t(y) = -t(-y)$ so

$$J = \int_{0}^{\infty} t(x)[f(x+\nu) - f(x-\nu)]\,dx.$$

Furthermore, since $\delta > 0$ and $\nu > 0$, the unimodality and symmetry of $f(x)$ imply that for all $x > 0$, $t(x) \leq 0$ and $f(x+\nu) - f(x-\nu) \leq 0$, so $J \geq 0$. To see that $J > 0$, consider the set $D_\nu = \{x: x \geq \nu,\ t(x) < 0\}$. Since $f(x)$ is positive and decreases to zero as $x \to \infty$, D_ν must have positive measure (otherwise $f(x)$ would be a constant). Also, since $0 < \delta < \nu$, x in D_ν implies that $f(x+\nu) - f(x-\nu) < 0$. Therefore,

$$J \geq \int_{D_\nu} t(x)[f(x+\nu) - f(x-\nu)]\,dx > 0.$$

(ii) Again by (i), (3.6) holds but without strict inequality, so (3.5) must be verified. Without loss of generality take $\theta_0 = 0$. Letting $h(x) = f'(x)/f(x)$, we find that

$$-g_1(\theta,0) = \int_{-\infty}^{\infty} h(x-\theta)f(x)dx$$

$$= \int_{-\infty}^{\infty} h(x)f(x+\theta)dx$$

$$= \int_{0}^{\infty} h(x)[f(x+\theta)-f(x-\theta)]dx,$$

since $h(x) = -h(-x)$. Suppose that $\theta > 0$. Then for all $x > 0$, $f(x+\theta) - f(x-\theta) \leq 0$ and $f'(x) \leq 0$ by symmetry and unimodality, so $g_1(\theta, 0) \leq 0$. To see that this inequality is actually strict, consider the set $B_\theta = \{x: x \geq \theta, f'(x) < 0\}$. Now, $f(x) > 0$ and $f'(x) \leq 0$ for all positive x, $f'(x)$ is continuous, and $f(x) \to 0$ as $x \to \infty$, so B_θ must have positive measure (otherwise $f(x)$ would be constant for $x \geq \theta$). Furthermore, by unimodality and symmetry, x in B_θ implies that $f(x+\theta) - f(x-\theta) < 0$, so

$$-g_1(\theta,0) \geq \int_{B_\theta} h(x)[f(x+\theta)-f(x-\theta)]dx > 0.$$

Lastly, if y^+ is locally dominated at θ, then $I_1^+(\theta,0) \leq g_1(\theta, 0)$ by Fatou's Lemma, which implies (3.5) and completes the proof.

We remark that when $p(x,\theta) = f(x-\theta)$, simpler versions of some of our earlier conditions can be stated easily. For example, G2b will be satisfied if $f'(x)/f(x)$ is uniformly continuous on $(-\infty,\infty)$, while L5b will hold if $(f''/f)-(f'/f)^2$ is uniformly continuous on $(-\infty,\infty)$, $\int f'' = 0$, and $\mathrm{Leb}\{x: f'(x)\neq 0\}>0$.

VI. TWO SPECIAL FAMILIES

Example 6.1. The Cauchy location-parameter family. Here, $p(x,\theta) = f(x-\theta)$ with $f(x) = \pi^{-1}(1+x^2)^{-1}$, x and θ in $(-\infty,\infty) \equiv \Omega$, and μ = Lebesgue measure. With the help of Proposition 5.2(ii) and the remark at the end of Section V, it is readily verified that conditions G1, G2b, L4, and L5b are satisfied: log f is twice continuously differentiable on $(-\infty,\infty)$, $(\log f)'$ and $(\log f)''$ are uniformly continuous on $(-\infty,\infty)$, $\int f'' = 0$, and $f' \neq 0$ a.e. Thus, the conclusions of Theorems 3.1(i) and 4.3 hold for the Cauchy location-parameter family. By the remark after Theorem 4.3 it follows that for any bounded interval $K \subset \Omega$, no matter how large, if $\theta_0 \in$ interior (K) then with probability 1 there is exactly one solution of the LEQ in K for a.e.n, this solution gives a relative maximum of the LF, and converges to θ_0. Furthermore, since the conditions of Wald (1949) and Perlman (1972) that guarantee the existence and strong consistency of the MLE are satisfied in the Cauchy case, the unique solution of the LEQ in K *must coincide* with the MLE for a.e.n.

Next, note that $g_1(\theta, \theta_0) \to 0$ as $\theta \to \pm\infty$ (by the Bounded Convergence Theorem), so condition G3 must fail and Theorem 3.1(ii) is not applicable. This does not guarantee the existence of multiple roots of the LEQ, of course, but direct examination of the LEQ shows that there may be as many as 2n-1 solutions, of which n may be relative maxima (cf. Barnett (1966)). Barnett's empirical study shows, in fact, that for $10 \leq n \leq 20$, multiple relative maxima occur in approximately 30% of all samples. If multiple solutions do occur infinitely often as $n \to \infty$, then the above discussion shows that all but one solution must approach $\partial\Omega \equiv \{\pm\infty\}$.

In a recent paper, Reeds (1979) has established the following remarkable result. Let $R_n \equiv R(x_1,\ldots,x_n)$ denote the number of relative maxima of the LEQ; R_n-1 is a random variable assuming values 0, 1, ..., n-1 and indicates the number of "false" relative maxima of the LF. Then R_n-1 converges in distribution to the Poisson (λ) distribution with $\lambda = \pi^{-1}$ as $n \to \infty$. In particular, the probability of at least one false relative maximum approaches $1-\exp(-\pi^{-1}) \approx 0.273$ as $n \to \infty$.

Copas (1975) has shown, however, that in the case of the *two-parameter* Cauchy family with both location and scale parameters, the LEQ's have a *unique* solution for all n and almost all sample sequences, and this solution gives the absolute maximum of the LF, i.e., the MLE.

Example 6.2. The bivariate normal distribution with unknown correlation but known means and variances. Take $x = (y, z)$, $\Omega = (-1, 1)$, $\theta = \rho$, and

$$p(x,\rho) = (1-\rho^2)^{-\frac{1}{2}} \exp\{-\tfrac{1}{2}(1-\rho^2)^{-1}(y^2-2\rho yz+z^2)\}.$$

The LEQ is a cubic equation in ρ (cf. Kendall and Stuart (1973), Example 18.3) which may have either one or three solutions in Ω. Let ρ_0 denote the true (but unknown) value of the correlation coefficient ρ. Then (cf. Kendall and Stuart (1973), Example 18.6)

$$(\partial/\partial\rho) \log p(x,\rho) = (1-\rho^2)^{-2}\{\rho(1-\rho^2)+(1+\rho^2)yz-\rho(y^2+z^2)\},$$

$$(\partial^2/\partial\rho^2)\log p(x,\rho)=(1-\rho^2)^{-3}\{(1-\rho^4)+2\rho(3+\rho^2)yz - (1+3\rho^2)(y^2+z^2)\},$$

and it is easily shown that the absolute values of these two functions are locally dominated on Ω. Therefore,

$$I_1(\rho,\rho_0) = g_1(\rho,\rho_0) = (1-\rho^2)^{-2}(1+\rho^2)(\rho_0-\rho),$$
$$I_2(\rho_0,\rho_0) = g_2(\rho_0,\rho_0) = -(1-\rho_0^2)^{-2}(1+\rho_0^2),$$

so that the family $\{p(x,\rho): -1<\rho<1\}$ is strongly K-L ordered and $I_2(\rho_0,\rho_0) < 0$. Thus conditions G1, G2a, L4, and L5a are satisfied (but not G2b or L5b), so that the conclusions of Theorems 3.1(i) and 4.3 hold.

In this example, conditions G3a and G3b fail, and it is difficult to determine whether or not G3 is satisfied. There is a simple alternate approach, however, that shows that the conclusion of Theorem 3.1(ii) holds. Let $y(x,\rho)$ be the function appearing in G3a, i.e., $y(x,\rho) = (\partial/\partial\rho) \log p(x,\rho)$, and define

$$y^*(x,\rho) = c(\rho)y(x,\rho),$$

where $c(\rho) = (1-\rho^2)^2$. Since

$$|y^*(x,\rho)| \leq 1+(y+z)^2,$$

it is immediate that $|y^*|$ is dominated on Ω, hence on U_δ^- and U_δ^+ for all $\delta > 0$. Furthermore,

$$\begin{aligned} g_1^*(\rho,\rho_0) &\equiv E_0[y^*(X,\rho)] \\ &= c(\rho)g_1(\rho,\rho_0) \\ &= (1+\rho^2)(\rho_0-\rho) \end{aligned}$$

is bounded away from 0 as $\rho\to\partial\Omega$. Therefore, condition G3a is satisfied with y replaced by y^*, hence so is G3 with the same replacement. By applying Lemma 2.5 with $(y,\Gamma) = ((y^*)^{\pm}, U_\delta^{\pm})$ for sufficiently small δ, we conclude that

$$P_0[\exists \text{ no solutions of the (LEQ)* in } U_\delta^- \cup U_\delta^+ \text{ a.e.n}] = 1 \qquad (6.1)$$

where (LEQ)* denotes the equation

$$c(\rho)(\partial/\partial\rho) \log L_n(\rho) = 0.$$

However, $c(\rho) > 0$ on $U_\delta^- \cup U_\delta^+$, so (6.1) remains valid with (LEQ)* replaced by LEQ. From this and (3.1) it follows that (3.2) is valid, so the conclusion of Theorem 3.1(ii) is satisfied as claimed.

With probability 1, therefore, there is exactly one solution of the LEQ in Ω for a.e.n, this solution is a relative maximum, and it converges to ρ_0 as $n \to \infty$. (Incidentally, this *proves* the existence, uniqueness, and strong consistency of the MLE in this example, where verification of Wald's (1949) conditions would be more difficult.) The uniqueness of the solution for a.e.n was demonstrated by Kendall and Stuart (1973) by direct examination of the (cubic) LEQ.

VII. IMPLICATIONS FOR THE ESTIMATION OF θ_0.

If the LEQ has multiple solutions for some or all n and $(x_1, \ldots, x_n)$, we are faced with the problem of choosing one of these solutions for our estimate of θ_0. If it is known that the MLE exists and is a strongly consistent estimator of θ_0 (as is almost always the case, since Wald's (1949) conditions are quite weak - however, see Kraft and LeCam (1956)), then one should choose that solution at which the LF is greatest, as this is the MLE. Here two steps are involved - finding the set S_n of all solutions of the LEQ, and evaluating the LF at each solution. (The usual Newtonian methods, based on successive iterations starting with a preliminary estimate, may converge to a solution other than the absolute maximum or may fail to converge at all - see Barnett (1966).) One or both of these steps may be difficult or lengthy to carry out. To insure that all solutions (and hence all maxima) are found, the entire possible range of solutions must be scanned. For example, in the case of the Cauchy distribution, there may exist up to 2n solutions, with possible range $[x_{min}, x_{max}]$. Since the Cauchy distribution has heavy tails this range may be quite wide, so the procedure of scanning this range for 2n possible solutions and evaluating the LF at each may be extremely laborious. (Barnett recommends the method of "false positions," which, although time-consuming, enables one to obtain (approximately) the MLE by a systematic scanning and evaluation process, and which avoids the possibility of failure of convergence. See also Richards (1967).)

If, therefore, we are concerned with a fixed sample size n (moderate) and our objective is to find the MLE, then in general we cannot avoid the necessity of a lengthy scanning procedure if multiple solutions of the LEQ occur. If, however, we are primarily concerned with large sample sizes and are satisfied to obtain an estimator which is asymptotically equivalent to the MLE, then simpler methods are available. For example, it is well-known that if $\bar{\theta}_n - \theta_0 = O_p(n^{-\frac{1}{2}})$ then the first Newton-Raphson iterate $\theta_n^{(1)}$, starting with $\tilde{\theta}_n$ as a preliminary estimate, is asymptotically equivalent to the MLE $\hat{\theta}_n$ in the (weak) sense that $\theta_n^{(1)} - \hat{\theta}_n = o_p(n^{-\frac{1}{2}})$.

We now propose three methods of a different sort, based on the results of Sections III and IV, which are aimed at simplifying the scanning procedure. These methods avoid the necessity of finding *all* solutions of the LEQ. The first makes use of a preliminary estimator $\bar{\theta}_n$ and only requires that we consider solutions "near" $\bar{\theta}_n$, while the other two methods require only that a single solution be found. (Numerical methods still may be needed to locate a solution.) These three methods provide estimators θ_n^* which are equivalent to the MLE in the very strong sence that $P_0[\theta_n^* = \hat{\theta}_n \text{ a.e.n}] = 1$.

First, suppose that the conclusions of Theorem 4.3 are true and that we have a preliminary estimate $\bar{\theta}_n$ such that $P_0[\bar{\theta}_n \to \theta_0] = 1$. Nothing need be assumed about the rate of convergence. Let $\theta_n^* \equiv \theta_n^*(x_1,\ldots,x_n)$ be that solution of the LEQ which is closest to $\bar{\theta}_n$. (In practice one would check that θ_n^* is a relative maximum, and if not, one might take the next closest solution; however, this is not necessary.) Then by Theorem 4.3,

$$P_0[\theta_n^* \text{ is the solution closest to } \theta_0 \ a.e.n] = 1 \tag{7.1}$$

and

$$P_0[\theta_n^* \to \theta_0] = 1 . \tag{7.2}$$

Furthermore, if it is assumed that the MLE $\hat{\theta}_n$ exists and is strongly consistent, then

$$P_0[\theta_n^* = \hat{\theta}_n \ a.e.n] = 1. \tag{7.3}$$

This method may greatly reduce the range which must be scanned for solutions. If one solution, say θ_n', is found by any method, then we need only scan the interval $[\bar{\theta}_n - d, \bar{\theta}_n + d]$, where $d = |\bar{\theta}_n - \theta_n'|$, to find θ_n^*. Note that the conclusions of Proposition 4.2 are not strong enough to guarantee the validity of (7.1) and (7.2). The present method was suggested by Barnett's empirical study for the Cauchy distribution, where he showed that if $\bar{\theta}_n$ is the sample median, then $P_0[\theta_n^* \neq \hat{\theta}_n] \leq .02$ for small n (n = 3), while $P_0[\theta_n^* \neq \hat{\theta}_n] \leq .001$ for moderate n ($13 \leq n \leq 19$).

If, in addition, the family $\{p(x,\theta): \theta \text{ in } \Omega\}$ is strongly K-L ordered and the conclusion of Theorem 3.1(i) is true, preliminary information of a much less precise nature is sufficient to simplify the scanning procedure. Suppose it is known only that θ_0 lies in the interior of a compact interval $K \subset \Omega$. Let θ_n^* be any solution of the LEQ lying in K, i.e., θ_n^* is chosen arbitrarily from $S_n \cap K$. (There may exist no solution in K, but with probability 1 there must be a unique solution in K for a.e.n). Then (7.1) and (7.2) follow from Theorem 3.1(i), and (7.3) from the (assumed) strong consistency of the MLE. This method is applicable, for example, in the case of the Cauchy distribution with an unknown location parameter - see Example 6.1.

Finally, if the conclusion of Theorem 3.1(ii) is also true, then *no* preliminary information is needed. Let θ_n^* be *any* root of the LEQ in Ω, i.e., θ_n^* is chosen arbitrarily from S_n. In this case we know that the MLE is strongly consistent, so (7.1), (7.2) and (7.3) again hold. This method is applicable in Example 6.2.

The main import of the second and third methods is that if n is large (how large?), once we have found a solution (by any means) in K or Ω, respectively, we need not worry that we have found the "wrong" solution. Of course, even if these methods are applicable, the first method might be preferred if a preliminary consistent estimate $\bar{\theta}_n$ is available.

VIII. CONCLUDING REMARKS

The basic local second-order condition L5 and global first order conditions G2-G3 are sufficient but not necessary for the conclusions of Theorems 4.3 and 3.1 to be valid. If the necessary smoothness and boundedness (dominance) conditions hold but L5, G2, and/or G3 fail because either $I_2(\theta_0, \theta_0) = 0$, $I_1(\theta_1, \theta_0) = 0$ for some $\theta_1 \neq \theta_0$, or $I_1(\theta, \theta_0)$ is not bounded away from 0 as $\theta \to \partial\Omega$, it may be possible to find a smooth one-to-one reparametrization $\phi \equiv \phi(\theta)$: $\Omega \to \Omega$ such that the family $\{p^*(x,\phi);\ \phi \text{ in } \Omega\}$ does satisfy L5, G2, and G3, where $p^*(x,\phi) = p(x, \theta(\phi))$. (Throughout this discussion we assume that $I_i(\theta,\theta_0) = g_i(\theta, \theta_0)$, $i = 1, 2$.)

For example, if $I(\theta, \theta_0)$ is a sufficiently smooth function of θ in a neighborhood of θ_0 and $I_2(\theta_0, \theta_0) = 0$, then

$$I(\theta,\theta_0) = -\alpha(\theta-\theta_0)^{2r} + O(|\theta-\theta_0|^{2r+1})$$

for some integer $r \geq 2$ and scalar $\alpha > 0$. Consider a local reparametrization $\phi = \phi(\theta)$ such that

$$\phi(\theta) = \begin{cases} \theta_0 + c\phi_r(\theta-\theta_0), & |\theta - \theta_0| \leq \delta \\ \theta, & |\theta - \theta_0| > \delta \end{cases} \tag{8.1}$$

for an appropriate $c > 0$ and sufficiently small $\delta > 0$, where

$$\phi_r(t) = \begin{cases} t^r & \text{if } r \text{ is odd,} \\ t^r & \text{if } r \text{ is even and } t \geq 0, \\ -t^r & \text{if } r \text{ is even and } t < 0. \end{cases}$$

Then if $r \geq 3$ (the case $r = 2$ requires special consideration, since $\phi_2''(0)$ does not exist),

$$I^*(\phi,\phi_0) \equiv I(\theta(\phi), \theta(\phi_0)) = -\alpha c^{-2}(\phi-\phi_0)^2 + O(|\phi-\phi_0|^{2+(1/r)}),$$

where $\phi_0 = \phi(\theta_0) = \theta_0$, so that

$$I_2^*(\phi_0,\phi_0) = -2\alpha c^{-2} < 0,$$

where I^*, I_1^*, and I_2^* are defined in terms of $p^*(x,\phi)$. This situation is illustrated by the example on page 18 of Pitman (1979).

Thus, if $I_2(\theta_0,\theta_0) < 0$ except for a finite or countable set of isolated values of θ_0 in Ω, a global reparametrization $\phi = \phi(\theta)$ can be found such that $I_2^*(\phi_0, \phi_0) < 0$ for every ϕ_0 in Ω, so that L5 will be satisfied (provided that dominance still holds) for the family $\{p^*(x,\phi)\colon \phi \text{ in } \Omega\}$. By (4.3), it follows that under suitable regularity conditions

$$\mu\{x\colon (\partial/\partial\phi)\log p^*(x,\phi) \neq 0\} > 0 \tag{8.2}$$

at each ϕ in Ω. If we now suppose that the original family $\{p(x,\theta)\}$ was K-L ordered but not strongly K-L ordered (i.e., $I_1(\theta_1,\theta_0) = 0$ for some θ_1, θ_0 in Ω with $\theta_1 \neq \theta_0$), then (8.2) suggests (but does not guarantee) that the new family $\{p^*(x,\phi)\}$ may in fact be strongly K-L ordered and satisfy G2. (This occurs in the example in Pitman (1979) mentioned above.) If $\{p(x,\theta)\} \equiv \{p^*(x,\phi)\}$ is a *complete* family of distribution, then (8.2) does guarantee at least that $I_1^*(\phi, \phi_0)$ has no "non-moving" zeroes, i.e., there can be no ϕ_1 in Ω such that $I_1^*(\phi_1,\phi_0) = 0$ for all ϕ_0 in Ω.

Next suppose that L5 and G2 hold but G3 fails because $I_1(\theta,\theta_0)$ is not bounded away from 0 as $\theta \to \partial\Omega \equiv \{a,b\}$. In this case, local reparametrizations $\phi_a = \phi_a(\theta)$ and/or $\phi_b = \phi_b(\theta)$ in neighborhoods of a and/or b sometimes can be found so that $I_1^*(\phi_a,\phi_0)$ and/or $I_1^*(\phi_b,\phi_0)$ are bounded away from 0 as $\phi_a \to a$ and/or $\phi_b \to b$. To suggest suitable reparametrizations, note that for $\phi = \phi_a$ or ϕ_b,

$$
\begin{aligned}
I_1^*(\phi,\phi_0) &\equiv E_0[(\partial/\partial\phi) \log p^*(X,\phi)] \\
&= E_0[(\partial/\partial\phi) \log p(X,\theta(\phi))] \\
&= E_0[\{(\partial/\partial\theta) \log p(X,\theta)\}/\phi'(\theta)]_{\theta=\theta(\phi)} \\
&= [I_1(\theta,\theta_0)/\phi'(\theta)]_{\theta=\theta(\phi)}.
\end{aligned} \tag{8.3}
$$

This suggests choosing $\phi_a(\theta)$ and/or $\phi_b(\theta)$ such that

$$
\begin{cases}
\phi_a'(\theta) = I_1(\theta,\theta_0) \\
\phi_b'(\theta) = -I_1(\theta,\theta_0)
\end{cases} \tag{8.4}
$$

in neighborhoods of a and/or b. If, then, $\phi = \phi(\theta)$ is a smooth one-to-one global reparametrization that coincides with $\phi_a(\theta)$ and $\phi_b(\theta)$ in neighborhoods of a and b, respectively (and such that $\phi'(\theta) > 0$ everywhere, as we assume throughout), we conclude that $I_1^*(\phi,\phi_0)$ does remain bounded away from 0 as $\phi \to \partial\Omega$. Thus, G3 will be satisfied for $\{p^*(x,\phi)\colon \phi \text{ in } \Omega\}$ (provided that dominance has not been destroyed).

The expression (8.3) suggests a weaker version of condition G3 that still suffices for (3.2) in Theorem 3.1(ii).

<u>*G3´*</u>. There exists $\delta_0 \equiv \delta(\theta_0) > 0$ and functions $c_a(\theta) > 0$, $c_b(\theta) > 0$ on $U_{\delta_0}^-$ and $U_{\delta_0}^+$, respectively, such that $c_a y^-$ and $c_b y^+$ are dominated by 0 on $U_{\delta_0}^-$ and $U_{\delta_0}^+$, where y^- and y^+ are as in G2.

Analogous versions of conditions G3a and G3b also can be stated. The expressions (8.3) and (8.4) suggest the following natural choices for $c_a(\theta)$ and $c_b(\theta)$.

$$\begin{cases} c_a(\theta) = 1/I_1(\theta,\theta_0) \\ c_b(\theta) = -1/I_1(\theta,\theta_0) \end{cases} \tag{8.5}$$

in neighborhoods of a and b, respectively. (Bear in mind that we are considering a strongly K-L ordered family, so $I_1 \neq 0$.) It would be of interest to determine whether or not the functions c_a and c_b in (8.5) *must* satisfy G3 whenever there exist some functions c_a and c_b that do so.

Condition G3′ was motivated by consideration of the situation where the functions y^- and/or y^+ in G3 might be dominated but not dominated by 0, so that $I_1(\theta,\theta_0)$ is not bounded away from 0 at $\partial\Omega$. On the other hand, in Example 6.2, $I_1(\rho,\rho_0)$ is bounded away from 0, but y^- and/or y^+ may not be dominated. This necessitated the introduction of the function $c(\rho)$, which is easily seen to fulfill the requirements imposed on the functions c_a and c_b in G3′, so that G3′ could have been used in that example as well to verify Theorem 3.1(ii). This dual success of G3′ in both situations suggest that G3′, rather than G3, may be the "right" condition for Theorem 3.1(ii), especially if the conjecture in the preceding paragraph proves to be true.

The reader is invited to consider the following modification of Pitman's example: take $\Omega = (0,\infty)$, μ = Lebesgue measure on $(-\infty,\infty)$, and

$$p(x,\theta) = (2\pi)^{-\frac{1}{2}}\exp\{-\tfrac{1}{2}(x-\log\theta)^2\}.$$

In this example both situations described in the preceding paragraph occur, one at each endpoint of Ω, and G3′ can be applied to show that the conclusion of Theorem 3.1(ii) is valid.

Note that in Example 6.1, the result of Reeds (1979) implies that condition G3′, as well as G3, must fail.

Finally, if $\{p(x,\theta): \theta \text{ in } \Omega\}$ is not K-L ordered, so that $I(\theta,\theta_0)$ has a relative maximum at some $\theta_1 \neq \theta_0$, then (assuming the necessary smoothness and dominance), the LEQ will have relative maxima near θ_0 and θ_1 for a.e.n and the conclusion of Theorem 3.1(i) will not hold. Furthermore, there can exist no *monotone* reparametrization $\phi = \phi(\theta)$ such that this conclusion is satisfied for $\{p^*(x,\phi)\}$. In such a case, *all* solutions of the LEQ must be determined and the LF evaluated at each relative maximum, in order to obtain a consistent root of the LEQ (assuming that the MLE is consistent). There may exist a *non-monotone* reparametrization, however, or a higher-dimensional reparametrization, such that the conclusions of Theorem 3.1(i) do hold (suitably modified if the reparametrization must be multidimensional). It would be of interest to obtain a general formula or algorithm for determining such reparametrizations, for then the simplified procedures given in Section VII for obtaining an appropriate root of the LEQ could routinely be applied.

This paper leaves several important questions open. First, one would like to know necessary and sufficient conditions, stated in terms of the original parametric family $\{p(x,\theta)\}$, for the existence of a suitable (possibly non-monotone) reparametrization $\phi = \phi(\theta): \Omega \to \Omega^*$, such that the new family $\{p^*(x,\phi): \phi \text{ in } \Omega^*\}$ satisfies the basic conditions L5, G2, and/or G3. Second, as already mentioned, an explicit procedure for determining a suitable reparametrization is required. (These two questions are related to the discussion following condition G3′.) A third question is that of extending the ideas in this paper to the multiparameter case, while a fourth is that of dropping the assumption that the observations are independent and identically distributed. Finally, the assumptions that Ω is open and that the support of $P_0 (\equiv P_{\theta_0})$ does not vary with θ_0 may be weakened.

A preliminary version of this paper appeared as Perlman (1969).

Acknowledgements. My warm thanks go to G. P. H. Styan who called this topic to my attention, to R. R. Bahadur and E. L. Lehmann for their continual interest and encouragement during the preparation of this paper, to Daijin Ko for several helpful discussions, and to Professor Chernoff, whose graduate course on large sample theory at Stanford prepared me for this study.

REFERENCES

Barndorff-Nielsen, O., and Blaesild, P. (1980). "Global Maxima, and Likelihood in Linear Models," Research Report No. 57, Department of Theoretical Statistics, University of Aarhus.

Barnett, V. D. (1966). "Evaluation of the Maximum Likelihood Estimator Where the Likelihood Equation Has Multiple Roots," *Biometrika 53,* 151-165.

Berk, R. (1972). "Consistency and Asymptotic Normality of MLE's for Exponential Models," *Ann. Math. Statist. 43,* 193-204.

Chernoff, H. (1954). "On the Distribution of the Likelihood Ratio," *Ann. Math. Statist. 25,* 573-578.

Copas, J. B. (1975). "On the Unimodality of the Likelihood for the Cauchy Distribution," *Biometrika 62,* 701-704.

Cramer, H. (1946). *Mathematical Methods of Statistics.* Princeton University Press, Princeton, New Jersey.

Haberman, S. J. (1979). "Log-Concave Likelihoods and Maximum Likelihood Estimation," Technical Report No. 99, Department of Statistics, University of Chicago.

Huzurbazar, V. S. (1948). "The Likelihood Equation, Consistency, and the Maxima of the Likelihood Function," *Ann. Eugenics 14,* 185-200.

Kendall, M. G. and Stuart, A. (1973). *The Advanced Theory of Statistics,* Vol. 2 (3rd ed.). Griffin, London.

Kraft, C. and LeCam, L. (1956). "A Remark on the Roots of the Maximum Likelihood Equation," *Ann. Math. Statist. 27,* 1174-1177.

LeCam, L. (1953). "On Some Asymptotic Properties of Maximum Likelihood Estimates and Related Bayes Estimates," *University of California Publications in Statistics 1,* 277-328.

Mäkeläinen, T., Schmidt, K., and Styan, G. (1981). "On the Existence and Uniqueness of the Maximum Likelihood Estimate of a Vector-Valued Parameter in Fixed Sample Sizes," *Ann. Statist. 9*, 758-767.

Perlman, M. D. (1969). "The Limiting Behavior of Multiple Roots of the Likelihood Equation," Technical Report No. 125, Department of Statistics, University of Minnesota.

Perlman, M. D. (1972). "On the Strong Consistency of Approximate Maximum Likelihood Estimators," *Proceedings of the Sixth Berkeley Symposium on Mathematical Statistics and Probability*, Vol. 1, 263-282.

Pitman, E. J. G. (1979). *Some Basic Theory for Statistical Inference*. Chapman and Hall, London.

Rao, C. R. (1972). *Linear Statistical Inference and Its Applications*. (2nd ed.) Wiley, New York.

Reeds, J. A. (1979). "Asymptotic Number of Roots of Cauchy Location Likelihood Equations," To appear.

Richards, F. S. G. (1967). "On Finding Local Maxima of Functions of a Real Variable," *Biometrika 54*, 310-311.

Scholz, F. W. (1981). "On the Uniqueness of Roots of the Likelihood Equations," Technical Report No. 14 , Department of Statistics, University of Washington.

Wald, A. (1949). "Note on the Consistency of the Maximum Likelihood Estimate," *Ann. Math. Statist. 20*, 595-601.

ON SOME RECURSIVE RESIDUAL RANK TESTS FOR CHANGE-POINTS[1]

Pranab Kumar Sen

Department of Biostatistics
University of North Carolina
Chapel Hill, North Carolina

I. INTRODUCTION

A general class of recursive residuals is incorporated in the formulation of suitable (aligned) rank tests for change-points pertaining to some simple linear models. The asymptotic theory of the proposed tests rests on some invariance principles for recursively aligned signed rank statistics, and these are developed. Along with the asymptotic properties of the proposed tests, allied efficiency results are studied.

Let $X_1,\ldots,X_n$ be n independent random variables (r.v.), taken at time-points $t_1<\ldots<t_n$, respectively, where X_i has an unknown, continuous distribution function (d.f.) F_i, defined on the real line E $(= (-\infty,\infty))$, for $i=1,\ldots,n$. In the simplest model, one may assume that $F_1=\ldots=F_n=F$ (unknown), and, based on $X_1,\ldots,X_n$, one may then like to draw statistical inference on suitable parameters (functionals) of the d.f. F. There are, however, problems in which a change of the d.f. (and hence, the

[1] *Work supported by the National Heart, Lung and Blood Institute, Contract NIH-NHLBI-71-2243-L from the National Institutes of Health.* *This research is dedicated to Professor Herman Chernoff on the occasion of his 60th birthday.*

ISBN 0-12-589320-5

the parameters) may occur at an unknown time-point [viz., Page (1957)], so that it may of some interest to test for such a possible change occurring at some unknown time point in (t_1, t_n). In a somewhat more general setup, one may conceive of the usual linear model:

$$F_i(x) = F(x - \beta_i' c_i),\ x \in E,\ i=1,\ldots,n\ , \tag{1.1}$$

where the c_i are q-vectors ($q \geq 1$) of known regression constants, the β_i are unknown regression parameters and F is an unspecified, continuous d.f.; the location model is a special case of (1.1) with $q = 1$ and $c_i = 1$, $\forall\ i \geq 1$. One may then conceive of the null hypothesis H_0 of the constancy of the regression relationships over time, i.e.,

$$H_0:\ \beta_1=\ldots=\beta_n=\beta \text{ (unknown)}, \tag{1.2}$$

and based on the given c_i and the observed X_i, one may then proceed to estimate the common β or to draw other statistical conclusions on it. But, a constancy of the regression relationships may not hold, and a change may occur at some unknown time-point, i.e., one may have

$$\beta_1=\ldots=\beta_m \neq \beta_{m+1}=\ldots=\beta_n,\ \text{for some } m:\ 1 \leq m < n. \tag{1.3}$$

As such, one may desire to test for the null hypothesis H_0 in (1.2) against the composite alternatives in (1.3), where m is unknown.

Testing procedures for a possible change in location or regression relationship occurring at an unknown time-point between consecutively taken observations have been proposed and studied by a host of workers; a recent bibliography by Hinkley (1980) and a somewhat specialized monograph by Hackle (1980) provide some detailed accounts of these developments. In the parametric case, test statistics are constructed from either residuals based on the terminal estimator of β or recursive residuals based on sequential estimators of β. An excellent account of this work is available with Brown, Durbin and Evans (1975). Some further recent studies in this direction (allowing F to be possibly unspecified) are due to Deshayes and Picard (1981) and Sen (1982a, b), among others. In the nonparametric case, the developments are mostly restricted to the location model ($q=1$, $c_i=1$, $\forall\ i\geq 1$) where recursive ranking [viz., Bhattacharya and Frierson (1981)] is adaptable, as is also the pseudo reduction to the two sample problem [viz., A. Sen and Srivastava (1975) and Sen (1978)]. Some ad hoc procedures are also due to Bhattacharyya and Johnson (1968), while some aligned rank tests based on terminal estimates are due to Sen (1977). Most of these procedures encounter difficulties when applied to the general model in (1.1) - (1.3); for such models, some nonrecursive residual rank procedures are discussed in Sen (1982b).

The object of the present study is to incorporate a general class of recursive residuals in the formulation of suitable rank order test statistics for some change-point problems pertaining to (1.1) - (1.3). For the sake of simplicity of presentation, these procedures are considered first for the location model in Section II; the case of the general model in (1.1) - (1.3) is then treated in Section III. Asymptotic properties of the proposed tests are studied in Section IV. Section V deals with some allied asymptotic efficiency results. The Appendix is devoted to the derivation of some results on recursive estimates.

II. RECURSIVE RESIDUAL RANK TESTS FOR THE LOCATION MODEL

We confine ourselves to the location model, for which, in (1.1), we have

$$F_i(x) = F(x-\theta_i), \quad x \in E,\ i=1,\dots,n \quad , \tag{2.1}$$

where $\theta_1,\ldots,\theta_n$ are the location parameters, and we want to test for

$$H_0: \quad \theta_1=\ldots=\theta_n = \theta(\text{unknown}), \tag{2.2}$$

against the composite alternative

$$H: \quad \theta=\ldots=\theta_m \neq \theta_{m+1}=\ldots=\theta_n, \text{ for some } m: 1 \leq m < n. \tag{2.3}$$

We assume that F is symmetric about 0, so that θ_i is the median of the d.f. F_i, for i=1,...,n. First, we proceed to estimate θ recursively as follows.

For every k ($\geq$ 1), let $U_{k1}<\ldots<U_{kk}$ be the ordered r.v.'s of a sample of size k from the uniform (0,1) d.f., $\phi^+ = \{\phi^+(u),\ 0 < u < 1\}$ be a non-constant, nondecreasing and square-integrable *score function*, generated by a skew-symmetric $\phi = \{\phi(u),\ 0 < u < 1\}$ (i.e., $\phi(u) + \phi(1-u) = 0,\ 0 < u < 1$) in the following way:

$$\phi^+(u) = \phi((1+u)/2), \quad 0 < u < 1, (\phi^+(0) = 0) \tag{2.4}$$

and let

$$a_k(i) = E\phi^+(U_{ki}), \quad i = 1,\ldots,k;\ k \geq 1. \tag{2.5}$$

Some other regularity conditions on ϕ will be introduced in Section IV. For every (k ($\geq$ 1) and b (ϵ E), let $R^+_{ki}(b)$ be the rank of $|X_i-b|$ among $|X_1-b|,\ldots,|X_k-b|$, for i=1,...,k, and let

$$T_k(b) = T(X_1-b,\ldots,X_k-b) = \Sigma_{i=1}^{k} \operatorname{sign}(X_i-b)a_k(R^+_{ki}(b)). \tag{2.6}$$

Note that $T_k(b)$ is $\searrow$ in b (ϵ E), and under (2.2), $T_k(\theta)$ has a specified distribution, symmetric about 0. Hence, based on $X_1,\ldots,X_k$, $\hat{\theta}_k$, the usual rank order estimator of θ, may be defined as

$$\hat{\theta}_k = \tfrac{1}{2}(\sup\{b: T_k(b)>0\} + \inf\{b: T_k(b)<0\}). \tag{2.7}$$

At the kth stage, we may define the (recursive) residuals as

$$\hat{X}_{ki} = X_i - \hat{\theta}_{k-1},\ i = 1,\ldots,k, \text{ for } 2 \leq k \leq n, \tag{2.8}$$

while $\hat{X}_{11}$ is conventionally taken as equal to 0. Let $\hat{R}^{+}_{ki} = R_{ki}(\hat{\theta}_{k-1})$ be the rank of $|\hat{X}_{ki}|$ among $|\hat{X}_{k1}|,\ldots,|\hat{X}_{kk}|$, for $i=1,\ldots,k$, and let

$$\hat{u}_k = \text{sign } \hat{X}_{kk} a_k(\hat{R}^{+}_{kk}), \text{ for } k \geq 2; \ \hat{u}_1 = 0. \tag{2.9}$$

We define the cumulative sums (CUSUM) for the recursive residual rank statistics in (2.9) by

$$\hat{U}_r = \Sigma_{k\leq r}\hat{u}_k \ , \text{ for } \ 1 \leq r \leq n. \tag{2.10}$$

It may be noted that in (2.8), it may not be necessary to employ the rank order estimates $\{\hat{\theta}_k, \ k\leq n\}$, in (2.7), for defining these residuals. We may, under fairly general regularity conditions, employ other recursive estimators of θ as well. This point will be elaborated in Section IV. Let

$$A^2 = \int_0^1 \phi^2(u)du = \int_0^1\{\phi^+(u)\}^2 du, \tag{2.11}$$

so that by assumption, $0 < A < \infty$. Define then

$$D_n^+ = n^{-\frac{1}{2}}A^{-1}\{\max_{1\leq r\leq n} \hat{U}_r\} \text{ and } D_n = n^{-\frac{1}{2}}A^{-1}\{\max_{1\leq r\leq n} |\hat{U}_r|\} \tag{2.12}$$

The proposed test for (2.2) against (2.3) is based on the statistic D_n^+ (for the one-sided alternative $\theta_m < \theta_{m+1}$) or D_n (for the two-sided one: $\theta_m \neq \theta_{m+1}$). Unlike the procedures considered by A. Sen and Srivastava (1975), Sen (1978) and Bhattacharya and Frierson (1981), the proposed tests may not be genuinely distribution-free under H_0 in (2.2). Nevertheless, they are asymptotically distribution-free and are easily extendable to the general model in (1.1) - (1.3), where the other procedures run into obstacles. This will be considered in the next section. The distribution theory of D_n^+ and D_n, under the null hypothesis as well as (local) alternatives, needed for the study of the (asymptotic) properties of the proposed tests will be considered in Section IV. It may be remakred that instead of the Kolmogorov-Smirnov type statistics in (2.12) one may also consider some Cramér-von Mises' type statistics based on the CUSUM's in (2.10), viz.,

$$V_n = n^{-2} A^{-2} \Sigma_{r=1}^{n} (\hat{U}_r^2), \tag{2.13}$$

or some weighted version of the same. In view of the invariance principles for the CUSUMs in (2.10), to be developed in Section IV, distributional results on such statistics would follow under the same set of regularity conditions, and hence, these details are omitted. Generally, D_n^+ (or D_n) has better (asymptotic) performance than V_n and intuitively more appealing too. Moreover, for small values of m, V_n may not perform that well.

III. RECURSIVE RESIDUAL RANK TESTS FOR THE REGRESSION MODEL

We consider here the general model in (1.1) and proceed to test for (1.2) against (1.3). As in Section II, we assume that the d.f. F is symmetric about 0, and define the scores $\{a_k(i)\}$ as in (2.4) - (2.5). Also, we would employ here recursive estimates of $\underset{\sim}{\beta}$ [under (1.2)] in the construction of residuals and aligned rank statistics.

Assuming (1.2) to be true, based on $X_1,\ldots,X_k$, let $\hat{\underset{\sim}{\beta}}_k$ be some suitable estimator of $\underset{\sim}{\beta}$. The estimator $\hat{\underset{\sim}{\beta}}_k$ may be quite arbitrary (e.g., least squares estimator, rank order estimator or some other robust estimator) and will be defined more formally in Section IV. Then, at the kth stage, we define the (recursive) residuals as

$$\hat{X}_{ki} = X_i - \hat{\underset{\sim}{\beta}}{}'_{k-1} \underset{\sim}{c}_i, \quad 1 \le i \le k, \text{ for } k=1,\ldots,n. \tag{3.1}$$

[Usually, for $k \le q$, the $\hat{X}_{ki}$ are all equal to 0.] Let then $\hat{R}_{ki}^+$ be the rank of $|\hat{X}_{ki}|$ among $|\hat{X}_{k1}|,\ldots,|\hat{X}_{kk}|$, for $i=1,\ldots,k$; $k \ge 1$. Further, as in (2.9), we define the residual signed rank scores by

$$\hat{u}_k = \text{sign}(\hat{X}_{kk}) a_k(\hat{R}_{kk}^+), \quad k=q+1,\ldots,n, \tag{3.2}$$

and, conventionally, we let $\hat{u}_k = 0$ for $k \le q$. Then the CUSUM's for the residual rank scores in (3.2) are

$$\hat{u}_r = \Sigma_{k \leq r} \hat{u}_k \quad , \quad \text{for } r=1,\ldots,n. \tag{3.3}$$

Finally, we define A^2 as in (2.11), and parallel to (2.12), we let

$$D_n^+ = (n-q)^{-\frac{1}{2}} A^{-1} \{\max_{r \leq n} \hat{U}_r\}, \tag{3.4}$$

$$D_n = (n-q)^{-\frac{1}{2}} A^{-1} \{\max_{r \leq n} |U_r|\}. \tag{3.5}$$

The proposed tests are based on the statistics D_n^+ and D_n. It may be noted that unlike the location model, here, the use of D_n^+ may only be advocated for certain cases, where under the alternative hypothesis, the $\underset{\sim}{\beta}'\underset{\sim}{c}_k$ are monotone. This may not generally be the case, and the two-sided statistic D_n is more generally applicable. The necessary distribution theory of D_n^+ and D_n will be studied in Section IV.

It may be noted that for the location model (2.1), under H_0 in (2.2), the X_i are independent and identically distributed (i.i.d.) r.v.'s, so that one may also use the sequential ranking scheme as in Bhattacharya and Frierson (1981), where the ranks R_{kk} (of X_k among $X_1,\ldots,X_k$), for different k, are stochastically independent, and R_{kk} assumes the values $1,\ldots,k$ with the equal probability k^{-1}, for $k \geq 1$. On the other hand, for the general linear model in (1.1), even under H_0 in (1.2), the X_i, though independent, are not identically distributed (unless $\underset{\sim}{\beta} = \underset{\sim}{0}$ or the $\underset{\sim}{c}_i$ are all equal), and hence, the stochastic independence and uniformity of the distributions of the R_{kk} may not hold. Thus, the procedure suggested by Bhattacharya and Frierson (1981) may not be generally applicable for the testing problem in (1.2) - (1.3). Further, the residuals in (3.1) are, in general, neither independent, nor (marginally) identically distributed, so that the exact distribution of D_n (or D_n^+) may be difficult to obtain even for small n and simple scores; in fact, the same generally depends on the underlying F. For this reason, for general linear models, exact distribution-free tests for change-points may not exist and one may have to be satisfied with ADF tests. We shall see in Section IV that under fairly general regularity conditions (on F, the c_i and the scores), some invariance principles hold for the CUSUM's in (2.10) or (3.3), and these provide the ADF structure of D_n^+ or D_n, where H_0 in (1.2) holds.

We conclude this section with the remark that, by construction, the proposed tests (based on the statistics D_n^+ and D_n) are quasi-sequential in character. If we denote the upper 100 % points of the null distributions of D_n^+ and D_n by $D_n^+(\alpha)$ and $D_n(\alpha)$, respectively, and, if we let $k_{n\alpha}^+ = (n-q)^{\frac{1}{2}} AD_n^+(\alpha)$ and $k_{n\alpha} = (n-q)^{\frac{1}{2}} AD_n(\alpha)$, then, we may define a stopping variable N as the minimum $r(\leq n)$ such that $\hat{U}_r$ exceeds $k_{n\alpha}^+$ (or $|\hat{U}_r|$ exceeds $k_{n\alpha}$, in the two-sided case), where N is set to be equal to n, if no such $r(\leq n)$ exists. Thus N is properly defined, is bounded from above by n, and based on this stopping variable, we have a stopping rule which permits early stopping of the trial. This characterization of the proposed tests enables one to use the recursive residual rank statistics for the sequential detection problem too. In a sequential detection problem, we have the same model as in (1.1) - (1.3), but, n is not specified in advance. Rather, as it usually happens in a continuous inspection plan, the trial is continued indefinitely if there is no change in the parameters, while, in case of a change occurring at any point, the goal is to detect the same as soon as possible and make corrections. This setup therefore requires that the probability of early stopping when there is no change (i.e., false alarm) should be very small (or more specifically, the expected stopping time in a false alarm case should be large) and the excess over the change-point time for a finite change-point should be small (in the sense that the expected value of the same is small). This formulation requires some other invariance principles for the recursive residual rank statistics, which will not be treated in the current paper. However, we may note that in the context of quality control, periodic inspections and adjustments are usually done routinely, and hence, in this setup, the sequential detection problem reduces to a change-point problem, where the proposed quasi-sequential procedures workout well.

IV. ASYMPTOTIC PROPERTIES OF D_n^+ AND D_n

We consider first some invariance principles for the CUSUM's in (2.10) or (3.3), when H_0 in (1.2) or (2.2) may or may not hold. First, consider the case where H_0 in (1.2) holds, and define $Y_i = X_i - \underset{\sim}{\beta}'\underset{\sim}{c}_i$, $i \geq 1$. Then, the Y_i are i.i.d.r.v. with the common d.f. F. Also, let R_{ki}^+ be the rank of $|Y_i|$ among $|Y_1|,\ldots,|Y_k|$, for $i=1,\ldots,k$; $k \geq 1$. Define then

$$u_k = \text{sign}(Y_k)a_k(R_{kk}^+),\ k \geq 1, \tag{4.1}$$

$$U_r = \Sigma_{k\leq r} u_k\ ,\quad \text{for}\quad r=1,\ldots,n. \tag{4.2}$$

Note that under H_0 in (1.2), (i) sign Y_k and R_{kk}^+ are independent, (ii) sign Y_k assumes the values ± 1 with equal probability ½, (iii) R_{kk}^+ assumes the values $1,\ldots,k$ with equal probability k^{-1}, and (iv) for different k, (sign Y_k, R_{kk}^+) (and hence, u_k) are stochastically independent of each other. Thus

$$E(u_k|H_0) = 0 \text{ and } E(u_k^2|H_0) = A_k^2 = \frac{1}{k}\Sigma_{i=1}^{k} a_k^2(i), \tag{4.3}$$

$$E(U_r|H_0) = 0 \text{ and } E(U_r^2|H_0) = \Sigma_{k=1}^{r} A_k^2,\ \forall\ r \geq 1, \tag{4.4}$$

where $A_k^2 \to A^2$ as $K \to \infty$. Hence, if we consider a stochastic process $Z_n^o = \{Z_n^o(t),\ 0 \leq t \leq 1\}$, by letting

$$Z_n^o(t) = n^{-\frac{1}{2}}A^{-1}U_k \text{ for } k \leq nt < k+1,\ k=0,\ldots,n\ , \tag{4.5}$$

then under H_0 in (1.2), by the stochastic independence of the u_k and (4.3) - (4.4),

$$Z_n^o \underset{\mathcal{D}}{\to} Z, \text{ in the } J_1\text{-topology on } D[0,1], \tag{4.6}$$

where $Z = \{Z(t): 0 \le t \le 1\}$ is a standard Wiener process on [0,1]. Side by side, we introduce the stochastic process $Z_n = \{Z_n(t): 0 \le t \le 1\}$ by letting

$$Z_n(t) = (n-q)^{-\frac{1}{2}} A^{-1} \hat{U}_k, \text{ for } \frac{k-q}{n-q} \le t < \frac{k-q+1}{n-q}, \; k=q,\ldots,n. \tag{4.7}$$

Note that by (3.4), (3.5) and (4.7),

$$D_n^+ = \sup\{Z_n(t): 0 \le t \le 1\} \text{ and } D_n = \sup\{|Z_n(t)|: 0 \le t \le 1\}. \tag{4.8}$$

Let us then define

$$D^+ = \sup\{Z(t): 0 \le t \le 1\} \text{ and } D = \sup\{|Z(t)|: 0 \le t \le 1\}. \tag{4.9}$$

It is well known that for every $\lambda \ge 0$,

$$P\{D^+ \le \lambda\} = 2\Phi(\lambda) - 1, \tag{4.10}$$

$$P\{D \le \lambda\} = \Sigma_{k=-\infty}^{\infty} (-1)^k \{\Phi((2k+1)\lambda) - \Phi((2k-1)\lambda)\} , \tag{4.11}$$

where Φ is the standard normal d.f. We shall show that under appropriate regularity conditions, when H_0 holds, as $N \to \infty$,

$$\rho(Z_n, Z_n^o) = \sup\{|Z_n(t) - Z_n^o(t)|: 0 \le t \le 1\} \underset{p}{\to} 0, \tag{4.12}$$

so that by (4.6) and (4.12),

$$Z_n \underset{\mathcal{D}}{\to} Z, \text{ in the } J_1\text{-topology on } D[0,1], \tag{4.13}$$

and hence, by (4.8), (4.9), (4.13) and (4.10) - (4.11), as $n \to \infty$,

$$P\{D_n^+ \le \lambda | H_0\} \to P\{D^+ \le \lambda\} = 2\Phi(\lambda) - 1, \tag{4.14}$$

$$P\{D_n \le \lambda H_0\} \to P\{D \le \lambda\} = \sum_{k=-}^{\infty} (-1)^k \{\Phi((2k+1)\lambda) - \Phi((2k-1)\lambda)\} . \tag{4.15}$$

Thus, ADF tests for H_0 in (1.2) against (1.3) may be based on D_n^+ or D_n, using the critical values of D^+ and D, respectively.

Looking at (3.2), (3.3), (4.1), (4.2), (4.5), (4.7) and (4.12), we gather that for proving (4.12), it suffices to show that under H_0,

$$\max_{k\leq n}\{n^{-\frac{1}{2}}|\Sigma_{i\leq k}(\hat{u}_i-u_i)|\} \xrightarrow{p} 0, \text{ as } n \to \infty. \tag{4.16}$$

For this purpose (as well as for studying the asymptotic nonnull distribution theory of D_n^+ and D_n), we introduce the following regularity conditions on F, the $\underset{\sim}{c}_i$ and the score function Φ. The d.f. F is assumed to have bounded and continuous first and second order derivatives [f(x) and f'(x), respectively] almost everywhere, and

$$I(f) = \int_{-\infty}^{\infty}\{f'(x)/f(x)\}^2 dF(x) < \infty. \tag{4.17}$$

Also, we assume that there exists a positive definite (p.d.) and finite matrix $\underset{\sim}{C}_0$, such that

$$n^{-1}\underset{\sim}{C}_n = n^{-1}\Sigma_{i=1}^{n}\underset{\sim}{c}_i\underset{\sim}{c}_i' \to \underset{\sim}{C}_0, \text{ as } n \to \infty, \tag{4.18}$$

$$\max_{1\leq k\leq n}\{\underset{\sim}{c}_k'\underset{\sim}{C}_0^{-1}\underset{\sim}{c}_k\} = O((\log n)^2). \tag{4.19}$$

[For the location model, $C_0 = 1$ and the left hand side of (4.15) is also equal to 1.] Note that (4.19) is weaker than the Hájek (1968) condition, but is more stringent than the classical Noether condition. Further, let $\phi^{(1)}$ and $\phi^{(2)}$ be the first and second derivatives of ϕ, and assume that there exist a generic constant K ($<\infty$) and a δ ($< 1/6$), such that

$$|\phi^{(r)}(u)| \leq K[u(1-u)]^{-r-\delta}, \quad 0<u<1, \quad r=0,1,2. \tag{4.20}$$

As in Sen (1980b), it is possible to replace $\delta < 1/6$ by $\delta < 1/4$, provided we assume that

$$\sup_x f(x)\{F(x)[1-F(x)]\}^{-\frac{1}{2}+\eta} < \infty \text{ for some } \eta < \infty. \tag{4.21}$$

Also, if $\phi^{(2)}$ is bounded a.e., then (4.19) may be replaced by the Noether condition: $\max\{\underset{\sim}{c}_k'\underset{\sim}{C}_n^{-1}\underset{\sim}{c}_k : 1 \leq k \leq n\} \to 0$ as $n \to \infty$. We may note that (4.20) holds for the Wilcoxon, Normal as well as all the other commonly adapted score functions. Finally, concerning the estimators $\{\hat{\beta}_k\}$ employed in (3.1), we assume that under H_0 in (1.2), for every $\varepsilon > 0$, there exists an integer k_0 (≥ 1), such that

$$P\{\max_{k_0 \leq k \leq n} (\log k)^{-1} k^{\frac{1}{2}} ||\hat{\beta}_k - \beta|| \geq 1\} < \varepsilon, \; \forall\, n \geq k_0. \tag{4.22}$$

Later on (in the appendix), we shall see that (4.22) holds under fairly general conditions.

Returning to the proof of (4.16), we may note first that by (4.20) and (2.4) - (2.5), for every k (≥ 1),

$$\max_{1 \leq i \leq k} |a_k(i)| = o(k^{\delta}), \tag{4.23}$$

so that by (2.9), (3.1), (3.2), (4.1) and (4.2), $|\hat{u}_k - u_k| = o(k^{\delta})$, with probability 1. Hence, we may always choose a sequence $\{k_n\}$ of positive integers, such that $k_n \nearrow \infty$ but $k_n^{1+\delta} n^{-\frac{1}{2}} \searrow 0$, as $n \to \infty$, and to prove (4.16), it suffices to show that for every $\varepsilon > 0$,

$$P_0\{n^{-\frac{1}{2}} \Sigma_{k_n \leq i \leq n} |\hat{u}_i - u_i| > \varepsilon\} \to 0, \text{ as } n \to \infty. \tag{4.24}$$

Let the $\hat{X}_{ki}$, Y_k, R^+_{ki} and $\hat{R}^+_{ki}$ be defined as in Sections III and IV, and let

$$B_{nk} = \{\max_{k \leq i \leq n} (\log i)^{-1} i^{\frac{1}{2}} ||\hat{\beta}_i - \beta|| \leq 1\} \;\; (k \leq n) \tag{4.25}$$

and B^c_{nk} be the complementary event to B_{nk}. Then, by (3.2), (4.1) and (4.25), for every $\varepsilon > 0$,

$$\begin{aligned} &P_0\{n^{-\frac{1}{2}} \Sigma^n_{i=k_n} |\hat{u}_i - u_i| > \varepsilon\} \\ &\leq P_0\{n^{-\frac{1}{2}} \Sigma^n_{k=k_n} |(\text{sign}\hat{X}_{kk} - \text{sign}Y_k) a_k(R^+_{kk})| > \varepsilon/2,\; B_{nk_n}\} \\ &\quad + P_0\{n^{-\frac{1}{2}} \Sigma^n_{k=k_n} |a_k(\hat{R}^+_{kk}) - a_k(R^+_{kk})| > \varepsilon/2,\; B_{nk_n}\} + P_0\{B^c_{nk_n}\}, \end{aligned} \tag{4.26}$$

where, by (4.22), the last term on the right hand side of (4.26) converges to 0 as $n \to \infty$. Note that by (4.18), (4.19), and (4.25), when B_{nk_n} holds, $|\text{sign}\hat{X}_{kk} - \text{sign}Y_k|$ may only be different from zero, when $|Y_k| \le dk^{-\frac{1}{2}}(\log k)^2$, for some finite d. Further, by the well-known results on the empirical d.f. (for the $|Y_i|$), we have on denoting by $F_k^*(x) = k^{-1}\Sigma_{i=1}^k I(|Y_i| \le x)$, $x \ge 0$, the sample d.f. and $F^*(x) = P\{|Y_k| \le x\} = F(x)-F(-x) = 2F(x)-1$, $x \ge 0$, that

$$P_0\{|k^{-1}R_{kk}^+ - F^*(|Y_k|)| \ge (2k^{-1}\log k)^{\frac{1}{2}}\}$$

$$\le P_0\{\sup_{x \ge 0}|F_k^*(x)-F^*(x)| \ge (2k^{-1}\log k)^{\frac{1}{2}}\} \tag{4.27}$$

$$\le 2\exp(-2k(2k^{-1}\log k)) = 2k^{-4}, \ \forall\, k \ge 2.$$

Also, $\phi^+(0) = 0$ and by the assumed boundedness of f, $F^*(dk^{-\frac{1}{2}}(\log k)^2) = O(k^{-\frac{1}{2}}(\log k)^2)$, $\forall\, k \ge 2$. Hence, by (4.20) and by some routine steps, we obtain that for every r: $0 \le r/k \le c < 1$, there exists a finite positive constant C, such that

$$\max_{1 \le i \le r}|a_k(i)| \le C(r/k), \ \forall\, k \ge k_0. \tag{4.28}$$

Note that for each k : $k_n \le k \le n$, $I_{B_{nk_n}}|(\text{sign}\hat{X}_{kk} - \text{sign}Y_k)$ $a_k(R_{kk}^+)|$ is bounded from above by $I_{B_{nk_n}}|a_k(R_{kk}^+)I(k^{-1}R_{kk}^+ \le F^*(|Y_k|)+$ $(2K^{-1}\log k)^{\frac{1}{2}} \cap I(|Y_k| \le dk^{-\frac{1}{2}}(\log k)^2) +$ $I_{B_{nk_n}}\max_{i \le k}|a_k(i)|I^C(k^{-1}R_{kk}^+ \le F^*(|Y_k|) +$ $(2k^{-1}\log k)^{\frac{1}{2}} \cup I^C(|Y_k| \le dk^{-\frac{1}{2}}(\log k)^2)$, where I^C stands for the complementary event. By (4.23), (4.27), (4.28) and the above inequality, we obtain that as $n \to \infty$,

$$E_0\{n^{-\frac{1}{2}}\Sigma_{k=k_n}^n I_{B_{nk_n}}|(\text{sign}\hat{X}_{kk} - \text{sign}Y_k)a_k(R_{kk}^+)|\}$$

$$\le n^{-\frac{1}{2}}\Sigma_{k=k_n}^n\{[O(k^{-\frac{1}{2}}(\log k)^2)]^2 + [o(k^\delta)O(k^{-4})]\} \tag{4.29}$$

$$= O(n^{-\frac{1}{2}}(\log n)^5) \to 0,$$

so that by (4.29) and the Chebyshev inequality, the first term on the right hand side of (4.26) converges to 0, as $n \to \infty$.

Consider now the sample d.f.'s $\hat{F}_k^*(x) = k^{-1}\Sigma_{i=1}^k I(|\hat{X}_{ki}|\leq x)$, $x\geq 0$, $k\geq 1$. Then, we may virtually repeat the proof of Theorem 3.1 of Ghosh and Sen (1972) and obtain that for every γ $(0<\gamma<\frac{1}{4})$ and h (which we take > 1), there exist positive constants K_1 and K_2 and an integer k_0 (≥ 1), such that under H_0 in (1.2), for every $k\geq k_0$,

$$P_0\{\sup_{x\geq 0} k^{\frac{1}{2}}|\hat{F}_k^*(x)-F_k^*(x)| > K_1 k^{-\gamma}(\log k)^2|B_{kk}\} \leq K_2 k^{-h}, \quad (4.30)$$

which for h>1, insures that

$$P_0\{\max_{k_n\leq k\leq n}\ \sup_{x\geq 0} k^{\frac{1}{2}+\gamma}(\log k)^{-2}|\hat{F}^*(x)-F_k^*(x)| > K_1|B_{nk_n}\} \to 0, \quad (4.31)$$

as $n\to\infty$. Also, note that $\{[F_k^*(x)-F^*(x):\ 0\leq x<\infty];\ k\geq 1\}$ is a reverse martingale (process) sequence, so that for every $\varepsilon>0$, $\{\sup_{x\geq 0}(F^*(x)[1-F^*(x)])^{-\frac{1}{2}+\varepsilon}|F_k^*(x)-F^*(x)|;\ k\geq 1\}$ is a reverse sub-martingale; by the use of the Hájek-Rényi-Chow inequality, we obtain that

$$P\{\max_{k_n\leq k\leq n}\ \sup_{x\geq 0} k^{\frac{1}{2}}(\log k)^{-1}|F_k^*(x)-F^*(x)|\{F^*(x)[1-F^*(x)]\}^{-\frac{1}{2}+\varepsilon}\geq 1\}$$

$$\leq k_n^{-1}(\log k_n)^{-2}E_0\{_{x\geq 0}k_n[F_{k_n}^*(x)-F^*(x)]^2\{F^*(x)\ 1-F^*(x)]\}^{-1+2\varepsilon}\}$$

$$+\ \Sigma_{k=k_n+1}^n\{(\log k)^{-2}\ -$$

$$-\ ((k-1)/k)(\log \overline{k-1})^{-2}\}E_0\{\sup_{x\geq 0} k[F_k^*(x)-F^*(x)]^2\{F^*(x)$$

$$[1-F^*(x)]\}^{-1+2\varepsilon}\}$$

$$= k_n^{-1}(\log k_n)^{-2}\ O(1) + \Sigma_{i=k_n+1}^n\ \left(\frac{1}{k((\log k)^2}\right) \to 0, \text{ as } n\to\infty, \quad (4.32)$$

where the penultimate step follows from the fact that the expectations in the preceding step are all bounded (uniformly in $k\geq k_n$) [see (7.4.54) - (7.4.55) of Sen (1981) in this respect]. Let us now define δ (> 0) as in (4.20), while in (4.31) - (4.32), we let $\varepsilon\in(0,\delta/2)$, and let

$$J_k = \{x: F^*(x)[1-F^*(x)] \geq k^{-1+\delta}(\log k)^2\},\ k \geq 2. \tag{4.33}$$

Then, from (4.32) and (4.33), we obtain that for every $\eta > 0$, as $n \to \infty$,

$$P_0\{\max_{k_n \leq k \leq n} \sup_{x \in J_k} |F_k^*(x)/F^*(x) - 1| \geq \eta\} \to 0, \tag{4.34}$$

$$P_0\{\max_{k_n \leq k \leq n} \sup_{x \in J_k} |\{1-F_k^*(x)\}/\{1-F^*(x)\} - 1| \geq \eta\} \to 0. \tag{4.35}$$

Now, the second term on the right hand side of (4.26) is bounded by

$$\begin{aligned} &P_0\{n^{-\frac{1}{2}}\Sigma_{k=k_n}^n (2 \max_{1 \leq i \leq k} |a_k(i)|) I(|Y_k| \notin J_k) > \varepsilon/4\} + \\ &P_0\{n^{-\frac{1}{2}}\Sigma_{k=k_n}^n I(|Y_k| \in J_k) |a_k(\hat{R}_{kk}^+) - a_k(R_{kk}^+)| > \varepsilon/4,\ B_{nk_n}\}. \end{aligned} \tag{4.36}$$

By (4.23) and (4.33), as $n \to \infty$,

$$\begin{aligned} &E_0\{n^{-\frac{1}{2}}\Sigma_{k=k_n}^n (2 \max_{1 \leq i \leq k} |a_k(i)|) I(|Y_k| \notin J_k)\} \\ &= 2n^{-\frac{1}{2}}\Sigma_{k=k_n}^n (\max_{1 \leq i \leq k} |a_k(i)|) P\{|Y_k| \notin J_k\} \\ &= 2n^{-\frac{1}{2}}\Sigma_{k=k_n}^n [o(k^{\delta})][O(k^{-1+\delta}(\log k)^2)] \\ &= o(n^{-\frac{1}{2}+2\delta}(\log n)^2) \to 0, \quad (\text{as } \delta < \tfrac{1}{4}) \end{aligned} \tag{4.37}$$

so that by (4.37) and the Chebyshev inequality, the first term of (4.36) converges to 0, as $n \to \infty$. For the second term, we make use of (4.20), (4.31), (4.32), (4.33) and (4.35), and obtain that when B_{nk_n} holds, with probability converging to 1 (as $n \to \infty$),

$$\begin{aligned} &n^{-\frac{1}{2}}\Sigma_{k=k_n}^n I(|Y_k| \in J_k) |a_k(\hat{R}_{kk}^+) - a_k(R_{kk}^+)| \\ &\leq cn^{-\frac{1}{2}}\Sigma_{k=k_n}^n I(|Y_k| \in J_k) k^{-\frac{1}{2}-\gamma}(\log k)^2[1-F^*(|Y_k|)]^{-1-\delta}, \end{aligned} \tag{4.38}$$

where C ($< \infty$) is a generic constant. Note that for $|Y_k| \in J_k$, $1-F^*(|Y_k|) \geq k^{-1+\delta}(\log k)^2$, so that on letting $\gamma = \delta + \eta$, $\eta > 0$, we have

$$k^{-\gamma}(\log k)^2[1-F^*(|Y_k|)]^{-1-\delta}I(|Y_k|\epsilon J_k)$$
$$\leq k^{-\gamma}(\log k)^2[1-F^*(|Y_k|)]^{-1+\eta}(k^{-1+\delta}(\log k)^2)^{-\delta} \tag{4.39}$$
$$= k^{-\eta-\delta^2}(\log k)^{2(1-\delta)}[1-F^*(|Y_k|)]^{-1+\eta}, \ \forall\ k \geq k_n.$$

By (4.39), the right hand side of (4.38) is bounded by

$$Cn^{-\frac{1}{2}}\Sigma_{k=k_n}^{n} k^{-\frac{1}{2}-\eta-\delta^2}(\log k)^{2(1-\delta)}[1-F^*(|Y_k|)]^{-1+\eta}. \tag{4.40}$$

Since (4.40) represents a positive r.v. whose expectation is

$$c\eta^{-1}n^{-\frac{1}{2}}\Sigma_{k=k_n}^{n} k^{-\frac{1}{2}-\eta-\delta^2}(\log k)^{2(1-\delta)} = O(n^{-\eta-\delta^2}(\log n)^2), \tag{4.41}$$

by the Chebyshev inequality, (4.40) converges to 0, in probability, as $n \to \infty$, which via (4.38) and (4.35) - (4.37) insures that (4.26) converges to 0 as $n \to \infty$. This completes the proof of (4.16), and hence, of (4.12).

Next, we proceed to study the non-null distribution theory of the proposed test statistics. In this context, we assume that in (1.3), the change-point τ (= τ_n) satisfy the condition that

$$t_{m_n} \leq \tau_n < t_{m_n+1} \quad \text{where } n^{-1}m_n \to \theta: 0 < \theta < 1, \tag{4.42}$$

as $n \to \infty$, that is, in the asymptotic case, a change point does not occur near the beginning or the end of the time period (t_1,t_n). Under (1.3), (4.41) and the regularity conditions assumed before, for $\underset{\sim}{\beta}_{m_n+1} = \underset{\sim}{\beta}_{m_n} + \underset{\sim}{\lambda}$, $\underset{\sim}{\lambda}$ ($\neq \underset{\sim}{0}$) fixed, it can be shown that the $\hat{u}_k$, $k>m_n$ are consistently shifted from the origin, and hence, by (3.3) - (3.5), D_n^+ or D_n will be $O_p(n^{\frac{1}{2}})$, and thus, by (4.14) - (4.15), the proposed tests will be consistent against $\underset{\sim}{\lambda}\neq\underset{\sim}{0}$. Thus, to study the asymptotic power properties, we confine ourselves to some local alternatives for which the asymptotic power does not converge to 1. With this in mind, we consider a sequence $\{K_n\}$ of alternative hypothesis, where under K_n,

$$\underset{\sim}{\beta}_1 = \ldots = \underset{\sim}{\beta}_{m_n} = \underset{\sim}{\beta}_{m_n+1} - n^{-\frac{1}{2}}\underset{\sim}{\lambda},\ \underset{\sim}{\beta}_{m_n+1} = \ldots = \underset{\sim}{\beta}_n, \tag{4.43}$$

where $\underset{\sim}{\lambda}$ ($\neq \underset{\sim}{0}$) is fixed and the m_n satisfy (4.41). Further, we strengthen (4.18) to

$$\lim_{m\to\infty} m^{-1}\underset{\sim}{C}_m = \underset{\sim}{C}_0 \text{ and } \lim_{m\to\infty} \frac{1}{m}\sum_{i=1}^{m}\underset{\sim}{c}_i = \bar{\underset{\sim}{c}} \tag{4.44}$$

both exist, where $\underset{\sim}{C}_0$ is p.d. Then

$$\underset{\sim}{C}^{-1}_{[nt]}\underset{\sim}{C}_{[ns]} \to (s/t)\underset{\sim}{I}, \text{ for every } 0\le s<t\le 1. \tag{4.45}$$

Under (4.17) - (4.19), the contiguity of the sequence of probability measures under $\{K_n\}$ with respect to those under H_0, follows then along the lines of the general results in Chapter VI if Hájek and Sidák (1967), so that proceeding as in Sen (1977, 1980a), we first extend (4.12) to that under $\{K_n\}$, (4.6) to that of a drifted Brownian motion under $\{K_n\}$, and finally, obtain the same result for $\{Z_n\}$. Thus, we obtain that under the regularity conditions assumed and $\{K_n\}$ in (4.41) - (4.42),

$$Z_n \underset{\mathcal{D}}{\to} Z + \xi, \text{ in the } J_1\text{-topology on } D[0,1], \tag{4.46}$$

where $\xi = \{\xi(t);\ 0\le t\le 1\}$ is specified by

$$\xi(t) = \begin{cases} 0, & 0\le t\le\theta \\ A^{-1}\gamma(\phi,F)(t-\theta)\underset{\sim}{\lambda}'\bar{\underset{\sim}{c}}, & \theta\le t\le 1, \end{cases} \tag{4.47}$$

where A is defined by (2.11) and

$$\gamma(\phi,F) = \int_{-\infty}^{\infty} (d/dx)\phi(F(x))dF(x) \quad (> 0). \tag{4.48}$$

[Note that by partial integration, $\gamma^2(\phi,F)\le A^2 I(f)<\infty$]. Thus, the asymptotic power of D_n^+, under $\{K_n\}$, is given by

$$P\{Z(t)+\xi(t)\ge D_\alpha^+ \text{ for some } t:\ 0\le t\le 1\} \tag{4.49}$$

where $\Phi(D_\alpha^+) = 1-\frac{1}{2}\alpha$; a similar expression holds for D_n.

V. ASYMPTOTIC RELATIVE EFFICIENCY RESULTS

For normal F, tests for change-points, relating to (1.1) - (1.3), based on recursive residuals are discussed in Brown, Durbin and Evans (1975). For F not necessarily normal, invariance principles for CUSUMs of such recursive residuals have recently been studied by Sen (1982a). It follows from the results in Section IV of Sen (1982a) that (4.14) - (4.15) hold for the parametric procedure based on the least squares recursive residuals, and also, (4.48) holds with the drift function $\xi = \{\xi(t),\ 0\le t\le 1\}$ replaced $\xi^* = \{\xi^*(t),\ 0\le t\le 2\}$, where

$$\xi^*(t) = \begin{cases} 0, & 0\le t\le \theta, \\ \sigma^{-1}(t-\theta)\underset{\sim}{\lambda}'\underset{\sim}{\bar{c}}, & \theta\le t\le 1, \end{cases} \tag{5.1}$$

where σ^2, the variance of F, is assumed to be finite; a similar result holds for the two-sided case too.

The two drift functions in (4.47) and (5.1) are proportional, i.e.,

$$\xi(t) = k\xi^*(t), \quad 0\le t\le 1, \text{ where } k^2=\sigma^2\gamma^2(\phi,F)/A^2. \tag{5.2}$$

Thus, as in Sen (1980a , 1982b), we may justify the use of the classical Pitman-efficiency results in this context too, and k^2 represents the asymptotic relative efficiency of the rank procedure relative to the least squares procedure. This agrees with the Pitman-efficiency of the rank test with respect to the Student t-test, [disucssed in detail in Chernoff and Savage (1958) and Puri and Sen (1971), among other places], and hence, the details are omitted here.

It may be noted that if instead of the recursive residuals, one would have used [as in Sen (1980a, 1982b)] aligned rank statistics based on the terminal estimator $\hat{\underset{\sim}{\beta}}_n$ of $\underset{\sim}{\beta}$, one has then the weak convergence to a Brownian bridge with a more complicated drift function (under $\{K_n\}$). In comparing such a test with the one considered here, it is, generally, not possible to adapt the measure of the Pitman-efficiency.

VI. APPENDIX

We proceed to verify here (4.22) for some typical estimators. First, consider the location model (2.1) - (2.3) and the rank estimates $\{\hat{\theta}_k\}$ in (2.7). In this case, under the assumed regularity conditions on ϕ and F, (4.22) follows directly from (10.3.39) through (10.3.45) of Sen (1981). For the general regression model, asymptotic theory of rank estimators of $\underset{\sim}{\beta}$ rests on an asymptotic linearity property of rank statistics in regression parameters, due to Jurečková (1969, 1971). Her results relate to the weak convergence properties under weaker regularity conditions, and, in view of Theorem A.4.1 of Sen (1981, p. 389), stronger conclusions, such as (4.22), hold under our assumed regularity conditions; similar results under more stringent regularity conditions are due to Ghosh and Sen (1972). Thus, (4.22) holds for the rank estimators under the assumed regularity conditions. We proceed to show that for the conventional least squares estimators too, (4.22) holds whenever $\sigma^2<\infty$. Towards this note that

$$
\begin{aligned}
&(\log k)^{-1}k^{\frac{1}{2}}||\hat{\underset{\sim}{\beta}}_k-\underset{\sim}{\beta}|| \\
&= (\log k)^{-1}k^{-\frac{1}{2}}||k\underset{\sim}{C}_k^{-1}\underset{\sim}{C}_k(\hat{\underset{\sim}{\beta}}_k-\underset{\sim}{\beta})|| \qquad (A.1)\\
&\le (\log k)^{-1}k^{-\frac{1}{2}}||\underset{\sim}{C}_k(\hat{\underset{\sim}{\beta}}_k-\underset{\sim}{\beta})||Ch_1(k\underset{\sim}{C}_k^{-1}),
\end{aligned}
$$

where Ch_1 stands for the largest characteristic root, and by (4.18), as $k \to \infty$, $Ch_1(k\underset{\sim}{C}_k^{-1}) \to [Ch_q\underset{\sim}{C}_0]^{-1}< \infty$. So, it suffices to show that under H_0 in (1.2), when $\sigma<\infty$, for every $c>0$

$$
P\{\max_{k_n\le k\le n}(\log k)^{-1}k^{-\frac{1}{2}}||\underset{\sim}{C}_k(\hat{\underset{\sim}{\beta}}_k-\underset{\sim}{\beta})|| > c\} \to 0, \qquad (A.2)
$$

as $n \to \infty$. Towards this, note that under H_0 in (1.2), for $\sigma<\infty$,

$$\{C_k(\hat{\beta}_k-\beta) = \Sigma_{i=1}^{k} c_i e_i;\ k\geq 1\} \text{ is a 0-mean martingale,} \tag{A.3}$$

where the e_i are i.i.d.r.v. with 0 mean and variance $\sigma^2<\infty$. As such, $\{Z_k = ||C_k(\hat{\beta}_k-\beta)||;\ k\geq 1\}$ is a nonnegative submartingale, where

$$EZ_k = (\Sigma_{i=1}^{k} c_i' c_i)\sigma^2 = \sigma^2 Tr(C_k),\ \forall\ k\geq 1. \tag{A.4}$$

Therefore, by the Hájek-Rényi-Chow inequality,

$$P_0\{\max_{k_n\leq k\leq n} (\log k)^{-1} k^{-\frac{1}{2}} ||C_k(\hat{\beta}_k-\beta)|| > c\}$$

$$\leq c^{-2}\sigma^{-2}\{n^{-1}(\log n)^{-2} TR(C_n) + \Sigma_{k=k_n}^{n-1}(Tr(C_k)) \cdot \tag{A.5}$$

$$\cdot \left[\frac{1}{k(\log k)^2} - \frac{1}{(k+1)(\log k+1)^2}\right]\},$$

where by (4.18), the right hand side of (A.5) is $O((\log k_n)^{-1})$ and converges to 0 as $k_n \to \infty$ (i.e., $n\to\infty$). Hence, (4.22) follows from (A.2) and (A.4).

VII. ACKNOWLEDGMENT

Thanks are due to the referee for his very critical reading of the manuscript.

VIII. REFERENCES

Bhattacharyya, G.K. and Johnson, R.A. (1968). "Nonparametric Tests for Shifts at an Unknown Time Point." *Ann. Math. Statist. 39,* 1731-1743.

Bhattacharya, P.K. and Frierson, D. (1981). "A Nonparametric Control Chart for Detection Small Disorders." *Ann. Statist. 9,* 544-554.

Brown, R.L., Durbin, J. and Evans, J.M. (1975). "Techniques for Testing Constancy of Regression Relationships Over Time (with discussions)". *J. Roy Statist. Soc. Ser. B 37,* 149-192.

Chernoff, H. and Savage, I.R. (1958). "Asymptotic Normality and Efficiency of Certain Nonparametric Test Statistics." *Ann. Math. Statist. 29,* 972-994.

Chernoff, H. and Zacks, S. (1964). "Estimating the Current Mean of a Normal Distribution Which is Subject to Change in Time." *Ann. Math. Statist. 35*, 999-1018.

Deshayes, J. and Picard, D. (1981). "Tests de Rupture de Regression: Comparison Asymptotic." *Teoria. Ver. Prem.*

Ghosh, M. and Sen. P.K. (1972). "On Bounded Length Confidence Interval for the Regression Coefficient Based on a Class of Rank Statistics." *Sankhyā, Ser. A 34*, 33-52.

Hackle, R. (1980). *Testing the Constancy of Regression Relationship Over Time.* Vandenhocck and Ruprecht, Gottingen.

Hájek, J. (1968). "Asymptotic Linearity of Simple Linear Rank Statistics Under Alternatives." *Ann. Math. Statist. 39*, 325-346.

Hájek, J. and Šidak, Z. (1967). *Theory of Rank Tests.* Academic Press, New York.

Hinkley, D.V. (1980). Bibliography of Articles on Change-Points. Tech. Report., Univ. of Minnesota.

Jurečková, J. (1969). "Asymptotic Normality of a Rank Statistic in Regression Parameter." *Ann. Math. Statist. 40*, 1889-1900.

Jurečková, J. (1971). "Nonparametric Estimates of Regression Coefficients." *Ann. Math. Statist. 42*, 1328-1338.

Page, E.S. (1955). "A Test for a Change in a Parameter Occurring at an Unknown Point." *Biometrika 42*, 523-527.

Page, E.S. (1957). "On Problems in Which a Change of Parameters Occurs at an Unknown Point." *Biometrika 44*, 248-252.

Puri, M.L. and Sen, P.K. (1971). *Nonparametric Methods in Multivariate Analysis.* Wiley, New York.

Sen, A. and Srivastava, M.S. (1975). "On Tests for Detecting Changes in Means." *Ann. Statist. 3*, 98-108.

Sen, P.K. (1977). "Tied-Down Wiener Process Approximations for Aligned Rank Order Processes and Some Applications." *Ann. Statist. 5*, 1107-1123.

Sen, P.K. (1978). "Invariance Principles for Linear Rank Statistics Revisited." *Sankhyā, Ser. A 40*, 215-236.

Sen, P.K. (1980a). "Asymptotic Theory of Some Tests for a Possible Change in the Regression Slope Occurring at an Unknown Time-Point." *Z. Warsch. Verw. Gebiete 52*, 203-218.

Sen, P.K. (1980b). "On Almost Sure Linearity Theorems for Signed Rank Order Statistics." *Ann. Statist. 8*, 313-321.

Sen, P.K. (1981). *Sequential Nonparametrics: Invariance Principles and Statistical Inference.* Wiley, New York.

Sen, P.K. (1982a). "Invariance Principles for Recursive Residuals." *Ann. Statist. 10*, 307-312.

Sen, P.K. (1982b). "Asymptotic Theory of Some Tests for Constancy of Regression Relationships Over Time." *Math. Operat. Statist. Ser. Statist 13*, 21-32.

Sen, P.K. (1982c). Tests for Change-Points Based on Recursive U-statistics. Inst. Statist., Univ. North Carolina, Mimeo Report No. 1402.

OPTIMAL UNIFORM RATE OF CONVERGENCE FOR NONPARAMETRIC ESTIMATORS OF A DENSITY FUNCTION OR ITS DERIVATIVES[1]

Charles J. Stone

Department of Statistics
University of California, Berkeley
Berkeley, California

I. INTRODUCTION

Let d be a positive integer and let $X_1, X_2, \ldots$ be independent $\mathbb{R}^d$ - valued random variables having common unknown density f, which is assumed to belong to a known collection $\mathcal{F}$ of density functions on $\mathbb{R}^d$. Let $T(f) = Qf$, where Q is a linear differential operator with constant coefficients of nonnegative integer order m. Let $\hat{T}_n$ denote an arbitrary estimator of T(f) based on $X_1, \ldots, X_n$. In this paper the optimal uniform rate of convergence of $\hat{T}_n$ to T(f) will be determined for various choices of $\mathcal{F}$.

Let $|| \; ||_\infty$ denote the usual L^∞ norm for functions on $\mathbb{R}^d$, defined by $||g||_\infty = \sup[|g(x)| : x \in \mathbb{R}^d]$. A sequence $\{b_n\}$ of (eventually) positive constants is called a *lower rate of convergence if*

$$\lim_n \inf_{\hat{T}_n} \sup_{\mathcal{F}} P_f(||\hat{T}_n - T(f)||_\infty \geq cb_n) = 1 \text{ for some } c > 0. \tag{1}$$

[1]*Research was supported by NSF Grant MCS 80-02732.*

ISBN 0-12-589320-5

(Here $\inf_{\hat{T}_n}$ means the infimum over all estimators of T(f) based on $X_1, \ldots, X_n$.) It is called an *achievable rate of convergence if there* is a sequence $\{\hat{T}_n\}$ of estimators such that

$$\lim_n \sup_F P_f(||\hat{T}_n - T(f)||_\infty \geq cb_n) = 0 \text{ for some } c > 0 . \quad (2)$$

It is called the *optimal rate of convergence* if it is both a lower and an achievable rate of convergence. If $\{b_n\}$ is the optimal rate of convergence and $\{\hat{T}_n\}$ satisfies (2), then $\{\hat{T}_n\}$ is said to be *asymptotically optimal.* (The conditions imposed by these definitions are stronger than the corresponding conditions in Stone (1980). They were formulated this way mainly because they could be verified in the present context.)

Let $\alpha = (\alpha_1, \ldots, \alpha_d)$ denote a d-tuple of nonnegative integers and set $[\alpha] = \alpha_1 + \ldots + \alpha_d$ and $\alpha! = \alpha_1! \cdot \ldots \cdot \alpha_d!$. For $x = (x_1, \ldots, x_d) \in \mathbb{R}^d$, set $|x| = (x_1^2 + \ldots + x_d^2)^{\frac{1}{2}}$ and $x^\alpha = (x_1^{\alpha_1} \cdot \ldots \cdot x_d^{\alpha_d})$. Let D^α denote the differential operator defined by

$$D^\alpha = \frac{\partial^{[\alpha]}}{\partial x_1^{\alpha_1} \cdot \ldots \cdot \partial x_d^{\alpha_d}} .$$

Then

$$Q = \sum_{[\alpha] \leq m} q_\alpha D^\alpha ,$$

where the q_α's are real constants such that $q_\alpha \neq 0$ for some α with $[\alpha] = m$.

Let k, β, and M denote real constants such that k is a nonnegative integer, $k \geq m$, $0 < \beta \leq 1$ and $M > 0$. Let F be the collection of k-times continuously differentiable probability densities f on $\mathbb{R}^d$ such that

$$|D^\alpha f(x_2) - D^\alpha f(x_1)| \le M|x_2 - x_1|^\beta \text{ for } [\alpha] = k \text{ and } x_1, x_2 \varepsilon \mathbb{R}^d. \quad (3)$$

Set $p = k + \beta$, $\gamma = 1/(2p + d)$ and $r = (p-m)/(2p+d) = (p-m)\gamma$.

Theorem 1. $\{(n^{-1} \log n)^r\}$ is the optimal rate of convergence.

Let K be an m-times differentiable function on $\mathbb{R}^d$ having compact support and such that $K^{(m)}$ is Hölder continuous,

$$\int_{\mathbb{R}^d} K(x)dx = 1, \quad (4)$$

and

$$\int_{\mathbb{R}^d} x^\alpha K(x)dx = 0 \quad \text{for} \quad 1 \le [\alpha] \le k. \quad (5)$$

(Note that if $k \ge 2$, then (4) and (5) together imply that K must take on both positive and negative values.) Let $\{\varepsilon_n\}$ be a sequence of real numbers such that

$$0 < \varepsilon_n \le 1 \quad \text{for } n \ge 1 . \quad (6)$$

Define K_n on $\mathbb{R}^d$ by

$$K_n(x) = \varepsilon_n^{-d} K(\varepsilon_n^{-1}x) \quad \text{for} \quad x \,\varepsilon\, \mathbb{R}^d . \quad (7)$$

Consider the kernel estimator $\hat{f}_n$ of f defined by

$$\hat{f}_n(x) = n^{-1} \sum_1^n K_n(x-X_i) \quad \text{for} \quad x \,\varepsilon\, \mathbb{R}^d \quad (8)$$

and the corresponding estimator T_n of T(f) defined by

$$\hat{T}_n(x) = Q\hat{f}_n(x) = n^{-1} \sum_1^n Q\, K_n(x-X_i) . \quad (9)$$

Theorem 2. Suppose that

$$\varepsilon_n / (\frac{\log n}{n})^{\gamma}$$

has a finite positive limit as $n \to \infty$. Then the sequence $\{\hat{T}_n\}$ defined by (9) is asymptotically optimal.

The proof that $\{(n^{-1} \log n)^r\}$ is a lower rate of convergence follows along the lines of Stone (1980, 1982). For a special case of this result see Khasminskii (1978). Theorem 2, which implies that $\{(n^{-1} \log n)^r\}$ is an achievable and hence an optimal rate of convergence will be proven in Section II. For weaker results see Singh (1981). There are analogues of Theorems 1 and 2 when the L^{∞} norm is replaced by the L^2 norm and $n^{-1} \log n$ is replaced by n^{-1} (see Bretagnolle and Huber (1979), and Müller and Gasser (1979)).

II. PROOF OF THEOREM 2

Throughout this section, (3)-(9) are assumed to hold. Given $f \in \mathcal{F}$, let $f_k(\cdot\,;x)$ denote the Taylor polynomial approximation to f of degree k about x. Then

$$f_k(x+y:x) = \sum_{[\alpha]\leq k} \frac{D^{\alpha} f(x)}{\alpha!} y^{\alpha} .$$

It follows from (3) and the integral form of the remainder term in Taylor's theorem that there is a positive constant M_1 such that if $f \in \mathcal{F}$ and $x \in \mathbb{R}^d$, then

$$|f(x+y)-f_k(x+y;x)| \leq M_1 |y|^p \quad \text{for} \quad |y| \leq 1. \tag{10}$$

Lemma 1. $\sup_{\mathcal{F}} ||D^{\alpha} f||_{\infty} < \infty$ *for* $[\alpha] \leq k$.

Proof. Since the functions in F are probability densities, it follows from (10) that

$$\sup_F \int_{|y|\leq 1} |f_k(y;0)|\ dy < \infty \ . \tag{11}$$

Suppose the lemma is false. Then

$$\sup_F \max_{[\alpha]\leq k} |D^{\alpha} f_k(0;0)| = \infty, \tag{12}$$

so there exist densities $f^{\nu} \in F$ for $\nu \geq 1$ such that

$$a_{\nu} = \max_{[\alpha]\leq k} |D^{\alpha} f_k^{\nu}(0;0)|$$

tends to infinity as $\nu \to \infty$. A subsequence of the polynomials $a_{\nu}^{-1} f_k^{\nu}(y;0)$ tends uniformly on $|y| \leq 1$ to a polynomial P on $\mathbb{R}^d$ of degree k such that

$$\max_{[\alpha]\leq k} |D^{\alpha} P(0)| = 1 \tag{13}$$

and, by (11),

$$\int_{|y|\leq 1} |P(y)|\ dy = 0. \tag{14}$$

Clearly, (13) and (14) yield a contradiction.

Write $Q = \sum_{0\leq j\leq m} Q_j$, where $Q_j = \sum_{[\alpha]=j} q_{\alpha} D^{\alpha}$.

Lemma 2. There is a positive constant M_2 *such that*

$$||E_f \hat{T}_n - T(f)||_{\infty} \leq M_2 \varepsilon_n^{p-m} \quad \textit{for} \quad n \geq 1 \textit{ and } f \in F.$$

Proof. If follows as in the proof of (3.5) of Stone (1980) that

$$E_f\hat{T}_n(x)-T(f)(x)= \int QK_n(-y)(f(x+y)-f_k(x+y;x))\,dy$$

$$= \sum_{j=0}^{m} \varepsilon_n^{-(d+j)} \int Q_jK(-\varepsilon_n^{-1}y)(f(x+y)$$

$$- f_k(x+y))\,dy \;.$$

Thus by (10),

$$|E_f\hat{T}_n(x)-T(f)(x)| \le M_1 \sum_{j=0}^{m} \varepsilon_n^{-(d+j)} \int |Q_jK(-\varepsilon_n^{-1}y)||y|^p\,dy$$

$$\le M_1 \sum_{j=0}^{m} \varepsilon_n^{(p-j)} \int |Q_jK(-y)||y|^p\,dy \;,$$

which yields the desired result.

Lemma 3. There are positive constants M_3 *and* λ *such that for* $n \ge 1$, $f \in F$ *and* $x_1, x_2 \in \mathbb{R}^d$

$$|\hat{T}_n(x_2) - \hat{T}_n(x_1)| \le M_3\, \varepsilon_n^{-(d+m+\lambda)}\, |x_2-x_1|^\lambda \tag{15}$$

and

$$|E_f\hat{T}_n(x_2)-E_f\hat{T}_n(x_1)| \le M_3\, \varepsilon_n^{-(d+m+\lambda)}\, |x_2-x_1|^\lambda . \tag{16}$$

Proof. Since K has compact support and $K^{(m)}$ is Hölder continuous, there are positive constants M_3' and λ such that

$$|Q_jK(x_2)-Q_jK(x_1)| \le M_3'|x_2-x_1|^\lambda \text{ for } 0\le j\le m \text{ and } x_1,x_2 \in \mathbb{R}^d .$$

Now

$$\hat{T}_n(x_2)-\hat{T}_n(x_1) = n^{-1}\sum_1^n[QK_n(x_2-X_i)-QK_n(x_1-X_i)]$$

$$= n^{-1}\sum_1^n\sum_{j=0}^m \varepsilon_n^{-(d+j)}[Q_jK(\varepsilon_n^{-1}(x_2-X_i))$$

$$- Q_jK(\varepsilon_n^{-1}(x_1-X_i))].$$

Consequently

$$|\hat{T}_n(x_2)-\hat{T}_n(x_1)| \leq M_3' \sum_{j=0}^m \varepsilon_n^{-(d+j+\lambda)}|x_2-x_1|^\lambda$$

$$\leq M_3'(m+1)\ \varepsilon_n^{-(d+m+\lambda)}\ |x_2-x_1|^\lambda.$$

This yield (15), from which (16) follows.

Given $x = (x_1, \ldots, x_d) \in \mathbb{R}^d$ and $\varepsilon > 0$, let $B_\varepsilon(x)$ denote the cube

$$\{x' = (x_1',\ldots,x_d') \in \mathbb{R}^d : |x_i'-x_i| \leq \varepsilon \text{ for } 1\leq i \leq d\}.$$

Let F_f denote the distribution having density f. Then

$$F_f(B_\varepsilon(x)) = \int_{B_\varepsilon(x)} f(x)\ dx,$$

so by Lemma 1 there is a positive constant M_4 such that

$$F_f(B_\varepsilon(x)) \leq M_4\varepsilon^d \text{ for } f \in \mathcal{F},\ \varepsilon>0 \text{ and } x \in \mathbb{R}^d.$$

Let $\hat{F}_n$ denote the empirical distribution of $X_1, \ldots, X_n$, so that

$$\hat{F}_n(B) = n^{-1}\ \#\{i : 1\leq i\leq n \text{ and } X_i \in B\},$$

where #A denotes the number of elements in the set A. The next result follows from (17) and the argument used to prove the convergence theorem for the empirical distribution in Chapter 12 of Breiman, et al (1983).

Lemma 4. There is a positive constant M_5 *such that*

$$\lim_n \sup_F P_f(\hat{F}_n(B_\varepsilon(x)) \geq \frac{3}{2} F_f(B_\varepsilon(x)) + M_5 \frac{\log n}{n}$$

$$\text{for some } x \in \mathbb{R}^d \text{ and } \varepsilon > 0) = 0.$$

Observe that

$$QK_n(x) = \sum_{j=0}^{m} Q_j K_n(x) = \sum_{j=0}^{m} \varepsilon_n^{-(j+d)} Q_j K(\varepsilon_n^{-1} x).$$

Thus there is a positive constant M_6 such that

$$|n^{-1} QK_n(x)| \leq M_6 \, n^{-1} \varepsilon_n^{-(m+d)} \quad \text{for } n \geq 1 \text{ and } x \in \mathbb{R}^d. \tag{18}$$

Set

$$\eta_n(x) = \{i : 1 \leq i \leq n \text{ and } X_i \in B_{\varepsilon_n}(x)\}$$

and $N_n(x) = \#\eta_n(x) = n\hat{F}_n(B_{\varepsilon_n}(x))$. Then $E_f N_n(x) = nF_f(B_{\varepsilon_n}(x))$.

Also

$$\hat{T}_n(x) = n^{-1} \sum_{1}^{n} QK_n(x - X_i) = \sum_{\eta_n(x)} n^{-1} QK_n(x - X_i). \tag{19}$$

Conditioned on $N_n(x)$, $\hat{T}_n(x)$ is the sum of $N_n(x)$ independent and identically distributed random variables each having mean $g_n(x;f)$, where

$$g_n(x;f) = \frac{\int_{B_{\varepsilon_n}(x)} n^{-1} Q K_n(x-y) f(y)\, dy}{\int_{B_{\varepsilon_n}(x)} f(y)\, dy} .$$

Thus (18) implies

$$|g_n(x;f)| \leq M_6\, n^{-2} \varepsilon_n^{-(m+d)} \quad \text{for } n \geq 1 \text{ and } x \in \mathbb{R}^d . \tag{20}$$

Also

$$E_f(\hat{T}_n(x) | N_n(x)) = g_n(x;f)\, N_n(x) \tag{21}$$

and

$$E_f \hat{T}_n(x) = g_n(x;f) E_f N_n(x) . \tag{22}$$

Lemma 5. Let M_7 be a positive constant and define c_n by

$$c_n = 2M_7 \left(\frac{\log n}{n}\right)^{\frac{1}{2}} \varepsilon_n^{-(2m+d)/2} \quad \text{for } n \geq 1. \tag{23}$$

If

$$\left(\frac{2M_4 M_6}{M_7}\right)^2 n\, \varepsilon_n^d \geq \log n ,$$

then, for $f \varepsilon F$ and $x \varepsilon \mathbb{R}^d$,

$$P_f(|\hat{T}_n(x) - E_f \hat{T}_n(x)| \geq c_n \text{ and } N_n(x) \leq 2M_4 n \varepsilon_n^d) \leq n^{-M_7^2/4M_4 M_6^2} .$$

Proof. By (18), (19) and Theorem 2 of Hoeffding (1963)

$$P_f(|\hat{T}_n(x)-g_n(x;f)N_n(x)| \geq \frac{c_n}{2} N_n(x))$$

$$\leq 2e^{-n^2c_n^2\varepsilon_n^{2(m+d)}/8M_6^2N_n(x)}$$

$$= 2n^{-M_7^2 n\varepsilon_n^d/2M_6^2N_n(x)}$$

$$\leq 2n^{-M_7^2/4M_4M_6^2} \quad \text{on } \{N_n(x) \leq M\, n\varepsilon_n^d\} .$$

Consequently

$$P_f(|\hat{T}_n(x)-g_n(x;f)N_n(x)| \geq \frac{c_n}{2} \text{ and } N_n(x) \leq 2M_4n\varepsilon_n^d)$$

$$\leq 2n^{-M_7^2/4M_4M_6^2} .$$

By (20) and (22)

$$|g_n(x;f)N_n(x) - E_f\hat{T}_n(x)| \leq M_6n^{-1}\varepsilon_n^{-(m+d)}|N_n(x)-E_fN_n(x)| .$$

Thus by Bernstein's inequality (see page 34 of Bennett (1962))

$$P_f(|g_n(x;f)N_n(x) - E_f\hat{T}_n(x)| \geq \frac{c_n}{2}) \leq 2e^{-q_n^2/(2EN_n(x)+q_n)}$$

$$\leq 2e^{-q_n^2/(2M_4n\varepsilon_n^d+q_n)} ,$$

where

$$q_n = \frac{n\varepsilon_n^{m+d}c_n}{2M_6} = \frac{M_7}{M_6}(\log n)^{\frac{1}{2}}(n\varepsilon_n^d)^{\frac{1}{2}} .$$

It follows from (24) that $q_n \leq 2M_4n\varepsilon_n^d$ and hence that

$$P_f(|g_n(x;f)N_n(x)-E_f\hat{T}_n(x)| \geq \frac{c_n}{2}) \leq 2e^{-q_n^2/4M_4n\varepsilon_n^d} = 2n^{-M_7^2/4M_4M_6^2}.$$

The desired conclusion follows immediately from these observations.

Lemma 6. Let c_n *be defined by* (23) *and suppose that*

$$\lim_n \frac{\log n}{n\,\varepsilon_n^d} = 0. \tag{25}$$

Set $A_n = \{x \in \mathbb{R}^d : F_f(B_{\varepsilon_n}(x)) \le c_n \varepsilon_n^{m+d}/4M_6\}$. *Then*

$$\lim_n \sup_F P_f(\sup[|\hat{T}_n(x)-E_f\hat{T}_n(x)| : x \in A_n] \ge c_n) = 0.$$

Proof. By (23), (25), and Lemma 4

$$\lim_n \sup_F P_f(\sup[\hat{F}_n(B_{\varepsilon_n}(x)) : x \in A_n] \ge \frac{c_n \varepsilon_n^{m+d}}{2M_6} = 0.$$

The desired result now follows from (18)-(20) and (22).

Lemma 7. Let M_8 *be a positive constant and set*
$B_n = \{x = (x_1, \ldots, x_d) \in \mathbb{R}^d : n^{M_8} x_i$ *is an integer for* $1 \le i \le d\}$.
If (23) *and* (25) *hold and* M_7 *is sufficiently large, then*

$$\lim_n \sup_F P_f(\sup[|\hat{T}_n(x)-E_f\hat{T}_n(x)| : x \in B_n] \ge c_n) = 0.$$

Proof. Recall than $\varepsilon_n \le 1$. Let n be sufficiently large so that $\log n \ge 1$ and $n\varepsilon_n^d \ge 1$. Then

$$\#(B_n - A_n)\,\frac{c_n\,\varepsilon_n^{m+d}}{4M_6} \le \sum_{x\varepsilon B_n} F_f(B_{\varepsilon_n}(x))$$

$$= \sum_{x\varepsilon B_n} \int_{B_{\varepsilon_n}(x)} f(y)\,dy$$

$$= \int_{\mathbb{R}^d} \#(B_{\varepsilon_n}(y)\cap B_n) f(y)\,dy$$

$$\le (2\varepsilon_n n^{M_8}+1)^d \, .$$

Consequently

$$\#(B_n - A_n) \le \frac{4M_6}{c_n\,\varepsilon_n^{m+d}}(2\varepsilon_n n^{M_8} + 1)^d$$

$$= \frac{2M_6}{M_7}\left(\frac{n}{\varepsilon_n^d \log n}\right)^{\frac{1}{2}}(2\varepsilon_n n^{M_8} + 1)^d$$

$$\le \frac{2M_6 n}{M_7}\,(2n^{M_8} + 1)^d .$$

Equation (26) now follows from Lemmas 4-6.

Lemma 8. If (23) *and* (25) *hold and* M_7 *is sufficiently large, then*

$$\lim_n \sup_F P_f(||\hat{T}_n - E_f\,\hat{T}_n||_\infty \ge 2c_n) = 0.$$

Proof. Let M_3 and λ be as in Lemma 3. Choose $M_8 > \lambda^{-1} + d^{-1}$. Let B_n be as in Lemma 7 and let M_7 be sufficiently large so that (26) holds. Let n be large enough so that $\log n \ge 1$, $n\varepsilon_n^d \ge 1$ and

$$n^{\lambda M_8} \geq \frac{M_3 d^{\lambda/2} n^{1+d^{-1}\lambda}}{M_7 2^{\lambda}} .$$

Suppose that

$$\sup[|\hat{T}_n(w) - E_f \hat{T}_n(w)| : w \in B_n] \leq c_n .$$

Choose $x \in \mathbb{R}^d$. There is a $w \in B_n$ such that $|x-w| \leq d^{\frac{1}{2}}/2n^{M_8}$.
By Lemma 3

$$|\hat{T}_n(x) - E_f \hat{T}_n(x)| \leq |\hat{T}_n(x) - \hat{T}_n(w)| + |\hat{T}_n(w) - E_f \hat{T}_n(w)| + |E_f \hat{T}_n(w) - E_f \hat{T}_n(x)|$$

$$\leq c_n + \frac{2M_3 \varepsilon_n^{-(d+m+\lambda)} d^{\lambda/2}}{2^{\lambda} n^{\lambda M_8}} \leq 2c_n,$$

from which the desired result follows.

With this preparation, it is easy to complete the proof of Theorem 2. It follows from Lemma 2 and Lemma 8 that if (25) holds, then for some positive number M_9

$$\lim_n \sup_F P_f(||\hat{T}_n - T(f)||_\infty \geq M_9(\varepsilon_n^{p-m} + (\frac{\log n}{n})^{\frac{1}{2}} \varepsilon_n^{-(2m+d)/2}) = 0.$$

If

$$\varepsilon_n^{p-m} = (\frac{\log n}{n})^{\frac{1}{2}} \varepsilon_n^{-(2m+d)/2}$$

or, equivalently, if

$$\varepsilon_n = (\frac{\log n}{n})^{\gamma},$$

then (25) holds and hence $\{\hat{T}_n\}$ is asymptotically optimal. Similarly, $\{\hat{T}_n\}$ is asymptotically optimal under the more general condition on $\{\varepsilon_n\}$ in the statement of Theorem 2.

REFERENCES

Bennett, G. (1962). "Probability Inequalities for the Sum of Independent Random Variables." *J. Amer. Statist. Assoc. 57,* 33-45.

Breiman, L., Friedman, J. H., Olshen, R. A. and Stone, C. J. (1983). *Tree-Structured Methods for Classification and Regression,* Wadsworth, Belmont.

Bretagnolle, J. and Huber, C. (1979). "Estimation des Densities: Risque Minimax." *Z. Warsch. verw. Gebiete 47,* 119-137.

Hoeffding, W. (1963). "Probability Inequalities for Sums of Bounded Random Variables." *J. Amer. Statist. Assoc. 58,* 13-30.

Khasminskii, R. Z. (1978). "A Lower Bound on the Risks of Nonparametric Estimates of Densities in the Uniform Metric." *Theor. Probability Appl. 23,* 794-798.

Müller, H. G. and Gasser, T. (1979). "Optimal Convergence Properties of Kernel Estimates of Derivatives of a Density Function," *Smoothing Techniques for Curve Estimation,* (T. Gasser and M. Rosenblatt, eds.), 144-154. Springer-Verlag, Berlin.

Singh, R. H. (1981). "Speed of Convergence in Nonparametric Estimation of a Multivariate μ-density and Its Mixed Partial Derivatives." *J. Statist. Plann. Inference 5,* 287-298.

Stone, C. J. (1980). "Optimal Convergence Rates for Nonparametric Estimators." *Ann. Statist. 8,* 1348-1360.

Stone, C. (1982). "Optimal Global Rates of Convergence for Nonparametric Regression. *Ann. Statistic 10,* 1040-1053.

RANKS AND ORDER STATISTICS

W. R. Van Zwet

Department of Mathematics
University of Leiden
Leiden, The Netherlands

I. INTRODUCTION

In their famous 1958 paper, Chernoff and Savage [4] proved the asymptotic normality of linear rank statistics for the two-sample problem under fixed alternatives. Such asymptotic normality proofs had been given before, but the degree of generality in this paper far surpassed these earlier efforts. The result was obtained for scores generated by a very smooth function J on (0,1) of controlled growth near 0 and 1 and for almost any fixed alternative. Classes of alternatives for which the convergence to normality is uniform were also investigated, thus extending the result to sequences of alternatives within such a class. The paper validated the normal approximation and the computation of asymptotic efficiencies for most two-sample rank statistics that one is likely to come across. It also struck terror into the hearts of graduate students at the time because of - what was then considered - its extreme technicality; in order to approximate the rank statistic by a sum of independent random variables no fewer than six remainder terms were shown

ISBN 0-12-589320-5

to tend to zero, each for its own particular reason. Unfortunately, the number of such remainder terms has increased monotonically over the years and nowadays authors in this area appear to need at least fifteen.

It is hard to overestimate the influence of the Chernoff-Savage paper. It started a steady stream of research resulting in a voluminous literature on the asymptotics of rank statistics. Many extensions to more general and more complicated rank tests were obtained and at the same time technical refinements have led to improved conditions. Even though contiguity arguments later took over part of the field, work along Chernoff-Savage lines is continuing to the present day.

Another one of Herman Chernoff's contributions to asymptotic statistics - and one that was almost as influential - is the 1967 paper by Chernoff, Gastwirth and Johns [3] on the asymptotic normality of linear functions of order statistics (LFO's). For uniform order statistics $U_{1:N} < U_{2:N} <\ldots< U_{N:N}$ and weights $a_{j,N}$ generated by a function J on (0,1), they prove asymptotic normality of $\Sigma a_{j,N}\ \psi(U_{j:N})$ for smooth ψ and under growth conditions on both J and ψ. The result provided normal approximations and an asymptotic theory for linear estimators. The proof is based on transforming to exponential order statistics and exploiting their very special structure. However, the authors point out that an alternative approach based on the methods of Chernoff and Savage would also have been possible.

The research on the asymptotics of LFO's that was initiated by the Chernoff-Gastwirth-Johns paper, is again quite substantial. Various techniques have been applied and have led to different and gradually improved sets of conditions and the end of this process doesn't yet seem to be in sight. It is interesting that there is a trade-off between assumptions on J and on ψ; one can either assume very little about J but a lot about ψ, or the other way round.

When one looks at the literature on the asymptotic normality of rank statistics and of LFO's, one can't help noticing a striking similarity of the techniques employed in the two areas. It was noted above that the Chernoff-Savage method is applicable to the study of LFO's too, but the similarity doesn't end there. Almost any technical device that has worked in one area, has worked for the other problem also. When viewing the research in the two areas, the image of two armies marching on parallel roads readily comes to mind. But this raises the further question whether perhaps the two seemingly very different problems of proving asymptotic normality for rank statistics and for LFO's, are essentially the same or at least more intimately connected than one would think at first sight. Or, in terms of our admittedly fanciful image: are the two armies perhaps going to the same place?

Basically, I believe they are. In this paper it will be shown that under very general conditions, asymptotic normality of a two-sample linear rank statistic under a fixed alternative follows from asymptotic normality of an appropriate LFO. Since the possibility of a result in the other direction won't even be considered in this paper, I can't possibly claim to have shown that the two problems are the same, but only that they are intimately related. Like everything else in this area, the proof of the result is highly technical; a part of it that can't possibly be of general interest, will be left to the interested reader with appropriate hints being given in the appendix. Problems concerning the uniformity of the convergence to normality of the rank statistic will be avoided by restricting attention to a single fixed alternative and a rapidly converging sample ratio. Finally, I should perhaps make it clear that I'm not advocating that one should use the result of this paper to prove asymptotic normality for a rank statistic by first proving it for the corresponding LFO. What motivates the result is not its possible application, but only the light it may throw on the connection between the two problems.

II. THE RESULT

Let F and G be two continuous distribution functions (d.f.'s) on the real line with densities f and g. For N = 1,2,..., consider independent random variables $X_{1,N}, X_{2,N}, \dots, X_{N,N}$ and assume that $X_{1,N}, \dots, X_{m_N,N}$ have common d.f. F and $X_{m_N+1,N}, \dots, X_{N,N}$ have common d.f. G. This is the two-sample situation with sample sizes m_N and $n_N = N - m_N$; $\lambda_N = n_N/N$ indicates the relative size of the second sample. Let $X_{1:N} < X_{2:N} < \dots < X_{N:N}$ denote the combined sample $X_{1,N}, \dots, X_{N,N}$ arranged in increasing order and define the antiranks $D_{1,N}, \dots, D_{N,N}$ by $X_{D_{j,N},N} = X_{j:N}$. The random variables

$$V_{j,N} = \begin{cases} 1 & \text{if } m_N + 1 \le D_{j,N} \le N \\ 0 & \text{otherwise} \end{cases}$$

indicate from which sample each of the ordered sample elements originates. For real numbers $a_{1,N}, \dots, a_{N,N}$ called scores, the two-sample linear rank statistic is defined by

$$T_N = \sum_{j=1}^{N} a_{j,N} V_{j,N}. \tag{2.1}$$

Suppose that $\lambda_N \to \lambda \in (0,1)$ as $N \to \infty$ and define

$$H = \lambda G + (1 - \lambda)F, \; h = \lambda g + (1 - \lambda)f, \tag{2.2}$$

$$\psi = \frac{\lambda g \circ H^{-1}}{h \circ H^{-1}}, \tag{2.3}$$

$$\pi_{j,N} = N. \int_{(j-1)/N}^{j/N} \psi(x)dx, \tag{2.4}$$

$$\bar{a}_N = \frac{\sum_{j=1}^{N} \pi_{j,N}(1 - \pi_{j,N}) a_{j,N}}{\sum_{j=1}^{N} \pi_{j,N}(1 - \pi_{j,N})}, \tag{2.5}$$

$$\tau_N^2 = \frac{1}{N} \sum_{j=1}^{N} \pi_{j,N}(1 - \pi_{j,N})(a_{j,N} - \bar{a}_N)^2 . \tag{2.6}$$

Let $U_1, U_2, \ldots$ be independent and identically distributed random variables with a common uniform distribution on (0,1) and let $U_{1:N} < U_{2:N} < \ldots < U_{N:N}$ denote the order statistics corresponding to $U_1, \ldots, U_N$. Define

$$L_N = \sum_{j=1}^{N} a_{j,N}\, \psi(U_{j:N}),$$
$$\hat{L}_N = \sum_{j=1}^{N} E(L_N | U_j) - (N - 1)\, E\, L_N , \tag{2.7}$$

and note that $\hat{L}_N$ is the L_2 - projection of L_N. Finally, let 1_A denote the indicator of a set A, let $\sigma^2(Y)$ denote the variance of a random variable Y and let $\xrightarrow{\mathcal{D}} N(0,1)$ denote convergence in distribution to the standard normal.

Theorem 2.1

Assume that $\lambda \in (0,1)$ *and* $\delta > 0$ *exist such that*

$$\lim_{N\to\infty} N^{1/2}(\lambda_N - \lambda) = 0, \tag{2.8}$$

$$\liminf_N \tau_N^2 > 0, \tag{2.9}$$

$$\limsup_N \frac{1}{N} \sum_{j=1}^{N} |a_{j,N}|^{2+\delta} < \infty. \tag{2.10}$$

If $\sigma^2(N^{-1/2} L_N)$ *is bounded and*

$$\frac{L_N - EL_N}{\sigma(\hat{L}_N)} \xrightarrow{\mathcal{D}} N(0,1), \tag{2.11}$$

then there exists a bounded sequence of positive numbers $\sigma_N \geq \tau_N$ *such that*

$$\frac{T_N - EL_N}{N^{1/2}\sigma_N} \xrightarrow{\mathcal{D}} N(0,1). \tag{2.12}$$

Some brief comments on assumptions (2.8) - (2.10) may be in order. First of all, (2.8) ensures an almost constant sample-ratio $\lambda_N = \lambda + \mathcal{O}(N^{-1/2})$ and together with the fact that F and G are fixed, this prevents uniformity problems. Another important aspect is that λ_N remains bounded away from 0 and 1 as $N \to \infty$. Without this, both $N^{-1/2}T_N$ and $N^{-1/2}L_N$ will degenerate as $N \to \infty$ and technical complications arise. Assumptions (2.9) and (2.10) together ensure that the scores $a_{j,N}$ are roughly of the order of 1 as $N \to \infty$, which is merely a norming convention. Apart from this, (2.9) prevents a more general kind of degeneration of $N^{-1/2}T_N$ which would occur if the $V_{j,N}$ would degenerate for certain indices j, whereas the scores $a_{j,N}$ would be almost constant for the remaining indices; of course (2.9) implies that $\lambda \in (0,1)$ in (2.8). Assumption (2.10) controls the growth of the scores.

III. PROOF

Recall (2.1) - (2.7) and define in addition

$$P_{j,N} = \psi(U_{j:N}), \quad P = P_N = (P_{1,N}, \ldots, P_{N,N}), \tag{3.1}$$

$$\omega(P) = N^{-1/2} \sum_{j=1}^{N} (P_{j,N} - \lambda_N), \tag{3.2}$$

$$\sigma^2(P) = \frac{1}{N} \sum_{j=1}^{N} P_{j,N}(1-P_{j,N}), \tag{3.3}$$

$$\bar{a}(P) = \frac{\sum_{j=1}^{N} P_{j,N}(1-P_{j,N})a_{j,N}}{\sum_{j=1}^{N} P_{j,N}(1-P_{j,N})}, \tag{3.4}$$

$$\tau^2(P) = \frac{1}{N}\sum_{j=1}^{N} P_{j,N}(1-P_{j,N})(a_{j,N}-\bar{a}(P))^2. \tag{3.5}$$

The following lemma will be the starting point of the proof.

Lemma 3.1. If (2.8) *and* (2.10) *are satisfied, then for every positive* δ *and* ε,

$$E\exp\{\text{it } N^{-1/2}T_N\} = E\,\frac{\{\lambda(1-\lambda)\}^{1/2}}{\sigma(P)}\exp\left\{-\frac{\omega^2(P)}{2\sigma^2(P)} + \right.$$
$$-\frac{1}{2}t^2\tau^2(P) - \text{it }\omega(P)\,\bar{a}(P) +$$
$$\left. + \text{it } N^{-1/2}\Sigma a_{j,N}P_{j,N}\right\}\cdot 1_{\{\tau^2(P)\geq\varepsilon\}} +$$
$$+ O(N^{1/2}P(\sigma^2(P)<\delta) + P(\tau^2(P)<\varepsilon)) + o(1)$$

as $N\to\infty$.

This lemma may be proved by modifying an argument in Bickel and Van Zwet [2]. Since this is a highly technical matter, the interested reader is referred to the appendix for details.

Define

$$S_N = \sum_{j=1}^{N}(\psi(U_j) - \lambda), \tag{3.7}$$

$$\sigma_0^2 = \lambda(1-\lambda)\int_{-\infty}^{\infty}\frac{f(x)g(x)}{h(x)}\,dx, \tag{3.8}$$

and let $\pi_{j,N}$, $\bar{a}_N$ and τ_N^2 be given by (2.4) - (2.6). The next lemma is needed to simplify (3.6).

Lemma 3.2. If (2.8) - (2.10) *are satisfied, then the following statements hold with probability* 1*:*

$$\lim_{N\to\infty} \frac{1}{N} \sum_{j=1}^{N} |P_{j,N} - \pi_{j,N}| = 0, \tag{3.9}$$

$$\lim_{N\to\infty} |\omega(P) - N^{-1/2} S_N| = 0, \tag{3.10}$$

$$\lim_{N\to\infty} \sigma^2(P) = \sigma_0^2 > 0, \tag{3.11}$$

$$\lim_{N\to\infty} |\bar{a}(P) - \bar{a}_N| = 0 \tag{3.12}$$

$$\lim_{N\to\infty} |\tau^2(P) - \tau_N^2| = 0. \tag{3.13}$$

Proof. Define the function ψ_N on $[0,1)$ by

$$\psi_N(t) = \psi(U_{j:N}) \quad \text{for} \quad \frac{j-1}{N} \leq t < \frac{j}{N}, \tag{3.14}$$

for $j = 1,\ldots,N$. Lemma 2.1 in Van Zwet [5] ensures that with probability 1 (w.p.1) ψ_N converges to ψ in Lebesgue measure. As $|\psi_N|$ and $|\psi|$ are bounded by 1 and

$$\frac{1}{N} \sum_{j=1}^{N} |P_{j,N} - \pi_{j,N}| \leq \int_0^1 |\psi_N(x) - \psi(x)| dx,$$

(3.9) follows. This implies that

$$\lim_{N\to\infty} |\sigma^2(P) - \frac{1}{N} \sum_{j=1}^{N} \pi_{j,N}(1-\pi_{j,N})| = 0$$

w.p.1. On the other hand, the strong law yields

$$\lim_{N\to\infty} \sigma^2(P) = \int_0^1 \psi(x)(1-\psi(x))dx = \sigma_0^2$$

w.p.1. Hence

$$\lim_{N\to\infty} \frac{1}{N} \sum_{j=1}^{N} \pi_{j,N}(1-\pi_{j,N}) = \sigma_0^2$$

and since (2.9) and (2.10) imply that the left-hand side is positive, (3.11) is proved. Because $\sigma_0^2 > 0$, (2.10) yields (3.12) and another application of (2.10) proves (3.13). As (3.10) is an immediate consequence of (2.8), the proof of the lemma is complete. □

Assumption (2.10) implies that on the set where $\tau^2(P) \geq \varepsilon$, $\sigma^2(P)$ is also bounded away from zero. Hence the expression following the expectation sign on the right in (3.6) is bounded and we may replace $\omega(P)$, $\sigma(P)$, $\bar{a}(P)$ and $\tau(P)$ is this expression by $N^{-1/2}S_N$, σ_0, $\bar{a}_N$ and τ_N, provided only that (2.8) - (2.10) hold. Take $\delta = \frac{1}{2}\sigma_0^2 > 0$ and $\varepsilon = \frac{1}{2}\liminf \tau_N^2 > 0$. Because of (3.11) and (3.13), combined with the fact that $\sigma^2(P)$ is a mean of independent, identically distributed and bounded random variables, one finds that

$$N^{1/2}\ P(\sigma^2(P) < \delta) + P(\tau^2(P) < \varepsilon) = o(1)$$

as $N \to \infty$. Hence lemmas 3.1 and 3.2 together yield

Lemma 3.3. If (2.8) - (2.10) *are satisfied, then, as* $N \to \infty$,

$$\phi_N(t) = E\exp\{itN^{-1/2}(T_N - EL_N)\} = \frac{\{\lambda(1-\lambda)\}^{1/2}}{\sigma_0}\exp\{-\tfrac{1}{2}\tau_N^2 t^2\}\cdot$$
$$\cdot\ E\exp\left\{-\frac{S_N^2}{2\sigma_0^2 N} - it\bar{a}_N N^{-1/2}S_N + itN^{-1/2}(L_N - EL_N)\right\} \tag{3.15}$$
$$+\ o(1).$$

The next step is to establish the asymptotic normality of the projection $\hat{L}_N$ of L_N.

Lemma 3.4. If (2.10) *is satisfied and* $\liminf_N \sigma^2(N^{-1/2}\hat{L}_N) >$ 0, *then*

$$\frac{\hat{L}_N - EL_N}{\sigma(\hat{L}_N)} \xrightarrow{\mathcal{D}} N(0,1)\ . \tag{3.16}$$

Proof. A straightforward calculation shows that

$$\hat{L}_N - EL_N = \sum_{i=1}^{N} \{\alpha_N(U_i) - \beta_N(U_i) + \gamma_N(U_i)\} \text{ where}$$

$$\alpha_N(u) = \sum_{j=1}^{N-1} \frac{N}{N-j} a_{j,N} E\, \psi(U_{j:N-1})(U_{j:N-1} - \frac{j}{N}) 1_{\{U_{j:N-1} \le u\}},$$

$$\beta_N(u) = \sum_{j=1}^{N-1} \frac{N}{j} a_{j+1,N} E\, \psi(U_{j:N-1})(U_{j:N-1} - \frac{j}{N}) 1_{\{U_{j:N-1} > u\}}, \tag{3.17}$$

$$\gamma_N(u) = \psi(u) \sum_{j=1}^{N} a_{j,N} \binom{N-1}{j-1} u^{j-1} (1-u)^{N-j}.$$

Because $|\psi| \le 1$ and $E\, U_{j:N-1} = j/N$,

$$|\alpha_N(u)| \le \left| \sum_{j=1}^{N-1} \frac{N}{N-j} a_{j,N} E\, \psi(U_{j:N-1}) \cdot \right.$$

$$\left. \cdot E(U_{j:N-1} - \frac{j}{N}) 1_{\{U_{j:N-1} \le u\}} \right| + \left| \sum_{j=1}^{N-1} \frac{N}{N-j} a_{j,N} E\{\psi(U_{j:N-1}) + \right.$$

$$\left. - E\, \psi(U_{j:N-1})\}(U_{j:N-1} - \frac{j}{N}) 1_{\{U_{j:N-1} \le u\}} \right| \le$$

$$\le \sum_{j=1}^{N-1} \frac{N}{N-j} |a_{j,N}| \left| E(U_{j:N-1} - \frac{j}{N}) 1_{\{U_{j:N-1} \le u\}} \right| +$$

$$+ \sum_{j=1}^{N-1} \frac{N}{N-j} |a_{j:N}| \sigma(\psi(U_{j:N-1}))\, \sigma(U_{j:N-1}).$$

Now $\sigma^2(U_{j:N-1}) = j(N-j)/\{N^2(N+1)\}$ and by lemma A2.1 in Albers, Bickel and Van Zwet [1]

$$P(\sigma^{-1}(U_{j:N}) |U_{j:N} - \frac{j}{N+1}| \ge t) \le 2 \exp\left\{ - \frac{3t^2}{6t+8} \right\}$$

for all $j = 1,\dots,N$, $N = 1,2,\dots$ and $t \ge 0$. It follows that

$$\left|E(U_{j:N-1} - \frac{j}{N})\, 1_{\{U_{j:N-1} \leq u\}}\right| \leq$$

$$\leq C\left(\frac{j(N-j)}{N^3}\right)^{1/2} \exp\left\{-\alpha \frac{|u-j/N|}{\{j(N-j)\}^{1/2}} N^{3/2}\right\}$$

for positive constants C and α. Hence

$$|\alpha_N(u)| \leq C \sum_{j=1}^{N-1} |a_{j,N}| \left(\frac{j}{N(N-j)}\right)^{1/2} \exp\left\{-\alpha \frac{|u-j/N|}{\{j(N-j)\}^{1/2}} N^{3/2}\right\} +$$

$$+ \sum_{j=1}^{N-1} |a_{j,N}| \left(\frac{j}{N(N-j)}\right)^{1/2} \sigma(\psi(U_{j:N-1})).$$

Take $\delta > 0$ as in assumption (2.10). There exist positive numbers $p \leq 2 + \delta$, $q < 2$ and $r > 2$ such that $p^{-1} + q^{-1} + r^{-1} = 1$ and repeated use of Hölder's inequality yields

$$|\alpha_N(u)| \leq C\, N^{-1/2}\left(\sum_{j=1}^{N-1} |a_{j,N}|^p\right)^{1/p}\left(\sum_{j=1}^{N-1} \left(\frac{j}{N-j}\right)^{q/2}\right)^{1/q} \cdot$$

$$\cdot\left[\left(\sum_{j=1}^{N-1} \exp\{-2\alpha r N^{1/2}|u-j/N|\}\right)^{1/r} + \left(\sum_{j=1}^{N-1} \sigma^r(\psi(U_{j:N-1}))\right)^{1/r}\right].$$

For ψ_N as in (3.14), we can argue as in the proof of lemma 3.2 to find that, as $N \to \infty$,

$$\frac{1}{N}\sum_{j=1}^{N-1} \sigma^r(\psi(U_{j:N-1})) \leq E \int_0^1 |\psi_{N-1}(x) - \psi(x)|^r dx = o(1)$$

because ψ_N and ψ are bounded and ψ_N converges to ψ in Lebesgue measure with probability 1. Bounding the other sums by the corresponding integrals or by using (2.10) we arrive at

$$\sup_{0<u<1} |\alpha_N(u)| = O(N^{-1/2+p^{-1}+q^{-1}})\{O(N^{(2r)^{-1}}) + o(N^{r^{-1}})\} =$$

$$= o(N^{1/2})$$

as $N \to \infty$. Similarly, $\sup|\beta_N(u)| = o(N^{1/2})$ and

$\sup|\gamma_N(u)| \leq \max_j |a_{j,N}| = o(N^{1/2})$ in view of (2.10). Since $\sigma(N^{-1/2}\hat{L}_N)$ is bounded away from zero, (3.17) and the central limit theorem yield (3.16). □

The next lemma deals with the asymptotic equivalence of L_N and $\hat{L}_N$.

Lemma 3.5. Suppose that $\liminf_N \sigma^2(N^{-1/2}\hat{L}_N) > 0$ *and* $\sigma^2(N^{1/2}L_N)$ *is bounded and that* (2.10) *and* (2.11) *hold. Then* $\sigma^{-1}(\hat{L}_N)(L_N-\hat{L}_N)$ *tends to zero in probability as* $N \to \infty$.

Proof. Write $Z_N = \sigma^{-1}(\hat{L}_N)(L_N-EL_N)$ and $\hat{Z}_N = \sigma^{-1}(\hat{L}_N)(\hat{L}_N-EL_N)$. The conditions of lemma 3.4 are satisfied so that we may assume that

$$E\, Z_N = 0, \quad E\, Z_N^2 \leq C, \quad Z_N \overset{\mathcal{D}}{\to} N(0,1),$$

$$E\, \hat{Z}_N = 0, \quad E\, \hat{Z}_N^2 = 1, \quad \hat{Z}_N \overset{\mathcal{D}}{\to} N(0,1)$$

for a positive constant C. We have to show that $\Delta_N = Z_N - \hat{Z}_N$ converges to zero in probability.

Because $\hat{Z}_N$ is the projection of Z_N we have $E\, \hat{Z}_N\Delta_N = 0$ for every N. Moreover, $E\Delta_N^2 \leq E\, Z_N^2 \leq C$ and since $E\, \hat{Z}_N^2 = 1$ and $\hat{Z}_N \overset{\mathcal{D}}{\to} N(0,1)$, the sequence $\{\hat{Z}_N^2\}$ is uniformly integrable. This implies that the sequence $\{\hat{Z}_N\Delta_N\}$ is uniformly integrable.

The sequence of joint distributions of $(Z_N, \hat{Z}_N, \Delta_N)$ is clearly tight. Take any weakly converging subsequence

$$(Z_{N_k}, \hat{Z}_{N_k}, \Delta_{N_k}) \overset{\mathcal{D}}{\to} (Z, \hat{Z}, \Delta), \text{ say.}$$

Obviously $Z = \hat{Z} + \Delta$ with probability 1, $E\, Z^2 = E\, \hat{Z}^2 = 1$ and as $\{\hat{Z}_N\Delta_N\}$ is uniformly integrable with $E\, \hat{Z}_N\Delta_N = 0$, we have $E\, \hat{Z}\Delta = 0$. It follows that $E\, \Delta^2 = E\, Z^2 - E\, \hat{Z}^2 = 0$ so that $\Delta_{N_k} \overset{\mathcal{D}}{\to} 0$. Hence $\Delta_N \overset{\mathcal{D}}{\to} 0$ and the lemma is proved. □

We are now in a position to prove the theorem.

Proof of Theorem 2.1. Assume first that $\liminf \sigma^2(N^{-1/2}\hat{L}_N) > 0$ so that the conclusion of lemma 3.5

holds. In view of (3.7), (3.17), the boundedness of ψ and the proof of lemma 3.4, we see that

$$\lim_{N\to\infty}[P(N^{-1/2}S_N \le x, N^{-1/2}(L_N - EL_N) \le y) - P(A_N \le x, B_N \le y)] = 0 \tag{3.18}$$

for all x and y, where A_N and B_N are jointly normally distributed with $E\,A_N = E\,B_N = 0$,

$$E\,A_N^2 = \int_0^1 \psi^2(x)dx - \lambda^2 = \lambda(1-\lambda) - \sigma_0^2, \tag{3.19}$$

$$E\,B_N^2 = \int_0^1 \{\alpha_N(x) - \beta_N(x) + \gamma_N(x)\}^2 dx, \tag{3.20}$$

$$E\,A_N\,B_N = \int_0^1 \psi(x)\{\alpha_N(x) - \beta_N(x) + \gamma_N(x)\}dx. \tag{3.21}$$

This is still true without the assumption that $\liminf \sigma^2(N^{-1/2}L_N) > 0$, the only difference being that (3.20) is now not necessarily bounded away from 0 for large N. To see this, note that if $\sigma^2(N^{-1/2}\hat{L}_N)$ (or any sub-sequence) tends to zero, then $N^{-1/2}(L_N - EL_N)$ (or its sub-sequence) will tend to zero in probability because of (2.11).

Since $\sigma_0^2 > 0$, assumption (2.10) ensures that $\bar{a}_N$ is bounded. It follows from lemma 3.3 that, for $N \to \infty$ and for every fixed t,

$$\begin{aligned}\phi_N(t) &= \frac{\{\lambda(1-\lambda)\}^{1/2}}{\sigma_0} \exp\{-\tfrac{1}{2}\tau_N^2 t^2\}.\\ &\quad\cdot E\exp\left\{-\frac{A_N^2}{2\sigma_0^2} - it\,\bar{a}_N A_N + it\,B_N\right\} + o(1) = \qquad (3.22)\\ &= \exp\{-\tfrac{1}{2}\sigma_N^2 t^2\} + o(1),\end{aligned}$$

where $\sigma_N \ge \tau_N$ and hence $\liminf \sigma_N > 0$ by (2.9). As $E\,B_N^2 = \sigma^2(N^{-1/2}\hat{L}_N) \le \sigma^2(N^{-1/2}L_N)$ and $\sigma^2(N^{-1/2}L_N)$ is bounded, it is easy to see that $\{\sigma_N\}$ is bounded. This completes the proof of the theorem. □

IV. APPENDIX

In this appendix we indicate how lemma 3.1 may be obtained by modifying an argument in Bickel and Van Zwet [2]. When referring to numbered formulas, lemmas etc. in that paper, we shall add an asterisk to avoid confusion; thus (2.10*) and lemma 2.3* refer to (2.10) and lemma 2.3 in Bickel and Van Zwet [2].

Since λ as defined in (2.5*) is the same as $\lambda_N = n_N/N$ in the present paper, we have to be careful to replace λ by λ_N in formulas such as (2.5*), (2.10*) and (2.14*). However, in (2.4*) we don't make this substitution, thus in effect replacing λ_N by its limiting value λ; note, however, that (2.3*) and (2.6*) remain valid. Similarly, at the beginning of section 3* we don't replace λ by λ_N in the definitions of H and h, thus making them coincide with definition (2.2) in the present paper. It is easy to check that lemma 3.1* remains valid with these modifications, i.e.

$$E \exp\{it\, N^{-1/2} T_N\} = \frac{E_H \nu(t,P) \exp\{it\, N^{-1/2} \Sigma a_j P_j\}}{2\pi\, N^{1/2}\, B_{N,n}(\lambda)}\,. \tag{A.1}$$

Next we need to establish conditions under which for every fixed t,

$$\nu(t,p) = \frac{(2\pi)^{1/2}}{\sigma(p)} \exp\left\{- \frac{\omega^2(p)}{2\sigma^2(p)} - \frac{\tau^2(p)t^2}{2} - i\, \omega(p)\bar{a}(p)t\right\} + o(1). \tag{A.2}$$

This is a weaker version of the conclusion of lemma 2.3* for which assumptions (2.21*) and (2.22*) would be needed. In the first place it is weaker because we are not concerned with values of $|t|$ tending to infinity with N, and therefore we can dispense with assumption (2.22*). Secondly, we don't need the asymptotic expansion for $\nu(t,p)$ established in lemma 2.3*, but only its leading term (A.2) and inspection of the proof of the lemma reveals that (2.21*) may be replaced by assumption (2.10) in the present paper and

$$\tau^2(p) \geq \varepsilon \qquad \text{for some } \varepsilon > 0. \tag{A.3}$$

Together, (2.10) and (A.3) guarantee the validity of (A.2).

If $\tau^2(p) < \varepsilon$, we can bound $|\nu(t,p)|$ as follows

$$|\nu(t,p)| \leq |\nu(0,p)| \leq 2\pi\cdot\min(N^{1/2},\sigma^{-1}(p)).$$

To see this, note that $|\rho(t,p)| \leq 1$ and $0 \leq c(p) \leq 1$ in (2.13*), that the first inequality in (2.24*) yields $|\psi(s,0,p)| \leq \exp\{-[\frac{1}{2} - (\pi^2/24)]\sigma^2(p)s^2\}$ for $|s| \leq \pi N^{1/2}$, and apply (2.11*). Hence, for positive ε and δ,

$$\begin{aligned} E_H|\nu(t,P)|\, 1_{\{\tau^2(P)<\varepsilon\}} &\leq E_H|\nu(t,P)|\, 1_{\{\sigma^2(P)<\delta\}} + \\ &+ E_H|\nu(t,P)|\, 1_{\{\tau^2(P)<\varepsilon,\sigma^2(P)\geq\delta\}} \leq \\ &\leq 2\pi N^{1/2} P_H(\sigma^2(P)<\delta) + 2\pi\delta^{-1/2} P_H(\tau^2(P)<\varepsilon). \end{aligned} \tag{A.4}$$

Assumption (2.8) of the present paper ensures that

$$\lim_{N\to\infty} 2\pi N^{1/2} B_{N,n}(\lambda) = \left[\frac{2\pi}{\lambda(1-\lambda)}\right]^{1/2} \in (0,\infty) \tag{A.5}$$

and, combining (A.1), (A.2), (A.4) and (A.5) we arrive at

$$\begin{aligned} E\exp\{it\,N^{-1/2}T_N\} &= E_H \frac{\{\lambda(1-\lambda)\}^{1/2}}{\sigma(P)} \exp\left\{-\frac{\omega^2(P)}{2\sigma^2(P)} + \right. \\ &\left. - \tfrac{1}{2}t^2\tau^2(P) - it\,\omega(P)\bar{a}(P) + it\,N^{-1/2}\Sigma a_j P_j\right\}\cdot 1_{\{\tau^2(P)\geq\varepsilon\}} + \\ &+ O(N^{1/2}P_H(\sigma^2(P)<\delta) + P_H(\tau^2(P)<\varepsilon)) + o(1), \end{aligned} \tag{A.6}$$

as $N\to\infty$, for every positive ε and δ. The assumptions needed to prove this are (2.8) and (2.10).

Finally we note that (2.4*) as modified above, implies that under H the vector $P = (P_1, \ldots, P_N)$ is distributed as $(\psi(U_{1:N}), \ldots, \psi(U_{N:N}))$ where ψ is given by (2.3) and $(U_{1:N}, \ldots, U_{N:N})$ are order statistics of a sample of size N from the uniform distribution on (0,1). Substituting this in (A.6) we obtain (3.6).

V. REFERENCES

[1] Albers, W., Bickel, P.J. and Van Zwet, W.R. (1976). "Asymptotic Expansions for the Power of Distribution Free Tests in the One-Sample Problem." *Ann. Statist. 4,* 108-156.

[2] Bickel, P.J. and Van Zwet, W.R. (1978). "Asymptotic Expansions for the Power of Distributionfree Tests in the Two-Sample Problem." *Ann. Statist. 6,* 937-1004.

[3] Chernoff, H., Gastwirth, J.L. and Johns, M.V. (1967). "Asymptotic Distribution of Linear Combinations of Functions of Order Statistics with Applications to Estimation." *Ann. Math. Statist. 38,* 52-72.

[4] Chernoff, H. and Savage, I.R. (1958). "Asymptotic Normality and Efficiency of Certain Nonparametric Test Statistics." *Ann. Math. Statist. 29,* 972-994.

[5] Van Zwet, W.R. (1980). "A Strong Law for Linear Functions of Order Statistics." *Ann. Probability 8,* 986-990.

IV. STATISTICAL GRAPHICS

M AND N PLOTS[1]

Persi Diaconis

Stanford University and
Stanford Linear Accelerator Center
Stanford, California

Jerome H. Friedman

Stanford Linear Accelerator Center
Stanford, California

I. INTRODUCTION

In this paper, we describe a method for scatterplotting four-dimensional data. The basic idea is a variation on a suggestion by Tukey and Tukey (1981): as shown in Figure 1, represent a four-dimensional observation p = (x,y,s,t) by a point in each of two coordinate systems. To show that the dots corresponding to (x,y) and (s,t) represent the same point p in four dimensions, connect them with a straight line. When many points are plotted, the larger number of lines can become visually confusing. The lines can be "thinned down" (details are given in Section III) so that the remaining lines give an accurate

[1]*Work partially supported by the Department of Energy under contract number DE-AC03-76SF00515.*
Work partially supported by National Science Foundation under grant MCS77-16974.

ISBN 0-12-589320-5

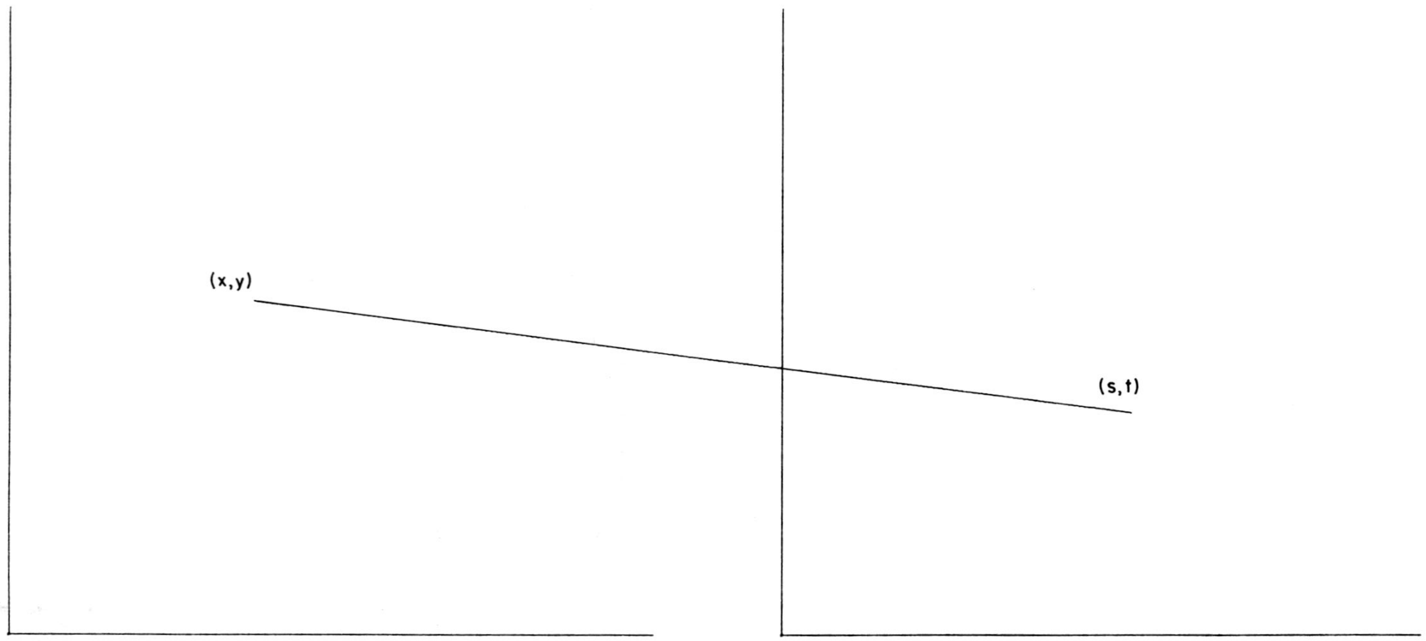

Figure 1

representation of which points in one plot correspond to those of the other plot. This picture is called a 2 and 2 plot. More general M and N plots are discussed in Section III.

Before a general discussion of the technique, we illustrate it by considering an example in some detail. These data are from a diabetes study of Reaven and Miller (1979). For each of 145 subjects in the study, five variables were measured. The variables, here somewhat crudely described, are:

1) relative weight
2) a measure of glucose tolerance
3) a second measure of glucose tolerance - *glucose area*
4) a measure of insulin secretion - *insulin area*
5) a measure of how glucose and insulin interact - *SSPG*.

Variables two and three, the two measures of glucose tolerance, exhibited a very high degree of linear association (R = 0.96) so that only variables one, three, four and five will be considered. Figure 2 shows a 2 and 2 plot of this data. The right-hand part has been rotated 180° to make the picture easier to view (this is discussed in Section III). We have found it useful to focus on each scatterplot separately, and then see what additional structure can be seen in the lines.

- In the left-hand plot, both variables are quite spread-out in their range. There are almost no points in the upper left-hand part of the plot, so thin people tend to have lower SSPG; medium weight and heavier people seem quite spread out in SSPG.
- In the right-hand plot, there is a good deal of structure. Glucose area seems tightly clustered around low values with some scattered high values. People with high values of glucose area are generally considered to be diabetics. The insulin area variable is more uniformly spread out with a high density of medium values. Subjects with very low values of insulin area seem to have higher values of glucose area.

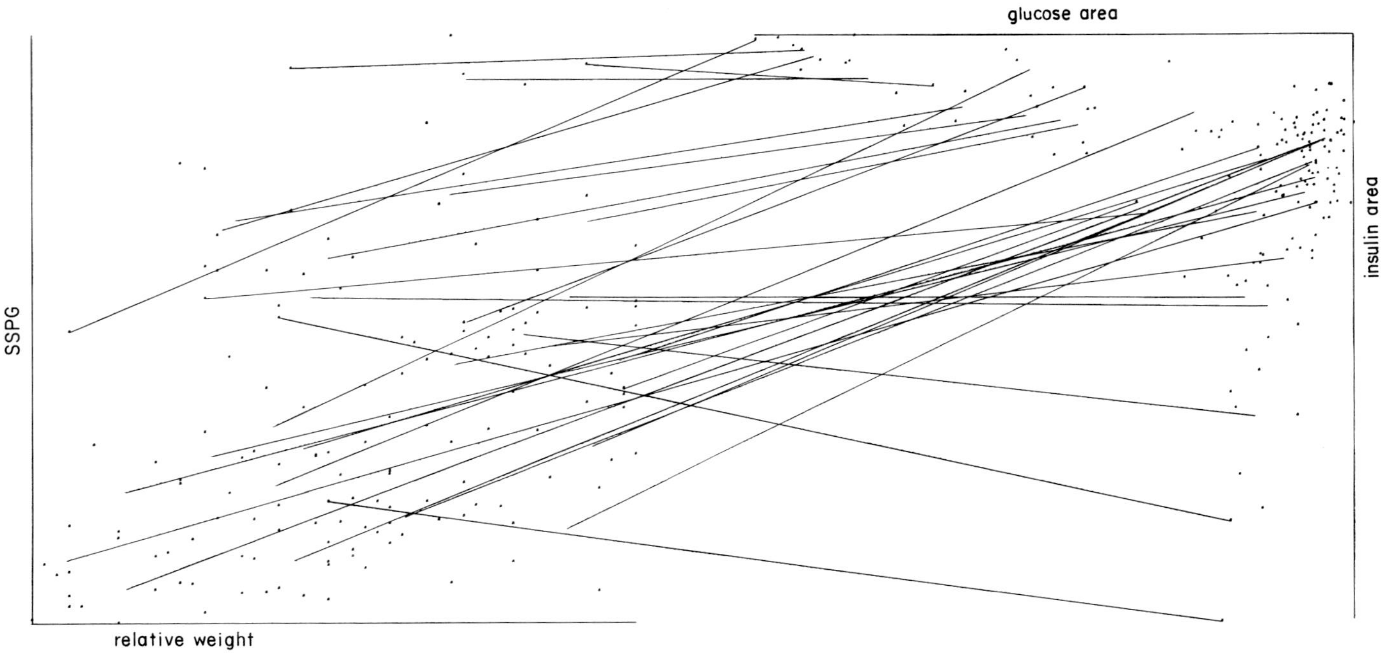

Figure 2

We next discuss the lines connecting the two plots. Consider first the lines emanating from the densist region of the right-hand picture. These lines move down to the left. They spread out on relative weight but seem to range over lower values of SSPG. This cluster of points represents "normal subjects" who have low values of glucose area and insulin area and low values of SSPG.

Next, consider the lines emanating from the points in the right-hand picture representing high values of glucose area. These lines fan out over medium values of relative weight and higher values of SSPG.

Finally, lines representing subjects with higher values of insulin area range over medium values of relative weight and medium values of SSPG. The last two inferences represent insights about genuinely four-dimensional aspects of this data.

The three groupings suggested above agree qualitatively with the groupings of Reaven and Miller (1979). One difference brought out in our analysis: relative weight seems to be a relevant factor. The interaction of weight and the other three variables is only made visible by four-dimensional graphics. Further pictures of this data set are in Section II.

In the diabetes example, 2 and 2 plots allows some higher dimensional aspects of the data set to be seen. The idea is easily generalized. Section II considers examples of general M and N plots. Section III explains an efficient thinning algorithm and ways of drawing M and N plots by hand.

The earliest reference to 2 and 2 plots we know of is Eckhart (1968). There is a good deal of literature on visualization of four dimensions. See Manning (1969), Brisson (1978), and Murrill (1980) for surveys. The best available reference to graphical methods for high-dimensional data is Gnanadesikan (1976).

II. EXAMPLES OF M AND N PLOTS

Conventional scatterplots use dots on a rectangular coordinate system to graphically represent k two-dimensional vectors. An M and N plot represents (M+N)-dimensional vectors by plotting M coordinates in one coordinate system and N coordinates in a second coordinate system. The two dots representing a point are connected by a straight line segment. Thus, scatterplots are 2 and 0 plots.

One-Dimensional Data

A 1 and 0 plot, the dot plot, is sometimes used to plot one-dimensional data, plotting each point as a dot. For example, Figure 3 shows a 1 and 0 plot of the glucose area variable in the diabetes example. Figure 3 shows the clustering around low values and the spreading at higher values reasonably well. A histogram might be a more informative picture for this data set.

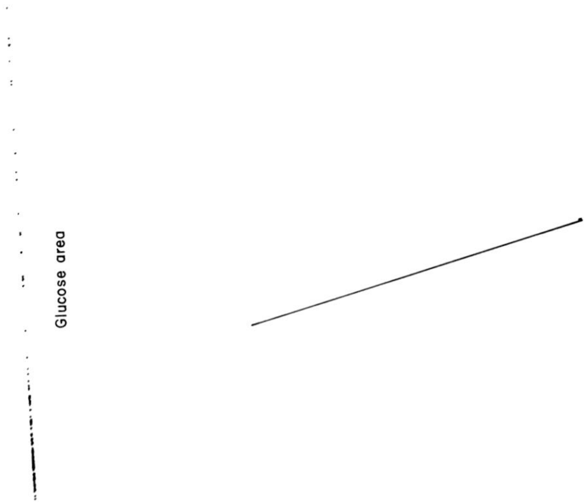

Figure 3

Two-Dimensional Data

There are two ways to draw M and N plots of two-dimensional data - the conventional scatterplot (a 2 and 0 plot), and the 1 and 1 plot. A 1 and 1 plot represents a two-dimensional point (x,y) by 2 dots and a line segment on a pair of parallel coordinate axis.

It is useful to look at 1 and 1 plots of familiar point clouds. For example, Figure 4 shows a cloud of points lying close to a straight line and the corresponding 1 and 1 plot.

Before considering other examples of 1 and 1 plots, we describe the connection between 1 and 1 plots and the set of lines in the plane. To avoid confusion, the line *segments* in 1 and 1 plots, such as Figure 4, will be called *segments* in what follows. Each segment can be thought of as a section of the (infinite) line in the plane that passes through that segment. Thinking of segments as lines helps in understanding the parallel segments in Figure 4.

Consider a collection of points in two dimensions: (x_1,y_1), $(x_2,y_2),\ldots,(x_n,y_n)$. It is easy to see that these points lie on a line if and only if the lines corresponding to these points in a 1 and 1 plot intersect at a point. As usual, parallel lines are thought of as intersecting at infinity.

One other aspect of 1 and 1 plots should be noted: the number of times that the segments cross in a 1 and 1 plot is equal to $\frac{1}{2}\binom{n}{2}$ $[1-\tau]$ where τ is Kendall's measure of association, τ computed for $(x_1,y_1),\ldots,(x_n,y_n)$. Thus, few crossings correspond to τ close to 1 and many corssings correspond to τ close to -1. The earliest reference we know for this is Griffin (1958).

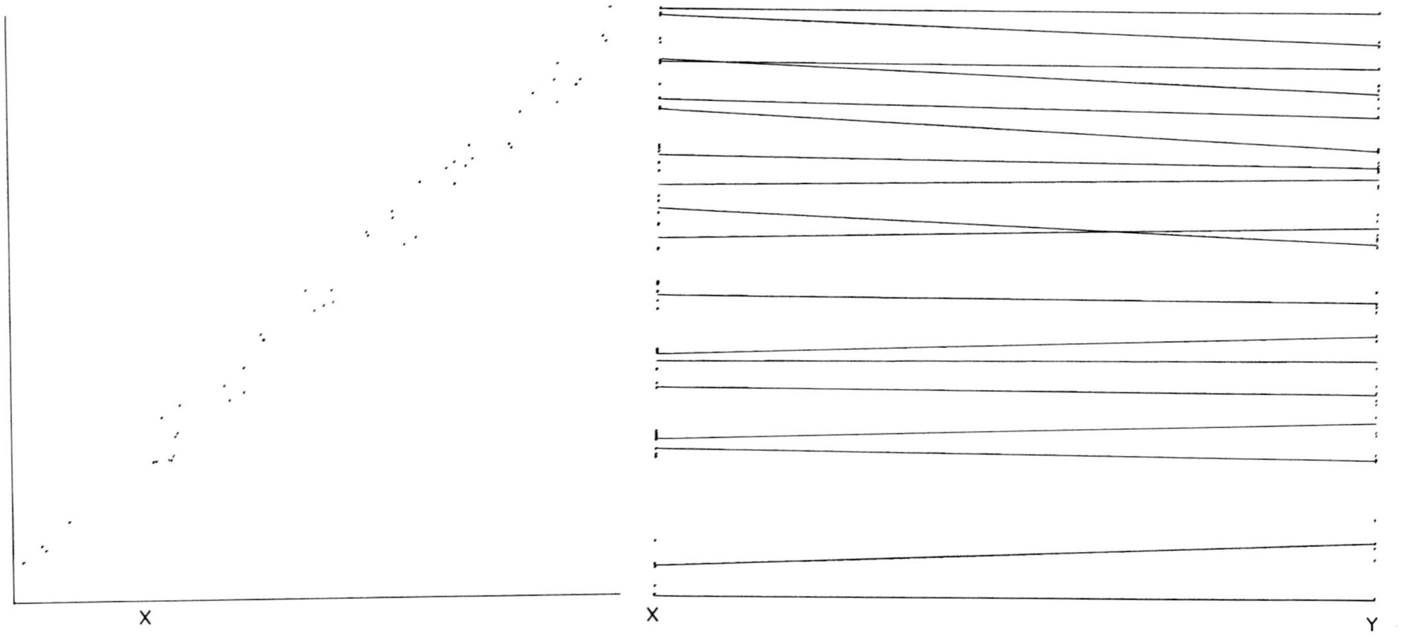

Figure 4

Three-Dimensional Data

It is not a straightforward task to make a scatterplot of three-dimensional data -- a 3 and 0 plot. Consider Figure 5 which is a scatterplot of three of the variables from the diabetes data set introduced in Section I.

The plot is not easy to make sense of because of the lack of perspective cues. A version of a three-dimensional scatterplot is at the center of the PRIM-9 plotting program (Fisherkeller, et. al. (1976)). Briefly, the idea is to show the collection of three-dimensional points as a rotating point cloud using a graphics display terminal. Points closer to the viewer rotate faster and parallax fools the eye into seeing the points as a three-dimensional point cloud. An artist's rendering of this display for the data of Figure 5 gives a very useful picture of this data set. Figure 6 is reproduced from Reaven and Miller (1979).

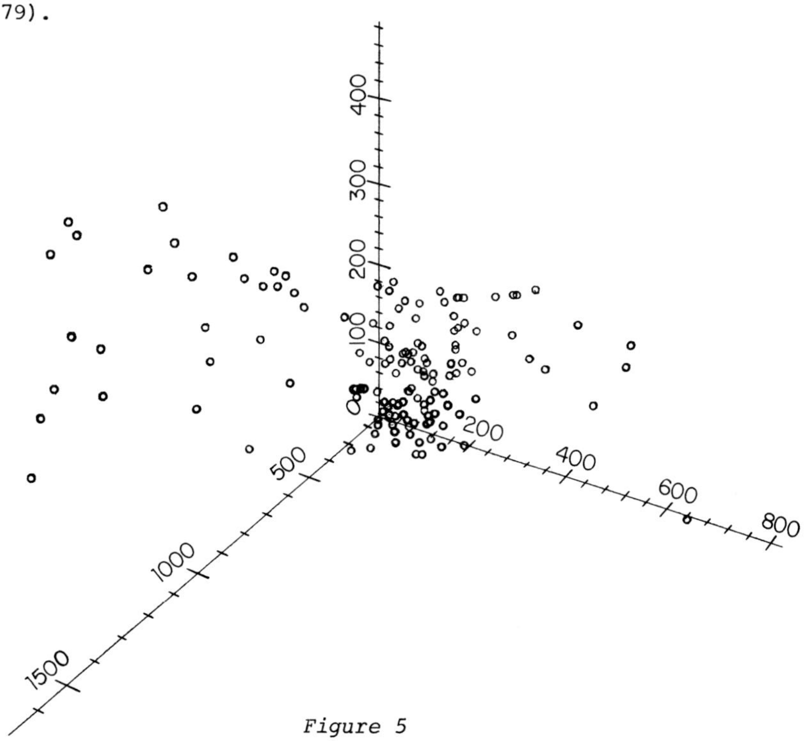

Figure 5

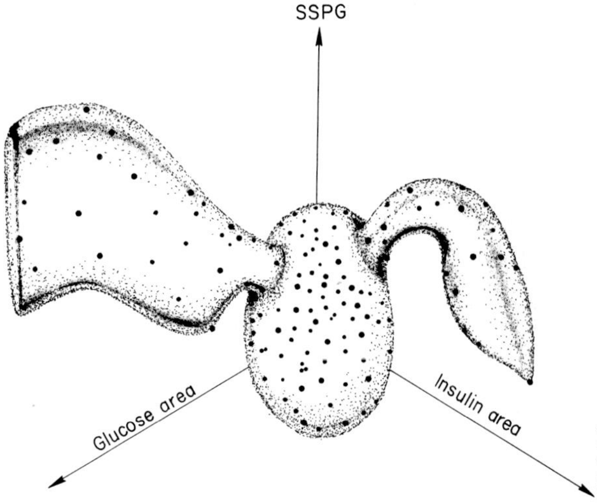

Artist's rendition of data as seen in three dimensions. View is approximately along 45° line as seen through Prim 9 program on the computer; coordinate axes are in the background

Figure 6

A 1 and 2 plot of the same three variables from the diabetes data set is shown in Figure 7. The right-hand scatterplot was discussed in Section I. The 1 and 0 plot of SSPG shows a high density of lower values. Looking at the lines, we see that points from the central cluster of "normal" patients have low values of SSPG. As one looks down in the right-hand plot, the values of SSPG increases, always remaining below the middle of the range. Looking left from the main cluster in the right-hand plot, the values of SSPG increase somewhat slowly. At the very top, there appears to be an interesting reversal -- the highest value of glucose area are connected to central values of SSPG.

The line segments in a 1 and 2 plot can also be viewed as a plot of lines in three dimensions by picturing the one-dimensional axis as a line lying above and parallel to the plane of the two-dimensional plot.

Another way to plot three-dimensional data is to make a (1,1,1) plot -- three distinct coordinate axes, each showing one dimension and a pair of line segments connecting each dot to the other two representing dots.

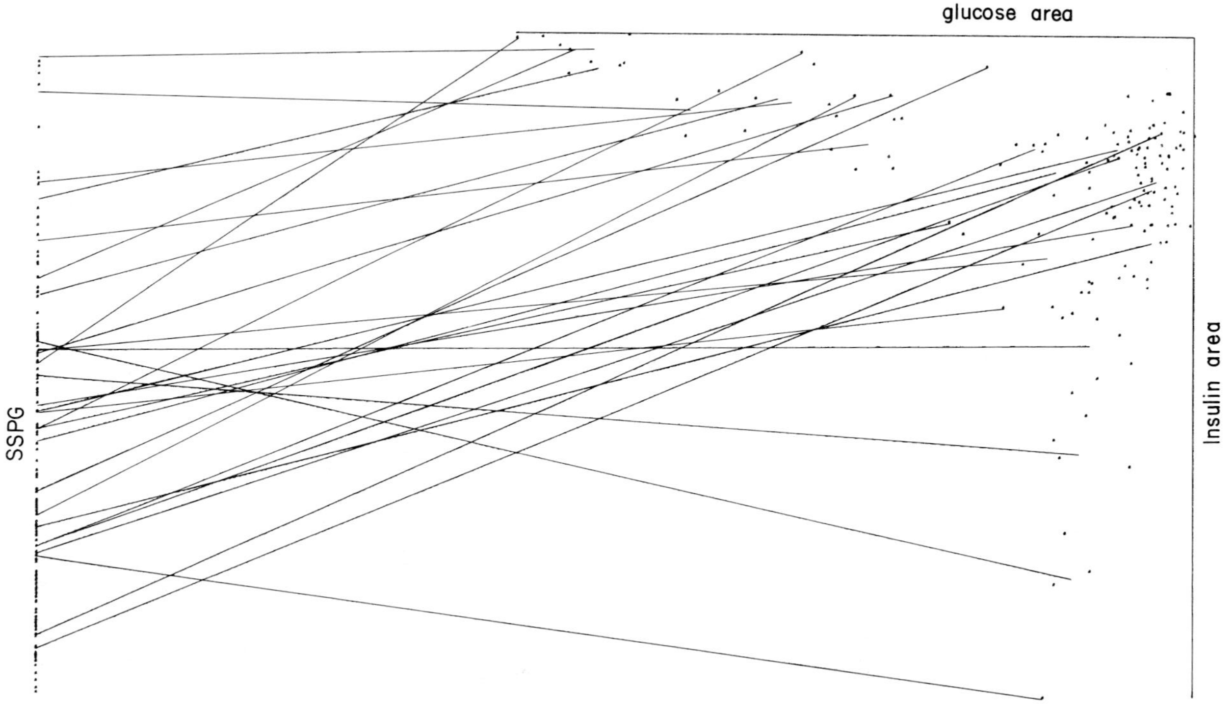

Figure 7

Four-Dimensional Data

We have already given an example of a 2 and 2 plot in Section I. We here discuss an interpretation of such plots as a picture of lines in three dimensions. The basic ingredients of a 2 and 2 plot are a pair of coordinate axes and a segment, see Figure 1.

Imagine the (S,T) plane as a plane parallel to but above the (X,Y) plane. A segment can then be pictured as part of a line running "down" from the (S,T) plane through the (X,Y) plane. Only that part of the line which lies between the planes is visible. In this way, the segments in a 2 and 2 plot can be thought of as lines in three-dimensional space. It is not hard to see that the set of all lines in three dimensions is a four-dimensional space. One argument considers a pair of parallel planes (like the (X,Y) and (S,T) planes described above). "Almost all" lines in the three-dimensional space pass through both planes and uniquely determine four coordinates. This omits lines parallel to the planes, but these form a lower dimensional surface. Hence, the lines in three-dimensional space form a four-dimensional space and can be used to picture four-dimensional data. A similar argument shows that the dimension of the set of lines in n-dimensional space is $2(n-1)$. Because we are using lines to represent our data, the geometry of the set of lines is of interest. For example, because of the way we have chosen to draw the line segments, a small change in the position of some four-dimensional points (x,y,s,t) can result in a large change in the plotted line: this happens if (x,y) and (s,t) are plotted physically close to each other (e.g., x,y is in the upper right quadrant and s,t in the upper left quadrant of Figure 1). It is amusing to show that the lines in two dimensions are topologically equivalent to a Mobius strip. Dan Asimov has shown that the lines in three dimensions are topologically equivalent to the usual tangent bundle of projective 2 space.

It is natural to try to find a set of coordinates for the lines in three dimensions which do not have the problem of omitting a low-dimensional set of lines. Several approaches are described in Chapter 1 of Jessop (1969). The most widely discussed coordinates - Plückers coordinates -- are not particularly suited to working with statistical data. While natural mathematically, Plückers coordinates use five coordinates to describe the four-dimensional set of lines.

We mention, in passing, that four-dimensional data can also be viewed using a (2,1,1) plot, a (1,1,1,1) plot or a (3,1) plot (a 3 and 1 plot would require a PRIM 9-like graphics device).

Higher-Dimensional Data and Multidimensional Scaling

With a PRIM 9-like graphical device, or an artist's rendering as in Figure 6, it is possible to draw 3 and 3 plots of six-dimensional data. Higher dimensional data can also be pictured by connecting together lower dimensional M and N plots. A more practical thought is to link the best two-dimensional projection with a plot of the output of one of the many nonlinear mapping or scaling algorithms. This gives a way of labeling the points of the resulting output. An example is given in Figure 8. This is a picture of the Reaven and Miller (1979) diabetes data. The right-hand plot is nonlinear mapping of the original five-dimensional data into the plane. The nonlinear mapping algorithm described in Friedman and Rafsky (1978) was used. This algorithm preserves 2n-1 of the $\binom{n}{2}$ interpoint distances. In this picture, the n-1 distances in the minimal spanning tree of the n points are preserved along with selected other distances. The right-hand plot has a dense area in the lower left-hand part and two "wings". The lines connect this plot to the ordinary projection of the data onto coordinates 2 and 4. The lines indicate that the dense part of the right-hand picture corresponds to "normal"

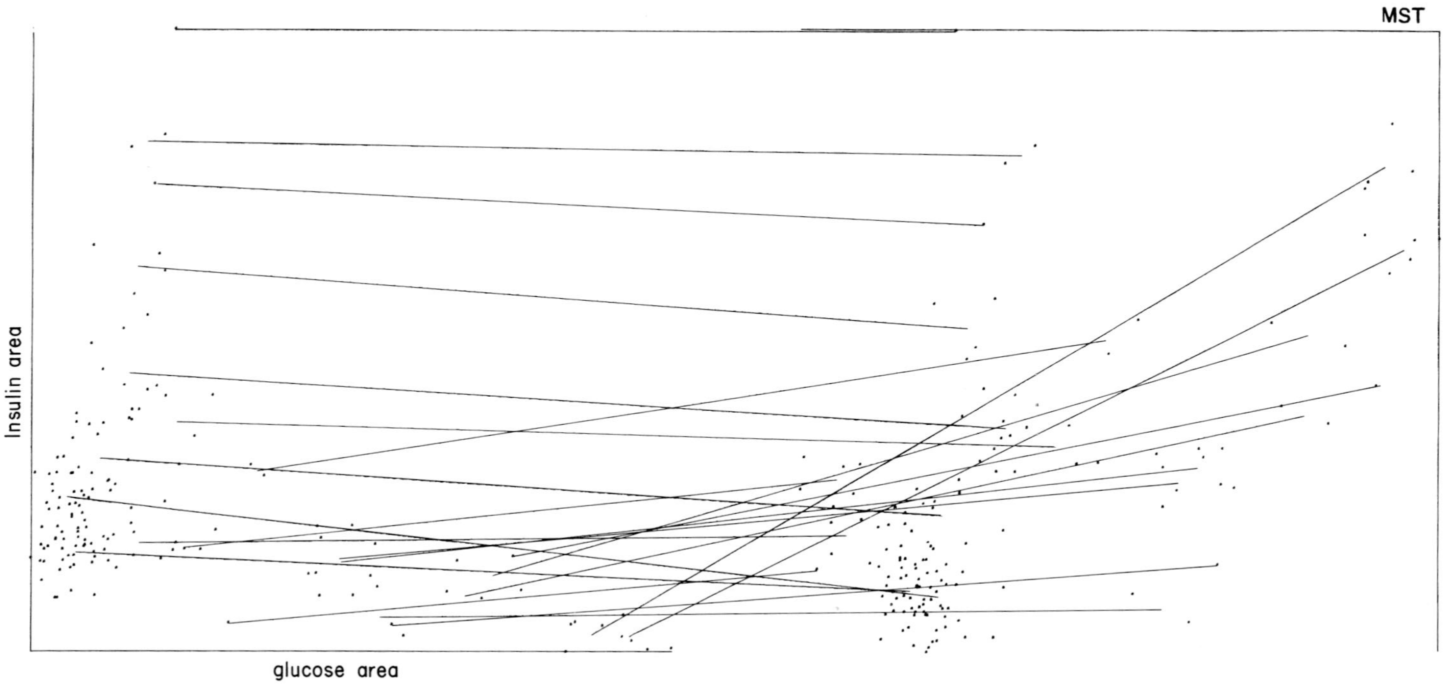

Figure 8

patients and the two "wings" correspond to the other two groups described previously. Looking more closely at the right-hand picture, it seems that the group of "normal" subjects may split into two groups - a dense lower group and a less dense upper group. An outlying observation shows up clearly in the line at the top of the plots.

One of the most popular methods of plotting high-dimensional data is Chernoff's faces. It is instructive to compare faces and M and N plots. Briefly, Faces seem useful for fairly high-dimensional data when there are not very many sample points; M and N plots seem useful for fairly low-dimensional data independent of the sample size.

III. SOME PRACTICAL DETAILS

In this section, we discuss thinning -- by hand or computer-- rotating, and some ideas for interactive implementation of M and N plots.

Thinning

To understand the need and results of thinning, consider the data set introduced in Figure 2. Figure 9 shows this picture with no thinning. Figure 10 shows this picture with very heavy thinning. Now the lines seem useless.

It is straightforward to make an M and N plot by hand. Simply draw a pair of coordinate axis and start plotting points. Lines can be drawn either at random (say, with probability 1 in 5) or more systematically. One systematic approach is to break the (X,Y) and (S,T) planes into boxes. The first time a point is plotted in two boxes, the connecting segment is drawn. Only one segment is drawn for each pair of boxes. Thinning at random causes the density of lines to be proportional to the density

Figure 9

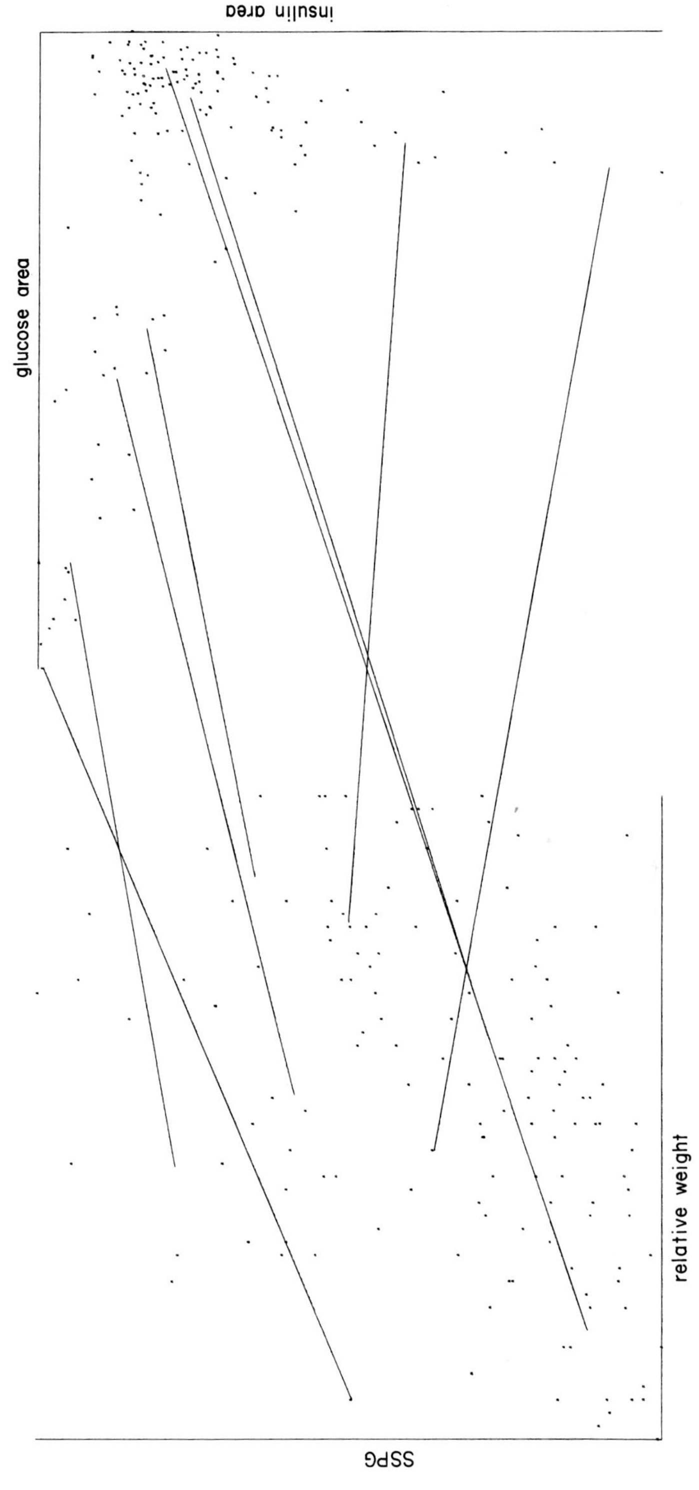

Figure 10

of points. Systematic thinning causes the density of lines to be proportional to volume. In either case, we recommend starting with a low density of lines and adding lines systematically as they seem useful.

We next describe an efficient thinning algorithm for use on a computer. This allows plots to be made rapidly in real time. The algorithm we prefer (based on some experimentation) causes the density of lines to be proportional to volume (not density). Here is a rough description: suppose we want to make a 2 and 2 plot of n points. Divide the four-dimensional space up into "boxes". A single connecting line is drawn for each non-empty box. The line for each box may be taken to be the line representing the (four-dimensional) average point for that box.

More generally, suppose that the data consists of p-dimensional vectors (or points) $\underset{\sim}{x}_1, \underset{\sim}{x}_2, \ldots, \underset{\sim}{x}_n$. For expository purposes, suppose that the scale is chosen so that the data lies between zero and one in each coordinate. Suppose that the p-dimensional unit cube is divided into "boxes" of side h in each dimension. This makes $J = (1/h)^p$ boxes in all. Each box can be indexed by a p-truple of integers. We will refer to the $\underset{\sim}{i}$-th box, and use lexographic order on the box labels. To determine what box a point $\underset{\sim}{x}$ is in requires checking p inequalities. With this notation we now present a semiformal description of our recommended algorithm. The main phase is to form a list containing

- the label of each non-empty box
- the sum of the vectors in that box
- the number of points in that box.

To begin, determine the box containing $\underset{\sim}{x}_1$. Call this $\underset{\sim}{i}_1$. The list begins

$$\underset{\sim}{i}\ ,\ \underset{\sim}{x}_1\ ,\ 1$$

the last component being a counter to indicate how many points are in box $\underset{\sim}{i}_1$. Next, consider $\underset{\sim}{x}_2$. If it is in box $\underset{\sim}{i}_1$, the first entry in the list is changed to

$$\underset{\sim}{i}_1,\ \underset{\sim}{x}_1 + \underset{\sim}{x}_2,\ 2$$

If $\underset{\sim}{x}_2$ is in box $\underset{\sim}{i}_2 \neq \underset{\sim}{i}_1$, the list contains

$$\underset{\sim}{i}_1,\ \underset{\sim}{x}_1,\ 1$$

$$\underset{\sim}{i}_2,\ \underset{\sim}{x}_2,\ 1.$$

Continue, for each point $\underset{\sim}{x}_j$ determine what box $\underset{\sim}{i}_j$ contains $\underset{\sim}{x}_j$. If the label $\underset{\sim}{i}_j$ appears in the list, add $\underset{\sim}{x}_j$ to the second component of that list entry and increment the counter in the third component of the entry by 1. If $\underset{\sim}{i}_j$ does not appear in the list, insert $\underset{\sim}{i}_j$ in the list by binary insertion (using lexographic order on the labels). After processing all the points, the list contains the labels of the non-empty boxes, the sum of the points in each box, and the number of points in each box.

To draw an M and N plot using this list, make a single pass through the list computing the k-dimensional average for each box, and draw the line corresponding to this average point. If the number of non-empty boxes is B, the algorithm may be seen to run in

$$O\{pn(1 + \log B)\} \text{ "operations".}$$

Rotation

Some of the earlier plots have been rotated to make them less confusing. The next example is a 1 and 2 plot to illustrate the usefulness of rotation. Figure 11 is a plot of 100 triples (x,y,z) where y and z were chosen independently and uniformly in [0,1] and x = y+z. Look at the outside edge of the right-hand plot. Notice how the lines move down in x as y or z decrease. Next, look from top to bottom on the right-hand picture along

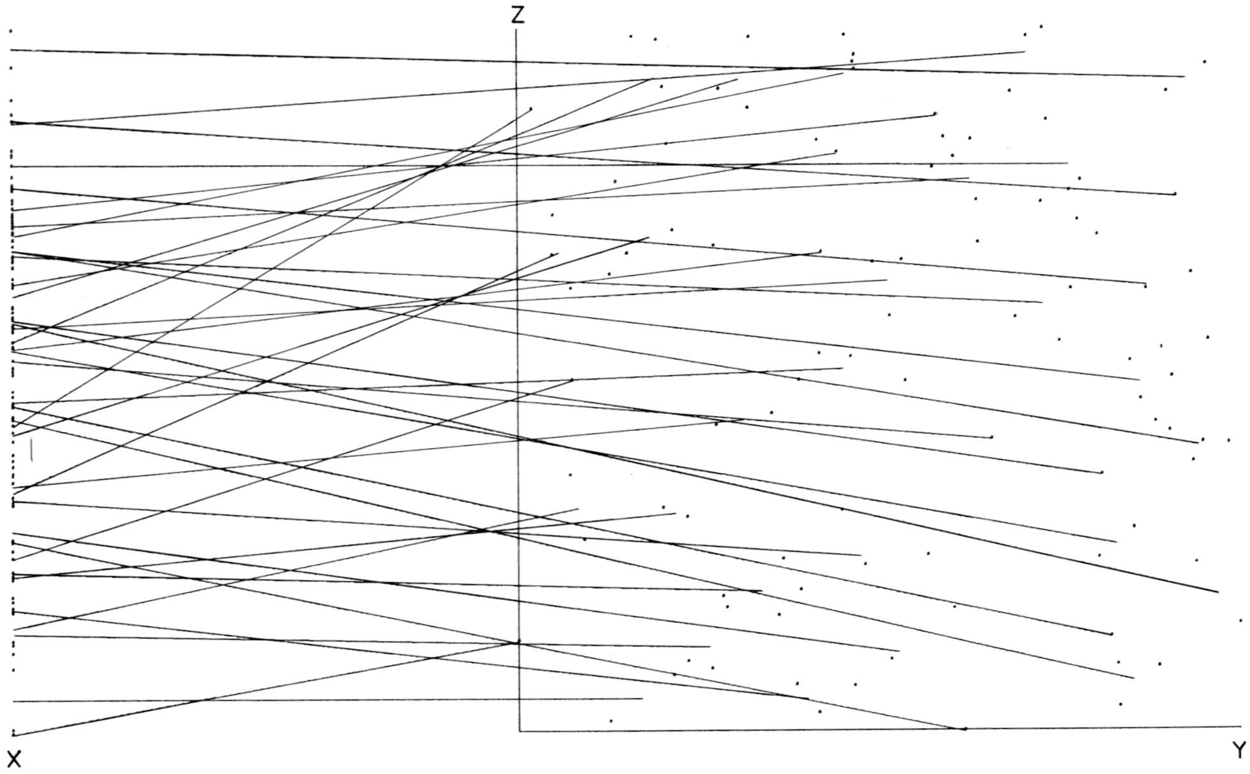

Figure 11

any fixed ℓ corresponding to y = constant. The lines in a neighborhood of ℓ have approximately constant slope. Recalling the discussion of 1 and 1 plots in Section 2, these observations suggest a linear relation between (y,z) and x.

Figure 11 is somewhat confusing to view because the lines change in slope and cross each other. Figure 12 shows a rotated 1 and 2 plot of the same data. Now the linear relation is striking.

Interactive Usage

When M and N plots are viewed on an interactive graphics device, the decisions about thinning and rotation are made by trial and error. A program making this easy to do was written at Stanford Linear Accelerator Center by Roger Chaffee. The program incorporates some other features which may be of interest.

- All points in the same box as a given point can be made brighter
- Lines can be easily added and deleted
- Lines can be made thicker to show another variable or the density of points in a box.

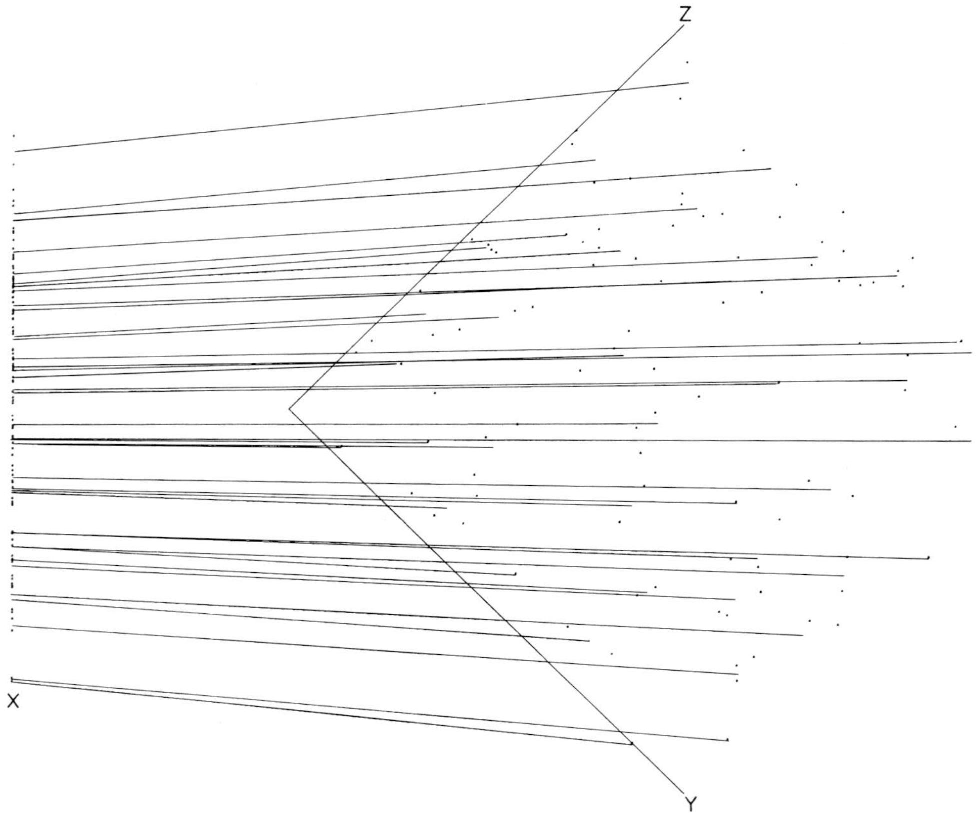

Figure 12

IV. ACKNOWLEDGMENT

We thank Roger Chaffee for extensive programming and helpful suggestions. We thank Dan Asimov, Rupert Miller, Colin Mallows, and Aiden Moran for their help. We are grateful to the editors and publishers of *Diabetologia* for permission to reproduce Figure 6.

V. REFERENCES

Brisson, D.W. (1978). Hypergraphics. *Visualizing Complex Relationships in Art, Science, and Technology.* Westview Press, Boulder, Colorado.

Eckhart, L. (1968). *Four-Dimensional Space.* (Translation by A.L. Bigelow and S.M. Slaby), Indiana University Press, Bloomington, Indiana.

Fisherkeller, M.A., Friedman, J.H., and Tukey, J.W. (1974). PRIM-9, An Interactive Multidimensional Data Display System. Stanford Linear Accelerator Pub-1408.

Friedman, J.H. and Rafsky, L.C. (1981). "Graphics for the Multivariate Two-sample Problem." *J. Amer. Statist. Assoc. 76,* 277-295.

Gnanandesikan, R. (1977). *Methods for Statistical Data Analysis of Multivariate Observations.* Wiley, New York.

Griffin, H.D. (1958). "Graphic Computation of Tau as a Coefficient of Disarray." *J. Amer. Statist. Assoc. 53,* 441-447.

Jessop, C.M. (1964). *A Treatise on the Line Complex.* Chelsea, New York.

Manning, H.P. (1960). *The Fourth Dimension Simply Explained.* Dover, New York.

Murrill, Malcom (1980). "On a Hyperanalytic Geometry for Complex Functions." *Amer. Math Monthly 8,* 7, 8-25.

Reaven, G.M. and Miller, R.G. (1979). "An Attempt to Define the Nature of Chemical Diabetes using a Multidimensional Analyses." *Diabetologia* 16, 17-24.

Tukey, P.A. and Tukey, J.W.T. (1981). Graphical Display of Data Sets in 3 or More Dimensions Chapter 10-12 of *Interpreting Multivariate Data,* V. Barnett, Ed. Wiley, New York.

INVESTIGATING THE SPACE OF CHERNOFF FACES

Robert J.K. Jacob

Computer Science and Systems Branch
Naval Research Laboratory
Washington, D.C.

and

Department of Statistics
George Washington University
Washington, D.C.

Since Herman Chernoff first proposed the use of faces to represent multidimensional data points (1971), investigators have applied this powerful graphical method to a variety of problems. With the method, each data point in a multidimensional space is depicted as a cartoon face. Variation in each of the coordinates of the data is represented by variation in one characteristic of some feature of the face. For example, one coordinate of the data might be represented by the curvature of the mouth, another by the size of the eyes, and so on. The overall multidimensional value of a datum is then represented by the expression on a single face. The strength of this method lies in the ability of an observer to integrate the elements of a face into a unified mental construct. With most other representations, it is more difficult for an observer to combine the coordinates of a data point to form a single mental impression.

Observers of Chernoff faces have, however, voiced one principal complaint: The perceived meaning of the resulting

ISBN 0-12-589320-5

face plots is dependent on the way that the coordinates of the data are assigned to the features of the face (e.g., Bruckner 1978, Fienberg 1979, Kleiner and Hartigan 1981). Each user of Chernoff faces claims to have developed an intuitively satisfying way to make this assignment for his own problem, but only a few investigators have attempted to find more general solutions. Chernoff and Rizvi (1975) provided an empirical measurement of the importance of the choice of this assignment. Tobler (1976) and Harmon (1977, 1978, 1981) both proposed criteria that an optimal data-to-face transformation should meet but did not provide methods for constructing transformations that meet their criteria. Kleiner and Hartigan (1981) proposed a method, based on clustering, for assigning data coordinates to the construction parameters of two new displays, designed specifically for their method. The same assignment procedure could usefully be applied to other displays including faces (Jacob 1981).

The basic problem is to find a transformation from the space of the data to be plotted to the space of possible face parameter vectors such that an observer's perception of the relationships among the resulting faces corresponds to the spatial configuration of the original data points. For example, given a set of data points in a Euclidean space, the assignment of data coordinates to face parameters should be chosen so that the similarity between any two plotted faces perceived by an observer reflects the Euclidean distance between the corresponding data points. The problem is that it is difficult to measure the spatial configuration formed by an observer's perception of similarities between faces. In principle, a multidimensional scaling procedure could be used to construct a description of an observer's "mental configuration" of a set of stimuli from a collection of measurements of similarities between all pairs of the stimuli (Torgerson 1952, 1958). However, the face plots are principally useful for data of high dimensionality, and, for such, a scaling procedure would require an impossibly large number of these similarity measurements.

A very simple approach is to look for "good" or "bad" regions of the space of face parameter vectors with respect to Euclidean data. A "good" region is one in which observers' perceptions of similarities between faces correspond well to the Euclidean distances between the points represented by the faces. Some good or bad regions may be found in an ad-hoc fashion, and then the bad ones improved by trial and error. Specifically, wherever uniform changes in a single face parameter produce widely differing changes in perceptions, a "bad" region is found. As Chernoff's (1973) geologist notes, a principal problem area is the face outline. It carries too much perceptual significance, partly because the placement of all the other features depends on the outline and partly because the parameter axes used to describe it do not map onto straight and orthogonal perceptual configurations. That is, as one moves along a straight line parallel to an axis in the space of face parameter vectors, there will be some points at which the perceptual effect of the move changes in kind, rather than degree. This problem was reduced for the outline by creating a single new parameter that controls a scaling function of the ratio of the eccentricities of the two ellipses in the outline and replacing the two original eccentricity parameters with it. These and other similar ad-hoc modifications resulted in a plotting program (Jacob 1976b) in which the outline is less prominent, the parameters that describe it are more nearly orthogonal, and the ill-behaved outer fringes of the space are avoided. As an example, Mezzich and Worthington (1978) used Chernoff's original face program in a careful comparative study of a number of multivariate methods. Faces did not yield particularly good performance in the comparison; but, in examining their faces, the outline seems to have been especially significant. Many of their data points lay in the "bad" region of the space of face parameter vectors, where small parameter variations cause disproportionate changes in the outline. Re-plotting their data with the new program provides a better representation of the underlying clusters.

I. AN EMPIRICAL METHOD FOR EVALUATING REGIONS OF THE SPACE OF FACE PARAMETER VECTORS

Beyond making ad-hoc improvements in the face parameters, a more general question arises: Is there some empirical way to identify "good" or "bad" regions of the space of face parameter vectors? This is, in a sense, the reverse of multidimensional scaling. Instead of using subjects' similarity judgments to infer the spatial configuration of the stimuli, this procedure starts by hypothesizing a particular configuration. Then, subjects' similarity judgments are compared to the distances derived from the hypothesized configuration. If there is a "good" region of the face space, where the subjective distances match the model, then the procedure has "found" the spatial configuration for that region. (If not, the hypothesized model may simply have been an inastute choice.) The advantage of this method is that considerably fewer distance observations are needed to confirm or refute an hypothesized model than to infer one. The disadvantage is, of course, that one must guess at a model rather than letting the data suggest one. Another disadvantage, as seen in the case presented here, is that even a large amount of data may not be sufficient to draw distinctions about regions within a many-dimensional space.

As in many conventional scaling experiments, subjects here

were not explicitly asked to rate the similarities between the stimuli, but such similarities were computed from their responses to a more familiar task. The data consist of subjects' responses to a categorization experiment. While they were originally obtained for a different purpose (Jacob, Egeth, and Bevan 1976), the responses are here used to investigate differences between regions within the face parameter space. Five widely-separated points *(prototypes)* were chosen from a 9-dimensional space. Then, a cluster of 10 more points *(deviants)* was generated randomly around each prototype, near it in Euclidean distance. Subjects were given the 5 prototypes and then instructed to assign each of the 50 deviants in turn to one of the prototypes ("the one they thought it belonged with"). These responses can be used to infer subjects' perceptual similarities between the 5 prototypes and the 50 deviants. This yields a total of 250 measurements-considerably less than the number of all possible pairs of stimuli in the set, which would be used for conventional scaling.

If many of the 24 subjects clustered a particular deviant with a particular prototype, then the perceptual similarity between those two points (combined over subjects) must be large-that is, greater than the similarity between that deviant and any of the other 4 prototypes. Since subjects did not differ markedly in their ratings, a combined measure for each of the 250 similarities was obtained by counting the number of times each deviant was judged to belong with each prototype. A linear regression was then performed to relate these similarities to the distances obtained by measuring the original points using some hypothesized distance metrics. Both the original Euclidean model

$$\sqrt{\sum_i (a_i - b_i)^2}$$

and a city block distance model

$$\sum_i |a_i - b_i|$$

were tested. The regressions predicted the subjects' similarity ratings from the hypothesized distances. Each yielded a correlation coefficient measuring how well the subjects' ratings corresponded to the model.

The simplest comparison that can be made is between the overall correlation coefficient for faces and that for the same experiment replicated using two other multivariate displays (polygons and digit matrices). Table I shows that faces fit either distance model better than polygons or digits.

One may also ask which model fits better-Euclidean or city block? It appears that the Euclidean model fits the responses *slightly* better for faces and polygons, but there is no difference for digits. In fact, for the particular 250 distance measurements in question, the two metrics give very similar relative distances, so it would be difficult to discriminate between them using this set of measurements.

TABLE I. Correlation Coefficients from Overall Linear Regressions

	Correlation coefficient for	
Display type	*Euclidean model*	*City block model*
Face	*0.556*	*0.511*
Polygon	*0.219*	*0.181*
Digits	*0.209*	*0.201*

Next, individual regions within the space may be considered. To do this, the 250 observed distances were divided into groups;

a regression was performed for each group; and the residuals for the groups were compared. Each of the original distance measurements was made between one of the 5 prototypes and one of the 50 deviants. One way to divide them is according to which of the 5 prototypes each distance is connected to. This gives 5 groups of 50 distances each. The average absolute residual for each of these, using a Euclidean distance model, is shown in Table II. To visualize the results, the actual prototypes are shown in Figure 1. For faces, ratings of distances from Prototype 2 appear least Euclidean, while those from Prototype 3 are most. One may hypothesize that Prototype 2 is the most distinctive face-that its location in the observers' perceptual configuration is further from the center than the location of the corresponding point in the 9-dimensional Euclidean data space. Prototype 2 is also the only face for which some of its deviants have mouths that extend beyond the face outline. Similar analyses may be made for the other display types. For polygons, observed distances from Prototype 1 fit the Euclidean model least well. Since it is the only regular polygon in the set, it is likely to be more perceptually distinctive than its location in the data space would indicate.

TABLE II. Residuals from Linear Regressions by Prototype[a]

	Prototype				
Display type	*1*	*2*	*3*	*4*	*5*
Face	*3.492*	*4.514*	*3.071*	*3.493*	*3.507*
Polygon	*4.099*	*3.515*	*3.143*	*3.392*	*3.491*
Digits	*2.928*	*3.669*	*2.644*	*2.908*	*2.801*

[a]*The similarity measurements from which these residuals are computed were integers between 0 and 24.*

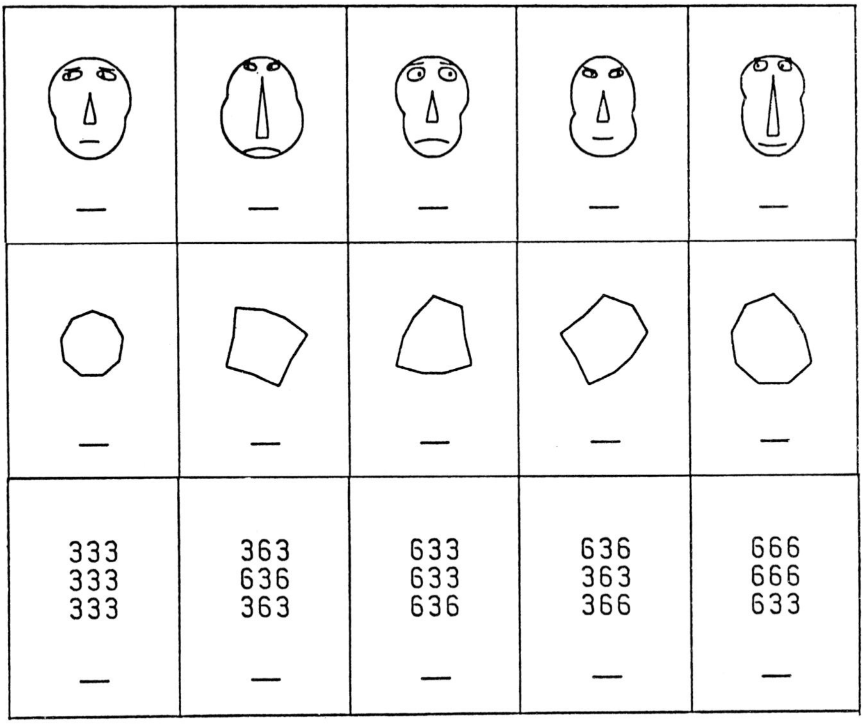

FIGURE 1. Five Prototypes Used for Distance Ratings

Aggregating all the distances *from* a particular prototype may be too gross a measure. Perhaps distances from some prototype to one region fit a Euclidean model well, and those from the same prototype to another region do not. The distances were therefore further divided so that all measurements from Prototype *i* to any of the 10 deviants around Prototype *j* were placed together in a single group. This yields 25 groups of 10 distances-one group for each pair that consists of a prototype and a cluster of 10 deviants. The residuals for the faces in Table III show some finer distinctions. The poor fit seen above for distances to Prototype 2 is attributable to the area immediately around Prototype 2; distances between Prototype 2 and the areas around the other prototypes are not particularly distorted. Prototype 5 exhibits a similar property. Thus, the regions of the face parameter space immediately surrounding these two points are relatively poor ones for representing Euclidean data.

TABLE III. Residuals from Linear Regressions by Prototype and Group

Display type	*From deviant group*	*To prototype* 1	2	3	4	5
Face	1	4.976	4.011	2.248	3.018	3.172
	2	4.532	7.862	3.403	3.056	2.914
	3	1.523	3.662	3.248	3.668	2.640
	4	2.909	4.349	3.290	4.510	2.437
	5	3.522	2.687	3.167	3.213	6.375
Polygon	1	4.937	2.656	2.906	2.821	2.611
	2	4.407	4.865	2.744	3.638	3.427
	3	5.226	2.243	4.969	3.145	2.148
	4	3.662	4.817	1.722	3.957	3.485
	5	2.265	2.995	3.374	3.401	5.784
Digits	1	3.021	3.766	1.987	3.633	3.206
	2	3.910	6.014	1.883	2.589	2.260
	3	3.338	2.720	3.286	2.228	3.454
	4	2.549	2.530	2.517	3.434	1.568
	5	1.825	3.315	3.547	2.659	3.514

Finally, a much finer measure may be considered. (This analysis was performed for the faces only.) Perhaps there are smaller good or bad regions of the space than the 5 or 25 considered above, and all distance measurements that *cross through* those regions of the space are distorted. To examine this, the 9-dimensional space was divided into uniform-size regions. Then, the average distortion for all distance measures that cross through a region was used as a measure of how well that region depicts Euclidean distances. The first step was to lay a grid over the 9-dimensional space to divide it into regions. The simplest grid, which divides each coordinate axis in half, was used; and this carves the space into 512 regions. (A finer grid would be difficult to support with only 250 observations. However, note that the two-level grid is not sufficient to test the hypothesis that regions near the center of the space are better than those near the edges, since all the regions are the same distance from the center.)

Each of the 250 subjective similarity observations was then compared to the similarity predicted by the hypothesized Euclidean metric. The absolute value of the resulting prediction error for each similarity observation provided a rating of the extent to which that observation fit the model. Next, a rating of the extent to which all the observations that cross through a particular region are Euclidean was needed. The rating for a region was defined as the mean of the ratings for all of the distance lines that traversed it, weighted by the portion of each distance line that lay in that region. This was approximated by dividing each distance line into 100 equal segments and assigning an equal share of the rating for that line to each of the regions that contained an endpoint of a segment. Of the 512 regions, only 189 are traversed by any of the distance lines. Hence this procedure yielded ratings of the "Euclidean-ness" of 189 out of the 512 regions of the 9-dimensional subspace of possible face parameter vectors that was used to represent the data. It is not easy to apprehend the resulting mass of numerical data. One approach was to divide the regions into groups according to their relative ratings and then cluster the regions within each group with other nearby regions. One then searches for a large agglomeration of good or bad regions that could be characterized more concisely than the collection of smaller regions; but such was not found when these data were clustered.

The problem of obtaining an overall picture of these detailed results clearly calls for a good statistical graphics technique! Figure 2 plots the center point of each of the 10 "most Euclidean" regions as a Chernoff face, followed by the 10 regions around the median, and the 10 worst regions. The rating for each region is shown below the face. Here the data are easy to understand, but what one learns is that the good and bad regions appear to be scattered fairly uniformly throughout the space. No one large area or axis direction appears to be particularly good or bad with respect to Euclidean distances. It is possible that the changes made to the face plotting program removed the most obvious non-Euclidean regions from the space, and a finer analysis requires more observations.

II. AN EMPIRICAL METHOD FOR ASSIGNING COORDINATES TO FACIAL FEATURES

Another approach to approximating a multidimensional scaling procedure is to perform it for some specific type of data. Instead of inferring the perceptual configuration of a set of faces from a collection of distance judgments, here one hypothesizes a set of axes for the perceptual configuration and asks subjects to rate the faces on the given axes. This is appropriate when a particular type of data have a known or well-established set of axes. The size of the scaling experiment

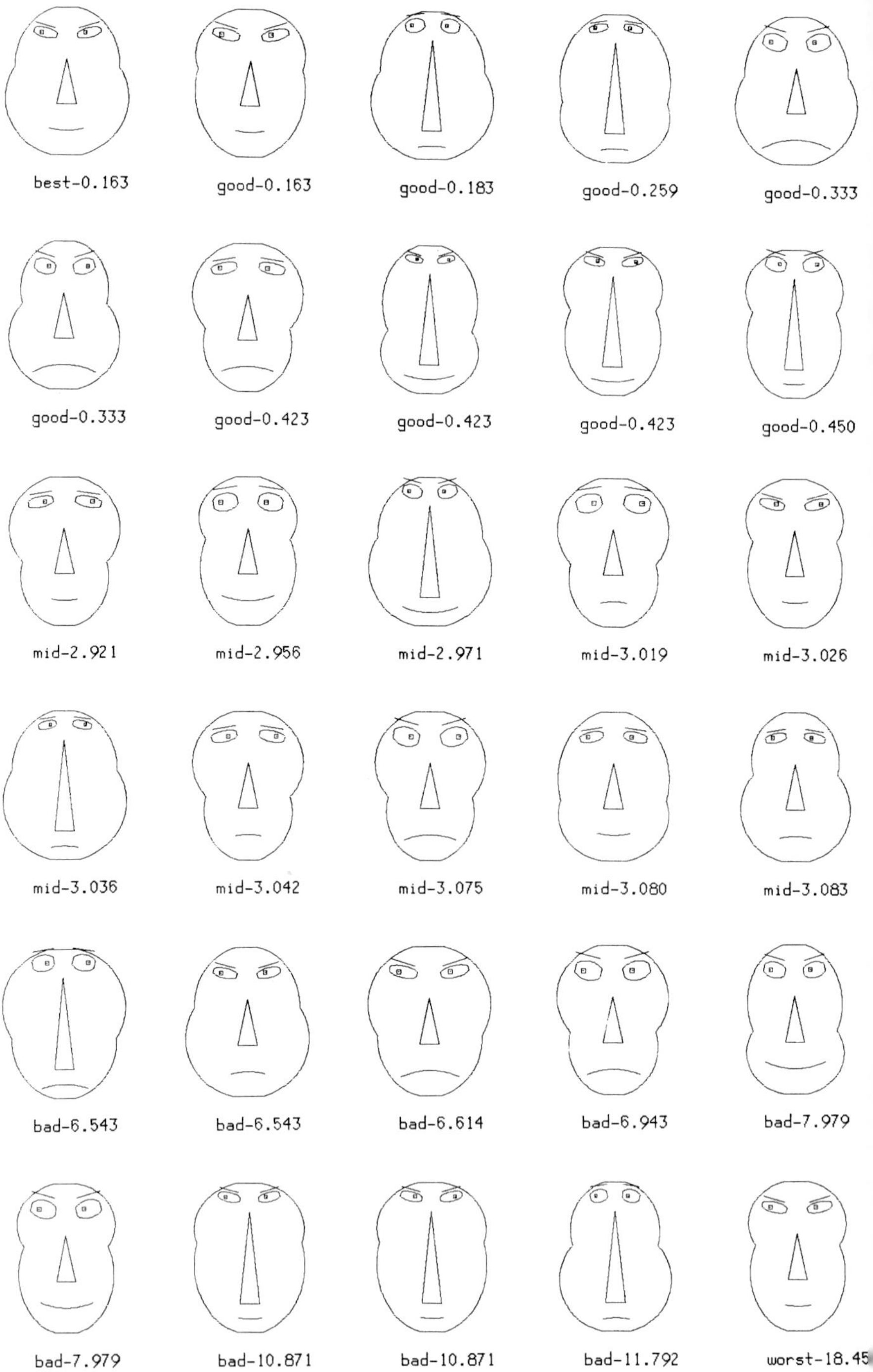

FIGURE 2. Facial Representation of the Centers of the Most, Median, and Least Euclidean Regions

that must now be performed is reduced by its square root, since only individual, rather than pairwise, ratings of the stimuli are needed. Each application of such a study yields a good mapping from one particular type of input data to faces. Obviously, it is most suitable where data of the same general type will be plotted again.

This technique was used to generate a data-to-face mapping for a type of psychiatric data. The resulting mapping was tested and found to be significantly more suggestive than other arbitrary mappings (Jacob 1978, 1976a). The data to be plotted consisted of the *Hypochondriasis, Depression, Paranoia, Schizophrenia,* and *Hypomania* scales of the Minnesota Multiphasic Personality Inventory (MMPI) psychiatric test. A naive approach would have been simply to assign the 5 scales arbitrarily to 5 facial features. Since the test was being given to many patients, however, it was worthwhile to conduct the scaling experiment to obtain a better mapping from the 5-dimensional space of MMPI scores to the 18-dimensional face parameter vector space. Once obtained, the same mapping could be used to plot the MMPI scores of new patients. It should provide a display of an MMPI score that intuitively suggests the meaning of the score, by exploiting observers' preconceptions or stereotypes. (The validity of these stereotypes is not at issue; it is only necessary that they be measurable and widely held.)

The first experiment attempted to find a linear transformation from the MMPI score space to the face parameter space. Unlike most applications of faces, in which each data coordinate is assigned to one face parameter, any linear transformation was permitted here. That is, each data coordinate could control any linear combination of all 18 face parameters. Subsequent analysis of the results for higher-order interactions showed the linear model to be an adequate approximation. A set of 200 faces was generated from uniform random face parameter vectors.

The subject then rated each face along the 5 MMPI axes. In effect, she was indicating what MMPI score she thought a person who looked like each of the faces might receive. A multiple linear regression of the face parameter vectors on the MMPI scores was computed, resulting in a regression equation that could then be used to produce a face for any given MMPI score.[1]

The resulting data-to-face transformation is described by an 18 by 5 matrix of regression coefficients. Again, the problem of interpreting it is solved with the faces themselves. To display the transformation, a series of face parameter vectors was computed, corresponding to equally-spaced points along the axes of the MMPI score space (i.e., the points represent patients who each have only one psychiatric disorder). Figure 3 shows the resulting plot; each row depicts a series of hypothetical patients with increasing amounts of a single disorder, from *0* (the first column), which represents an inverse amount of the disorder, through *1*, representing no disorder (the origin of the MMPI space), to *4*, representing a large, extrapolated amount of the disorder. These faces (particularly those in the column labeled *3*) appear to resemble common stereotypes of the personality traits they are purported to represent. In fact, the subject had not rated faces like these; most of the faces in the original stimuli had been reported to have more than one disorder. The scaling procedure generated an intuitively appealing linear transformation from MMPI scores to faces in an objective manner.

The resulting transformation can be used to plot new MMPI scores as faces. In a subsequent experiment, 30 subjects were each given 50 questions, each consisting of a text description of a patient with an hypothetical MMPI score vector and 5 faces, one of which was generated from this transformation. Subjects chose the "correct" face with significant ($p < 0.0005$) accuracy.

[1] *If the predicted face parameters are compared to the original parameters, the mean squared error over all components of all the vectors is 0.075, where vector components were between 0 and 1.*

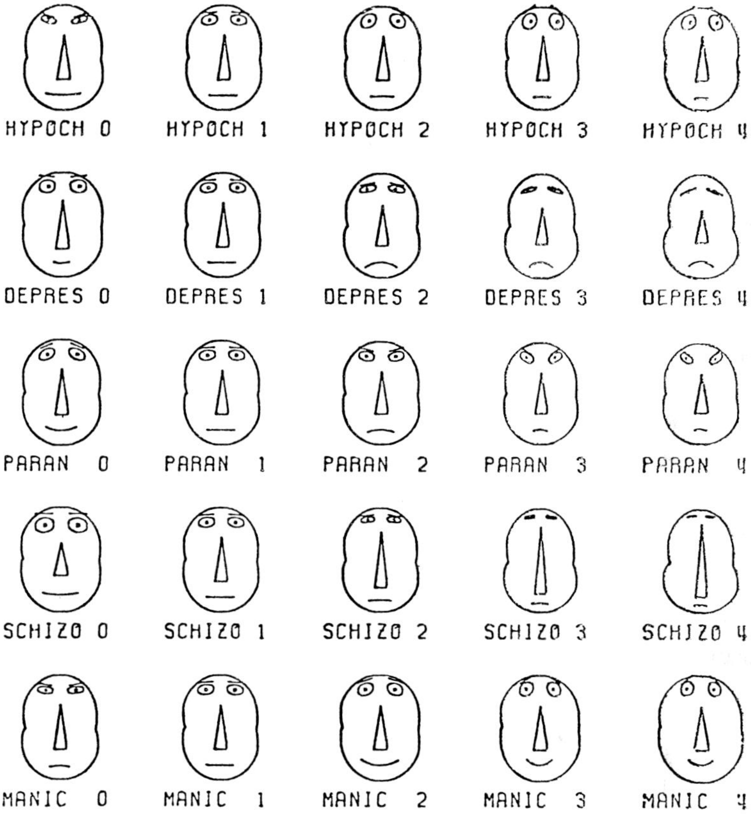

FIGURE 3. Facial Representation of MMPI Scores

These results can provide insight into the face parameter space in another way. The facial representations of all possible MMPI scores comprise a 5-dimensional subspace of the 18-dimensional space of all possible face parameter vectors. There remains an orthogonal 13-dimensional subspace of facial variation. Variation in this subspace should have little effect on the MMPI-related meaning of a facial expression. To see this subspace, Figure 4 shows faces corresponding to 5 arbitrarily-selected, mutually-orthogonal axes in this 13-dimensional subspace. (As before, each row of faces represents movement along one axis; and here the center of each row corresponds to the origin of the space.) Examination of faces in this subspace suggests that all facial variation could be partitioned into two distinct orthogonal subspaces. One (the 5-dimensional range of the transformation from MMPI scores to faces) depicts an *emotional* component of facial expressions, and the other (the remaining orthogonal subspace) depicts an *identification* component, representing variations that help distinguish the faces of individuals from one another but transmit little emotional content. In fact, most of the variation in the former space is in the facial features that a person can move (eyes, mouth) to indicate emotion, while that of the latter is in the immovable features (nose, outline) that distinguish individuals from one another.

To test the hypothesis that variation in the orthogonal subspace carries little psychological significance, 15 faces were generated. Five contained random variation in the MMPI space plus no variation in the orthogonal non-MMPI space, while the remaining 10 varied randomly only in the non-MMPI space. A group of 32 subjects rated the MMPI-varying faces significantly ($p < 0.0005$) more psychologically disturbed than the others. Thus the space of the faces appears to be divisible into two orthogonal subspaces, each of which carries a distinct type of facial information.

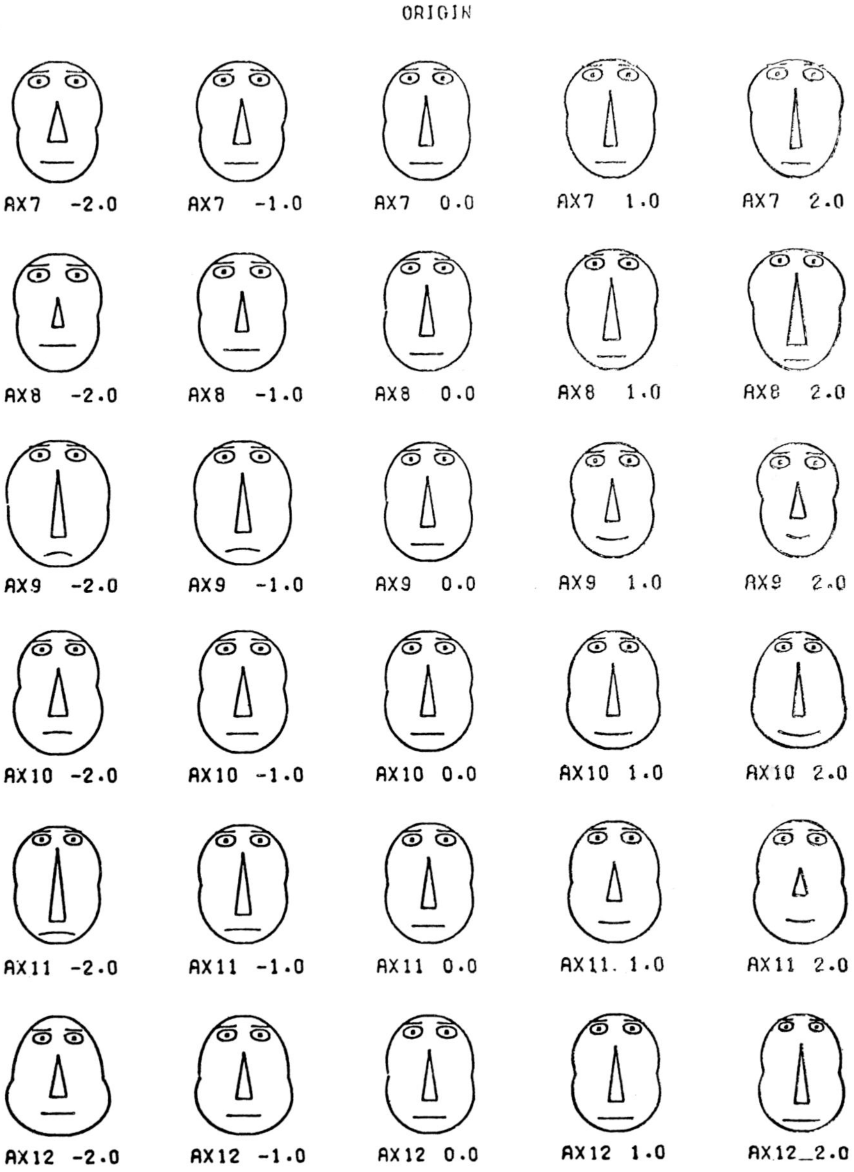

FIGURE 4. Facial Representation of the Axes of the Orthogonal Subspace

III. CONCLUSIONS

Using Chernoff faces to their best advantage to represent multivariate data requires that the perceptual configuration in which the faces are perceived resemble the configuration of the original data points. This means observers' judgments of similarities between faces should reflect the distances between the original data points. Multidimensional scaling solves this problem in principle, but it is not practical because of the dimensionality of the spaces. Three other approaches were considered:

- Intuitively obvious deficiencies in the way face parameters map onto faces can be corrected by trial and error, as was shown for the face outline parameters.
- Regions of the space of possible face parameter vectors can be evaluated according to the degree to which subjective distance judgments within each region match a particular distance model, as was attempted for the 9-dimensional data with the Euclidean distance model. If "bad" regions of the space can be identified from this procedure, one could then avoid them when producing face plots.
- A special case of the scaling paradigm can be used for any specific type of data, as was shown for the MMPI data. The result was a demonstrably mnemonic transformation from data to faces and also the identification of two basic components of facial variation.

ACKNOWLEDGMENTS

Professor William H. Huggins provided inspiration and insight for much of this work, while the author was a student at the Johns Hopkins University. Professors Howard Egeth and William Bevan at Johns Hopkins also provided a great deal of valuable guidance.

Portions of this research were supported by a contract between the Johns Hopkins University and the Engineering Psychology Programs, Office of Naval Research; and by the U.S. Public Health Service Hospital in Baltimore.

And, of course, Herman Chernoff is father to all of this research. Happy Birthday!

REFERENCES

Bruckner, L.A. (1978). On Chernoff Faces, pp. 93-121, *Graphical Representation of Multivariate Data,* ed. P.C.C. Wang, Academic Press, New York.

Chernoff, H. (1971). The Use of Faces to Represent Points in n-Dimensional Space Graphically. Technical Report No. 71, Dept. of Statistics, Stanford University.

Chernoff, H. (1973). "The Use of Faces to Represent Points in k-Dimensional Space Graphically." *J. Amer. Statist. Ass. 68,* 361-368.

Chernoff, H. and Rizvi, M.H. (1975). "Effect on Classification Error of Random Permutations of Features in Representing Multivariate Data by Faces." *J. Amer. Statist. Ass. 70,* 548-554.

Fienberg, S.E. (1979). "Graphical Methods in Statistics." *Amer. Statist. 33,* 165-178.

Harmon, L.D. and Hunt, W.F. (1977). "Automatic Recognition of Human Face Profiles." *Computer Graphics and Image Processing 6,* 135-156.

Harmon, L.D., Kuo, S.C., Ramig, P.F., and Raudkivi, U. (1978). "Idenification of Human Face Profiles by Computer." *Pattern Recognition 10,* 301-312.

Harmon, L.D., Khan, M.K., Lasch, R., and Ramig, P.F. (1981). "Machine Identification of Human Faces." *Pattern Recognition 13,* 97-110.

Jacob, R.J.K., Egeth, H.E., and Bevan, W. (1976). "The Face as a Data Display." *Human Factors 18,* 189-199.

Jacob, R.J.K. (1976a). Computer-produced Faces as an Iconic Display for Complex Data. Doctoral dissertation, Johns Hopkins University. University Microfilms Inc. Order No. 76-22926.

Jacob, R.J.K. (1976b). PLFACE Program, Available from the author upon request.

Jacob, R.J.K. (1978). Facial Representation of Multivariate Data, pp. 143-168, *Graphical Representation of Multivariate Data,* ed. P.C.C. Wang, Academic Press, New York.

Jacob, R.J.K. (1981). "Comment on Representing Points in Many Dimensions by Trees and Castles." *J. Amer. Statist. Ass. 76,* 270-272.

Kleiner, B. and Hartigan, J.A. (1981). "Representing Points in Many Dimensions by Trees and Castles." *J. Amer. Statist. Ass. 76,* 260-269.

Mezzich, J.E. and Worthington, D.R.L. (1978). A Comparison of Graphical Representations of Multidimensional Psychiatric Diagnostic Data, pp. 123-141, *Graphical Representation of Multivariate Data,* cd. P.C.C. Wang, Academic Press, New York.

Tobler, W.R. (1976). The Chernoff Assignment Problem. Unpublished paper.

Torgerson, W.S. (1952). "Multidimensional Scaling: I. Theory and Method." *Psychometrika 17,* 401-419.

Torgerson, W.S. (1958). *Theory and Methods of Scaling,* Wiley, New York.

ON MULTIVARIATE DISPLAY

Howard Wainer[1]

Educational Testing Service
and
Bureau of Social Science Research
Princeton, New Jersey

Iconic metaphorical methods for the storage and display of multivariate data have been in use for centuries, from the pre-Columbian Quipu (Figure 1) to Chernoff's Face (1973) and its subsequent modifications (Wakimoto, 1977; Flury and Riedwyl, 1981). The Quipu is a data display made out of rope that used the number, position, color and type of know to allow the Peruvian Incans to record a wide variety of statistical data (Locke, 1923). Modern displays employ similar iconic devices, usually, however, on a two-dimensional surface rather than hanging freely on a wall or slung over one's shoulder. Despite the apparent variety, the study of the history of multivariate displays reveals more similarities than differences in the aspects of the display characteristics utilized. A major contribution to this ancient methodology was made by Herman Chernoff when he emphasized, through his now famous Face, the usefulness

[1] *This research was partially supported by the Program Statistics Research Project of the Educational Testing Service, and the National Science Foundation (Grant #SES80-08481) to the Bureau of Social Science Research, Howard Wainer, Principal Investigator.*

ISBN 0-12-589320-5

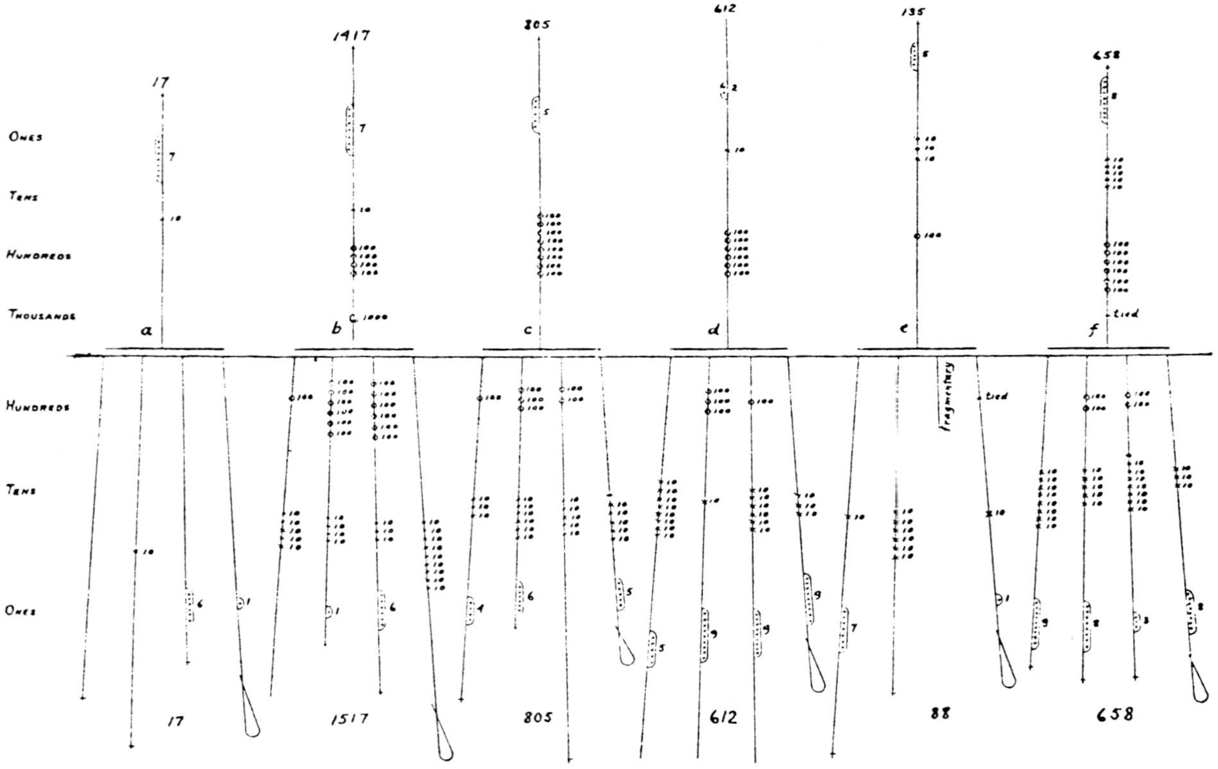

FIGURE 1. An example of the Incan Quipu. The pendant strands are grouped in fours, each group tied with a top strand. The top strand sums the numbers on the pendant strands (Locke, 1923).

of a display icon which is memorable in its totality. This important characteristic of a display was not emphasized before, and its highlighting by Chernoff has increased our understanding of what makes a useful display.

A wide variety of techniques for the display of multivariate displays has developed, and the resulting multiple-bar charts, subdivided bilateral frequency diagrams, perspective drawings, stereograms, contour plots, shaded maps, etc., became familiar to the readers of 19th Century statistical atlases. It is interesting to note that among Francis Galton's other accomplishments was a monograph (1863) describing graphical "methods for mapping the weather", many of which are now the standard. Karl Pearson followed Galton's example and "devoted considerable attention to graphics" (Kruskal, 1975, p. 31). This devotion took a backseat to the development of the mathematical side of statistics during subsequent years, but the past decade has seen a renewed effort in the devlopment of multivariate displays. It would seem that this effort was spurred by the increasing popularity of multivariate statistical procedures and the availability of inexpensive plotters and good software.

Broadly available computing hardware and software made complex analyses easy for everyone, yet their very complexity removed them further from the intuition of the user. This increased the possibility of erroneous or inappropriate usage going undetected. Two avenues to guard against this were explored. One was the explosion of interest in robust statistical procedures that yielded protection from many sorts of unforeseen events; the second was the variety of graphical methods that helped to bring into consonance the user's intuition of the multivariate data structure with the results provided by the analytic model.

Many of the techniques developed for looking at multivariate data derived from the same basic notion; an icon of many parts is used to represent a multivariate point, the size and/or shape of each part of the icon representing one component of the data. Herman Chernoff, whom we now honor, was instrumental in this development. His notions about the necessity of a representative icon having a memorable perceptual gestalt joined comfortably with the ideas being expressed by Bertin (1967)[2] regarding the mental process of "visual selection". Together they represent an important contribution to the emerging theory of data display.

In this essay, I would like to discuss and illustrate several multivariate display methodologies. We start our examination of multivariate display methods with the table. Although the table (like the Quipu) has traditionally been a medium for data storage more than display, it can also serve a display purpose. Ehrenberg (1977) has presented a variety of suggestions that allow a table to better communicate the data structure contained in it. Now that data storage is primarily accomplished through electronic records, the communicative role of a table must be considered seriously indeed. Among these suggestions are:

1) Massive rounding of table entries. The human eye and mind do not easily perceive differences in numbers with more than two digits. Often rounding to two meaningful digits aids perception enormously.

2) Ordering rows and columns of the table by some aspect of the data. We have previously (Wainer, 1978) referred to this as the principle of, "I'm not interested in Alabama first". Too often we see data tables ordered alphabetically. This aids in locating particular entries for large tables, but does not simplify the perceptual structure of the table. For smallish

[2] *It is important to note that this classic work, previously only available in French and German, has recently appeared in an English translation, published by the University of Wisconsin Press.*

tables (10-20 rows) alphabetizing is not too crucial. In situations where it is, we can utilize the techniques employed in the 19th Century statistical atlases; the data were ordered by some meaningful aspect, and an alphabetical index to aid in locating entries of interest was provided.

3) Bordering the table with summary values to aid in comparisons. Adding a robust summary value like a median to each column or row (when summarizing is appropriate) facilitates comparisons within that column or row. It also helps to answer preliminary questions that one has when faced with the data.

4) Choosing scales wisely. Scale variables so that comparisons that are properly made can be made easily. Those comparisons that ought not be made, should not be included in the same table.

5) Using space wisely. If some data entries are different from the rest by construction (summary values rather than actual data), physically separate them a bit from the rest. Group together those that are similar. Grouping columns closer together facilitates visual comparisons.

Following these simple and commonsense rules, we will find that the tables thus prepared are easier to understand (and lose little in terms of their accuracy). They will also lead us toward a recently developed multivariate display technique. Rather than deal with display methods in the abstract, we will work within the context of a concrete example. Within this context we will describe several display methods and illustrate them on the same data set.[3] The data consist of seven variables for 10 states (from the *Statistical Abstract of the United States: 1977)* and were gathered as part of an effort to determine what is the worst American State.

[3] *This section is drawn from Wainer and Thissen, 1981.*

Tables

Table I shows data on ten states for the seven variables indicated. This form of the table is not recommended, but is probably most common. The states are ordered alphabetically; the numbers are given in varying accuracy (population is given to the nearest thousand, life expectancy to the hundredth of a year [4 days!]). It is clear than although the accuracy represented may be justified by the data, it is not the most useful way to show the data for communicative nor exploratory purposes.

Following Ehrenberg's (1977) suggestions, the data were rounded, states were reordered by an aspect of the data (life expectancy), column medians were calculated and shown, and space was used to block the table, which Ehrenberg (personal communication) suggests facilitates 'reading across'. In addition, the variables were reordered, bringing to the left those variables whose direct relationship with quality of life was clear (i.e., lower homicide rate is better), and moving to the right those variables whose relationship to quality of life was uncertain (i.e., is it better or worse to have warm weather?).

Viewing Table II tells us much more than was immediately evident from the original data table--that the two southern states are substantially inferior to the other eight on the first five social indicators. We see that California seems to have a relatively high homicide rate and per capita income. Further we get an impression of the overall levels of each of these indicators from the summaries at the bottom of the table, thus adding meaning to any particular entry. We lose the ability to locate immediately any particular state, which does not seem too important in a short list, but which may become more important with larger tables. Ehrenberg (1977) argues that most tables are looked at by individuals who are not naive with respect to their contents, and so have a reasonably good idea about the approximate position of each member. Thus if

TABLE I. Excerpted From "Worst American State: Revisted" Data

State	Population (1000's)	Average per capita income ($)	Illiteracy rate (% pop.)	Life expectancy	Homicide rate (1000)[a]	Percent high school graduates	Average # days year below freezing
Alabama	3,615	3,624	2.1	69.05	15.1	41.3	20
California	21,198	5,114	1.1	71.71	10.3	62.6	20
Iowa	2,861	4,628	.5	72.56	2.3	59.0	140
Mississippi	2,341	3,098	2.4	68.09	12.5	41.0	50
New Hampshire	812	4,281	.7	71.23	3.3	57.6	174
Ohio	10,735	4,561	.8	70.82	7.4	53.2	124
Oregon	2,284	4,660	.6	72.13	4.2	60.0	44
Pennsylvania	11,860	4,449	1.0	70.43	6.1	50.2	126
South Dakota	681	4,167	.5	72.08	1.7	53.3	172
Vermont	472	3,907	.6	71.64	5.5	57.1	168

[a] "Homicide Rate/1000" is actually combined murders and non-negligent manslaughter per 1000 population.

TABLE II. *Table I Rounded With Rows and Columns Reordered and Closed Up*

State	Life expectancy	Income (100's)	% High school graduates	Homicide rate	Illiteracy rate	Population (100,000's)	Days below freezing
Iowa	73	46	59	2	.5	29	140
Oregon	72	47	60	4	.6	23	44
South Dakota	72	42	53	2	.5	7	172
California	72	51	63	10	1.1	212	20
Vermont	72	39	57	6	.6	5	168
New Hampshire	71	43	58	3	.7	8	174
Ohio	71	46	53	7	.8	107	124
Pennsylvania	70	44	50	6	1.0	119	126
Alabama	69	36	41	15	2.1	36	20
Mississippi	68	31	41	12	2.4	31	50
Median	72	44	55	6	.8	26	125

one looked at the bottom of a list for a state one expected to find there, and only later discovered that it was somewhere else, this would be useful and surprising information. He claims that the losses associated with not having the table in alphabetical order are modest (for most applications) compared to the gains in potential for internal comparisons.

Table II provides us with some insights into the data structure. One of these is that the variables are not oriented in a way that allows easy comparison. For example, we see that when life expectancy is long, homicide rate seems low. Similarly, "% High School Graduates" seems negatively related to the illiteracy rate--a comforting expected result. By reordering the variables as well as we can manage so that they 'point in the same direction', we will gain additional insight. We also make use of this transformation step to make the univariate distributions less skewed. Thus we transform the variables as follows:

Life expectancy--untransformed

Homicide rate (per thousand, denoted P)--transform to non-homicide measure by using $f(P) = -5 \log [P/(1-p)]$

Income--use square root of income

% High School Graduates--untransformed

Illiteracy rate--use literacy measure = 1 $\sqrt{(\text{illiteracy rate})}$

Population--use $\log_e$ (Population) x 10

Days/year below freezing--untransformed.

The resulting Table III is not as close to intuition as Table II because of the transformations, but is a useful next analysis step. While the transformed numbers may be further from intuition than the originals, we can see the same structure that was evident in Table II, and the problem of scale becomes clearer still. In a true multivariate situation, like this one, making comparisons across variables is difficult. Some transformation is required.

TABLE III. The Same Data as in Table II After Transformations

State	Life expectancy	Income	% High school graduates	Homicide measure	Literacy measure	Population	Days below freezing
Iowa	73	68	59	30	141	14.9	140
Oregon	72	68	60	27	129	14.6	44
South Dakota	72	65	53	32	141	13.4	172
California	72	72	63	23	95	16.9	20
Vermont	72	63	57	26	129	13.1	168
New Hampshire	71	65	58	29	120	13.6	174
Ohio	71	68	53	24	112	16.2	124
Pennsylvania	70	67	50	25	100	16.3	126
Alabama	69	60	41	21	69	15.1	20
Mississippi	68	56	41	22	65	14.7	50
Median	71.5	66	55	25.5	116	14.8	125

Inside-Out Plots

The first step toward alleviating the problem of comparing across variables is to center the table by column. we can do this easily and robustly by subtracting out column medians and displaying the residuals. This is shown in Table IV. We note more clearly that the bottom of the table tends to have many negative residuals, the top many positive ones. This reassures us that the variables have been oriented correctly, and that they do, in fact, form a positive manifold in the variable space, with the possible exceptions of population and temperature.

Even after column-centering the table, we see that the residuals are still quite different and require some rescaling. An easy and robust way to do this is to divide each column by the median of the absolute value of each residual in that column. This is the well-known MAD estimator of spread. When we do this (Table V), we have finally arrived at a table that, although the numbers are removed from their original meaning (i.e., an entry of .3 under 'coldness' for Iowa is a long way from the original '140 days a year below freezing'), they now allow us to compare the relative standing of each state across variables. This was not possible in the original metric. If the aim of the original table was not to facilitate these comparisons, why were these data included in the table in the first place?

Now that the states are comparable across variables (the entries can be thought of as proportional to robust z-scores), we can row-center the table by calculating and subtracing out row medians. These medians are shown flanking Table V and provide one measure to answer the original question, "What is the worst American state"? Our answer, based upon only these 10 states, points to Mississippi as being without many serious rivals to the lamentable preeminence of the Worst American State.

TABLE IV. *The Same Data as in Table III Column-Centered by Subtracting Out Medians*

State	Life expectancy	Income	% High school graduates	Non-homicide measure	Literacy measure	Population	Days below freezing
Iowa	1.5	2	4	4.5	25	.1	15
Oregon	.5	2	5	1.5	13	- .2	- 81
South Dakota	.5	- 1	- 2	6.5	25	-1.4	47
California	.5	6	8	-2.5	-21	2.1	-105
Vermont	.5	- 3	2	.5	13	-1.7	43
New Hampshire	- .5	- 1	3	3.5	4	-1.2	49
Ohio	- .5	2	- 2	-1.5	- 4	1.4	- 1
Pennsylvania	-1.5	1	- 5	- .5	-16	1.5	1
Alabama	-2.5	- 6	-14	-4.5	-47	.3	-105
Mississippi	-3.5	-10	-14	-3.5	-51	- .1	- 75
Median	71.5	66	55	25.5	116	14.8	125
MAD	.5	2	4.5	3	18	1.3	45

TABLE V. *The Same Data as in Table IV Column-Centered and Column-Standardized by the MAD*

State	Life exceptancy	Income	% High school graduates	Non-homicide measure	Literacy measure	Population	Days below freezing	Row median
Iowa	3	1.0	.9	1.5	1.4	.1	.3	1.0
Oregon	1	1.0	1.1	.5	.7	- .2	-1.8	.7
South Dakota	1	- .5	- .4	2.2	1.4	-1.1	1.0	1.0
California	1	3.0	1.8	- .8	-1.2	1.6	-2.3	1.0
Vermont	1	-1.5	.4	.2	.7	-1.3	1.0	.4
New Hampshire	-1	- .5	.7	1.2	.2	- .9	1.1	.2
Ohio	-1	1.0	- .4	- .5	- .2	1.1	0	- .2
Pennsylvania	-3	.5	-1.1	- .2	- .9	1.2	0	- .2
Alabama	-5	-3.0	-3.1	-1.5	-2.6	.2	-2.3	-2.6
Mississippi	-7	-5.0	-3.1	-1.2	-2.8	- .1	-1.7	-2.8
Median	71.5	66	55	25.5	116	14.8	125	
MAD	.5	2	4.5	3	18	1.3	45	

We can see the relative position of the various states rather well by their ordering and spacing in Table V. If we use space more conspicuously to relate the relative position of the ten states overall, we arrive at the diagram shown in Figure 2. This is a back-to-back stem-and-leaf display (Tukey, 1977) in which the row medians are written down linearly in the center (the stem), and the states which have these medians

IOWA, SOUTH DAKOTA, CALIFORNIA	1.0	IA SD CA
OREGON	.6	OR
VERMONT, NEW HAMPSHIRE	.2	VT NH
OHIO, PENNSYLVANIA	-.2	OH PA
	-.6	
	-1.0	
	-1.4	
	-1.8	
	-2.2	
ALABAMA	-2.6	AL
MISSISSIPPI	-3.0	MS

FIGURE 2. Back-to-back stem-and-leaf display of the Worst American State effects showing both state names and their two-letter abbreviations.

are written down adjacent to the appropriate median (the leaves).[4] This can be usefully thought of as turning a table inside out, for we ordinarily have labels on the OUTSIDE of a table and numbers on the INSIDE. For a stem-and-leaf display we move the numbers outside to make them labels, and the labels inside next to the appropriate number. This concept of inside-out plotting will be even more useful shortly.

Figure 2 shows the overall position of each of the ten states. What remains when the data matrix in Table V is row-centered by subtracting out these overall positions is what is of interest next. Table VI shows the final matrix. It has been column-centered by removing column medians, column-scaled by dividing each column by the median absolute deviation from the column median, and last, row-centered by subtracting out the row medians. All of this was done after the variables were oriented in the same direction and transformed to be as symmetric as possible (within generous limits). Table VI shows the resulting residual matrix.

Glancing over Table VI our eyes note several unusually large entries. It is hard to see much order in the residuals. Can we display this in a way that makes the features of interest more obvious? Ramsay (1980) has suggested that plotting Table VI inside-out is a way that answers this question straightaway. To do this, we prepare stem-and-leaf diagrams of each column of this table separately, but since they are on comparable scales we can plot them side-by-side on a common stem (see Figure 3). Thus turning this table inside out--placing the labels inside and the numbers outside--we immediately see details of structure that add to the states' overall ranking. We see that Pennsylvania seems to have a shorter life expectancy

[4] *The values plotted here were arrived at with somewhat more precision than what was shown in Table V, thus the slight discrepancies. Obviously, our advocacy of extreme rounding for presentational purposes does not extend into intermediate calculations.*

TABLE VI. The Same Data as in Table V, Column-Scaled and Doubly Centered by Medians--A Scaled Residual Matrix

State	Life exceptancy	Income	% High school graduates	Non-homicide measure	Literacy measure	Population	Days below freezing
Iowa	2.0	0	- .1	.5	.4	- .9	- .7
Oregon	.3	.3	.4	- .2	0	- .9	-2.5
South Dakota	0	-1.5	-1.4	1.2	.4	-2.1	0
California	0	2.0	.8	-1.8	-2.2	.6	-3.3
Vermont	.6	-1.9	0	- .2	.3	-1.7	.6
New Hampshire	-1.2	- .7	.5	1.0	0	-1.1	.9
Ohio	- .8	1.2	- .2	- .3	0	1.3	.2
Pennsylvania	-2.8	.7	- .9	0	- .7	1.4	.2
Alabama	-2.4	- .4	- .5	1.1	0	2.8	.3
Mississippi	-4.2	-2.2	- .3	1.6	0	2.7	1.1

	Life Expec- tancy	Income	% HS Graduates	Non- Homicide Measure	Literacy Measure	Popu- lation	Days Below Freezing
2.8						AL MS	
2.4							
2.0	IA	CA		NH			
1.6				MS			
1.2		OH		SD AL		PA OH	MS
0.8	VT	PA	CA				NH
0.4	OR	OR	OR NH	IA	IA SD VT	CA	AL VT
0.0	SD CA	IA	IA VT	OR VT PA	OR NH OH AL MS	OH	PA OH SD
-0.4		AL	OH AL MS	OH			
-0.8	OH	NH	PA		PA	IA OR	IA
-1.2	NH					NH	
-1.6		SD	SD			VT	
-2.0		VT		CA		SD	
-2.4	AL	MS			CA		OR
-2.8	PA						
-3.2							
-3.6							CA
-4.0							
-4.4	MS						

FIGURE 3. Inside-out display of residuals in Table VI. Plot allows viewer to center attention on unusual state-variable interactions by showing unusual residuals.

than its overall rank would predict--perhaps looking at infant mortality rates and black lung statistics would be helpful to understand this. We note that California's literacy measure seems unusually low when one considers its high overall ranking, suggesting an interpretation based upon its large Spanish-speaking population.

The plotting of multivariate residuals inside-out follows logically from the basic premises of tabular display espoused by Ehrenberg. They allow the tabular format to communicate information and provide that most valuable of graphic services--"force us to notice what we never expected" (Tukey, 1977). We discover that many of the findings seen clearly in Figure 3 were visible in Table II, now that we know to look for them.

This display has the added advantage of getting better as the data set (number of observations) gets larger. We are typically only interested in observations with large residuals; therefore when the data bunch up in the middle (as would be the case with a larger data set), we can safely ignore them. Further, this display method works well when data are not scored in the same way, though it does take some preprocessing to get the data ready for plotting. That includes a variety of two-way scatter plots to determine the best orientation of each variable. In this instance we ordered 'number of days below freezing' in the same direction as literacy not because of any love for cold weather, but rather because it was in this direction that it related to all the variables that are easily oriented.

The operations described previously involved with the transformation of variables prior to, or in the midst of, preparing a multivariate display are common to all display methodologies. One must almost always make a sequence of displays--each one telling how the next should be scaled. Univariate stem-and-leaf diagrams were used to help determine the transformations needed for symmetry. Bivariate plots aid in determining the correct orientation of variables and point toward unusual points that might confuse solely analytic transformation methods. However, while the analyses are done sequentially, the wise analyst often returns to previous displays with newfound knowledge, to redisplay the data after subsequent analyses pointed toward an important or useful transformation. In addition to providing additional information about the data set, this technique also reduces the effect that the *particular* sequence used has had on the results.

Function Plots

Andrews (1972) developed a scheme for multivariate display that is quite different from most methods used previously. His method involves calculating a periodic function of k-Fourier components for each k-dimensional data point. The value of each Fourier component of that function is determined by the value of its associated variable. The strength of this technique is that it allows the inclusion of many variables but limits the number of data points that can be effectively shown; with more than 10 or 20 points the plot can get too busy to be helpful. Another disadvantage is that such plots are quite far removed from the data, so that even though one can see which points tend to group together, one must look to other display methods to better understand why.

Shown in Figure 4 is a function plot of the Worst American State data. We note the similarity of Alabama and Mississippi and the close matching of Pennsylvania and Ohio. If one chooses the plots with care, such characteristics as 'overall height' and 'general jiggliness' can have particular importance.

As with almost any innovative methodology, experience with its use is required before the subtleties of its character reveal themselves. The potential value of function plots does

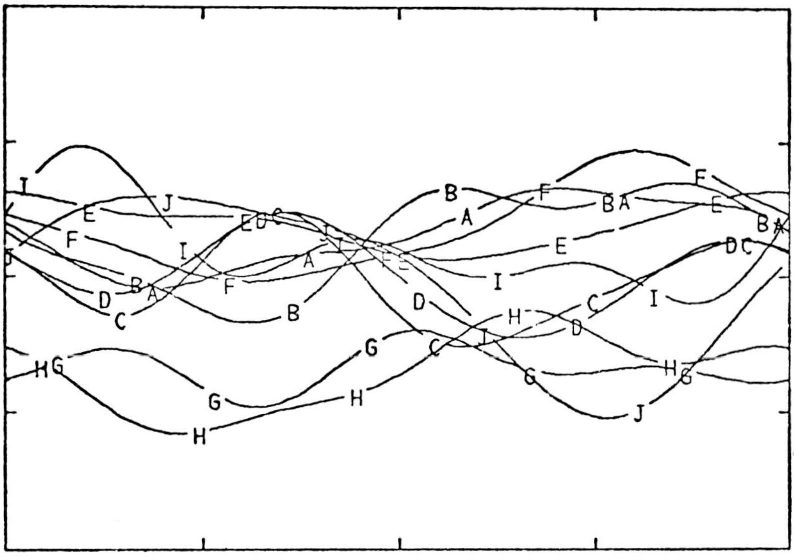

Function Plot Prepared by David Andrews

Legend

- A New Hampshire
- B Vermont
- C Pennsylvania
- D Ohio
- E Iowa
- F South Dakota
- G Alabama
- H Mississippi
- I Oregon
- J California

FIGURE 4. Function plot of Worst American State data (plot prepared by David Andrews).

not always manifest itself upon first acquaintance. Tales are told in the hallways of Bell Laboratories about a contest to discover the secret of a data set constructed especially for this purpose. There several of the talented statisticians examined the problem under the illumination of their favorite display/analysis methods. The only one to come up with the correct solution was David Andrews, who used function plots.

The apportioning of credit for this victory between the tool and the craftsman is difficult, yet it does supply support for the inclusion of function plots within the toolbox of the serious data analyst.

Polygons

A more traditional icon for multivariate display is a polygon formed by k-secting a circle with k-radii each associated with a particular variable. The length of each radius is proportional to the value of the variable for the data point to be displayed, and these various points are then connected to yield an irregularly shaped polygon. This technique is more than a century old and provides an often useful display. Usually each variable is scaled separately, so that the maximum value of each variable touches the circle. Sometimes this is varied to make a regular polygon with all radii touching the circle a standard (i.e., this could represent the condition for the entire U.S.). Than any point that stuck out past the circle's perimeter would represent 'greater than average'. We choose the former convention in our display of the Worst American State data shown in Figure 5. Note the convention of orienting all variables so that 'bigger = better' is quite useful. Thus states that are in the running for the title of 'Worst' have smallish polygons, whereas those at the other end look much larger. Note further that it is relatively easy to discover (using the legend) what are each state's strong and weak points. The shortcomings of this method are:

1) Producing a legible display with many more than seven variables becomes difficult.

2) The shape of the polygons is arbitrary, because of the arbitrariness of the order of the variables around the circle.

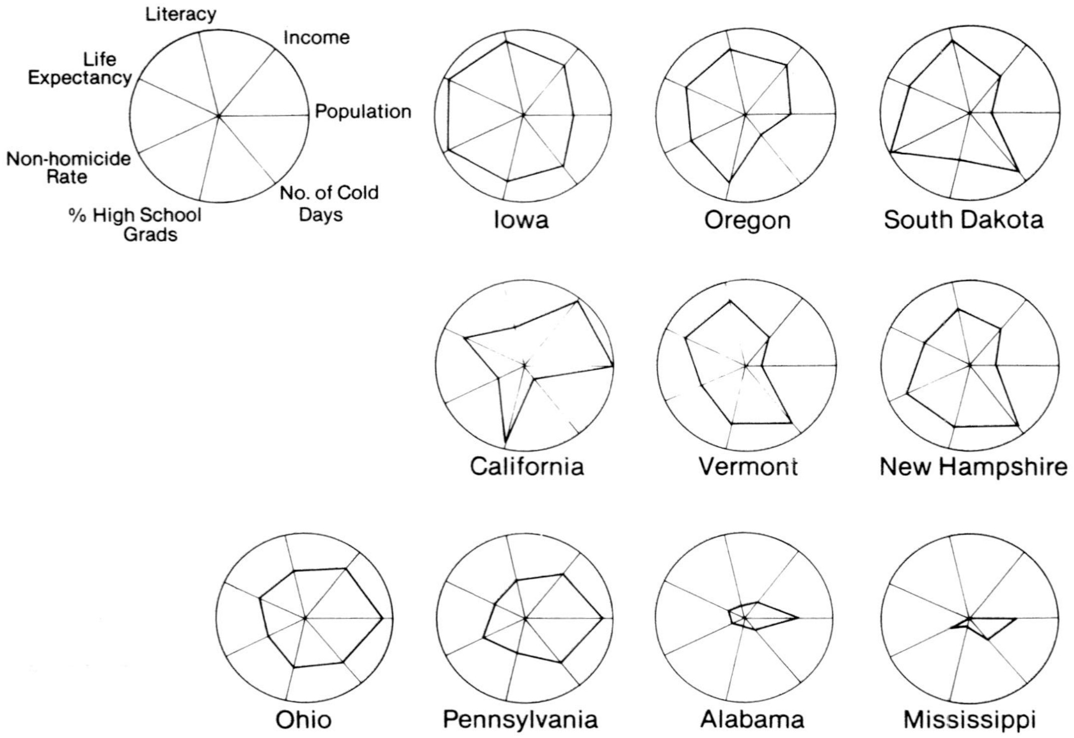

FIGURE 5. Polygon plot of Worst American State data.

3) Odd shaped polygons are not memorable, so one is hard pressed to remember any details about a state other than its gross structure (if it was small or large, regular or irregular--this latter bit of structure can be varied by reordering the variables).

4) The legend is crucial to the display. It seems difficult to keep in mind which corner refers to which variable. Take away the legend and the interpretability of the display shrinks.

Similarities in shape seem to be perceived irrespective of differences in orientation. Thus Vermont and Oregon seem similar (except for rotation). Of course, this similarity is to some extent false, due just to the particular ordering of the variables.

Despite these shortcomings, the polygon can be a useful display.

Returning to our investigation into the Worst American State and Figure 5, we see: Alabama and Mississippi standing out, the similarities between Ohio and Pennsylvania (among the causes of this similarity is the relatively low life expectancy), and Iowa emerging as a contender for the other end of the spectrum.

Trees

In the display methods so far discussed the viewer is able, more or less well, to visually cluster the multivariate observations (states) through the similarity of their representative icons. Clustering of the variables (state characteristics) is not easily done. This inability to represent simultaneously the clustering of variables and of observations I view as a weakness of the display. Kleiner and Hartigan (1981) solved this problem in a simple and ingenious way. They noted that a tree is a common icon to represent a hierarchical cluster structure, and that the structure of the tree is reasonably well determined (Johnson, 1967; Gruvaeus and Wainer, 1972). They reasoned that one could use a variation of the tree structure obtained from the inter-observation distances as a basic icon. This single tree would provide the shape of the icon, but the size of each branch of the tree would be determined by the value of the variables represented by that branch. Thus each observation would be represented by a tree with the same general structure, but with different size branches.

In Figure 6 we see a tree derived from the Worst American State data (averaged across all ten states). We see immediately that there are two major groupings of variables: one includes life expectancy, percent of high school graduates and income; and the other includes homicide rate, literacy and temperature. The populations of the states appear to be a relatively isolated variable. In Figure 7 this tree is distorted to reflect each state's multivariate structure. We note California's asymmetry with warm weather and high homicide rate shrinking the left cluster; the symmetric structure of Vermont, New Hampshire and South Dakota is similar to that of Iowa, but their small populations separate them. Mississippi and Alabama are, again, shown to be distinctive.

We can conclude that the tabular scheme seems better at supplying details, for clustering states function plots win out, but clustering of variables is done well only by trees. Polygons seem to provide adequate performance in displaying qualitative detail and clustering observations, and so is a reasonable compromise candidate. It is also easier to use within the context of mass communication than the other three methods. It provides a more eye-catching image than a table and is easier to explain than a tree or function plot. Inside-out plots are a curious mixture of tabular and graphic display. They provide a clear view of the best unidimensional ordering, as well as highlighting deviations from this order.

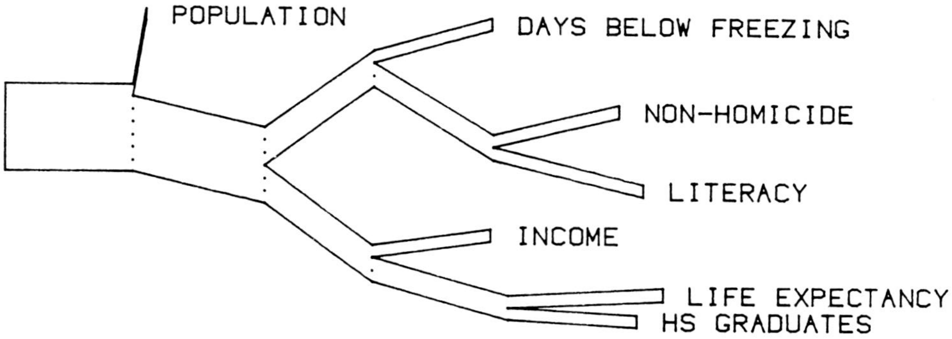

FIGURE 6. Cluster diagram of the variables considered in the search for the Worst American State. This tree forms the basic template for the trees in Figure 19 (plot prepared by Beat Kleiner).

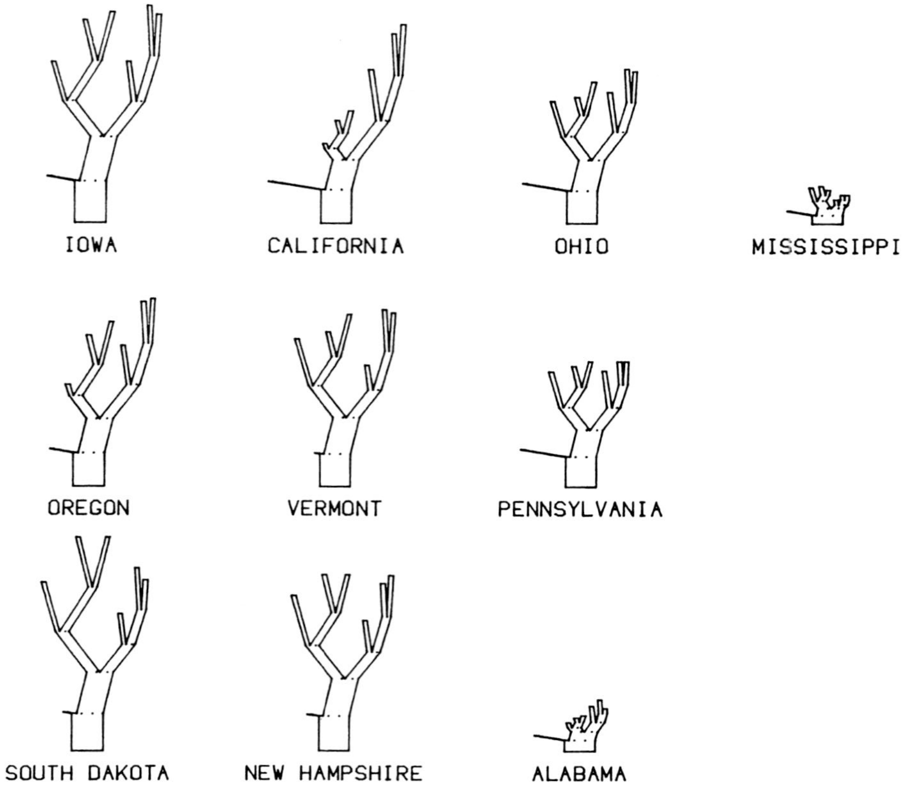

FIGURE 7. *Tree display of Worst American State data (plot prepared by Beat Kleiner).*

Faces

I previously referred to the polygon display as "a more traditional icon". Let me take this opportunity to explain that remark a bit further, and thus better set the stage for the discussion of the Chernoff Face. A survey of multivariate icons reveals that many of them share a common structure. Typically, each icon is made up of many parts, each of which varies in size and shape as a function of the data. Thus the final appearance of the whole icon reflects contributions of the various parts. Polygons obviously fit this mold, as do Anderson's (1960) *metroglyphs*, Bertin's (1977) *graphical matrices*, Kleiner and Hartigan's (1981) *trees*, as well as more ancient types like the quipu and the various visual metaphors employed by cartographers. To some extent so too does the Chernoff Face, with an important variant. In the Face construction the size, shape and orientation of each feature is related to a characteristic of the data, yet the icon is conceived of as a whole and is intended to be perceived from the outset as a gestalt--not as the sum of its parts.

Before we discuss this icon in greater detail, let us first look at its use within the context of our example--the Worst American State. The first step in displaying data with faces (after the sorts of orientations and scalings described previously) is the assignment of variable to feature. This is not always easy nor is it unimportant. Jacob (1978; 1981) has suggested several empirical aids to this assignment problem; they supplement but do not supplant displayer wisdom. The assignment plan we arrived at (after several tries) was:

1) Population => the number of faces/state--The number of faces is proportional to the log of the population. We used one large face in each state for easier identification and as many identical small faces as required. The size of these latter faces was kept small so as to allow us to fit them within the confines of the state boundaries.

2) Literacy rate => size of the eyes (bigger = higher)

3) % HS graduates => slant of the eyes (the more slanted the better)

4) Life expectancy => the length of the mouth (the longer the better)

5) Homicide rate => the width of the nose (the wider the nose, the lower the homicide rate)

6) Income => the curvature of the mouth (the bigger the smile, the higher the income)

7) Temperature => the shape of the face (the more like a peanut, the warmer; the more like a football, the colder)

8 and 9) Longitude and latitude => the X and Y position of the face on the coordinate axes of the paper represents the position of the state.

Thus we tried to use sensible visual metaphors for representing each variable; 'Bigger = Better' was the general rule when normative direction was clear. In the case of a variable (such as weather) where desirability could not be determined easily, we used an aspect of the face that is not ordered. To show the adaptability of the FACE scheme to larger data sets, we prepared a plot involving all fifty states. Similar plots could have been prepared for most other meta-iconic schemes (STARs, TREEs, etc.). The resulting plot is shown in Figure 8.

A viewing of the map reveals many things. First, we can see the relative density of population in the East at a glance. Next, note the temperature gradient as one goes south from North Dakota to Texas. The Pacific current is also evident. Looking more closely we see that the deep South looks very homogeneous and low on all variables of quality of life used.

Note further that three New England states, Vermont, New Hampshire and Maine, look the same: they are generally high on education variables, as well as having low homicide rates, but appear to have low incomes. Massachusetts seems to have a

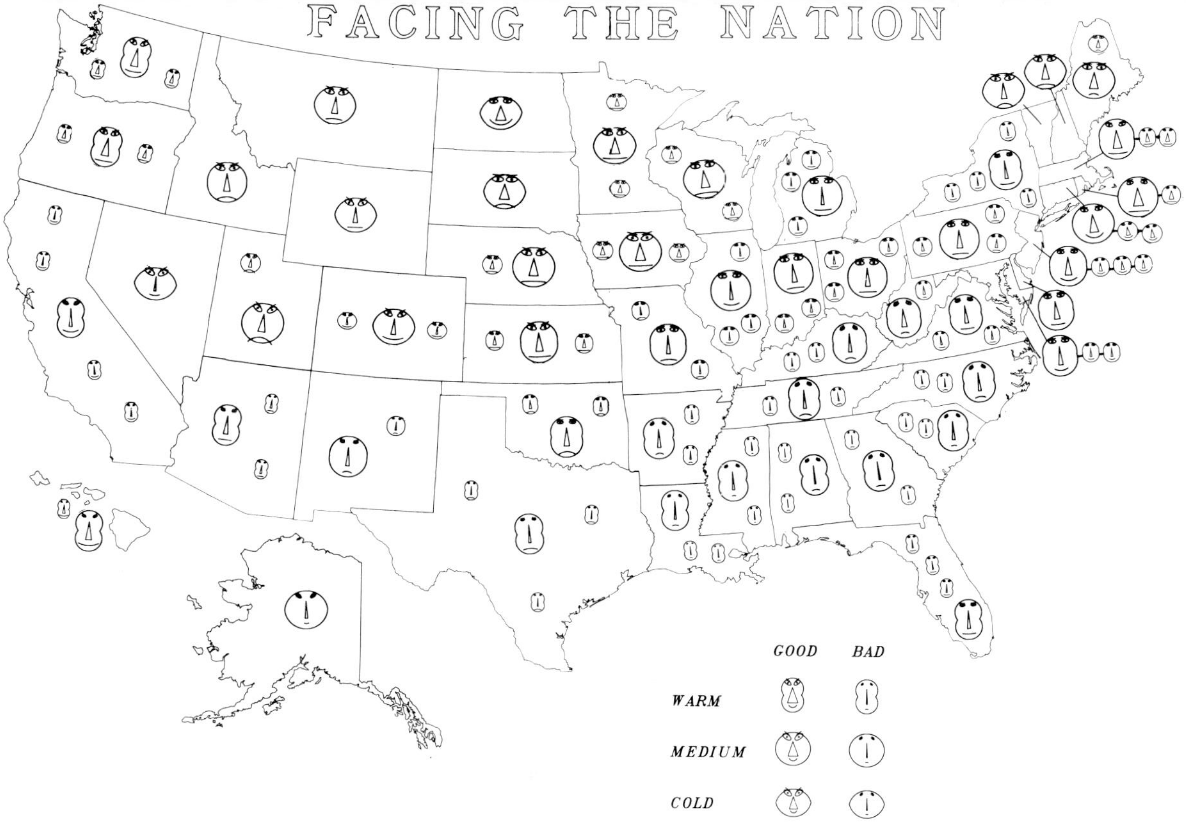

FIGURE 8. The Worst American State data shown for all 50 states as Chernoff Faces.

somewhat lower literary rate but higher per capita income. Connecticut and New Jersey seem remarkably similar with higher incomes still. We also see a clustering of Mid-East states on all variables (Pennsylvania, Ohio, Michigan and Indiana) with the border states of Virginia, West Virginia and Kentucky falling somewhere between their rural neighbors and their more industrial ones.

The Mid-West looks homogeneous with the rather surprising difference between North and South Dakota in income. A check back to the original data indicates that this is not a data entry error. There is a substantial difference between these two neighbors. We also note that except for the cold weather, North Dakota seems a pleasant place to live. Utah has an interesting structure, being very high on all variables except income, which may reflect the influence of the Mormon church on the state. There are many other interesting similarities that might be noted, and the reader is invited to study the map more closely. As a last comment, we draw the reader's attention to the similarity between California and Hawaii. We see benign weather, long life, high income, low homicide rate, high proportion of high school graduates, but a low literacy rate. This reflects either their high proportion of non-English speaking inhabitants, or a propensity for lotus eating so common in paradise.

The map in Figure 8--Facing the Nation--has several unique characteristics. The human face quality of the display icon makes the display memorable. Without having the map in front of you, it is easy to characterize from memory the sad squinty-eyed face of the South. Similarly, the broad faced stolid Mid-Westerner emerges clearly from memory, as do the smiling faces of Hawaii and California. Once one remembers the correspondences between facial features and statistical variables, the attaching of meaning to the visual memory becomes easy.

Quantitative details are lost, but general structure remains. Visual clustering in k-space is done almost without noticing. Care must be taken in the assignation of features, for some seem to have a more profound effect upon perception than others. Yet, with the addition of a little wisdom a useful display appears.

Extensions

Chernoff's original formulation has been adapted, modified and improved upon. Flury and Riedwyl (1981) developed a modified face that is a more realistic version with much greater detail (see Figure 9). There is less implicit covariation among the features than in the Chernoff version, and the two sides of the face can have features assigned to them independently, thus allowing the possibility of an asymmetric face.

In an experiment that Flury and Riedwyl ran, they found that their version of the face (used either symmetrically or asymmetrically) provided for more accurate judgments as to which pairs of twins were monozygotic or dizygotic.

Wakimoto (1977) in a further extension of the Chernoff idea used an entire body plot to represent various physiological measurements (see Figure 10). This may be a useful form for certain kinds of applications, but aside from providing room for a few more parameters, does not seem to improve on the original formulation. It is just another realistic icon.

This leads to an interesting question. Wakimoto's body plots use the characteristics of the body to represent the physiological characteristics. Thus a fat figure implies an overweight person, small eyes represent poor eyesight, a tall figure represents a tall person, etc. This means that we have an excellent visual metaphor, with the legend able to be learned with little effort. Should this concept be generalized? If we want to represent bank notes (as in Flury and Riedwyl), why use faces? Why not

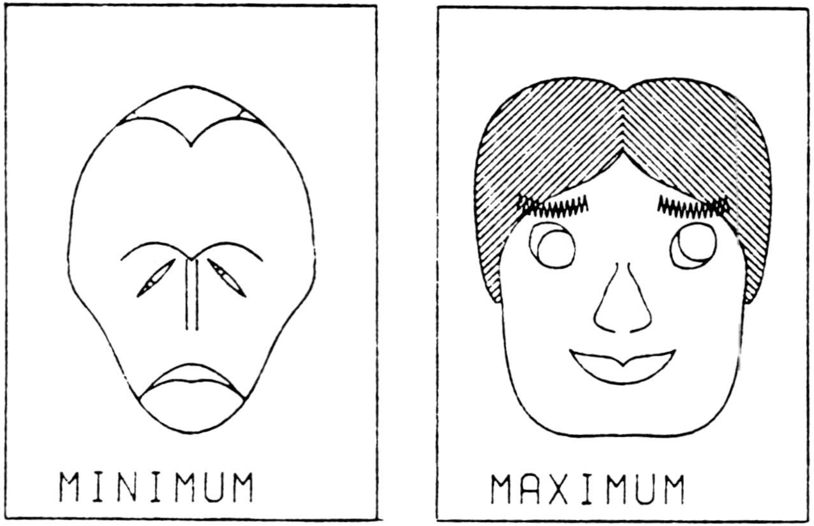

FIGURE 9. *Two Flury and Riedwyl asymmetric faces plotted symmetrically. One face shows each feature at its minimum value, the other shows all features at their maximum.*

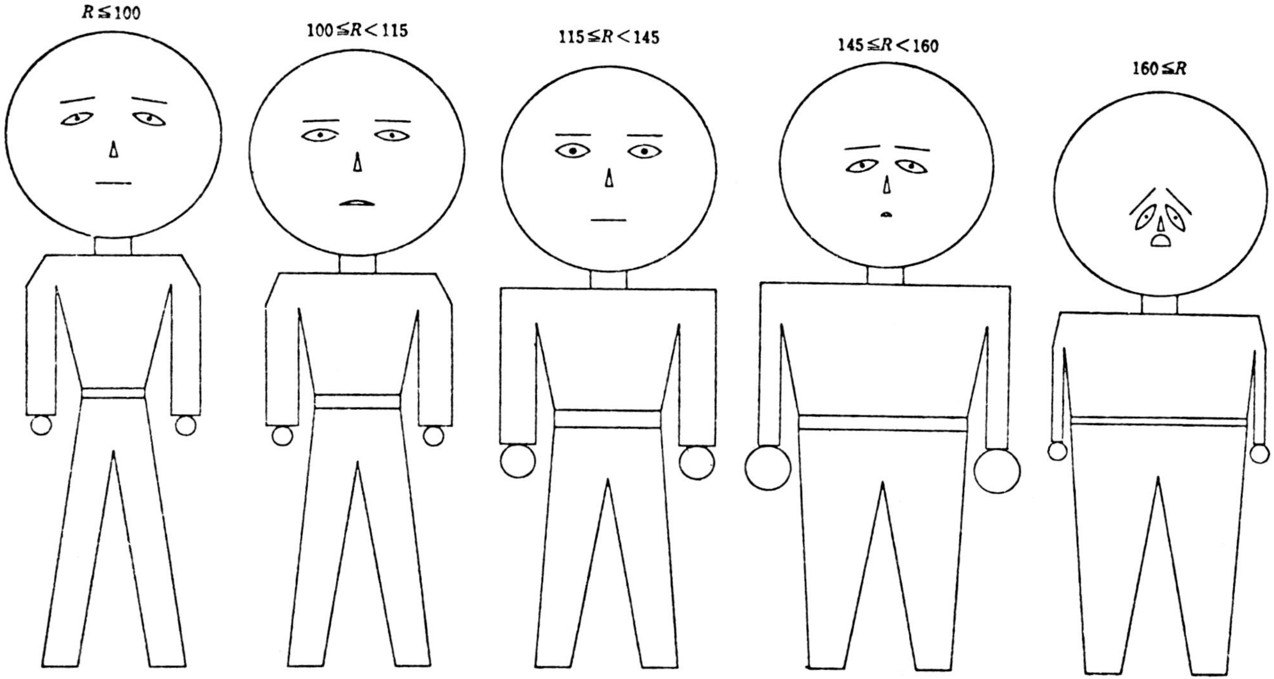

FIGURE 10. Wakimoto's Deshi, which uses a whole body to convey multivariate information.

cartoon bank notes with the features measured enhanced in some way? If we want to represent various characteristics of housing, why not use cartoon houses? Previously, I suggested (Wainer, 1979) that an icon like the Wabbit (see Figure 11) be used on cereal boxes to convey nutrient information to children. In this instance the characteristics of the Wabbit's face would be related to the vitamins whose deficiency affects them.

Obviously, such methods have much to recommend them, yet they are not practical as a general approach. The programming necessary to draw a specialized icon for a single application is too costly in time and resources to justify itself unless this use is to be routinized. Thus the power of the face as a generalizable multivariate display--it provides a memorable display for data from a variety of origins. It may not be optimal in any given circumstance, but may provide (as Mosteller and Tukey [1968] describe the Jackknife--another general technique) 'the broad usefulness of a technique as a substitute for specialized tools that may not be available' (p. 134).

Conclusions

In this essay we focussed on a few methods that are either exemplary of a class of others or are sufficiently useful and unique to warrant inclusion. The use of several of these methods was illustrated on social indicators data gathered to compare the standings of states in modern America.

To summarize what was illustrated in this essay we noted:

1) The table did a pretty good job once it was redone, although the full (50 x 7) data set for all states would have been less transparent. Enhanced tables, from which main effects have been removed do better. Still better for large tables are

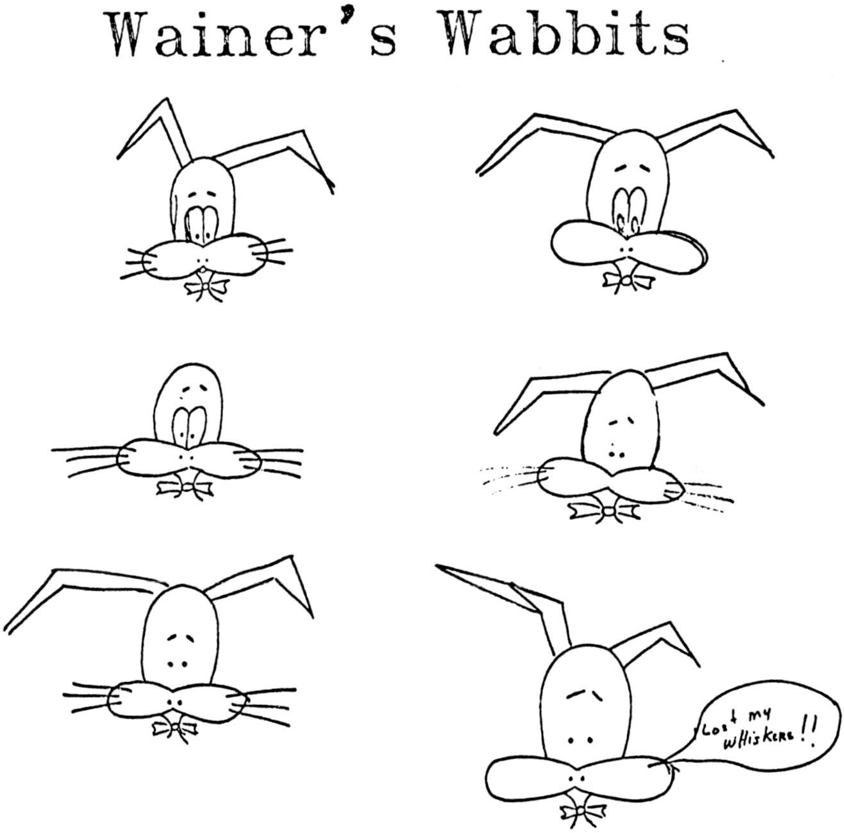

FIGURE 11. Wainer's Wabbits shown here in several variations. These were originally recommended as a display icon to communicate multivariate nutritional information to children.

inside-out plots which convey the detail of a table along with a more visually compelling image of the distributions and clusterings of the data.

2) Function plots are sufficiently far from convention that considerable user experience seems to be required before being able to utilize fully the information carried in them. They do seem to easily provide clustering information, although they can be limited in the number of data points that they can convey and still allow the extrication of identification information of individual points. Their strength is the number of variables that can be portrayed.

3) Polygon plots, as well as other sorts of iconic displays based upon non-metaphorical icons, are useful for qualitative conveyance of information and are made more useful still if the icons are displayed in a position that is meaningful. This was illustrated in Figure 8, when the geographic location provided information about the identification of each icon that was far more visually compelling than merely tagging each with the associated state's name. This led us to an important conclusion about multivariate plotting methods--do not waste the two dimensions of the plane. Early plotting schemes showed just two variables by using the two dimensions of the plane to represent them. Multivariate display methods allow many variables to be shown, yet we would be wise not to waste the physical dimensions, for they provide one of the most compelling of visual metaphors.

4) Trees are the only display method that do a good job at depicting the covariance structure among the variables. Their perceptual complexity diminishes their usefulness as a memorable icon, but they are a new idea--still under development.

5) Faces (and subsequent versions including nicer faces, bodies, and wabbits) allow the display of complex data in a

memorable way. There are some problems with their use, particularly those associated with the assignation of features. This problem is eased through the use of clustering schemes like those used by Kleiner and Hartigan (1981) in the formation of their Trees and Castles. Further details of this point are contained in Jacobs' (1981) discussion of their paper.

Seeing vs. Reading a Display

A recurring issue in discussions of display methodologies relates to the kinds of cognitive processes that ought to be brought into play when confronted with a display. Bertin (1977) contends that a display is most efficient when we can 'see' the data structure. This entails the use of a display in which the data are represented by a compelling visual metaphor. Shading (darker = more) and stereograms (higher = more) are two examples of such metaphors. Displays that are less efficient need to be 'read'. Colors used for shading often need to be read (i.e., red = most, blue = medium, yellow = least). Obviously, when a display needs to be read, we both perceive less efficiently and remember more poorly. The great power of Chernoff's invention is that it allows the seeing of quite complex data structures.

Tables can be made more 'seeable' through Ehrenberg's guidelines, but this rests to some extent on the iconic characteristics of the numbers themselves. It is apparent that two-digit numbers look bigger than one-digit ones, and thus we can 'see' that 123 is bigger than 62 and is greater than 8. Confusion manifests itself when the numbers being compared are of the same order. Here we are helped because the numbers themselves are somewhat ordered--1 looks smaller than 2, and 3 looks smaller than 8. This relationship is not perfect--7 looks smaller than 6--yet it does help. Bachi's (1968) Graphical Rational Patterns are an attempt to make the numbers more graphical. This same conception is behind tallying schemes

(/, //, ///,...) as well as quipu accounting. Thus though numerical representation can be made somewhat graphical and hence 'seeable', it is important that we are aware of their limitations.

If a display must be read, is that bad? This question seems to be often misunderstood. I believe that a graphic display can serve many purposes. Sometimes we want a display to show general structure quickly and efficiently, and provide a memorable image that can subsequently be retrieved and utilized. This certainly is one of the more important uses of graphical display, and one on which we have concentrated most of our attention. There are others. We note that communication with prose tends to be linear--the storyteller starts at the beginning and proceeds to the end in a manner that appears to him to be logical. Yet stories, or even 'how-to' directions can be told in many ways. If one was to prepare directions as to how to get up in the morning and get to work successfully, there are many possible orders that could be specified--but only one would be. Such is the nature of linear presentation. John Tukey acknowledges this in his classic text (1977) when he points out alternative orders in which the book can be profitably read--some chapters are in series, others in parallel.

A powerful use of a graphic display is to present information in a non-linear way. Thus exploded diagrams of automobile transmissions show clearly which pieces go where and indicate clearly which orders of assembly are possible and which are not. Similarly, the complex charts of population often found in statistical atlases provide many stories of immigration trends. These can be read and studied from left to right, top to bottom, or in many other nonlinear ways and provide fresh insights. I do not mean to imply that such complex charts are superior to telling any particular tale--rather they are a reasonably compact way of providing the raw material for many stories.

Such displays have their place and should not be denigrated when this is the purpose for which they are employed. Our primary concern in this essay is with displays that aid in seeing structure, but we would have been remiss in not mentioning this other role.

An Appreciation

A glance over the history of display indicates the many contributors who have conceived important and useful methods to allow us to better look at our data. I am indeed grateful to Herman Chernoff who has allowed our data to look back at us.

Acknowledgments

I would like to thank Albert D. Biderman for important references, stimulating discussions, and pointed criticism, all of which added considerably to whatever value this essay may have. In addition this paper has profited from the comments on an earlier draft by: William Angoff, Andrew Ehrenberg, Mary Elsner, Norman Fredrikson, Beat Kleiner, and John Tukey.

References

Anderson, E. (1957). A Semigraphical Method for the Analysis of Complex Problems. *Proceedings of the National Academy of Sciences 13*, 923-927; reprinted in *Technometrics 2*, 3, 387-392 (1960).

Andrews, D.F. (1972). "Plots of High-Dimensional Data." *Biometrics 28*, 125-136.

Bachi, R. (1968). *Graphical Rational Patterns: A New Approach to Graphical Presentation of Statistics.* University Press, Jerusalem, Israel.

Bertin, J. (1977). *La graphique et le traitement graphique de l'information.* Flammarion, Paris.

Bertin, J. (1973). *Semiologie Graphique.* Mouton-Gautier, The Hague, 2nd ed.

Bertin, J. (1967). *Semiologie Graphique.* Gauthier-Villars, Paris.

Chernoff, H. (1973). "The Use of Faces to Represent Points in K-Dimensional Space Graphically." *J. Amer. Statist. Assoc. 68,* 361-368.

Ehrenberg, A.S.C. (1977). "Rudiments of Numeracy." *J. Roy. Statist. Soc. Ser. A* 140, 277-297.

Flury, B. and Riedwyl, H. (1981). "Graphical Representations of Multivariate Data by Means of Asymmetrical Faces." *J. Amer. Statist. Assoc. 76,* 757-765.

Galton, F. (1863). The Weather of a Large Part of Europe, During the Month of December 1861. *Meteorographica or Methods of Mapping the Weather.* London and Cambridge.

Gruvaeus, G.T. and Wainer, H. (1972). "Two Additions to Hierarchical Cluster Analysis." *British J. Math. Statist. Psych. 25,* 200-206.

Jacob, R.J.K. (1978). Facial Representation of Multivariate Data. In *Graphical Representation of Multivariate Data,* (ed.) P.C.C. Wang, 143-168. Academic Press, New York.

Jacob, R.J.K. (1981). "Comment on Trees and Castles." *J. Amer. Statist. Assoc. 76*: 374, 270-272.

Johnson, S.C. (1967). "Hierarchical Clustering Schemes." *Psychometrika 32,* 241-254.

Kleiner, B. and Hartigan, J.A. (1981). "Representing Points in Many Dimensions by Trees and Castles." *J. Amer. Statist. Assoc. 76:* 374, 260-269.

Kruskal. W.H. (1975). Visions of Maps and Graphs. *Auto Carto II: Proceedings International Symposium Computer-Assisted Cartography.* Census Bureau, Washington, D.C.

Locke, L.L. (1923). The Ancient Quipu or Peruvian Knot. *The American Museum of Natural History: Science Education.* Library of Congress, Smithsonian, 12-73.

Mosteller, F. and Tukey, J.W. (1968). Data Analysis: Including Statistics. *The Handbook of Social Psychology,* (eds.) G. Lindsey and E. Aronson, 80-203. Addison-Wesley, Reading, Massachusetts.

Ramsay, J.O. (1980). *Inside-Out Displays and More.* Presented at a Symposium on Multivariate Data Displays. Psychometric Society Meeting, Iowa City.

Tukey, J.W. (1977). *Exploratory Data Analysis.* Addison-Wesley, Reading, Massachusetts.

Wainer, H. (1979). The Wabbit: An Alternative Icon for Multivariate Data Display. BSSR Technical Report, 547-792.

Wainer, H. (1978). Graphical Display. Presented at European Mathematical Psychological Association, Uppsala, Sweden.

Wainer, H. and Thissen, D. (1981). "Graphical Data Analysis." *Annual Review of Psychology 32,* 191-241.

Wakimoto, K. (1977). A Trial of Modification of the Face Graph Proposed by Chernoff: Body Graph. *Quantitative Behavior Science (Kodo Keiryogaku) 4,* 67-73.

V. OTHER TOPICS

MINIMAX ESTIMATION OF THE MEAN OF A NORMAL DISTRIBUTION SUBJECT TO DOING WELL AT A POINT

P. J. Bickel[1]

Department of Statistics
University of California
Berkeley, California

SUMMARY

We study the problem: Minimize $\max_\theta E_\theta(\delta(X)-\theta)^2$ subject to $E_0\delta^2(X) \leq 1-t$, $t > 0$, when $X \sim N(0,1)$. This problem arises in robustness questions in parametric models (Bickel (1982)). We

(1) Partially characterize optimum procedures.

(2) Show the relation of the problem to Huber's (1964) minimax robust estimation of location and its equivalence to a problem of Mallows on robust smoothing.

(3) Give the behaviour of the optimum risk for $t \to 0, 1$ and (4) Study some reasonable suboptimal solutions.

[1] *This research was supported in part by the Adolph C. and Mary Sprague Miller Foundation for Basic Research in Science and the U.S. Office of Naval Research Contract No. N00014-75-C-0444 and N00014-80-C-0163.*

The results of this paper constituted a portion of the 1980 Wald Lectures.

ISBN 0-12-589320-5

I. THE PROBLEM

Let $X \sim N(\theta,\sigma^2)$ where we assume σ^2 is known and without loss of generality equal to 1. Let δ denote estimates of θ (measurable functions of X), and

$$M(\theta,\delta) = E_\theta(\delta-\theta)^2$$

$$M(\delta) = \sup_\theta M(\theta,\delta)$$

For $0 \leq t \leq 1$, let

$$\mathcal{D}_t = \{\delta: M(0,\delta) \leq 1-t\}$$

and

$$\mu(t) = \inf\{M(\delta): \delta\varepsilon \mathcal{D}_t\}.$$

By weak compactness an estimate achieving $\mu(t)$ exists. Call it δ_t^*. Of course, $\mu(0) = 1$ and

$$\delta_0^* = X$$

while $\mu(1) = \infty$ and

$$\delta_1^* = 0.$$

Our purpose in this paper is to study δ_t^* and μ_t and approximations to them, based on a relation between the problem of characterizing μ and δ^* and Huber's classical (1964) minimax problem.

The study of δ^* and μ can be viewed as a special case, when the prior distribution is degenerate, of the class of restricted Bayes problems studied by Hodges and Lehmann (1952) and the subclass of normal estimation problems studied by Efron and Morris (1971).

Our seemingly artificial problem is fundamental to the study of the question: In the large sample estimation of a parameter η in the presence of a nuisance parameter θ, which we believe to be 0, how can we do well when $\eta = 0$ at little expense if we are wrong about η? This question is discussed in Bickel (1982).

The paper is organized as follows. In section II we sketch the nature of the optimal procedures, establish the connection to robust estimation and introduce and discuss reasonable sub-optimal procedures. In sections III and IV we give asymptotic approximations to $\mu(t)$ and δ_t^* for t close to 0 and 1. Proofs here are sketched with technical details reserved for an appendix labeled (A) which is available only in the technical report version of this paper.

II. OPTIMAL AND SUBOPTIMAL PROCEDURES AND THE CONNECTION TO ROBUST ESTIMATION OF LOCATION

For $0 \le \lambda \le 1$ let,

$$M_\lambda(\delta) = (1-\lambda)m(\delta) + \lambda M(0,\delta)$$

$$\rho(\lambda) = \inf_\delta M_\lambda(\delta)$$

and let δ_λ be the estimate which by weak compactness achieves the inf. By standard arguments (see e.g. Neustadt (1976)) $\forall\ 0 < t < 1$ there exists $0 < \lambda(t) < 1$ such that

$$\delta_t^* = \delta_{\lambda(t)}, \qquad \mu(t) = \rho(\lambda(t)). \tag{2.1}$$

Given a prior distribution P on R define the Bayes risk of δ by

$$M(P,\delta) = \int M(\theta,\delta)\, P(d\theta)$$

and its risk,

$$R(P) = \inf_\delta M(P,\delta).$$

By arguing as in Hodges and Lehmann (1952) Thms. 1, 2 and using standard decision theoretic considerations,

$$\begin{aligned}\rho(\lambda) &= \inf_\delta \sup\{M(P,\delta) : P \in P_\lambda\} \\ &= \sup\{R(P) : P \in P_\lambda\}\end{aligned} \tag{2.2}$$

where P_λ is the set of all prior distributions P on $[-\infty, \infty]$ such that $P = (1-\lambda)K+\lambda I$ where K is arbitrary and I is point mass at 0. In fact, there exists a proper least favorable distribution $P_\lambda \in \mathcal{P}_\lambda$ against which δ_λ is necessarily Bayes. The distribution P_λ is unique and symmetric about 0. Unfortunately it concentrates on a denumerable set of isolated points. This fact as well as the approximation theorems which represent the only analytic information we have so far on $\mu(t)$, δ_t^* are related to the "robustness connection" which we now describe.

If ψ denotes functions from R to R, P, F, K probability distributions on $[-\infty, \infty]$ and * convolution let

$$\mathcal{F}_{\lambda o} = \{F : F = P * \Phi,\ P \in \mathcal{P}\}$$

$$= \{F : F = (1-\lambda)K*\Phi + \lambda\Phi,\ K \text{ arbitrary}\}$$

$$I(F) = \int \frac{[f'(x)]^2}{f(x)}\, dx$$

if F has an absolutely continuous density f with derivative f'.

$$= \infty \quad \text{otherwise.}$$

If $I(F) < \infty$ let

$$V(F,\psi) = \int (\psi^2(x)\ f(x) + 2\psi(x)\ f'(x))dx \tag{2.3}$$

if $\int \psi^2(x)\ F(dx) < \infty$

$$= \infty \quad \text{otherwise.}$$

By integration by parts if ψ is absolutely continuous and $\int|\psi'(x)|F(dx) < \infty$

$$V(F,\psi) = \int \psi^2(x)F(dx) - 2\int \psi'(x)F(dx). \tag{2.4}$$

Given δ define

$$\psi(x) = x - \delta(x). \tag{2.5}$$

Then, it is easy to show by direct computation

$$M(P,\delta) = 1 + V(P*\Phi,\ \psi) \tag{2.6}$$

a formula due to Stein (Hudson (1978)) if P is a point mass.

By minimizing (2.6) we get

$$R(P) = 1 - I(P*\Phi) \tag{2.7}$$

achieved when

$$\psi = \frac{-f'}{f}$$

where f is the density of $P*\Phi$, a special case of an identity of Brown (1971).

Standard minimax arguments yield that if $\mathcal{F}$ is any convex weakly closed set of distributions on $[-\infty, \infty]$ with finite Fisher information then

$$V(F_o, \psi_o) = \sup_{\mathcal{F}} V(F, \psi_o) = \inf_{\psi} V(F_o, \psi) = I(F_o)$$

where F_o minimizes I(F) over $\mathcal{F}$ and

$$\psi_o = \frac{-f_o'}{f_o} \tag{2.8}$$

and f_o is the density of F_o. Specializing to $\mathcal{F}_{\lambda o}$ we obtain

$$\rho(\lambda) - 1 = - I(F_{\lambda o}),$$
$$\delta_\lambda(x) = x + \frac{f'_{\lambda o}}{f_{\lambda o}}(x) \tag{2.9}$$

where $F_{\lambda o}$ is the least favorable distribution in $\mathcal{F}_{\lambda o}$ and $f_{\lambda o}$ is its density. The characterization of P_λ we mentioned follows immediately from (2.9) and theorem 2 of Bickel and Collins (1982).

For $\mathcal{F}$ as above, Huber (1964) essentially considered the game (with "Nature" as player I) and payoff (to I),

$$W(F,\psi) = \int \psi^2(x)f(x)dx/(\int\psi(x)\ f'(x)dx)^2.$$

Here ψ is the score function of an (M) estimate and W its asymptotic variance under F. (Huber restricted ψ, for instance to continuously differentiable functions with compact support, redefining the denominator of W to be $\int \psi'(x)\ F(dx)$ and permitting $I(F) = \infty$. But this seems inessential.) Here again the game has a value,

$$W(F_o,\psi_o) = \text{Sup}_{\mathcal{F}}\ W(F,\psi_o) = \inf_{\psi} W(F_o,\psi) = I^{-1}(F_o)$$

where F_o, ψ_o are the same strategies as for the payoff V.

$\mathcal{F}_{\lambda o}$ arose in the context of Huber's game in connection with robust smoothing, Mallows (1978), (1980). He posed the problem of minimizing I(F) for $F \in \mathcal{F}_{\lambda o}$ and conjectured that K_λ corresponding to the optimal P_λ concentrates on $\{kh: k = \pm1,\pm2,\ldots\}$, for some $h > 0$, and assigns mass

$$K_\lambda\{kh\} = \tfrac{1}{2}\lambda(1-\lambda)^{|k|-1} \quad \forall k.$$

As of this writing it appears that this conjecture is false although a modification of D. Donoho in which the support is of the form $\{\pm(a+kh): a,h > 0,\ k = 0,1,\ldots\}$ may be true.

The Efron-Morris Estimates

Let

$$\mathcal{F}_{\lambda 1} = \{F: F = \lambda\Phi + (1-\lambda)G,\ G \text{ arbitrary}\}. \tag{2.10}$$

$\mathcal{F}_{\lambda 1}$ is Huber's (1964) contamination model. As Huber showed, the optimal $F_{\lambda 1}$ has $-\frac{f'}{f}$ of the form

$$\begin{aligned} \psi_m(x) &= x, && |x| \le m \\ &= m \operatorname{sgn} x, && |x| > m. \end{aligned} \tag{2.11}$$

The estimate corresponding to ψ_m in the sense of (2.5) is given by

$$\begin{aligned} \bar{\delta}_m(x) &= 0, && |x| \le m \\ &= x - m \operatorname{sgn} x, && |x| > m. \end{aligned} \tag{2.12}$$

This is a special case of the limited translation estimates proposed by Efron and Morris (1971) as reasonable compromises between Bayes and minimax estimates in the problem of estimating θ when θ has a normal prior distribution. We will call $\bar{\delta}_m$ the E-M estimate. Since $\bar{\delta}_m$ is not analytic it cannot be optimal. Nevertheless it has some attractive features.

The M.S.E. of δ_m is given by

$$M(\theta,\bar{\delta}_m) = 1 + m^2 + (\theta^2-(1+m^2))(\Phi(m+\theta)+\Phi(m-\theta)-1)$$
$$- ((m-\theta)\Phi(m+\theta) + (m+\theta)\Phi(m-\theta)).$$

Since $-2\psi_m' + \psi_m^2$ is an increasing function of $|x|$ we remark, as did Efron and Morris, that $M(\theta,\bar{\delta}_m)$ is an increasing function of $|\theta|$ with

$$M(\bar{\delta}_m) = M(\infty,\bar{\delta}_m) = 1 + m^2.$$

For fixed λ the $m(\lambda)$ which minimizes $M_\lambda(\bar{\delta}_m)$ is the unique solution of the equation

$$2\Phi(m) - 1 + \frac{2\Phi(m)}{m} = \lambda^{-1}. \tag{2.13}$$

This is also the value of m which corresponds to F_{λ_1}. We deduce the following weak optimality property: Let ψ correspond to δ by (2.5) in the following.

$$\mathcal{D}_\infty = \{\delta: E_\theta|\delta'(X)| < \infty, \forall\theta;\ \lim_{|x|\to\infty}[\psi^2(x)-2\psi'(x)]$$
$$= \sup_x[\psi^2(x)-2\psi'(x)]\}$$

$\mathcal{D}_\infty$ is a subclass of estimates which achieve their maximum risk at $\pm\infty$.

Theorem 2.1. If $m(\lambda)$ is given by (2.13) then $\bar{\delta}_{m(\lambda)}$ is optimal in $\mathcal{D}_\infty$, i.e.

$$M_\lambda(\bar{\delta}_{m(\lambda)}) = \min\{M_\lambda(\delta): \delta \in \mathcal{D}_\infty\}.$$

Proof. By (2.6) and (2.4) if $E_\theta|\delta'(X)| < \infty$, $\forall\theta$,

$$M_\lambda(\delta) - 1 = \sup\{V(F,\psi) : F \in \mathcal{F}_{\lambda o}\} \le \sup\{V(F,\psi):$$
$$F \in \mathcal{F}_{\lambda 1}\} \le \sup_x\{\psi^2(x) - 2\psi'(x)\}.$$

These inequalities become equalities for $\delta \in \mathcal{D}_\infty$ by letting $|\theta| \to \infty$ in $M(\theta,\delta)$. The result follows from the optimality property of $\mathcal{F}_{\lambda 1}$.

The Pretesting Estimates

There is a natural class of procedures which are not in $\mathcal{D}_\infty$ and are natural competitors to the E-M estimates. A typical member of this class is given by

$$\tilde{\delta}_m(x) = 0, \qquad |x| \leq m$$

$$= x, \qquad |x| > m.$$

Implicitly, in using $\tilde{\delta}_m$ we test H: $\theta=0$ at level $2(1-\Phi(m))$. If we accept we estimate 0, otherwise we use the minmax estimate of X. We call these pretesting estimates. The ψ function corresponding to $\tilde{\delta}_m$ is of the type known as "hard rejection" in the robustness literature.

Comparison of E-M and Pretesting Estimates

Hard rejection does not work very well--nor do pretesting estimates. Both the E-M and pretesting procedures have members which are approximately optimal for λ close to 1 or what amounts to the same, t close to 1. However, the pretesting procedures behave poorly for λ (or t) close to 0. This is discussed further in sections III and IV. The following table gives the maximum M.S.E. of $\bar{\delta}$ and $\tilde{\delta}$ which have M.S.E. equal to 1-t at 0 as a function of t. The E-M rules always do better for the values tabled, spetacularly better in the ranges of interest. This is consistent with results of Morris et al. (1972) who show that Stein type rules render pretesting type rules inadmissible in dimension 3 or higher.

Notes:

(1) The connection between restricted minmax and more generally restricted Bayes and robust estimation was independently discovered by A. Marazzi (1980).

(2) Related results also appear in Berger (1982). His approach seems related to that of Hampel in the same way as ours is to that of Huber.

TABLE I. Maximum M.S.E. and Change Point m as a Function of M.S.E. at 0 for $\bar{\delta}_m$ and $\tilde{\delta}_m$

$M(0,\delta)$	$M(\delta)$		m	
	E-M	Pretest	E-M	Pretest
.1	2.393	3.626	1.180	2.500
.2	1.756	2.839	.869	2.154
.3	1.452	2.383	.672	1.914
.4	1.275	2.058	.525	1.716
.5	1.164	1.805	.405	1.538
.6	1.092	1.597	.303	1.367
.7	1.046	1.418	.215	1.193
.8	1.018	1.262	.137	1.002
.9	1.004	1.124	.065	.728
1.0	1.000	1.000	.000	.000

III. THE BEHAVIOUR OF $\mu(t)$ FOR SMALL t

Let

$$\Delta(t) = \mu(t) - 1$$

Theorem 3.1. As $t \to 0$

$$\Delta(t) = o(t^2) \tag{3.1}$$

but

$$t^{-(2+\varepsilon)}\Delta(t) \to \infty \tag{3.2}$$

for every $\varepsilon \to 0$.

Notes:

(1) The E-M rule $\bar{\delta}_m$ with $M(0,\bar{\delta}_m) = 1\text{-}t$ has $M(\bar{\delta}_m) = = 1 + \frac{\pi}{2} t^2 + o(t^2)$. However our proof of (3.2) suggests that the asymptotic improvement over this rule is attainable only by very close mimicking of the optimal rule. This does not seem worthwhile because of the oscillatory nature of the optimal rule.

(2) The pretesting rule $\tilde{\delta}_m$ with $M(0,\tilde{\delta}_m) = 1\text{-}t$ has $M(\tilde{\delta}_m) = 1 + \Omega(t)$. This unsatisfactory behaviour is reflected in Table I.

We need

Lemma 3.1. Let $\lambda(t)$ be as in (2.1). Then λ is continuous.

Proof. By the unicity of δ_λ and weak compactness, $M(0,\delta_\lambda)$ is continuous and strictly decreasing in λ on $[0,1]$. The lemma follows.

Lemma 3.2. As $\lambda \to 0$

$$\lambda^{-2}(1-\rho(\lambda)) \to \infty. \tag{3.3}$$

Lemma 3.3. As $\lambda \to 0$

$$\lambda^{-2+\varepsilon}(1-\rho(\lambda)) \to 0 \tag{3.4}$$

for every $\varepsilon > 0$.

Proof of Theorem 3.1 from Lemmas 3.1-3.3

Claim 3.1. For any sequence $t_k \to 0$ let $\lambda_k = \lambda(t_k)$ so that,

$$1 - \rho(\lambda_k) = \lambda_k t_k - (1-\lambda_k)\Delta(t_k). \tag{3.5}$$

Then, by lemma 3.1, $\lambda_k \to 0$ and

$$\begin{aligned}\lambda_k^{-2}(1-\rho(\lambda_k)) &\le \lambda_k^{-2} \max_x\{\lambda_k x - (1-\lambda_k)\frac{\Delta(t_k)}{t_k^2} x^2\} \\ &= 0(t_k^2/\Delta(t_k)).\end{aligned} \tag{3.6}$$

By (3.3), $\frac{t_k^2}{\Delta(t_k)} \to \infty$ and (3.1) follows.

Claim 3.2. Note that

$$1 - \rho(\lambda) \ge \max_t[\lambda t - (1-\lambda)\Delta(t)] \tag{3.7}$$

If $\Delta(t_k) \le C\, t_k^{2+\varepsilon}$ for some $C \le \frac{1}{2}$, $\varepsilon > 0$, $t_k \to 0$ put $\lambda_k = t_k^{1+\varepsilon}$ to get

$$1 - \rho(\lambda_k) \ge t_k^{2+\varepsilon}(1+o(1)) \ge \lambda_k^{2-\frac{\varepsilon}{1+\varepsilon}}(1 + o(1))$$

a contradiction to (3.4).

Proof of Lemmas 3.1-3.3

The proof proceeds via several sublemmas.

Lemma 3.4. Let $\{\nu_n\}$ be a sequence of Bayes prior distributions on R and let δ_n be the corresponding Bayes estimates. Suppose that, as $n \to \infty$,

$$M(\delta_n) \to 1. \tag{3.8}$$

Then

$$\delta_n(x) \to x \quad \text{a.e.} \tag{3.9}$$

$$\frac{\nu_n(I_1)}{\nu_n(I_2)} \to 1 \tag{3.10}$$

for any pair of intervals I_1, I_2 of equal length.

Proof. By Sacks' (1963) theorem, there exists a subsequence $\{n_k\}$ such that $\{\delta_{n_k}\}$ converge regularly to δ and

$$\delta_{n_k}(x) \to \delta(x) \quad \text{a.e.}$$

where δ is generalized Bayes with respect to ν (σ finite) such that for some sequence $\{a_k\}$

$$\nu_{n_k}/\nu_{n_k}(-a_k, a_k) \to \nu \tag{3.11}$$

weakly. But, by (3.8) and regular convergence $M(\delta) \leq 1$ and hence δ is minmax. Therefore, $\delta(x) = x$ a.e. and (3.9) follows by Sacks' theorem. Since δ is generalized Bayes with respect to ν, ν must be proportional to Lebesgue measure and (3.10) follows from (3.11).

Lemma 3.5. If $\lambda \to 0$, and $P_\lambda = (1-\lambda)K_\lambda + \lambda I$

$$\frac{1}{\lambda}\int_{-\infty}^{\infty} \phi(\theta)K_\lambda(d\theta) \to \infty.$$

Proof. Since $M(\tilde{\delta}_\lambda) \to 1$

$\{P_\lambda\}$ satisfies (3.10) as $\lambda \to 0$.

Therefore for all $a > 0$

$$\frac{P_\lambda[0,a)}{P_\lambda(0,a)} \to 1.$$

Hence,

$$\lambda = o(K_\lambda(0,a)). \tag{3.12}$$

By the same argument

$$\frac{K_\lambda(0,a)}{K_\lambda(0,1)} \to a, \ a \leq 1$$

and hence

$$[K_\lambda(0,1)]^{-1} \int_0^1 \phi(\theta)K_\lambda(d\theta) \to \int_0^1 \phi(\theta)d\theta. \tag{3.13}$$

The lemma follows from (3.12) and (3.13).

Proof of Lemma 3.2. We compute the Bayes risk of a reasonable E-M estimate, viz. $\bar{\delta}_\lambda$. We claim that

$$\lambda^{-2}(M(P_\lambda,\bar{\delta}_\lambda) - 1) \to -\infty \tag{3.14}$$

Since $\rho(\lambda) = M(P_\lambda) \le M(P_\lambda,\bar{\lambda})$ the lemma will follow. To prove (3.14) apply Stein's formula to get

$$M(P_\lambda,\bar{\delta}_\lambda) - 1 = \int_{-\infty}^{\infty} E_\theta(\bar{\delta}_\lambda(X)-X)^2 P_\lambda(d\theta)$$

$$+ 2\int_{-\infty}^{\infty} E_\theta(\bar{\delta}_\lambda(X) - 1)P_\lambda(d\theta).$$

The expression in (3.15) is bounded by

$$\lambda^2 - 2\int_{-\infty}^{\infty} [\Phi(\lambda-\theta) - \Phi(-\lambda-\theta)]P_\lambda(d\theta)$$

$$\le \lambda^2 - 2(1-\lambda)\int_{-\infty}^{\infty} [\Phi(\lambda-\theta) - \Phi(-\lambda-\theta)]K_\lambda(d\theta).$$

Since $\Phi(\lambda-\theta) - \Phi(-\lambda-\theta) \ge \lambda\phi(\theta)$ for $\lambda \le \sqrt{2 \log 2}$ we can apply Lemma 3.5 to conclude that

$$\lambda^{-2}\int_{-\infty}^{\infty} [\Phi(\lambda-\theta) - \Phi(-\lambda-\theta)]K_\lambda(d\theta) \to \infty$$

and claim (3.14) and the lemma follow.

We sketch the proof of Lemma 3.3. Details are available in (A).

Proof of Lemma 3.3. It suffices for each $\varepsilon > 0$ to exhibit a sequence of prior distributions $\bar{P}_\lambda$ such that

$$\lambda^{-2+\varepsilon}(1 - R(\bar{P}_\lambda)) \to 0. \tag{3.16}$$

By Brown's identity, claim (3.16) is equivalent to

$$\int \frac{[f'_\lambda]^2}{f_\lambda} = o(\lambda^{2-\varepsilon}) \tag{3.17}$$

for f_λ the density of $\bar{P}_\lambda * \Phi$. Here is the definition of $\bar{P}_\lambda$. Let

$$e_\tau(x) = \frac{1}{\pi\tau}(1 + (\frac{x}{\tau})^2)^{-1}.$$

Write ϕ_σ, (Φ_σ) for the normal $(0,\sigma^2)$ density (d.f.). Given $k \geq 1$, let h be a (C^∞) function from R to R such that

$$|h(x)| \leq c_r(1 + |x|^r)^{-1} \text{ for all } r > 0,\ x, \text{ some } c_r. \tag{3.18}$$

$$\int_{-\infty}^{\infty} h(x)dx = 1$$

$$\int_{-\infty}^{\infty} x^j h(x)dx = 0, \quad 1 \leq j \leq 2k-1.$$

An example of h satisfying these conditions is

$$h(x) = \frac{1}{2\pi i} \int_{-\infty}^{\infty} \exp\{-itx-t^{2k}\}dt.$$

Let

$$h_\sigma(x) = \frac{1}{\sigma} h(\frac{x}{\sigma}).$$

Set, for c_k in (3.18),

$$\sigma = 2\pi \lambda m c_k, \tag{3.19}$$

and

$$m = \lambda^{-(2k+2)/(2k+3)}. \tag{3.20}$$

If $\overline{P}_\lambda = (1-\lambda) L_\lambda + \lambda I$ define $\overline{P}_\lambda$ by the density of L_λ,

$$\ell_\lambda = (e_m - \lambda h_\sigma)(1-\lambda)^{-1}$$

for $\lambda \leq 1$ and $\sigma \leq 1$. By construction L_λ is a probability measure (see (A)).

Let

$$g_\lambda = e_m * \phi$$

$$q_\sigma = \phi - h_\sigma * \phi$$

Then,

$$\int \frac{[f_\lambda']^2}{f_\lambda} = \int \frac{[g_\lambda']^2}{f_\lambda} + 2\lambda \int \frac{g_\lambda' q_\sigma'}{f_\lambda} + \lambda^2 \int \frac{[q_\sigma']^2}{f_\lambda}.$$

It is shown in (A) that,

$$\int \frac{[g_\lambda']^2}{f_\lambda} = O(m^{-2}) \tag{3.21}$$

$$\int \frac{g'_\lambda q'_\sigma}{f_\lambda} = 0(m^{-1}) \tag{3.22}$$

$$\int \frac{[q'_\sigma]^2}{f_\lambda} = 0(m\sigma^{2k}). \tag{3.23}$$

We combine (3.21) - (3.23) to get

$$\int \frac{[f'_\lambda]^2}{f_\lambda} = 0(m^{-2}+\lambda m^{-1}+m\lambda^2\sigma^{2k}) = 0(\lambda^{2-2(k+3)^{-1}}).$$

Since k is arbitrary we have proved (3.17) and the lemma.

The theorem is proved.

IV. THE BEHAVIOUR OF $\mu(t)$ FOR t CLOSE TO 1

We sketch the proof of,

Theorem 4.1. As $t \to 1$

$$\mu(t) = 2|\log(1-t)|(1+o(1)). \tag{4.1}$$

If $\delta_t \in \mathcal{D}_t$ and

$$\delta_t(x) = 0, \quad |x| \le g(t)$$

$$\sup\{|\delta_t(x)-x|: \quad |x|>g(t)\} = o(g(t)) \tag{4.2}$$

$$g(t) = \sqrt{2|\log(1-t)|}(1 + o(1)).$$

then

$$M(\delta_t) = \mu(t)(1 + o(1)). \tag{4.3}$$

Note. It is easy to see that both E-M and pretest estimates which are members of $\mathcal{D}_t$ satisfy (4.2) and are optimal in this sense. The approximation (4.1) is thus crude and not practically useful.

Lemma 4.1. As $\lambda \to 1$

$$\rho(\lambda) = 2(1-\lambda)|\log(1-\lambda)|(1+o(1)). \tag{4.4}$$

Moreover if $\{\delta_\lambda\}$ is any sequence of estimates such that

$$\begin{aligned} \delta_\lambda(x) &= 0, \quad |x| \le c(\lambda) \\ |\delta_\lambda(x)-x| &\le b(\lambda), \quad |x| > c(\lambda) \end{aligned} \tag{4.5}$$

where

$$\begin{aligned} c(\lambda) &= [2|\log(1-\lambda)|]^{1/2}(1+o(1)) \\ b(\lambda) &= o(c(\lambda)) \end{aligned} \tag{4.6}$$

then

$$M(0,\delta_\lambda) = \frac{2}{\sqrt{\pi}}(1-\lambda)|\log(1-\lambda)|^{1/2}(1+o(1)) \tag{4.8}$$

$$M(\delta_\lambda) = 2|\log(1-\lambda)|(1+o(1)) \tag{4.9}$$

and hence

$$M_\lambda(\delta_\lambda) = \rho(\lambda)(1+o(1)).$$

Proof. We establish the lemma by

(i) For every $\gamma > 0$ exhibiting $\tilde{P}_\lambda$ such that

$$R(\tilde{P}_\lambda) \ge 2(1-\lambda)|\log(1-\lambda)|(1-\gamma)(1+o(1)).$$

(ii) Showing that δ_λ given in (4.5) satisfy (4.8) and

$$M(\delta_\lambda) \le 2|\log(1-\lambda)|(1+o(1)).$$

Since, by (4.8),

$$M(0,\delta_\lambda) = o((1-\lambda)|\log(1-\lambda)|)$$

and

$$R(\tilde{P}_\lambda) \le \rho(\lambda) \le (1-\lambda)M(\delta_\lambda) + \lambda M(0,\delta_\lambda)$$

the lemma will follow. Here is $\tilde{P}_\lambda$. Let,

$$\varepsilon = 1 - \lambda \tag{4.10}$$

$$a = a(\varepsilon) = \sqrt{|2\log\varepsilon|(1-\gamma)}, \quad \gamma > 0 .$$

Let $\tilde{P}_\lambda$ put mass $\frac{\varepsilon}{2}$ at $\pm a$, and λ at 0. The calculations establishing (i) and (ii) are in (A).

Proof of Theorem 4.1. Putting $\lambda = t$ we must have,

$$\rho(t) \leq (1-t)(\mu(t)+t).$$

Therefore, by Lemmas 3.1 and 4.1, as $t \to 1$,

$$\mu(t) \geq |2 \log(1-t)|(1+o(1)).$$

By (4.8) and (4.9) we can find members of $\mathcal{D}_t$ with maximum risk $|2 \log(1-t)|(1+o(1))$ and the theorem follows.

V. ACKNOWLEDGMENT

A brief but stimulating conversation with P. J. Huber was very helpful.

VI. REFERENCES

Berger, J. (1982). Estimation in Continuous Exponential Families: Bayesian Estimation Subject to Risk Restrictions and Inadmissibility Results. Statistical Decision Theory and Related Topics III. S. Gupta and J.O. Berger Eds. Academic Press, New York.

Bickel, P.J. (1982). Parametric Robustness and Prestesting. Submitted to *J. Amer. Statist. Assoc.*

Bickel, P.J. and Collins, J. (1982). "Minimizing Fisher Information Over Mixtures of Distributions." *Sankhya*, to appear.

Brown, L.D. (1971). "Admissible Estimators, Recurrent Diffusions and Insoluble Boundary Value Problems." *Ann. Math. Statist. 42*, 855.

Efron, B. and Morris, C. (1971). "Limiting the Risk of Bayes and Empirical Bayes Estimates: Part I. The Bayes Case." *J. Amer. Statist. Assoc. 66*, 807.

Hodges, J.L. and Lehmann, E.L. (1952). "The Use of Previous Experience in Reaching Statistical Decisions." *Ann. Math. Statist. 23*, 396-407.

Huber, P.J. (1964). "Robust Estimation of a Location Parameter." *Ann. Math. Statist. 35*, 73.

Hudson, H.M. (1978). "A Natural Identity for Exponential Families with Applications in Multiparameter Estimation." *Ann. Statist. 6*, 473.

Mallows, C.L. (1978). Problem 78.4. *S.I.A.M. Review.*

Marazzi A. (1980). Robust Bayesian Estimation for the Linear Model. Tech. Report E.T.H. Zürich.

Morris, C.N., Radhadrishnan, R. and Sclove, S.L. (1972). "Non-optimality of Preliminary Test Estimators for the Mean of a Multivariate Normal Distribution." *Ann. Math. Statist. 43,* 1481.

Neustadt, L.W. (1976). *Optimization.* Princeton University Press.

Port, S. and Stone, C. (1974). "Fisher Information and the Pitman Estimator of a Location Parameter." *Ann. Statist. 2,* 25.

Sacks, J. (1963). "Generalized Bayes Solutions in Estimation Problems." *Ann. Math. Statist. 34,* 751.

SOME NEW DICHOTOMOUS REGRESSION METHODS

William DuMouchel

Statistics Center
Massachusetts Institute of Technology
Cambridge, Massachusetts

Christine Waternaux

Biostatistics Department
Harvard School of Public Health
Boston, Massachusetts

I. INTRODUCTION

This paper illustrates some new methods in dichotomous regression. Taken together, they increase the power of such models and make them easier to fit and interpret. Our context is the prediction of a dichotomous variable Y, based on the row vector $\underset{\sim}{X}$, which is conditioned upon and taken as fixed. We do not consider here the situation in which Y is determined by design and one observes the distributions of $\underset{\sim}{X}$ given Y, as happens, e.g. in an epidemiology case-control study.

The function $P = P(Y = 1|\underset{\sim}{X})$ will be called the regression of Y on X and we discuss the fitting of models $P(Y=1|\underset{\sim}{X}) = F(\underset{\sim}{X}\beta)$, where F is a cumulative distribution function and $\underset{\sim}{X}\beta$ is a linear function of $\underset{\sim}{X}$. For fixed F, the column vector β is estimated by maximum likelihood. An analysis of longitudinal medical data

ISBN 0-12-589320-5

from critically ill patients in a surgical intensive care unit (ICU) illustrates the techniques. Section II describes the data and discusses the models used. Section III discusses the analogy between dichotomous regression and normal-theory ordinary least squares (OLS) regression, and introduces a statistic, the likelihood-equivalent error rate, which has R^2-like properties. Section IV uses this statistic in a method of assessing goodness of fit. Section V assesses the adequacy of asymptotically normal approximations upon which the data analysis depends. Finally, Section VI discusses the importance of comprehensive, user-oriented computer output as an aid to interpretation.

II. MODELS FOR PREDICTING THE OUTCOME OF INTENSIVE CARE

Data from 156 patients who entered an ICU following surgery are used to develop a method for identifying those critically ill patients who are most likely to fail to recover in spite of all the resources and support provided by the intensive care unit. Dr. David Cullen, of the Department of Anaesthesia at the Massachusetts General Hospital, collected data from approximately 200 patients who were admitted to the Gray Recovery Room between March 1977 and December 1978. Patients entered the study if they were classified as class IV critically ill patients (requiring constant physician and nursing care). Values of clinical indicators such as platelet counts, state of consciousness, etc., and of the intensity of care were recorded on each of days 1, 2, 3, 4, 7, and 14 after entry of the study (Cullen et al, 1982).

Only patients who remained in intensive care until the measurements at day 2 are considered for this analysis. We here concentrate on predicting short-term outcome, defined as the outcome of the time period until the next scheduled data collection, using as predictors the data collected in the two most recent observation periods. We define 5 possible time intervals for each patient as the time periods between successive measurements. Interval one is the time between measurements on days 2 and 3, and so forth. Interval four is the time between the measurements on days 7 and 14, and interval five is the time from day 14 until final follow-up within a year from first admission to ICU. During the first four intervals, "failure" is defined as death, but for patients who enter the last interval, "failures" include those who at final follow-up are still severely disabled, and excludes those who die from causes other than those which brought them to the ICU. Patients who recover enough during an interval to be removed from the ICU do not contribute data to later intervals. Table I lists the variables used in the present analysis and describes their marginal distributions. Initially, we combine the data from all intervals and consider the sample size to be N = 481 patient-intervals in the ICU.

Choosing a Functional Form for F

Conditional on the past values of $\underset{\sim}{X}$, the outcomes of each patient-interval are assumed to be independent Bernoulli trials with probability of failure $P = F(\underset{\sim}{X}\beta)$. The inverse cumulative, F^{-1}, has been called the "link" function (Nelder and Wedderburn, 1972) because it links P and $\underset{\sim}{X}$, in the sense of specifying that $F^{-1}(P)$ is linear in $\underset{\sim}{X}$. The logistic link, $F^{-1} = \log P/(1-P)$ is the "natural" link for the binomial distribution, since it defines the natural parameter when the binomial is viewed as a member of the exponential family of distributions. But this mathematical statistics argument should not preclude the investigation of other models in applications. For example, with these longitudinal data, assuming that the log of the hazard rate is constant within each interval and equal to $\underset{\sim}{X}\beta$ yields the link $F^{-1} = \log(-\log(1-P))$. The equivalence between similar survival models and log-linear models is discussed by

TABLE I. Description of Variables (N = 481)

OUTCOME (of interval)		*1 = Failure*	*(18%)*
		0 = No Failure	*(82%)*
INTERVAL		*1 = Day 2*	*(32%)*
		2 = Day 3	*(26%)*
		3 = Days 4-7	*(22%)*
		4 = Days 8-14	*(14%)*
		5 = Days 15+	*(6%)*
SOC	*State of Consciousness*		
		1 = awake	*(70%)*
		2 = obtunded	*(25%)*
		3 = coma	*(5%)*
DIAL		*1 = on dialysis*	*(7%)*
		2 = no dialysis	*(93%)*
LOGURIN	*$\log_e$ urine output in ml. (= 0 if urine = 0)* ($\bar{x} = 6.9$, $s = 1.5$)		
BIL	*Bilirubin test* ($\bar{x} = 6.6$, $s = 9.7$)		
BASE	*Base excess* ($\bar{x} = 3.1$, $s = 4.3$)		
DPLAT	*Increase in Platelet count in thousands since previous measurement* ($\bar{x} = -0.2$, $s = 75.6$)		
DTIST	*Increase in Total TISS points (a measure of ICU resource utilization) since previous measurement.* ($\bar{x} = -2.5$, $s = 8.7$)		

Laird and Olivier (1981). Table II lists the seven models which were considered for this analysis, together with rationales supporting the use of each. None of these rationales are compelling enough for the adoption of a particular functional form, but each rationale has enough appeal to warrant an exploratory fit, provided that the appropriate computer programs are available.

Table III displays the asymptotically normal test statistics, $\hat{\beta}/\hat{SE}(\hat{\beta})$, for each of the estimated coefficients in each of the models of Table II. The estimated standard errors are computed from the Fisher information matrices. The five coefficients for INTERVAL are constrained so that their weighted average is zero, as are the three SOC coefficients. The Table also presents the likelihood ratio chi-squared statistic for testing the hypothesis that $P(Y = 1|\underset{\sim}{X})$ depends only on INTERVAL. Since $\chi^2(8df) > 65$ in all seven models, there is no doubt that the clinical predictors have a strong effect. The final row of Table III shows the relative maximized likelihood for each model, taking the logit model's value as 1.

Table III shows that the models agree well on the relative significance of the predictors. The log, probit, and log hazard rate models agree especially well with each other. Five of the seven χ^2 values are in the range $74.6 \leq \chi^2 \leq 76.5$, while the Cauchy model ($\chi^2 = 65.6$) fits poorly and the linear model ($\chi^2 = 80.3$) fits best by the criterion of χ^2. We do not have a formal procedure for choosing among models. Pregibon (1980) proposes a method for estimation of the link function in a more restricted context. A Bayesian who is willing to assume 7 identical vague prior distributions for β, conditional on each model, could obtain a Bayes factor for each model by

TABLE II. Models Used

Model	link function: $F^{-1}(P)=X\beta$	Rationale(s)
logit	$\log P/(1-P)$	most common; simple sufficient statistics; numerically stable; related to discriminant analysis.
probit	$\Phi^{-1}(P)$	historical; assumes a normal distribution of individual tolerances.
linear	P	simplicity; MLE version of OLS; ease of interpretation of coefficients.
log hazard rate	$\log(-\log(1-P))$	survival curve interpretation: the hazard rate is $\exp(X\beta)$. Also called complementary log-log or extreme value model.
Cauchy	$\tan \pi(P-1/2)$	robustness; the heavy tails of the Cauchy distribution diminish the influence of extreme X-values.
complementary log	$-\log(1-P)$	simple multiplicative model for survival rate.
log	$\log P$	simple multiplicative model for failure rate.

TABLE III. Ratios $\hat{\beta}/\hat{SE}(\hat{\beta})$ for the Seven Models of Table II

VARIABLE		MODEL					
	Logit	Probit	Linear	log hazard	Cauchy	-log(1-P)	log P
CONSTANT	1.7	1.5	1.0	1.8	3.2	4.9	0.3
INTERVAL = 1	-3.4	-3.7	-8.3	-3.4	-2.6	-5.6	-3.4
2	-3.0	-3.3	-7.3	-2.8	-2.2	-5.0	-2.6
3	1.9	1.9	1.4	1.9	1.5	.8	1.9
4	5.4	5.5	5.0	5.8	4.5	4.0	6.8
5	6.4	6.4	9.4	7.3	5.1	3.6	9.2
SOC = 1	-1.4	-1.4	-2.6	-1.3	-1.4	-1.7	-1.4
2	.3	.2	- .2	.2	.6	- .2	.0
3	2.7	2.9	3.0	2.7	1.9	2.7	2.7
DIAL = 1	-2.4	-2.3	-2.5	-2.8	-3.3	-2.1	-3.3
LOGURIN	-4.0	-3.9	-4.4	-4.2	-4.7	-3.6	-4.4
BIL	2.8	3.0	4.4	2.9	2.1	3.3	3.0
BASE	-3.8	-3.7	-4.6	-4.2	-3.8	-3.4	-5.2
DPLAT	-2.3	-2.3	-2.5	-2.4	-1.9	-1.3	-3.5
DTIST	2.3	2.6	4.4	2.4	1.9	4.3	3.0
χ^2 (8df)[a]	74.7	76.5	80.3	75.0	65.6	74.7	74.6
relative maximized likelihood[b]	1	2.5	16.	1.2	.011	1.0	.95

[a] Likelihood ratio statistic to test whether P depends only on INTERVAL.
[b] Taking Logit as standard, relative maximized likelihood = exp $\frac{1}{2}(\chi^2 - 74.7)$.

integrating out the respective likelihood functions in 13 dimensions. We make our choice based on the following considerations: (a) The log hazard rate model is *a priori* appealing for this data since we are in effect modeling a survival curve.[1] (b) Only the linear model leads to a much higher maximized likelihood function than the log hazard model. (c) The linear model allows predicted probability of failures equal to zero or one. In fact, out of 481 cases, 136 have $\hat{p} = 0$, while 4 have $\hat{p} = 1$, where $\hat{p} = F(\underset{\sim}{X}\hat{\beta}) = \max(0, \min(1, \underset{\sim}{X}\hat{\beta}))$ for the linear model. It does not make sense to claim that so many ICU patients were certain to survive the next interval. In constrast, every $\hat{p} \geq .010$ for the log hazard model, although even here 5 cases have $\hat{p} \geq .9998$. But there are often cases with poor prognosis in the ICU; the main purpose of this study is to learn to identify them. (One price which the Cauchy model pays for its presumed robustness in an inability to make extreme predictions. All of the Cauchy fitted values satisfied $.041 \leq \hat{p} \leq .942$.) (d) Finally, Section 5 will cast doubt on the estimated standard errors and the asymptotic normality of $\hat{\beta}$ for the linear model applied to this data.

In the rest of this paper, we will illustrate our techniques using the log hazard model: $P(Y = 1|\underset{\sim}{X}) = 1 - \exp\{-\exp(\underset{\sim}{X}\beta)\}$.

III. ANALOGIES TO NORMAL REGRESSION AND R^2.

This section discusses analogies between dichotomous regression and the ordinary least squares (OLS) regression based on normally distributed residuals. In OLS, the MLE of β is found by minimizing $\sum(Y - X\beta)^2$, while dichotomous regression

[1] *In particular the effect of having intervals of varying lengths can be easily represented: if the kth interval has length Δ_k, then the log of the hazard rate during this interval is $\underset{\sim}{X}\beta - \log \Delta_k$.*

maximizes the log-likelihood

$$L(\beta) = \sum_{i=1}^{n} [(Y_i = 1)\log p_i + (Y_i = 0)\log(1 - p_i)], \text{ where}$$

$p_i = P(Y_i = 1 | \underset{\sim}{X} = \underset{\sim}{X}_i, \beta)$, and $(Y_i = y)$ is the indicator function of the corresponding event. However, interpreting the maximized value $L(\hat{\beta})$ is problematic. The OLS statistic $\hat{\sigma}_e = [N^{-1}\sum(Y - X\hat{\beta})^2]^{\frac{1}{2}}$ estimates the standard deviation of the prediction error, an important parameter for interpreting the normal model. In the case of the binomial distribution, the prediction error variance, $p(1-p)$ is less interesting since the distribution of Y is already completely specified by p. The quantity $-2L(\hat{\beta})$ has been called the *deviance* (see, e.g. Baker and Nelder (1978)) and is often used in likelihood ratio chi-squared tests. One attempt to form a descriptive statistic from the dichotomous regression deviance is that of Goodman (1970) who defines an R^2-type statistic $R_G^2 = [L(\hat{\beta})-\hat{L}_0]/[\hat{L}_2-\hat{L}_0]$. In this expression, $\hat{L}_0$ is the log-likelihood function maximized under the constraint that every $p_i = p$, while $\hat{L}_2$ is the maximum of the log-likelihood defined by the "saturated model", which is an extended model $P(Y = 1|\underset{\sim}{X}) = F(\underset{\sim}{X}\beta + \underset{\sim}{Z}\gamma)$, where $\underset{\sim}{Z}\gamma$ represents a large number, say m_2, of interaction terms involving the components of $\underset{\sim}{X}$. However, if one assumes that the original model with $\gamma = 0$ holds in the population, but that Y and $\underset{\sim}{X}$ are not independent, then it is easily shown that $R_G^2 \to 1$ as $N \to \infty$ with probability one, so that R_G^2 is not estimating an interesting population parameter.

Efron (1978) also considers general measures of residual variation for dichotomous data. His emphasis is on the mathematical properties such measures should possess, while our emphasis is on the intuitive interpretation of such measures.

For Bernoulli trials with constant probability p of sucess, p itself is perhaps the most intuitive measure of predictive error. The quantity min(p, 1-p) represents the minimum error rate for predicting Y, achievable by always predicting Y = 1 if $p > \frac{1}{2}$, and always predicting Y = 0 if $p \leq \frac{1}{2}$. When the p_i differ, the quantity $N^{-1}\sum_i \min(p_i, 1-p_i)$ represents the smallest error rate that can be expected, and the quantity $N^{-1}\sum_i (Y_i = 1)(P_i \leq \frac{1}{2}) + (Y_i = 0)(P_i > \frac{1}{2})$ (the quantities in parentheses are indicator functions) represents the misclassification rate for this sample. There is an extensive literature on the estimation of misclassification error rates; see c.g. Hills (1966). However, if these error rates are estimated by substituting $\hat{p}_i = P(Y = 1|X_i, \hat{\beta})$ in one of the above expressions, a desirable analogy to normal regression fails: choosing $\hat{\beta}$ to maximize the likelihood does not necessarily minimize the measure of predictive error. We next define a statistic, the likelihood-equivalent error rate (LE error rate) which does have this feature.

First, let $L_0(\hat{p}) = N\hat{p}\log \hat{p} + N(1-\hat{p})\log(1-\hat{p})$ be the *maximized* simple binomial log-likelihood function from a sample with observed proportion $\hat{p}$.

The (estimated) LE error rate in a dichotomous regression is defined as the unique solution, $0 \leq \hat{\alpha} \leq \frac{1}{2}$, to

$$L_0(\hat{\alpha}) = L(\hat{\beta});\ \text{or}$$

$$N\hat{\alpha}\log\hat{\alpha} + N(1-\hat{\alpha})\log(1-\hat{\alpha}) = \sum_i[(Y_i=1)\log \hat{p}_i + (Y_i=0)\log(1-\hat{p}_i)]. \tag{1}$$

Thus $\hat{\alpha}$ may be interpreted as the error rate (estimated to be) achievable in a dichotomous population without predictors having the same entropy as this sample conditioning on the predicators. For example, the ICU data has proportion .181 failures (Y = 1) and proportion .819 successes (Y = 0), so the marginal or null hypothesis error rate is α_0 = min(.181, .819) = .181. When the $\hat{\beta}$ for the log hazard model (model 4 of Table II) is the MLE, the solution to (1) is $\hat{\alpha}$ = .099. If the size of the likelihood function is taken as the measure of predictive power, then the use of these predictors reduces the error rate from that of an 18-82 distribution to that of a 10-90 distribution, a reduction of R^2_α = (18.1 - 9.9)/18.1 = 45%. No other estimate of β would lead to a lower LE error rate than $\hat{\beta}$ does, nor to a higher R^2_α, the percent of LE error rate explained by $\underset{\sim}{X}$.

Just as $N^{-1} \sum (Y - \underset{\sim}{X}\hat{\beta})^2$ is a biased estimate of σ^2_e in a normal regression, so $\hat{\alpha}$ and R^2_α are unduly optimistic estimates here, especially if m, the degrees of freedom in β, is large. The usual OLS correction, replacing N^{-1} by $(N-m)^{-1}$ in the formula for $\hat{\sigma}^2_e$, has a counterpart for dichotomous regression. It would be preferable to base $\hat{\alpha}$ on $L(\beta_0)$ rather than $L(\hat{\beta})$, where β_0 is the true value of β. Since $2(L(\hat{\beta}) - L(\beta_0))$ has an approximate χ^2 distribution, with mean m, an *adjusted* LE error rate, α_{adj}, can be defined by the solution to: $L_0(\alpha_{adj}) = L(\hat{\beta}) - m/2$. (Take $\alpha_{adj} = \frac{1}{2}$ if this equation has no solution, i.e. if $L(\hat{\beta}) < N \log \frac{1}{2} + m/2$.)

Our ICU data analysis yields the following results for the log hazard rate model:

Hypothesis	*df*	*LE error rate*	*"adjusted" LE error rate*
H_0: *Y, X independent*	*1*	*.181*	*.182*
H_1: *INTERVAL only*	*5*	*.138*	*.141*
H_2: *Full model*	*13*	*.099*	*.105*

The full model (using all the variables in Table I) explains (.181 - .099)/.181 = 45% of the LE error rate under H_0, but adjusted for degrees of freedom the percentage is (.185 - .105)/.185 = 42%. For comparison, the sum of squared residuals for H_0 and H_2 are 71.3 and 48.1 respectively, a 33% reduction in sum of squares. We can also define a "partial" R^2_α. For example, the medical predictors explain 28% (=(.138 - .099)/.138) of the LE error rate under H_1, or 26% after adjusting for degrees of freedom.

IV. A SCREENING PROCEDURE FOR GOODNESS OF FIT

This section describes a screening procedure which is useful for detecting ill-fitting regions of the design space. Before describing the procedure, we mention problems with some of the common methods for measuring goodness of fit to dichotomous regressions.

If each X-variable takes on only a few discrete values, then the standard methods of analysis of contingency tables may be used, (e.g. Bishop, Fienberg, and Holland 1974) in which Pearson or likelihood ratio goodness of fit test is performed. In addition, one can define residuals based on the observed and expected count for each cell, and examine them in an exploratory manner. But these methods are not applicable if some of the X's are continuous or if the sample size is such that most of the cells have 0 or 1 observation, as is common if there are many predictors.

Alternatively, by analogy with stepwise regression, one can fit many larger models, each formed from the one under investigation by the addition of one or a few interaction variables, etc. But fitting so many models, each requiring the use of an iterative maximization algorithm in many dimensions, is time consuming, and one would like guidance as to which variables are most likely to need transformations or be involved in interaction terms.

Data analysis should routinely include graphical methods. Although they may still be useful, the standard diagnostic graphics of normal theory regression - histograms of residuals, plots of residuals versus fitted values or versus the predictors, and the newer techniques of Belsley, Kuh, and Welsch (1980) - do not transfer very well to the dichotomous regression situation, especially when the fitted p_i are near 0 or 1. The extreme skewness of the dichotomous residuals, and the fact that this skewness changes with the values of the p_i, make the plots hard to interpret. Also, as shown in Pregibon (1981) the "influence" of observation i having $Y_i = 1$ looks large whenever the fitted p_i is small, even though the position of point i in the $\underset{\sim}{X}$-space is not extreme or unusual in the sample.

Finally, some goodness of fit tests compare observed and expected counts of Y = 1 across a partition of the design space. Tsiatis (1981) shows that such a test can be constructed by adding a set of dummy variable predictors, one for each subset of the partition, and testing their joint significance by an efficient scores test. But comparisons of (observed-expected) counts over a subset of the sample can be insensitive. To take an extreme example, the three pairs of points $\{(p_1, Y_1), (p_2, Y_2)\} = \{(.1, 0), (.9, 1)\}, \{(.5, 0), (.5, 1)\}$, or $\{(.9, 0), (.1, 1)\}$ all look like perfect fits by this criterion. A measure like $\sum(Y_i - \hat{p}_i)^2$ is more sensitive than $\sum(Y_i - \hat{p}_i)$ for picking out ill-fitting subsets of the sample. This motivates the following goodness of fit procedure, which is based on comparing likelihood equivalent error rates over various partitions of the sample.

Given a particular partition of the sample into k subsets, $S_1, \ldots, S_K$, let q_k, $k = 1, \ldots, K$, be the observed proportion of Y = 1 in subset k, and define

$$\alpha_k^M = \min(q_k, 1 - q_k)$$

to be the "marginal" LE error rate within subset k. This is to be compared to the LE error rate achievable when the estimates $\hat{p}_i = F(\underset{\sim}{X}_i\hat{\beta})$ are used to predict the observations in S_k, namely $\hat{\alpha}_k$, where $0 \leq \hat{\alpha}_k \leq \frac{1}{2}$ and

$$N_k\hat{\alpha}_k \log\hat{\alpha}_k + N_k(1-\hat{\alpha}_k)\log(1-\hat{\alpha}_k) = \sum_{i \in S_k} [(Y_i=1)\log\hat{p}_i + (Y_i=0)\log(\ -\hat{p}_i)] \ .$$

(N_k is the sample size in subset k.)
If there are strong predictors in the model used to estimate the $\hat{p}_i$, then one expects $\hat{\alpha}_k << \alpha_k^M$. However, if the model does not predict well within S_k, then $\hat{\alpha}_k$ may be near to or even greater than α_k^M, since $\hat{\beta}$ and the $\hat{p}_i$ have been estimated from the whole sample, not just S_k.

Table IV shows how the computer program DREG (Dichotomous REGression - see DuMouchel, 1981) screens for poor-fitting subsets of the ICU data. The method (choosing a partition and comparing α_k^M to $\hat{\alpha}_k$, $k = 1, \ldots, K$) is repeated for eight different partitions of the sample, each partition defined by intervals of one of the X-variables used in the model. For categorical predictor variables (i.e., X's which take on only the nominal values $k = 1, 2, \ldots, K$, and for which dummy variables are generated to compute $\underset{\sim}{X}\beta$) the partition is defined by the subsets $S_k = \{i | X_i = k\}$. For example, the variable INTERVAL specifies which of the 5 time periods is being used, and the first few lines of Table IV are based on a partition of the data by time period. The first two columns of the table show the N_k and q_k for each subset. Thus, there were 156 patients at risk in time category 1 (day 2) and 106 patients at risk at the start of time category 3 (days 4-7). From column 2, $q_1 = .10$ of the initial 156 patients "failed" immediately and $q_3 = .18$ of the 106 patients at risk in interval 3 failed. Columns 3-5 of Table IV show α_k^M, $\hat{\alpha}_k$, and $\%RED = 100\% \times (\alpha_k^M - \hat{\alpha}_k)/\alpha_k^M$. Note that %RED ranges from 31% to 56% for time intervals 1, 2, 4, 5, but in interval 3, $\alpha_3^M = .18$, $\hat{\alpha}_3 = .17$, %RED = 7%. The use of the other 7 predictors, all strongly significant in the sample as a whole, hardly helps predict these 106 cases at all, if predictive success is measured by the LE error rate.

TABLE IV. DREG Goodness of Fit Output for the ICU Data and the log Hazard Rate Model.

GOODNESS OF FIT ANALYSIS BY EACH PREDICTOR VARIABLE

STRATUM ASSOC-		OBS	LIKE-EQUIV ERROR RATE		
IATED WITH	N	P(Y=1)	MARGINAL	MODEL2	%RED
15 Interval					
CAT. 1	156	.1026	.1026	.0625	39.1
CAT. 2	124	.0645	.0645	.0422	34.7
CAT. 3	106	.1792	.1792	.1671	6.8
CAT. 4	67	.3731	.3731	.2561	31.4
CAT. 5	28	.6786	.3214	.1398	56.5
OVERALL	481	.1809	.1379	.0991	28.1
2 soc					
CAT. 1	336	.1429	.1429	.0875	38.7
CAT. 2	121	.2479	.2479	.1104	55.5
CAT. 3	24	.3750	.3750	.2563	31.7
OVERALL	481	.1809	.1729	.0991	42.7
14 dial					
CAT. 1	35	.2571	.2571	.1260	51.0
CAT. 2	446	.1749	.1749	.0971	44.5
OVERALL	481	.1809	.1799	.0991	44.9
21 logurin					
V LO	48	.3958	.3958	.1318	66.7
LOW	87	.3678	.3678	.2148	44.6
HIGH	340	.1000	.1000	.0725	27.5
V HI	6	.3333	.3333	.2013	39.6
OVERALL	481	.1809	.1506	.0991	34.2
8 bil					
V LO	0				
LOW	341	.1290	.1290	.0753	41.6
HIGH	85	.2353	.2353	.1708	27.4
V HI	55	.4182	.4182	.1743	58.3
OVERALL	481	.1809	.1643	.0991	39.7
12 base					
V LO	65	.4154	.4154	.2022	51.3
LOW	183	.1530	.1530	.0850	44.4
HIGH	160	.1312	.1312	.0801	38.9
V HI	73	.1507	.1507	.1068	29.1
OVERALL	481	.1809	.1652	.0991	40.0
29 dplat					
V LO	38	.2105	.2105	.1837	12.7
LOW	215	.1767	.1767	.0788	55.4
HIGH	176	.1420	.1420	.1018	28.3
V HI	52	.3077	.3077	.1298	57.8
OVERALL	481	.1809	.1761	.0991	43.7
24 dtist					
V LO	65	.1231	.1231	.0771	37.4
LOW	191	.1728	.1728	.0937	45.8
HIGH	161	.1615	.1615	.0833	48.4
V HI	64	.3125	.3125	.2019	35.4
OVERALL	481	.1809	.1750	.0991	43.4

Following up this clue, Table V shows the coefficients and standard errors of $\hat{\beta}$ when the model is fitted separately to the 3 subsets corresponding to INTERVAL = (1,2), 3, (4,5). (The ordinal variable SOC was treated as a continuous variable in these runs in order to save on degrees of freedom.) Note that the variables SOC, DIAL, LOGURIN, and DTIST tend to diminish in importance as the time spent in intensive care increases, while BIL and DPLAT show the opposite trend, and the coefficient for BASE shows no trend. As evidenced by the standard errors in Table V, there is only mild evidence for these trends. A multivariate test on the difference between $\beta_{1,2}$ and $\beta_{4,5}$ for these 7 predictors can be based on the statistic $(\hat{\beta}_{1,2} - \hat{\beta}_{4,5})'(\hat{S}_{1,2} + \hat{S}_{4,5})^{-1}(\hat{\beta}_{1,2} - \hat{\beta}_{4,5}) = 12.4$ where $\hat{S}_{1,2}$ and $\hat{S}_{4,5}$ are the estimated covariance matrices from the first and third regressions of Table V. Since $P(\chi^2(7df)>12.4) = .09$, there is evidence of a trend in β, but the post-hoc nature of this analysis makes verification on new data necessary.

Going back to Table 4, the partitions based on the other predictors do not lead to indications of poor fitting subsets, since %RED > 10% in the rest of the table. Experience with this procedure in a large number of data sets indicates that %RED < 0 is a reliable indicator (The range 0 < %RED < 10% is indicative, but not so definite.) that an interaction term or variable transformation will improve the model, at least when N_k is as large as 50 or 100 and when there are strongly significant predictors among those not involved in the partition under consideration. The strength of predictors other than those involved in the partition is shown by the lines of Table 4 labeled "OVERALL", which display α^M, the LE error rate based on using subset membership itself as a predictor, and $\hat{\alpha}$ from the full model. If the "OVERALL %RED" is less than 20% or so, then the individual %RED values tend to be

TABLE V. Coefficients and Asymptotic Standard Errors From Separate Dichotomous Regressions (log hazard rate model).

Predictor	Intervals[c] 1 and 2		Interval[c] 3		Intervals[c] 4 and 5	
Constant	-9.7	(4.2)	-3.4	(2.2)	-5.7	(2.2)
INTERVAL[b]	.2	(.5)			1.6	(.5)
SOC[b]	.8	(.3)[a]	.1	(.4)	- .2	(.4)
DIAL[b]	4.8	(2.2)[a]	2.6	(1.5)	.2	(.8)
LOGURIN	- .5	(.1)[a]	- .5	(.2)	- .2	(.2)
BIL	.01	(.02)	.03	(.02)	.05	(.02)[a]
BASE	- .14	(.04)[a]	- .03	(.07)	- .09	(.04)[a]
DPLAT	- .003	(.003)	- .002	(.004)	.006	(.003)[a]
DTIST	.05	(.02)[a]	.02	(.03)	.02	(.02)
LE error rate						
interval only	.085		.179		.356	
full model	.051		.156		.198	
χ^2 (7df)[d]	50.1		8.1		29.2	

[a] significant effect at asymptotic .05 level

[b] original variable used rather than dummy variables

[c] coefficients are based on prediction of outcomes which fall into the indicated time-interval(s) indicated

[d] likelihood ratio chi-squared statistic testing the hypothesis that Y depends only on INTERVAL

less informative. The partition used when X is continuous is the four intervals defined by the points $\bar{x} - s$, $\bar{x}$, $\bar{x} + s$, where s is the sample standard deviation of X. In an earlier run with a somewhat different set of predictors, the value of %RED was negative for URINE > $\bar{x} + s$, leading us to use the variable LOGURIN instead of URINE. Although the mathematical theory of this procedure has not yet been fully developed, the method is an effective exploratory tool which is cheap to compute. Note that the computation of observed minus expected counts of Y = 1 is not nearly as powerful when the partitions are defined by single X variables. In fact, if the logistic model is used, partitions based on a categorical predictor will have the observed count equal to the expected count for every subset of the partition, because the MLE procedure fits the corresponding 2-way X-Y marginals exactly.

V. EXAMINATION OF THE LIKELIHOOD SURFACE

While normal-theory regression has a simple small-sample mathematical theory, inferences from dichotomous regressions are usually based on the asymptotic theory of the maximum likelihood estimate for large samples. Is the present sample size large enough? Attempts to answer the question from a classical statistical veiwpoint usually require simulation or the bootstrap method (Efron, 1982). The Bayesian answer requires the numerical integration of the joint posterior distribution, over the space of all nuisance parameters, in order to compare the resulting marginal posterior distribution with that predicted by asymptotic theory. The following procedures is in the spirit of the latter approach, but requires much less computation.

Bayesian large-sample theory (Lindley, 1980) predicts that the posterior distribution of β will approximate a multivariate normal density with mean $\hat{\beta}$ and covariance matrix I^{-1}, where I is the Fisher information, if the prior information about β is small compared to I. Even when there is substantial prior information, the likelihood function $\exp(L(\beta))$ is predicted to be approximately proportional to the same normal density. A comparison of $L(\beta)$ to its approximating quadratic form, namely $\tilde{L}(\beta) = L(\hat{\beta}) - \frac{1}{2}(\beta-\hat{\beta})' I(\beta-\hat{\beta})$ will show how well the likelihood conforms to the asymptotic theory. When m, the dimensionality of β, is large, a complete comparison of $L(\beta)$ and $\tilde{L}(\beta)$ requires excessive computing. But it is not too burdensome to compute $L(\beta)$ in each of the 2m directions away from $\hat{\beta}$ corresponding to positive and negative deviations $\beta_j - \hat{\beta}_j$, $j = 1, \ldots, m$, of the components of the vector β. The program DREG has an option to produce a table of 2m values of $\exp\{L(\beta) - L(\hat{\beta})\}$ for values of β such that $\exp\{L(\hat{\beta})\} - \tilde{L}(\hat{\beta})\} = .05$. That is, the *j*th pair of β-vectors are defined by

$$\begin{cases} \beta_j = \hat{\beta}_j \pm [(2 \log 20)/I_{jj}]^{\frac{1}{2}} \\ \beta_{j'} = \hat{\beta}_{j'} \qquad j' \neq j \end{cases}$$

where I_{jj} is the *j*th diagonal element of I.

Table VI shows these values of $\exp\{L(\beta) - L(\hat{\beta})\}$ for the ICU data. The first two columns use the linear model while the last two columns show the corresponding output for the log hazard rate model. Remembering that values near .05 indicate agreement with the asymptotic theory, the likelihood based on the log hazard rate model seems quite in agreement with its normal approximation in all but 2 of the directions chosen. The likelihood drops off somewhat faster than expected when the coefficient for DIAL is moved toward 0 and also when the coefficient of LOGURIN is moved away from 0. By contrast, the

TABLE VI. *Likelihood Surface Exploration for the Log Hazard Rate and Linear Models (Models 3 and 4 of Table II).*

Parameter Corresponding to	linear model decrease β	linear model increase β	log hazard rate model decrease β	log hazard rate model increase β
constant	.488	.563	.063	.040
INTERVAL 1	.577	.718	.090	.030
2	.148	.398	.109	.020
3	.010	.112	.076	.022
4	0	.074	.070	.030
SOC = 1	.483	.588	.070	.036
2	.436	.478	.070	.034
DIAL	0	.132	.071	.011
LOGURIN	.383	.322	.007	.048
BIL	.365	.418	.069	.036
BASE	.186	.391	.053	.074
DPLAT	0	.284	.041	.048
DTIST	0	.255	.052	.029

Tabled values are $\exp\{L(\beta) - L(\hat{\beta})\}$ *for values of* β *identical to* $\hat{\beta}$ *in all but one component, which is either decreased (cols 1, 3) or increased (cols 2, 4) so that asymptotic theory predicts* $\exp\{L(\beta) - L(\hat{\beta})\} = .05$.

behavior of the linear model likelihood function does not at all agree with its Fisher information approximation. In most directions $L(\beta)$ has not decreased nearly as much as predicted, but in 4 directions the likelihood drops all the way to zero where its normal approximation is .05. The fact that this model allows fitted probabilities to take on the values $\hat{p}_i = 0$ or 1 means that the likelihood function can be very ill behaved. For the linear model, minus the second derivative of the log likelihood $L(\beta)$ with respect to β_j is

$$\frac{-\partial^2 L(\beta)}{\partial \beta_j^2} = \sum_{0<\underset{\sim}{X}_i\beta<1} x_{ij}^2 \left[\frac{(Y_i=1)}{p_i^2} + \frac{(Y_i=0)}{(1-p_i)^2}\right]$$

and its expectation is

$$I_{jj} = \sum_{0<\underset{\sim}{X}_i\beta<1} x_{ij}^2/p_i(1-p_i)$$

where $p_i = F(\underset{\sim}{X}_i\beta) = \max(0, \min(1, \underset{\sim}{X}_i\beta))$ and x_{ij} denotes the jth element of $\underset{\sim}{X}_i$. The presence of p_i and $(1-p_i)$ in the denominators of these expressions causes them to change quickly when β changes so that some of the $\underset{\sim}{X}_i\beta$ are near 0 or 1. In this case $L(\beta)$ will not be well approximated by a quadratic, and the maximum likelihood estimate will not be approximately normally distributed except for very large samples. This leads to severe difficulties in basing inferences on the linear model for moderate sample sizes, and provides another reason for prefering the log hazard rate model for this data. The latter model has

$$I_{ij} = \sum_i x_{ij}^2(1-p_i)[\log(1-p_i)]^2/p_i$$

and is much better behaved numerically.

The examination of the likelihood function as a standard practice helps avoid unwarranted reliance on asymptotic theory. In addition, information like that in Table VI provides a way to improve the usual interpretation of asymptotic tests or credible intervals. For example, from the entry in column 4 of the DIAL row of Table VI, we infer that DIAL is even less likely to be a nonimportant effect than the corresponding test statistic in Table III ($\hat{\beta}/\hat{SE} = -2.8$) would imply.

Sprott (1980) and Lindley (1980) each propose, from different perspectives, methods involving the third derivatives of the log-likelihood function. Application of their proposals to the dichotomous regression framework might improve upon our qualitative results.

VI. DISCUSSION AND CONCLUSION

Now that several statistical computer program packages have dichotomous regression commands, its use is becoming more common. Years ago, a program that could successfully maximize the likelihood function in many dimensions and print out the estimated coefficients and standard errors was thought to be a sufficient achievement. Today a program which does only that is woefully inadequate. Our ICU analysis has required the fitting of a wide variety of functional forms for the link function, the availability of intuitive statistics to describe the explanatory power of $\underset{\sim}{X}$, special procedures to screen for bad fitting subsets of the data and for poorly behaved likelihood functions, and the ready availability of fitted probabilities and residuals for each case. In addition, it helps to have comprehensive, user-oriented output to help make interpretation easier. For example, since nominal-scale predictors having 3 or more categories are very common in these models, it is important that the program be able to handle them with ease. It should generate coefficients, standard errors and other output for every category to avoid a distracting focus on the one omitted dummy variable[2], and it should provide an overall χ^2 statistic for the significance of the variable as a whole. Similarly, the ability to treat the linear and quadratic components of a continuous predictor as a single two-degree-of-freedom predictor is helpful, as is the ability of the input to and output from the program to easily handle interaction terms involving predictors each of which have multiple degrees of freedom. (The package GLIM described in Baker and Nelder, 1978 is outstanding in this respect.)

The fact that dichotomous regression coefficients are usually measured in nonintuitive units (logistic, etc.) makes it hard to get a feel for the practical effect that varying a particular X-variable has on the $\hat{p}$. The program should produce a table of adjusted means (i.e., standardized rates) so that these effects are clearly shown. The so-called "direct standardization" to the distribution of the sample is easily interpreted and is computed as follows: Let X denote one of the predictors and $\underset{\sim}{X}^*$ denote the remaining predictors. Let $\hat{\beta}_1$ and $\hat{\beta}^*$ denote the respective coefficients. Suppose we want to express the effect of changing X from x_1 to x_2. We define for k = 1, 2:

$$\hat{p}(x_k) = N^{-1} \sum_i F(x_k\hat{\beta}_1 + \underset{\sim}{X}_i^*\hat{\beta}^*).$$

Thus $\hat{p}(x_k)$ estimates the average P(Y = 1) if every case had its X changed to x_k. Table 7 shows a table of standardized rates for the ICU data. The values of x_k chosen for the continuous predictors are the 10*th* and 90*th* percentiles of X.

[2] *In the DREG program, the coefficients associated with each category of a nominal-scale variable are constrained so that their weighted average is zero, where the categories are weighted by their sample sizes.*

TABLE VII. Standardized Rates for the ICU Data.

Variable	*Value*	$\hat{p}$ *(Failure)*
INTERVAL	*1*	*.080*
	2	*.078*
	3	*.207*
	4	*.395*
	5	*.774*
SOC	*1*	*.166*
	2	*.182*
	3	*.363*
DIAL	*1*	*.064*
	2	*.199*
LOGURIN	*5.2*	*.267*
	8.0	*.129*
BIL	*1*	*.156*
	19	*.222*
BASE	*-2*	*.118*
	8	*.251*
DPLAT	*-66*	*.214*
	84	*.147*
DTIST	*-13*	*.140*
	9	*.227*

Finally, we need to develop more graphical aids to the use of dichotomous regression. One useful display is an estimate of the operating characteristic curve, or, equivalently, the sensitivity-specificity curve, (Weinstein and Fineberg, 1980). Figure 1 shows the curve computed for our ICU analysis.

In conclusion, the preceding analyses have led to the following model for the log of the hazard rate in the ICU at time t:

$$\log h(t, \underset{\sim}{X}) = h_0(t) + \begin{matrix} -.10\ (SOC=1) \\ .04\ (SOC=2) \\ 1.19\ (SOC=3) \end{matrix} + \begin{matrix} -1.56\ (DIAL=1) \\ .12\ (DIAL=2) \\ \end{matrix}$$

$$- .39\ LOGURIN + .029\ BIL - .11\ BASE$$

$$- .004\ DPLAT + .033\ DTIST,$$

where $h_0(t)$ is constant within each interval.
The sign and the magnitude of all the $\hat{\beta}$'s are consistent with current medical knowledge about critically ill patients. As discussed in Cullen et al (1982), these predictors were chosen only after extensive analyses on a larger set of variables. If confirmation of the model on an independent sample is achieved, these results can help the physician allocate the limited resources of the ICU. In addition, the outcomes of intensive care in different hospitals can be compared. Until now such comparisons have been hampered by the lack of standard methods of adjusting for differences in the distribution of severity of critical illness in hospitals.

ACKNOWLEDGMENT

We would like to thank Dr. David Cullen of the Department of Anaesthesia, Massachusetts General Hospital, for allowing the use of his data in this paper.

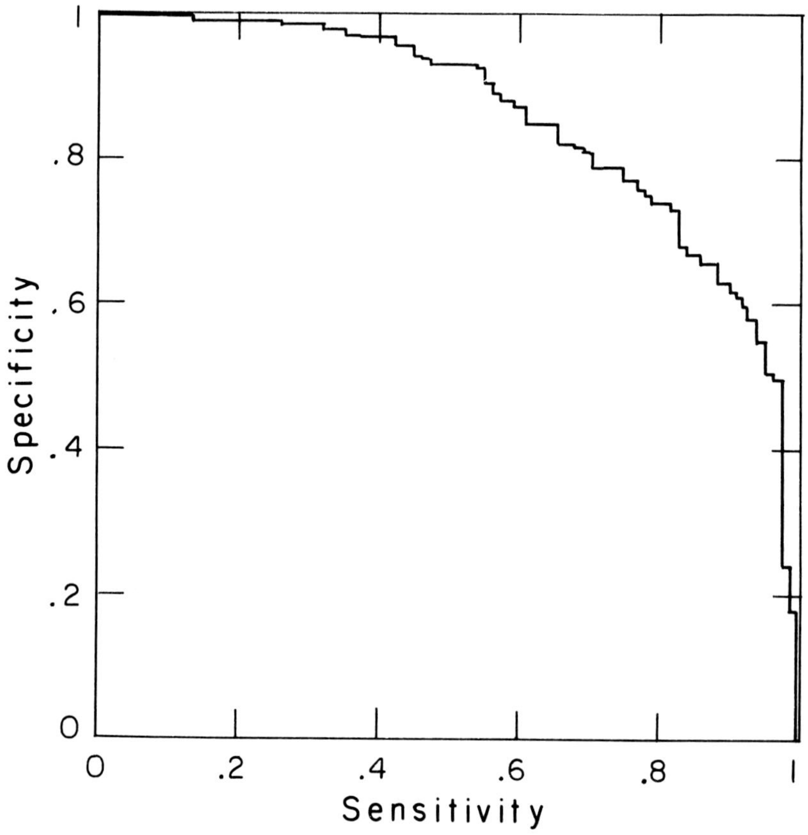

FIGURE 1. Sensitivity-Specificity Curve for the ICU Data.

Each of the 481 patient-intervals is predicted to be a failure (or nonfailure) if the corresponding $\hat{p}$ *is greater (or less) than* p_0. *Sensitivity (the proportion of failures correctly identified) and specificity (the proportion of nonfailures correctly identified) are each functions of* p_0.

REFERENCES

Baker, R. J. and Nelder, J. A. (1978). *The GLIM Manual-Release 3*, Distributed by Numerical Algorithms Group, Oxford, England.

Belsley, D. A., Kuh, E., and Welsch, R. E. (1980). *Regression Diagnostics: Identifying Influential Data and Sources of Collinearity*, Wiley, New York.

Bishop, Y., Fienberg, S., and Holland, P. (1975). *Discrete Multivariate Analysis*, M.I.T. Press, Cambridge, MA.

Cullen, D. et al (1982). "The Outcome of Intensive Care in Critically Ill Patients," Manuscript in preparation.

DuMouchel, W. (1981). "Documentation for DREG - A Dichotomous Regression Program," M.I.T. Statistics Center Technical Report # NSF-27, Cambridge, Masschusetts.

Efron, B. (1978). "Regression and ANOVA with Zero-One Data: Measures of Residual Variation," *JASA 73*, 113-121.

Efron, B. (1982). *The Jackknife, The Bootstrap and Other Resampling Plans*, CBMS-NSF Regional Conference Series in Applied Mathematics, Society for Industrial and Applied Mathematics, Philadelphia.

Goodman, L. A. (1970). "The Multivariate Analysis of Qualitative Data: Interactions Among Multiple Classifications, *JASA 65*, 226-256.

Hills, M. (1966). "Allocation Rules and Their Error Rate," *JRSS B 28*, 1-30.

Laird, N. and Olivier, D. (1981). "Covariance Analysis of Censored Survival Data Using Log-Linear Analysis Techniques," *JASA 76*, 231-240.

Lindley, D. V. (1980). "Approximate Bayesian Methods," in *Bayesian Statistics: Proc. of First International Meeting*, 223-237, J. M. Bernardo, et al, eds., University Press, Valencia, Spain.

Nelder, J. A. and Wedderburn, R. W. M. (1972). "Generalized Linear Models," *JRSS A, 135*, 370-384.

Pregibon, D. (1980). "Goodness of Link Tests for Generalized Linear Models," *Appl. Statist. 29*, 15-24.

Pregibon, D. (1981). "Logistic Regression Diagnostics," *Annals of Statistics, 9*, 705-724.

Sprott, D. A. (1980). "Maximum Likelihood in Small Samples: Estimation in the Presence of Nuisance Parameters," *Biometrika 67*, 515-23.

Tsiatis, A. A. (1980). "A Note on a Goodness of Fit Test for the Logistic Regression Model," *Biometrika 67*, 250-51.

Weinstein, M. C. and Fineberg, H. V. (1980). *Clinical Decision Analysis*, W. B. Saunders, Philadelphia.

THE APPLICATION OF SPLINE FUNCTIONS TO THE PHARMACOKINETIC ANALYSIS OF METHOTREXATE INFUSED INTO MALIGNANT EFFUSIONS[1]

Stephen B. Howell
Richard A. Olshen
John A. Rice

Cancer Center,
Laboratory for Mathematics and Statistics,
and Department of Mathematics
University of California
La Jolla, California

I. INTRODUCTION

The human body contains cavities inside which are various organs. The 3 cavities of special interest to us are the pleural (which contains the lungs), the peritoneal (or abdominal), and the pericardial (which contains the heart). The cavities are lined with membranes, which face them and which weep fluid into the cavities. In normal individuals this fluid is absorbed by the lymph system. However, in the cavities of some individuals with intracavitary tumors the fluid is not absorbed, and thus there arise what are termed malignant effusions. Lymph nodes can be plugged by cancer cells which have been sloughed by a tumor. A tumor itself can be growing on the membranes,

[1] *Supported by National Cancer Institute Grants CA-23100 and CA-23334, National Science Foundation Grants MCS-7901800 and MCS-7906228, and National Institutes of Health Grant CA-26666.*

ISBN 0-12-589320-5

thereby inflaming them and limiting the above mentioned absorption. Also, fluid can collect in the cavities from a generalized edema. Pharmacologic modelling performed by Dedrick et al. (1978) suggested that administration of MTX by the intracavitary route would result in much greater drug exposure for the intracavitary tumor than administration of the same dose via the intravenous route. Since even small amounts of MTX reaching the blood can be very toxic, in this study we not only administered MTX by the intracavitary route, but also infused a neutralizing agent into the blood that was capable of blocking the action of MTX (Howell et al., 1981). The Phase I clinical trial from which our data were drawn involved 18 cancer patients with malignant effusions: 12 of these were located in the pleural cavity, 4 in the peritoneal cavity, and 2 in the pericardial cavity. Six patients had lung cancer, 5 breast, 2 ovarian, 1 pancreatic cancer, 2 cancer of unknown origin, and 2 patients had melanoma. The details of the patient characteristics and study design have been published elsewhere (Howell et al., 1981). The MTX was infused into the cavity at a dose of 30 mg/m^2 of body surface area per day in 10 ml/hr of fluid through one catheter, and in most cases 2 ml samples were removed periodically via another catheter for measurement of drug concentrations.

Models and Problems of Goodness of Fit

The principal goal of this paper is to give methods for inferring the goodness of fit of certain models for the concentrations of MTX in the study just described. Most models of such concentrations are of a "compartmental" nature (see Rubinow (1973), Endrenyi (1981), and Dedrick et al. (1978)). The most general (open) model which we deemed suitable for our study is given in Figure 1. If changes in concentration are governed by linear, first order kinetics, then for $t \geq 0$ the vector of concentrations is the solution to a system of differential equations of the form

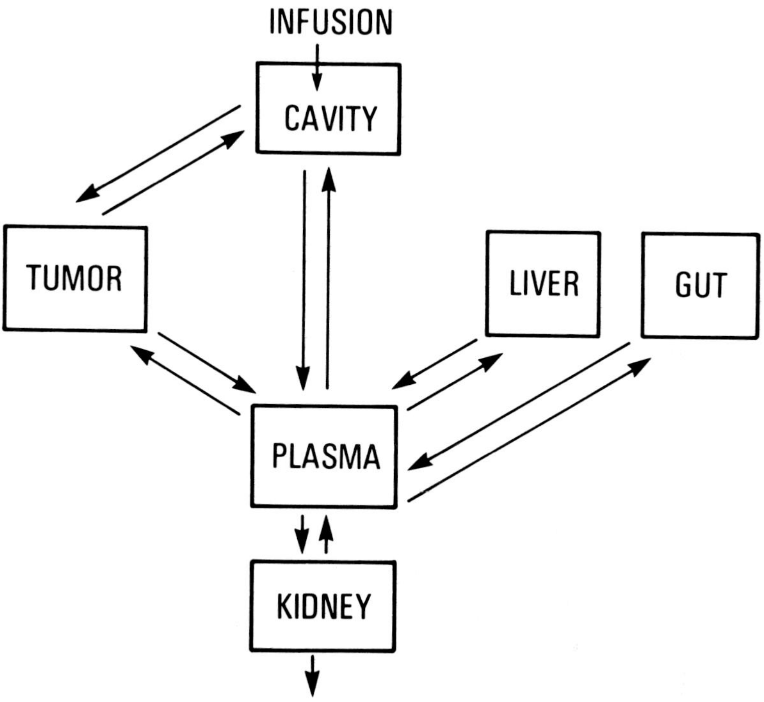

FIGURE 1

$$\dot{\underline{C}}(t) = \underline{M}\,\underline{C}(t) + \underline{I},\ \underline{C}(0) = \underline{0}, \tag{1.1}$$

where the dot denotes differentiation with respect to time. There are 6 coordinates to $\underline{C}$, one for each compartment. All entries of the vectors $\underline{I}$ are 0 except for a positive entry in the component which corresponds to the cavitary infusion rate. The (i,j) entry of the 6 x 6 constant matrix $\underline{M}$ is the proportionality constant for the rate of flow from compartment i to compartment j. Note that $M_{ii} \leq 0$ and $M_{ij} \geq 0$ for $i \neq j$. The roots of $\underline{M}$ for such systems of equations have non-positive real parts (Rubinow, 1973, p. 28). In practice, the imaginary parts of the roots are nearly always 0 (Rubinow, loc. cit.), and we shall suppose that to be the case here; moreover, we shall suppose that the roots are distinct. If the foregoing assumptions hold, then the entries of $\underline{C}(t)$ are nonnegative and are linear combinations of $\{e^{-\lambda_i t} : -\lambda_i$ is a root of $\underline{M}\}$ and a constant. (See Bellman, 1970, Chapters 10 and 11).

Many pharmacokinetic models include consideration of drug metabolism based on Michaelis-Menton theory (see Rubinow (1973), Endrenyi (1981), and Dedrick et al. (1978)). However, in our study we deemed these inappropriate since the extent of MTX metabolism to products not measured by the drug assay we employed was negligible.

II. ANALYSIS AND RESULTS

The models we ultimately study can be motivated by a much simpler model than those which correspond to Figure 1. It assumes that the tumor and fluid (that is, the cavity) are one compartment; this assumption is in force the remainder of the paper. Temporarily, we also assume that the plasma and kidney compartment are one, and further we ignore the liver and gut. The resulting model is basically the first model of Dedrick et al. (1978).

The system is described by the equations

$$V_p \dot{C}_p(t) = PA[C_f(t) - C_p(t)] - K C_p(t) \qquad (2.1)$$

$$V_f \dot{C}_f(t) = PA[C_p(t) - C_f(t)] + I \quad , \qquad (2.2)$$

where V_p and V_f are the volume of distribution of the plasma and fluid; $C_p(t)$ and $C_f(t)$ are the concentrations in plasma and fluid at time t: PA is the permeability-area product; I is the infusion rate; and K is an excretion rate. (That (2.1) and (2.2) ought to contain the same rate constant PA is a consequence of protein binding studies.) As was indicated, in this model the plasma and the rest of the body are considered as one compartment; if the model were to allow for more compartments, equation (2.1) would contain additional terms, but if we assume that the only significant transfer of MTX from the fluid was via the plasma, then equation (2.2) would remain unchanged. Note that equation (2.2) implies that at equilibrium $C_f - C_p = I/PA$. The solution of the system given by equations (2.1) and (2.2) with initial conditions $C_f(0) = C_p(0) = 0$ is necessarily of the form

$$C_f(t) = A_1(1 - e^{-\alpha_1 t}) + A_2(1 - e^{-\alpha_2 t}) \tag{2.3}$$

$$C_p(t) = B_1(1 - e^{-\alpha_2 t}) + B_2(1 - e^{-\alpha_1 t}), \tag{2.4}$$

where the parameters of equations (2.3) and (2.4) are uniquely determined from those of equations (2.1) and (2.2). As was mentioned, $-\alpha_1$ and $-\alpha_2$ are the roots of the matrix which corresponds to $\underline{M}$. In this case the roots of $\underline{M}$ are real and unequal.

In a conventional analysis one attempts to fit equations of the form of equations (2.3) and (2.4) to the observations and then to estimate the parameters of equations (2.1) and (2.2). The fitting is often accomplished by nonlinear least squares, which entails the use of iterative estimates of the parameters. This process can be quite difficult since the function being minimized is typically not convex; frequently several different starting values must be tried. If more than two compartments are important, there will be more than two exponential terms in equations (2.3) and (2.4), even though equation (2.2) remains unchanged. Fitting several exponentials to limited and noisy data is especially difficult when the number must be determined from the data.

The method we discuss below focuses directly on the kinetics described by equation (2.2) by examining the concentrations and their derivatives through time. The method is data-analytic in nature--it provides a preliminary and informal graphical method for ascertaining the validity of the linear kinetic model. Although our immediate objective is not the estimation of parameters, but an examination of goodness-of-fit, the method does yield an estimate of PA if the model is reasonable.

To motivate the method, note that if one had very accurate measurements taken continuously in time, $\dot{C}_p$ and $\dot{C}_f$ could be calculated from the data, and the parameters V_f and PA could be determined from equation (2.2) and knowledge of the infusion rate by the linear relation between $\dot{C}_f$ and $C_p - C_f$. If the observed relation were not linear, the validity of the model would be called into question. It might be possible, for example, that PA is not constant but is a function of $C_p - C_f$.

Since we did not have such measurements, we first smoothed the data, producing differentiable curves, and then calculated differences and derivatives of these curves. There are many methods of carrying out such smoothing; we chose to smooth with spline functions since they have good approximation properties and since excellent computer programs are available for computing and fitting them (see for example, de Boor, (1978). Spline smoothing has also been applied to pharmacokinetic data by Wold (1971).

We next briefly introduce some terminology and facts about spline functions. Following de Boor, we note that spline functions are functions which are polynomials over intervals between breakpoints, and which join smoothly accross the breakpoints. For example, given $\tau_1 < \tau_2 < \ldots < \tau_k$, a cubic spline with this breakpoint sequence consists of different cubic polynomials in each interval (τ_i, τ_{i+1}). (For technical reasons, we distinguish breakpoints from "knots"; τ_1 and τ_k will typically be "multiple knots" and the interior breakpoints will be simple knots (de Boor, 1978)). These polynomials are constrained to have two continuous derivatives across the breakpoints. Similarly, one can define splines of any order; linear splines are continuous piecewise linear functions; quadratic splines are piecewise quadratic functions which are one time continuously differentiable; and so forth.

Given data $y(t_i) = g(t_i) + \text{error}$, $i = 1,\dots,n$, we may try to smooth them and obtain an estimate of the unknown function g and its derivatives by choosing a relatively small number of breakpoints $\tau_1,\dots,\tau_p$, $p < n$, and by finding the spline function s(t) with those breakpoints which minimizes $\sum_{i=1}^{n} (y(t_i) - s(t_i))^2 w_i^2$, where the w_i's are weights which reflect the accuracy of the data. In fitting a cubic spline in this manner, there are p + 2 unknown parameters which must be determined by linear least squares. Since the fitting procedure is linear, the computations are inexpensive and quite easy to perform.

A particularly convenient representation for spline functions with a given breakpoint sequence is afforded by the B-spline basis (de Boor, 1978); if B_j denotes the jth B-spline, then an arbitrary spline can be represented as $s(t) = \sum_{j=1}^{p+k-1} \beta_j B_j(t)$, where k is the degree of the spline function. An example of a B-spline basis is shown in Figure 2. Thus, to fit a spline to data, $\beta_1,\dots,\beta_{p+k}$ are chosen to minimize

$$S(\beta_1,\dots,\beta_{p+k}) = \sum_{i=1}^{n} [y(t_i) - \sum_{j=1}^{p+k-1} \beta_j B_j(t_i)]^2 , \qquad (2.5)$$

which is a linear least squares problem. Once the β's are thus determined, the smoothing function $s(t) = \Sigma\beta_j B_j(t)$ and its derivatives $s^{(\ell)}(t) = \Sigma\, \beta_j B_j^{(\ell)}(t)$ $(\ell \leq k - 2)$ can be evaluated easily. We mention that asymptotic properties of this sort of nonparametric regression have been investigated by Agarwal and Studden (1980); they note that if p = p(n) grows at the

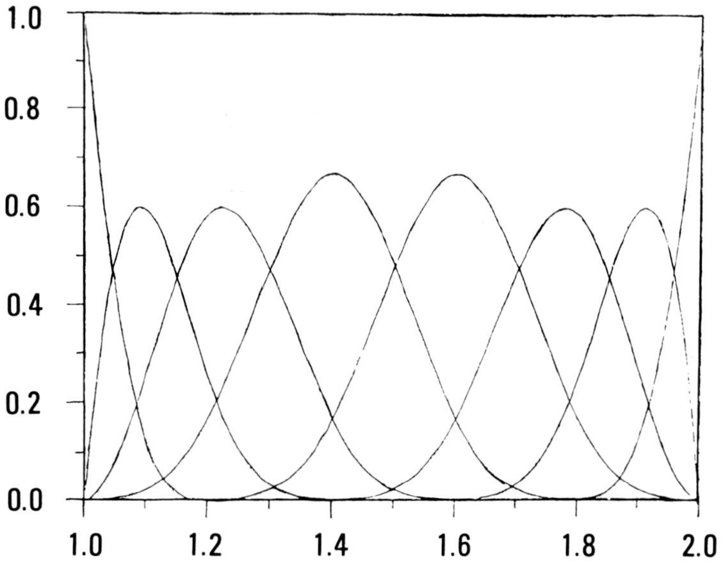

FIGURE 2

proper rate and the τ's are suitably chosen, then the integrated mean squared error tends to zero at the fastest rate possible for nonparametric estimators.

Spline approximations work well when the data have locally polynomial behavior. Since pharmacokinetic data typically vary over several orders of magnitude, it is necessary to transform the data prior to smoothing. Estimates may be then back-transformed. The transformations we used were of the form

$$u_i = \log(A_p\, C_p(t_i) + 1); \tag{2.6}$$

$$v_i = \log(A_f\, C_f(t_i) + 1); \tag{2.7}$$

$$y = \log(t + 1). \tag{2.8}$$

The constant 1 is added to the argument of the logarithm so that at time t = 0, u = v = y = 0 ($C_p(0) = C_f(0) = 0$). This linear constraint was incorporated in the spline smoothing. The constants A_p and A_f were chosen so that when u versus y and v versus y were graphed, there was a smooth transition in the behavior of the graph from y = 0 to small y. Our computer program allowed us to choose A_p and A_f interactively. For example, for a given A_p, a plot of u_i versus y_i was displayed on a CRT screen; the program then inquired whether we wished to try another A_p, and so forth. As a rule of thumb, A_p, for example, was chosen to be of the same order of magnitude as $1/C_p(t_1)$.

The spline fitting was done in a similar interactive manner. After choosing to fit either cubic or quadratic splines, we tried various breakpoint sequences and examined plots of the splines until we settled on a satisfactory smoothing. Generally, using too few knots causes the smoothing curve to systematically err in approximating the general trend of the points, whereas if too many knots are used, the spline oscillates as it tracks random fluctuations in the data. It is effective to place knots in regions where the function changes rapidly. To avoid overfitting and unnecessary inflation of variance, we tried to keep the number of breakpoints as small as possible; since we wished to estimate derivatives, we preferred to err on the side of oversmoothing than of undersmoothing. The smoothing procedure is admittedly subjective--knots were added and their locations varied until there were no blatant patterns in the resuduals. Aside from the determination of knot location, the procedure is similar to that used in determining the order of a polynomial regression without carrying out formal F-tests. We found, however, the independent analyses of different users of the program were in qualitative agreement, and we note that the initial information we sought was qualitative in nature.

We could, perhaps, have used some sort of cross-validation to produce the smoothing automatically. However, relatively little is known about the behavior of cross-validation in this context. The sample sizes are fairly small (10 - 20 points), and we do not understand clearly the relative influences of different points on the number of knots and their locations and on the coefficients of the B-spline smoothing. We note that the problem of optimal knot location is non-linear and that there may be more than one local minimum. Thus, although a fit determined by cross-validation may have a certain appealing "objectivity", the fit will not necessarily be better than one determined in a more subjective, interactive manner.

After this fitting one has constructed smoothing curves $\hat{u}(y)$ and $\hat{v}(y)$ to the points (u_i, y_i) and (v_i, y_i). From the B-spline representation of these curves $d\hat{u}/dy$ and $d\hat{v}/dy$ can be easily evaluated. We then have

$$\hat{C}_f(t) = \frac{\exp[\hat{v}(y(t))]-1}{A_f} \tag{2.9}$$

$$\frac{d}{dt}\hat{C}_f(t) = \frac{d\hat{v}}{dy}\frac{dy}{dt}\frac{\exp[\hat{v}(y(t))]}{A_f} \tag{2.10}$$

and similar expressions for $\hat{C}_p(t)$ and $\frac{d}{dt}\hat{C}_p(t)$.

Our data were fairly noisy, with a coefficient of variation in the range 25 - 50%, and we typically had a relatively small number of observations. In order to assess the accuracy of the results of smoothing, it was very useful to carry out simulations in which we generated values according to equations (2.3) and (2.4), added noise, and carried out the analysis. Figures 3 through 7 show the results of a simulation in which the coefficient of variation was 50%.

Figures 3 and 4 show the smoothing with quadratic splines of the plasma and fluid points on the transformed scales given by equations

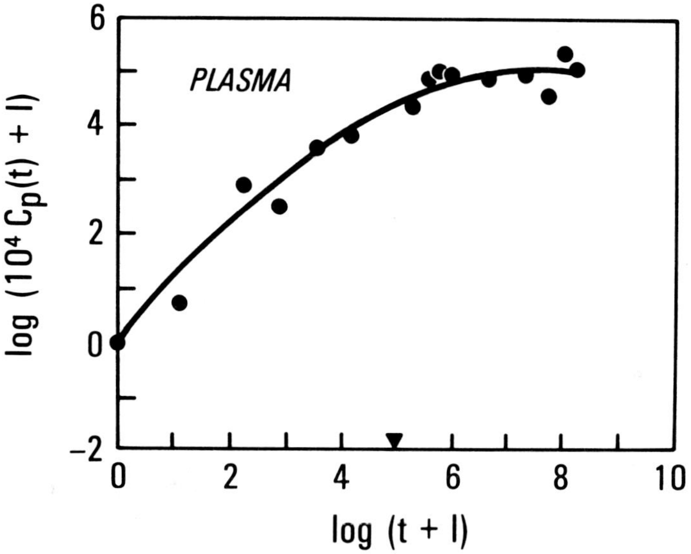

FIGURE 3

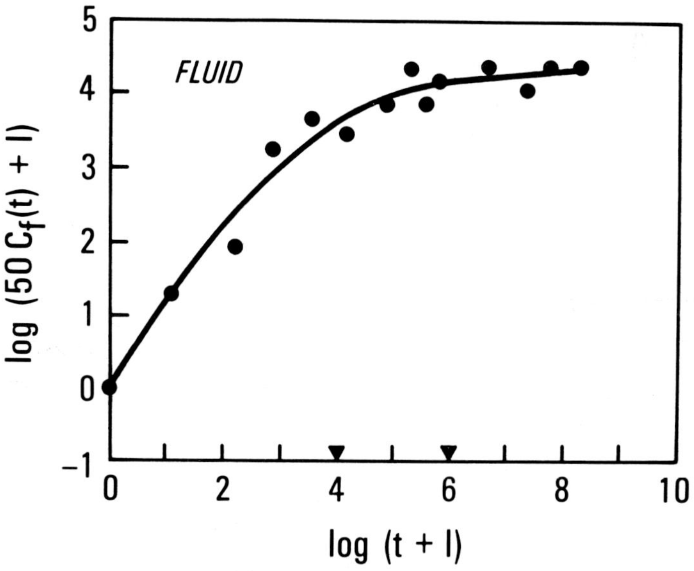

FIGURE 4

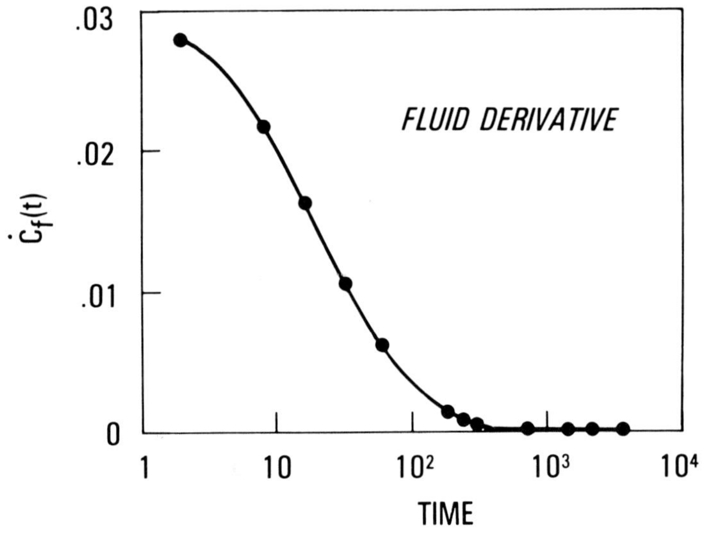

FIGURE 5

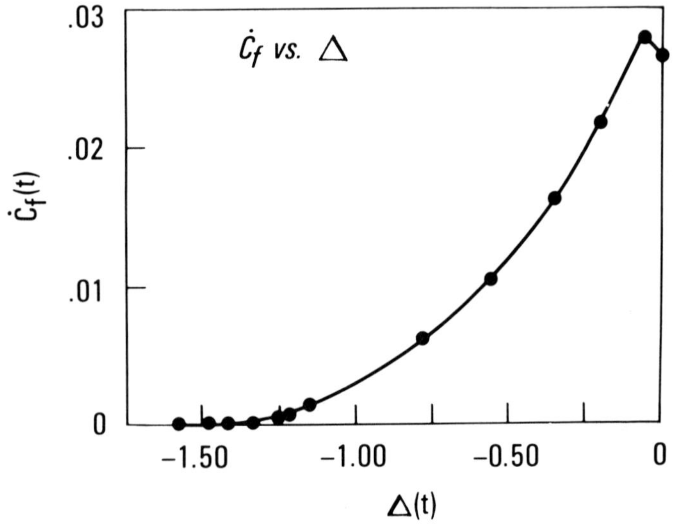

FIGURE 6

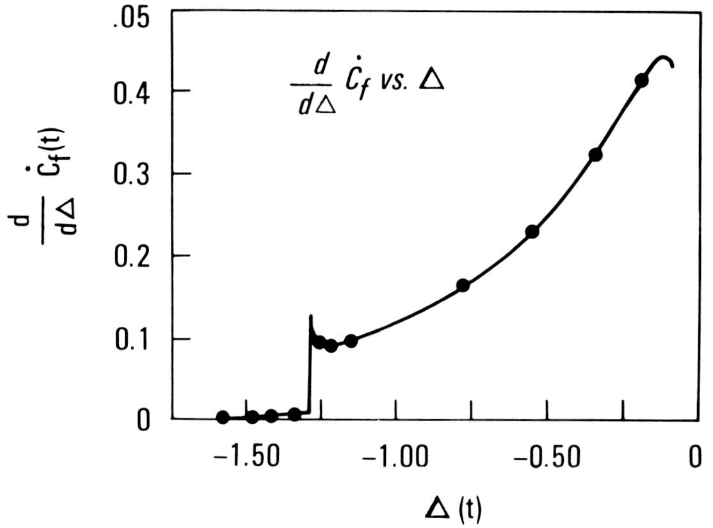

FIGURE 7

(2.6) through (2.8). Figure 5 is graph of $d\hat{C}_f/d\ t$ versus $\Delta(t) = C_p(t) - C_f(t)$. The model predicts this to be linear, and with the exception of early and late times, the graph qualitatively agrees. (In general, the derivative is not estimated well at early and late times). The constant PA/V_f is the slope of this line; we could have estimated this by fitting a straight line to the curve of Figure 6, but instead we calculated and plotted in Figure 7 the local slope

$$\frac{d}{d\Delta}\left(\frac{d}{dt}\hat{C}_f(t)\right) = \frac{d^2}{dt^2}\hat{C}_f(t)\left(\frac{d\Delta}{dt}\right)^{-1}. \qquad (2.11)$$

The "true value" is .016, and Figure 7, gives us an indication of how accurate we might expect to be with this method of analysis.

Figures 8 through 12 show the results of a similar analysis done on data from an actual patient who had a pleural effusion. The model appears reasonable. The slight peak in Figure 10 is an artifact of the smoothing; it does not show up consistently as different spline schemes are used. Figure 11 shows a relation between $\dot{C}_f$ and Δ which is, over all, qualitatively in agreement with the model. As shown in Figure 12, the slope is approximately 2×10^{-3}, which we take as an estimate of PA/V_f.

Figures 13 - 15 show results from another patient who had a pleural effusion. Figures 13 and 14 show nothing suspicious, but Figure 15, a graph of $\frac{d}{dt}\hat{C}_f$, is completely unexpected. The derivatives *increase* initially, behavior which is completely inconsistent with equation (2.2) or any reasonable modification of it. From equation (2.2), $\ddot{C}_f = PA(\dot{C}_p - \dot{C}_f)$, and in any model of the form (1.1) with our initial conditions $\dot{C}_p(0) = 0$. Furthermore, 5 of the 9 patients who underwent pleural infusions showed this sort of phenomenon, as did all three patients who underwent peritoneal infusions. (The data on the single evaluable patient with a pericardial infusion conformed to the model.) Initially, it was very difficult for us to conceive how this could possibly be taking place. One possible explanation might be that the infusion rate was not constant, but increased with time; however, we are quite certain that this was not the case. The most likely explanation is that the methotrexate did not disperse homogeneously throughout the fluid, and that the catheters by which fluid was extracted for assay were located in regions of relatively low concentrations. Although mechanical activity should enhance mixing in the pleural cavity, structural aspects may cause low concentrations in some regions. Nonconstant PA that decreased with increasing $C_f - C_p$ could not alone account for the observed character of the $\frac{d}{dt}\hat{C}_f$ curves, but would produce an effect in that direction.

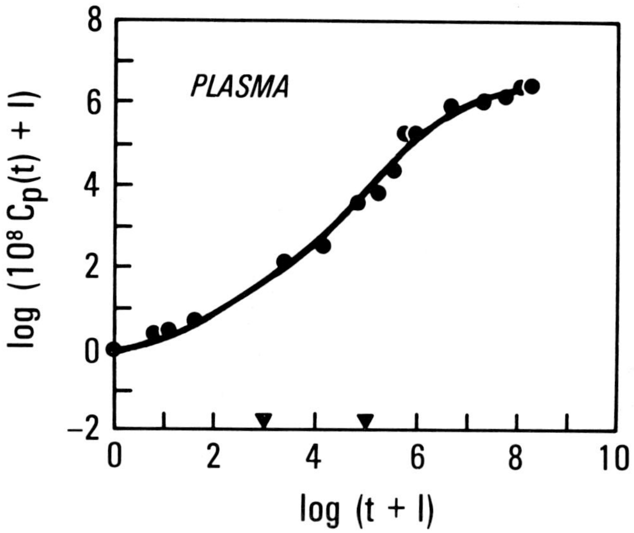

FIGURE 8

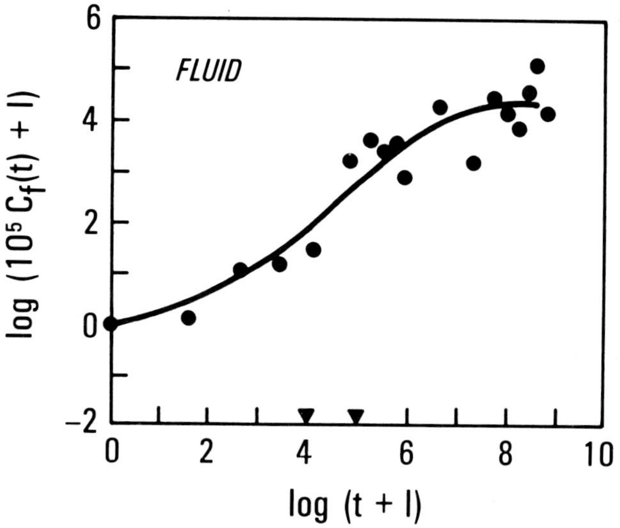

FIGURE 9

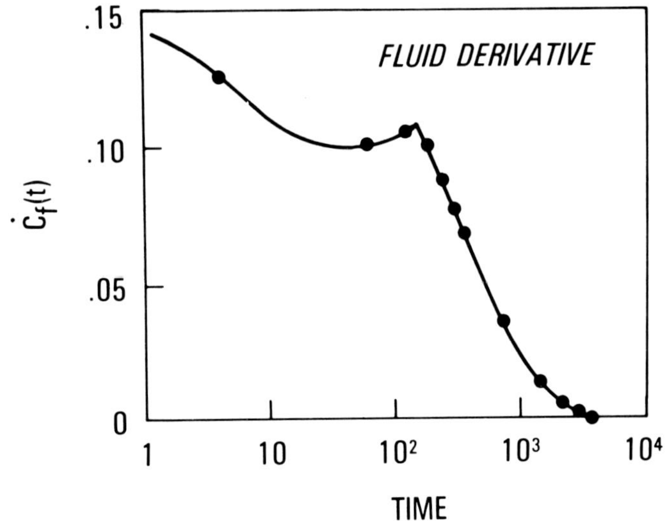

FIGURE 10

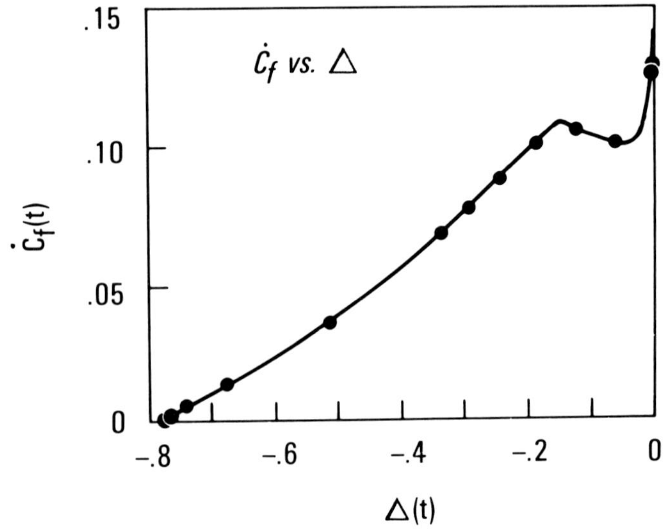

FIGURE 11

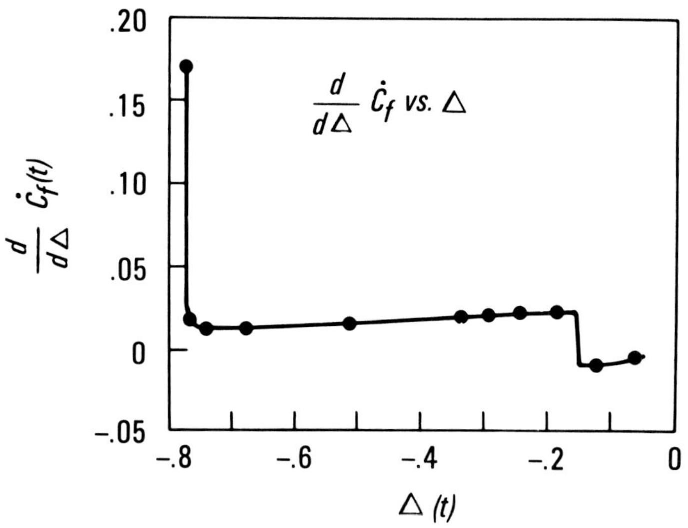

FIGURE 12

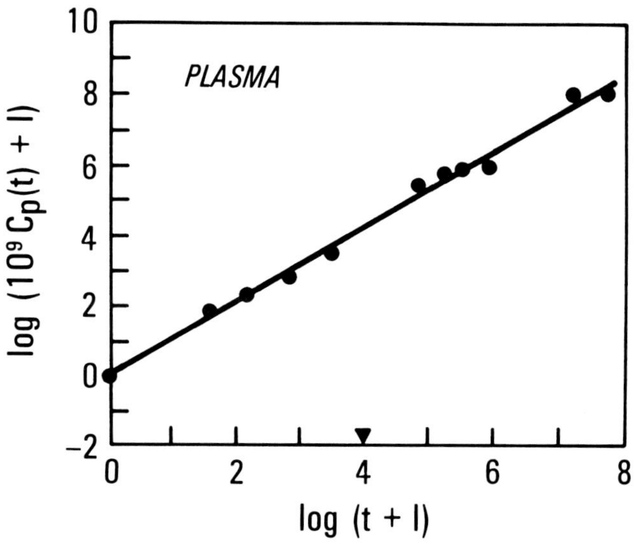

FIGURE 13

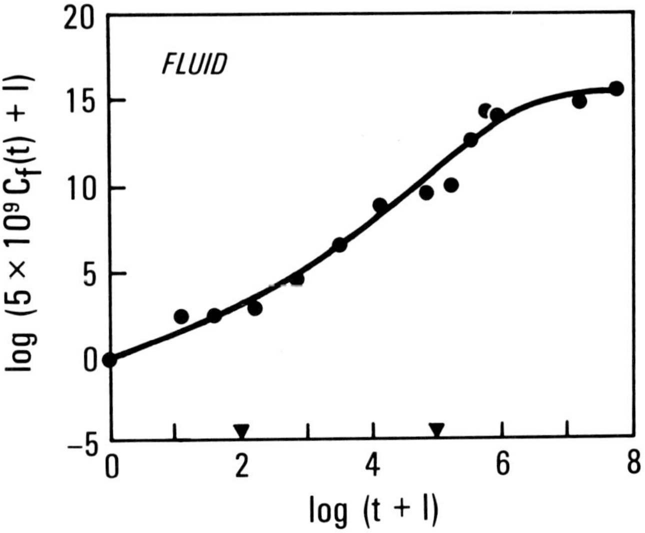

FIGURE 14

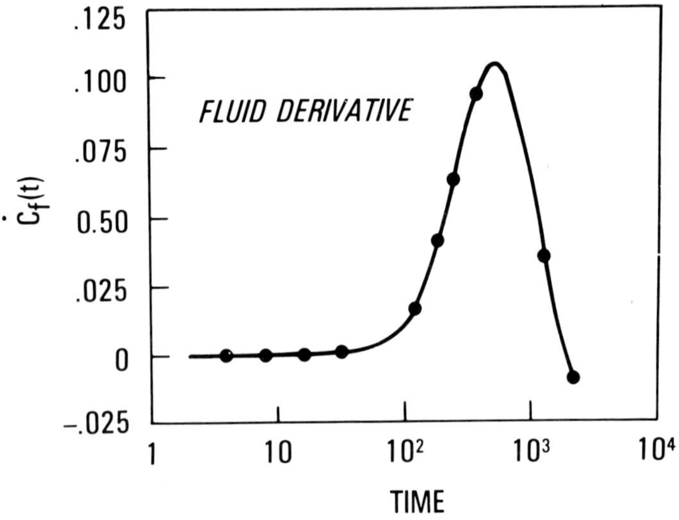

FIGURE 15

Another possible explanation is that the volume of fluid decreases with time, giving rise to an apparent increase in the infusion rate. If we let V_f be a function of time and differentiate equation (2.2) we obtain

$$\dot{V}_f \ddot{C}_f + \dot{V}_f \dot{C}_f = PA(\dot{C}_p - \dot{C}_f). \tag{2.12}$$

Evaluating this at t = 0 and using the initial conditions $V_f \dot{C}_f = I$, $\dot{C}_p = 0$, we obtain

$$V_f \ddot{C}_f = - \frac{I}{V_f} (PA + \dot{V}_f) . \tag{2.13}$$

Thus $\ddot{C}_f(0) > 0$ if $\dot{V}_f < -PA$, so that a leakage rate greater in magnitude than PA would account for the initial increase in $\dot{C}_f$. However, leakage rates of this magnitude do not seem plausible since the cavity would have been evacuated in a relatively short time.

For the 5 patients for whom the model fit we calculated PA both from the equilibrium concentrations and from the derivative calculated from the smoothed curves. The agreement is fair but not excellent; we do not believe that the data allow very precise determinations of PA because of the noise and the small number of observations. The comparisons are shown in Table 1.

TABLE I. Comparison of PA from Spline Fit and from Equilibrium Values (mℓ/min)

Patient	*(PA Equilibrium)*	*PA (Spline)*
1	6.6×10^{-2}	1.0×10^{-1}
2	1.2	6.0×10^{-2}
3	3.0	2.6×10^{-1}
4	2.2×10^{-1}	1.7×10^{-1}
5	3.1	3.9

III. DISCUSSION

We have discussed and illustrated how spline functions may be used in smoothing intrinsically variable data arising from pharmacokinetic experiments. This technique can be especially useful in the initial or exploratory stages of the analyses of such data and can be helpful in ascertaining the suitability of tentatively held models. Unexpected behavior of the data can be made apparent by these techniques.

By providing continuous values of the concentrations and their derivatives through time, it is possible to examine the plausibility of a first order linear kinetic model. If PA changes as a function of the concentration difference between the fluid and plasma, it is possible to examine this relationship.

In our investigations, that aspect of a compartmental model described by equation (2.2) was found to be reasonable in about half the patients. In these cases we were able to derive estimates of PA without making the assumption that plasma was the only other important compartment, for equation (2.2) only entails the assumption that the sole route of significant transfer of methotrexate from the fluid occurs via the plasma. Possible communication between the plasma and other compartments is not relevant.

The technique revealed that the data were inconsistent with the model of equation (2.2) for many of the patients. By examining the smoothed concentration derivatives we reached the tentative conclusion the misfit was due to systematic sampling error.

IV. ACKNOWLEDGMENTS

We gratefully acknowledge Shri Apta Good for writing the Fortran programs.

V. REFERENCES

Agarwal, G.G. and Studden, W.J. (1980). "Asymptotic Integrated Mean Squared Error Using Least Squares and Bias Minimizing Splines." *Ann. Statist. 8*, 1307-1325.

Bellman, R. (1970). *Introduction to Matrix Analysis*, McGraw-Hill, New York.

de Boor, C. (1978). *A Practical Guide to Splines*, Springer-Verlag, New York.

Braun, M. (1978). *Differential Equations and Their Applications*, Springer-Verlag, New York.

Dedrick, R.L., Myers, C.E., Bungay, P.M. and DeVita, V.T. (1978). "Pharmacokinetic Rationale for Peritoneal Drug Administration in the Treatment of Ovarian Cancer." *Cancer Treatment Reports 62*, 1-11.

Endrenyi, L. (editor) (1981). *Kinetic Data Analysis*, Plenum, New York.

Howell, S.B., Chu, B.C.F., Wung, W.E., Metha, B.M. and Mendelsohn, J. (1981). "Long-Duration Intracavitary Infusion of Methotrexate with Systemic Leucovorin Protection in Patients with Malignant Effusions." *J. Clinical Investigation 67*, 1161-1170.

Rubinow, S. I. (1973). *Mathematical Problems in the Biological Sciences*, Society for Industrial and Applied Mathematics, Philadelphia.

Wold, S. (1971). "Analysis of Kinetic Data by Means of Spline Functions." *Chemica Scripta. 1*, 97-102.

SELECTING REPRESENTATIVE POINTS IN NORMAL POPULATIONS

S. Iyengar

Department of Mathematics and Statistics
University of Pittsburgh

H. Solomon

Department of Statistics
Stanford University
Stanford, California

The representation of a continuous random variable by several discrete points occurs often in applied probability problems. Quantization is the term applied to this procedure and optimal quantizers have been sought by a number of investigators. This requires defining suitable measures for the error inherent in the procedure and then constructing quantizing procedures that minimize the expected error. While the problems that motivate quantization are far ranging, the mathematization leading to solutions is essentially always the same.

Most efforts are devoted to one dimensional random variables. Obviously two, three, and higher dimensional variables can lead to more intractability but in this paper we will explore some special cases in two and three dimensions. The loss function we employ is that of mean square error. Zador [1963, 1982] explored the multivariate normal random variable and its quantization by a random choice of representative points. He does not restrict himself to mean square error;

ISBN 0-12-589320-5

rather, he defines error as the sth power of the distance between the random variable and its quantization and derives results about its asymptotic properties.

The IEEE very recently published a special issue on the topic of quantization [1982] that collected quite recent work and work by Zador and other investigators reported but not published as much as 25 years ago. The papers in the issue arise out of an electrical engineering and information theory framework and ignore the efforts of workers in other disciplines who in turn are unaware of the work of these authors.

For the one-dimensional normal random variable situation, there are papers by Bofinger [1970] who studies the question of grouping a continuous bivariate normal by selecting intervals on the marginals that would provide the maximum possible correlation between the marginal variables and Sitgreaves [1961] who arrived at the same bivariate model as Bofinger in connection with a psychometric query on optimal test items for an achievement test. In each case, the univariate normal is quantized in an optimal manner. Maximizing the correlation is equivalent to minimizing mean square error in those cases. Previous workers also are Cox [1957], and Anderberg [1973], each of whom seeks to sectionalize or quantize the univariate normal for subsequent data analysis. Recently Fang and He [1982], motivated by clothing size category representations provide a detailed analysis for the univariate normal and give tables of representative points for N = 1,2,3,...,31. In an earlier paper in the electrical engineering literature, Max [1960] gives representative points for N = 1,2,3,...,26. When tabled values of optimal representative points and interval endpoints are listed there is consensus among the investigators where values can be compared.

A rather early paper on quantization is by Steinhaus [1956]. In that paper, he demonstrates the necessary (but not sufficient) conditions for optimal quantization, namely, that the optimal representative points are given by $q_i = E(X|X\varepsilon Q_i)$ when mean squared error is the loss function; and the optimal regions are nearest neighbor regions, namely

$$Q_i = \{x: |x-q_i| \leq \min_{\leq i \leq j} |x-q_j|\}.$$

Let us look into the computation of the optimal quantization of a continuous random variable, X. That is, we divide the real line into N disjoint intervals $\{Q_i\}_1^N$, pick a representative point $q_i \varepsilon Q_i$, and define $Q(X) = q_i$ whenever $X \varepsilon Q_i$. The loss of information is indicated by, say $(X-Q(X))^2$, and we wish to minimize $E(X-Q(X))^2$ by choosing $\{Q_i\}_1^N$ and $\{q_i\}_1^N$ appropriately. We now describe and compare the methods proposed by Lloyd [1957, 1982] and Zador [1963, 1982], and our modification of Zador's method.

Lloyd notes that given the intervals $\{Q_i\}_1^N$, the optimal representative points are given by the centroids $q_i = E(X|X\varepsilon Q_i)$. This is, of course, a consequence of the fact that we use mean squared error; if, instead we used $E|X-Q(X)|$ as our criterion, the optimal q_i would be given by the conditional median. He also notes that given $\{q_i\}_1^N$, the optimal intervals are just the nearest neighbor regions, $Q_i = \{x: |x-q_i| \leq \min_{1 \leq i \leq 1} |x-q_j|\}$. We have already noted that Steinhaus lists these two conditions.

These two necessary conditions for an optimal quantization suggest an iterative procedure. In particular, we start with points $\{q_i^{(1)}\}$ and define the corresponding optimal intervals $\{Q_i^{(1)}\}$; then we let $q_i^{(2)} = E(X|X\varepsilon Q_i^{(1)}),\ldots$. We repeat this procedure until we have convergence. One importance question, then, is when does the procedure converge? Lloyd presents a simple example to show that when the density of X is bimodal, then the iterative procedure may converge to a local minimum and not a global one. Kieffer [1982] has the following positive result: if the density of X is a log-concave which is not piecewise affine on $\mathbb{R}$, then this iterative procedure converges to the unique optimal points at an exponential rate; that is, if $q^{(N)} = \{q_i^{(N)}\}_{i=1}^N$, then $||q^{(N)}-q^*|| < \alpha^N$ for all large N where q* is the optimal quantization and $0 < \alpha < 1$. When $X \sim N(0,1)$, Lloyd gives a table of optimal points for N = 2,4,8,16 and the corresponding mean square errors. Our experience has shown that the initial points should be chosen symmetrically about zero, else the procedure converges much more slowly. For a table of the optimal representation points and errors, see the papers by Lloyd, Max, and Fang and He.

For future reference we write out the mean squared error when $X \sim N(0,1)$. Let $\phi(x)$ and $\Phi(x)$ be the standard normal density and distribution functions, respectively, and assume that $q_1 < q_2 < \ldots < q_N$. Then

$$\begin{aligned} E(X-Q(X))^2 &= 1-2EXQ(X) + EQ(X)^2 \\ &= 1+2q_1\phi\left(\frac{q_1+q_2}{2}\right) + 2\sum_{2}^{N-1} q_i\left\{\phi\left(\frac{q_i+q_{i+1}}{2}\right) - \phi\left(\frac{q_i-q_{i-1}}{2}\right)\right\} \\ &\quad - 2q_N\phi\left(\frac{q_{N-1}+q_N}{2}\right) + q_1^2\Phi\left(\frac{q_1+q_2}{2}\right) + q_N^2\Phi\left(\frac{q_N+q_{N-1}}{2}\right) \\ &\quad + \sum_{2}^{N-1} q_i^2\left\{\Phi\left(\frac{q_i+q_{i+1}}{2}\right) - \Phi\left(\frac{q_i+q_{i-1}}{2}\right)\right\} . \end{aligned}$$

Zador proposed a random quantizer: the q_i are chosen randomly according to some density g. The mean squared error is then a random variable and the problem now is to choose g so that some aspect of the distribution of the mean squared error is optimized. Zador shows that N^2 times the mean squared error has a limiting distribution and he computes the mean of this limit; $N^2(MSE_g) \xrightarrow{d} Z_g$ and $EZ_g = \mu_g$. He then shows that by choosing $g(x) = \phi^{1/3}(x)/\int\phi^{1/3}(t)dt$, μ_g is minimized. It is clear that the optimality criterion chosen by Zador is quite distinct from Lloyd's criterion. Random quantization in one dimension is not a crucial issue. This is because optimal quantization typically involves only one-dimensional integrals whose computation is efficient. Optimal quantization in higher dimensions, though, rapidly becomes expensive, and it is here that random quantization could be valuable. However, since we know many results for the one-dimensional case, it is of interest to see how random quantization compares in that case. In the one-dimensional case, we now propose several improvements over Zador's scheme.

First, choosing the asymptotically optimal $g(x) = (1/\sqrt{3})\phi(x/\sqrt{3})$ may not do well for finite N. In fact, we shall show that in one case, choosing $g_\sigma(x) = \frac{1}{\sigma}\phi(\frac{x}{\sigma})$ with $\sigma = .545$ yields substantial improvement. Second, it seems intuitively clear that the optimal points $\{q_i\}$ ought to be located symmetrically about zero. Zador's scheme does not guarantee this symmetry, but it is fairly easy to modify it to do so. To illustrate these modifications, we do the analytically tractable cases, N=2 and N=3.

When N=2, assuming $q_1 < q_2$, we have that

$$E(X-Q(X))^2 = 1+q_1^2\Phi(\frac{q_1+q_2}{2}) + q_2^2\Phi(-\frac{q_1+q_2}{2}) - 2(q_2-q_1)\phi(\frac{q_1+q_2}{2})$$

$$= 1-2(q_2-q_1)\phi(\frac{q_1+q_2}{2})+q_2^2-(q_2-q_1)(q_2+q_1)\Phi(\frac{q_1+q_2}{2}) .$$

Under Zador's scheme, we generate Y_1, Y_2 i.i.d. $N(0,\sigma^2)$ and set $q_1 = Y_{(1)}$, $q_2 = Y_{(2)}$. Then the range (q_2-q_1) is clearly independent of the mean $(q_1+q_2)/2$. Now let X_1, X_2 be i.i.d. $N(0,1)$. Then

(i) $q_2 \overset{d}{=} \sigma X_{(2)} \Rightarrow Eq_2^2 = \sigma^2 EX_{(2)}^2 = 2\sigma^2\int t^2\Phi(t)\phi(t)dt - \sigma^2.$

(ii) $E(q_2-q_1) = \sigma E|X_2-X_1| = 2\sigma/\sqrt{\pi}.$

(iii) $E\phi(\frac{q_1+q_2}{2}) = E\phi(\frac{\sigma}{\sqrt{2}} X) = 1/\sqrt{\pi(2+\sigma^2)} .$

$$\text{(iv)} \quad E(q_1+q_2)\Phi\left(\frac{q_1+q_2}{2}\right) = 2E\,\frac{\sigma}{\sqrt{2}}\,X\Phi\left(\frac{\sigma}{\sqrt{2}}X\right) = \frac{\sigma^2}{\sqrt{\pi(2+\sigma^2)}}\,.$$

Collecting terms, we see that under Zador's scheme, $MSE(\sigma) = E(X-Q(X))^2 = 1 + \sigma^2 - \frac{2\sigma}{\pi}\sqrt{2+\sigma^2}$. The asymptotically optimal σ^2 is 3, for which the mean square error is 1.53. Straightforward numerical work shows, however, that MSE (σ) is minimized for $\sigma_{min} = .545$ and MSE(.545) = .77, which is almost half of $MSE(\sqrt{3})$.

It is also of interest to know if we get much improvement if we force the random points to be symmetric about the origin. In this case we generate $Y \sim N(0,\sigma^2)$ and let $q_1 = -|Y|$, $q_2 = |Y|$. For this situation, the mean squared error is $E(1+Y^2-4|Y|/\sqrt{2\pi}) = 1+\sigma^2 - \frac{4}{\pi}\sigma$, which is minimized for $\sigma = \frac{2}{\pi}$ and the minimum value is $1 - \frac{4}{\pi^2} = .59$. A similar computation for N=3 (in which case $q_1 = -|Y|$, $q_2 = 0$, and $q_3 = |Y|$) shows that the mean square error assumes a minimum value of .41 for $\sigma = 1.32$.

We thus have the following short table:

Mean Squared Error

N	*Optimal*	*Zador,* $\sigma^2=3$	*Zador* σ^2_{min}	*Symmetrized*
2	.3634	1.53	0.77	0.59
3	.1902	.86[a]	—	0.41

[a] *Simulation*

The results for N = 4 are analytically intractable. Also, the .86 entry above is a simulation result while all others are exact. The optimal results for N=2, N=3 are known, Gray and Karnin [1982].

Before we turn to higher dimensional problems, we describe some difficulties in the one dimensional normal case. One might naively expect that the symmetry of the distribution requires that the points be located symmetrically about zero. The conclusion is indeed true for the normal, but the reason is not symmetry alone. To illustrate the difficulties here, we consider the cases N=2 and N=3 for an arbitrary symmetric density f.

When N=2, we have, say

$$Q_{ab}(x) = \begin{cases} a & \text{if} \quad x \leq \dfrac{a+b}{2} \\ \\ b & \text{if} \quad x \geq \dfrac{a+b}{2} \end{cases}$$

and we seek a and b to minimize $E(X-Q_{ab}(X))^2$. Differentiating this expression with respect to a and b and setting the partial derivatives equal to zero, we get

$$\int_{-\infty}^{\frac{a+b}{2}} (x-a)f(x)dx = 0$$

and

$$\int_{\frac{a+b}{2}} (x-b)f(x)dx = 0.$$

If we let $h(a,b) = \int_{-\infty}^{(a+b)/2} (x-a)f(x)dx$, then the two equations can be rewritten as $h(a,b) = h(-b,-a) = 0$. We can now say that if (a^*,b^*) provides an optimal quantization, then so does $(-b^*,-a^*)$: this seems to be the only consequence of symmetry. In order to say that (a^*,b^*) lie symmetrically about zero, we must have $h(a,b) = 0$ has a unique solution. One simple sufficient condition for this is that log f be strictly concave (see Fleischer [1964], Trushkin [1982]). We conjecture that a weaker sufficient condition is that f be unimodal and strictly decreasing from the mode. Notice that one solution is always $a^* = E(X|X<0)$ and $b^* = E(X|X>0)$.

If we require three points, one might invoke a symmetry argument to say that one of the points be at the origin. However, the following intuitively clear example shows that this is not the case. Consider the following class of symmetric bimodal densities:

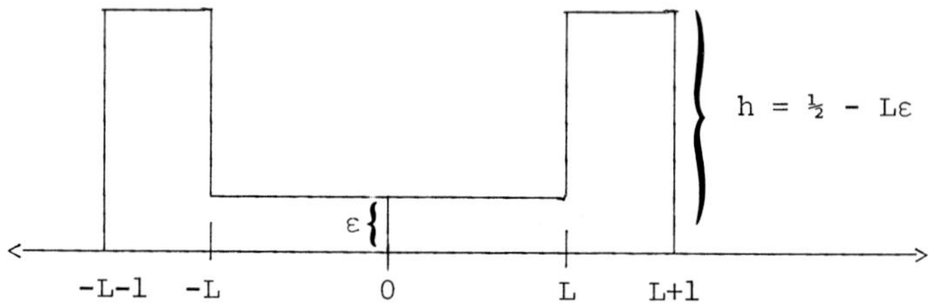

If ε is very small, then there is virtually no mass between $-L$ and L. Thus, if we put one of the points at zero, we are wasting it. If we put two points in the right mode and one in the left, we capture much more information in the random variable. Of course, in this case, we can reflect the asymmetrical quantization without changing the mean square error. We omit the details of such a counterexample. The non-optimality of the symmetric quantizer is a feature of an odd number of points.

The quantization of random vectors in $\mathbb{R}^d$ is in many ways a much more difficult problem than that of ordinary random variables. Even the "simplest" case of $X \sim N(0,I)$ presents many difficulties, as we shall see. First of all, whenever the random variable has a spherically symmetric density, any quantization can be rotated without changing the mean square error. More precisely, we have the following lemma, which is a generalization of the lemma of Gray and Karnin [1982].

Lemma. Suppose X has density $f(x) = g(x'x)$ and $Q(X)$ is any quantizer. Then the family of quantizers $\{Q_\Gamma(X)\}_\Gamma$ where $Q_\Gamma(X) = \Gamma' Q(\Gamma X)$ and $\Gamma'\Gamma = I$ have the same mean square error.

Proof. Clearly $X \overset{d}{=} \Gamma X$. Thus,

$$E|X-\Gamma' Q(\Gamma X)|^2 = E|\Gamma'(\Gamma X-Q(\Gamma X))|^2 = E|\Gamma X-Q(\Gamma X)|^2 = E|X-Q(X)|^2.$$

Thus, we should not consider the quantizations $Q(X)$ and $\Gamma' Q(\ X)$ as distinct.

As in the one-dimensional case, any quantization has two components, the subsets $\{Q_i\}_1^N$ of $\mathbb{R}^d$ and the representative points $\{q_i\}_1^N$ of each subset. Because we use mean square error, we again have the following necessary conditions for the optimality of a quantizer:

(i) the representative point must be the centroid of the respective subset: $q_i = E(X|X\varepsilon Q_i)$.

(ii) Q_i is determined by the nearest neighbor rule:

$$Q_i = \{x: ||x-q_i|| \leq \min_j ||x-q_j||\}.$$

It is clear that if Q(X) satisfies (i) and (ii), then so does $\Gamma' Q(\Gamma X)$. Thus, $\{\Gamma' Q(\Gamma X)\}_\Gamma$ are all fixed points of Lloyd's algorithm. This is not the real source of the difficulty, however. In general, there will be several distinct local minima. For example in the normal case, if d=2 and N=4, Gray and Karnin give the following configurations which are fixed points of Lloyd's algorithm.

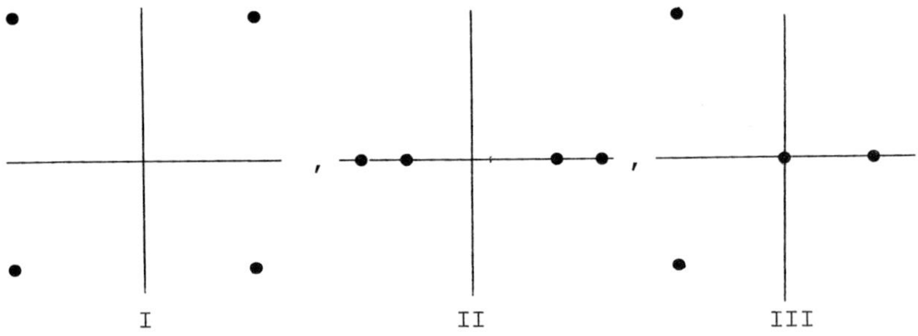

They conjecture that these three are the only fixed points of the algorithm. Standardizing the error - that is, considering $\frac{1}{d} E||X-Q(X)||^2$, Gray and Karnin show that the average errors are .3634, .5588, and .4102 for I, II, and III respectively. We call configuration I the product quantizer since it is the Cartesian product of the optimal one-dimensional quantizer with itself. Quantizer II performs rather poorly. Quantizer III has the intuitive appeal that one point is located at the origin, which is the mode of the distribution. Gray and Karnin comment that it "was a surprise to us that the distortion resulting from [code III] was so much larger than that of [code I]."

However, the intuition that says that one representative point should be the origin because that is the location of the mode of the distribution can be very misleading. Consider, for instance, the problem of quantizing $X \sim N_2(0,I)$ with three points. Two configurations immediately come to mind:

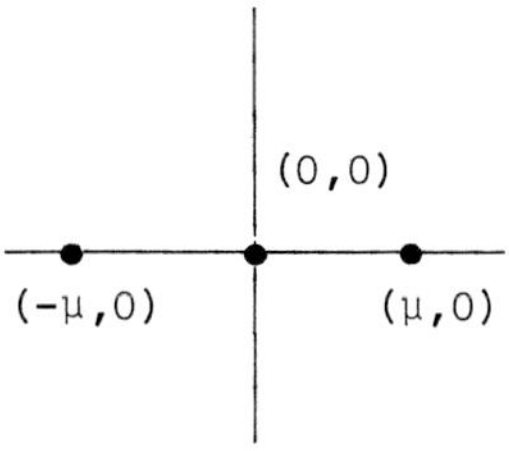

and

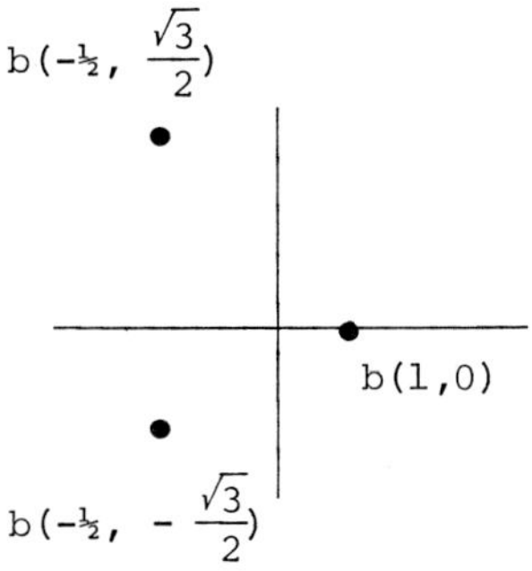

In the first case, the mean square error as a function of μ is

$$\begin{aligned} \mathrm{MSE}_1(\mu) &= E||X-Q(X)||^2 = E||X||^2 + E||Q(X)||^2 - 2EX'Q(X) \\ &= 2+2\mu^2\Phi(-\frac{\mu}{2})\ 4\int_{\infty}^{\infty}\int_{\mu/2}^{\infty} x\mu\phi(x)\phi(y)\,dx\,dy \\ &= 2+2\mu^2\Phi(-\frac{\mu}{2}) - 4\mu\phi(\frac{\mu}{2}). \end{aligned}$$

To find the μ that minimizes this, we set $\mathrm{MSE}_1(\mu) = 0$ to get $R(\frac{\mu}{2}) = \frac{1}{\mu}$, where $R(x) = \Phi(-x)/\phi(x)$ is Mills' ratio (for a more thorough discussion of Mills' ratio, see Iyengar [1982]). Straightforward numerical work shows that the minimizing μ is 1.224 and that the average noise is $\frac{1}{2}\,\mathrm{MSE}_1(1.224) \simeq .5951$.

The noise for the second configuration is

$$\begin{aligned} \mathrm{MSE}(b) &= 2 + b^2 - 2EX'Q(X) \\ &= 2 + b^2 - 2(3)\int_0^{\infty}\int_{-x\sqrt{3}}^{x\sqrt{3}} xb\ \phi(y)\phi(x)\,dy\,dx \\ &= 2 + b^2 - 3\sqrt{3}\ \phi(0)b. \end{aligned}$$

In this case our optimizing b is easily seen to be $\frac{3\sqrt{3}}{2}\phi(0)$, so that the average mean squared error is $1-27/16\pi \simeq .4629$. It is now clear that the second configuration is considerably better than the first one, even though the first one has a point at the origin. (Note that the two configurations do satisfy the two necessary conditions for optimality; we omit a formal proof.)

Quantization of a standard normal in three dimensions provides some new interesting twists. If we use eight points, one obvious choice is the product quantizer; the interesting result here is that the product quantizer can be improved upon. Indeed, Gray and Karnin give three different configurations that beat the product code. In one, there is a point at the origin, two points lie symmetrically on a line orthogonal to a plane formed formed by a pentagon whose vertices are the other five points. Another quantizer, suggested by N.J.A. Sloane to Gray and Karnin [1982] was obtained by rotating the top square of the product quantizer by 45° to obtain a twisted cube. Gray and Karnin report simulation results to show that these quantizers are superior. We did a straightforward (but very expensive) numerical integration to get the following results listed below. Notice that the best of these three does not have a representative point at the origin.

Mean Square Error

Quantizer	*Simulation (Gray,Karnin)*	*Numerical Integration*	*True*
Product	-	*.3635*	$1-\frac{2}{\pi} = .3634$
Pentagon-origin-poles	*.3590*	*.3585*	-
Twisted cube	*.3573*	*.3581*	-

REFERENCES

Anderberg, M.R. (1973). *Cluster Analysis for Applications.* Academic Press, New York and London.

Bofinger, E. (1980). "Maximizing the Correlation of Grouped Observations." *J. Amer. Statist. Assoc. 65,* 1632-1638.

Cox, D.R. (1957). "Note on Grouping." *J. Amer. Statist. Assoc. 52,* 543-547.

Fang, K.-T. and He, S.-D. (1982). The Problem of Selecting a Given Number of Representative Points in a Normal Population and a Generalized Mills' Ratio. Department of Statistics Technical Report, Stanford University.

Fleischer, P.E. (1964). "Sufficient Conditions for Achieving Minimum Distortion in a Quantizer." *IEEE Int. Conv. Rec.* Part 1, 104-111.

Gray, R. and Karnin, E. (1982). "Multiple Local Minima in Vector Quantizers." *IEEE Trans. Inf. Theory,* March 1982, 256-261.

Iyengar, S. (1982). On the Evaluation of Certain Multivariate Normal Probabilities. Department of Statistics Technical Report No. 322, Stanford University.

Kieffer, J. (1982). "Exponential Rate of Convergence for Lloyd's Method I." *IEEE Trans. Inf. Theory,* March 1982, 205-210.

Lloyd, S. (1957). Least Squares Quantization in PCM, draft manuscript, Mathematical Research Department, Bell Laboratories, New Jersey.

Lloyd, S. (1982). "Least Squares Quantization in PCM." *IEEE Trans. Inf. Theory,* March 1982, 129-136.

Max, J. (1960). "Quantizing for Minimizing Distortion." *IRE Trans. Inform. Theory,* IT6, 7-12.

Sitgreaves, R. (1961). Optimal Test Design in a Special Situation. *Studies in Item Analysis and Prediction,* H. Solomon, ed., 129-145. Stanford University Press, Stanford, California.

Steinhaus, H. (1957). "Sur la Division des Corps Materials en Parties." *Bull. Acad. Pol. Sci. Ser. Math. CL. III, IV,* 12, 801-804.

Trushkin, A. (1982). "Sufficient Conditions for Uniqueness of a Locally Optimal Quantizer." *IEEE Trans. Inf. Theory,* March 1982, 187-198.

Zador, P. (1963). Development and Evaluation of Procedures for Quantizing Multivariate Distributions, Ph.D. dissertation, Stanford University, Stanford, California.

Zador, P. (1982). "Asymptotic Quantication Error of Continuous Signals and the Quantization Dimension." *IEEE Trans. Inf. Theory,* March 1982, 139-148.

LEAST INFORMATIVE DISTRIBUTIONS

E. L. Lehmann

Department of Statistics
University of California
Berkeley, California[1]

I. INTRODUCTION

It was shown by Kagan, Linnik and Rao (1973) that among all distributions with given variance and satisfying certain regularity assumptions, the normal distribution minimizes the Fisher information. It is thus the distribution for which in large samples the task of estimating the location parameter is the hardest.

The purpose of the present note is to point out a formulation in which this assertion holds also for small samples, and without the regularity assumptions required for the asymptotic result. Section IV discusses a somewhat analogous property of the exponential distribution and the last section suggests some generalizations.

Consider a sample $\underline{X} = (X_1, \ldots, X_n)$ from a location family $F(x - \xi)$ and the problem of estimating ξ with a loss function $L(d,\xi) = \rho(d - \xi)$. We shall restrict attention to equivariant estimators, ie. estimators δ satisfying

$$\delta(x_1 + c, \ldots, x_n + c) = \delta(x_1, \ldots, x_n) + c. \qquad (1)$$

[1] *Research partially supported by NSF Grant MCS79-03716.*

ISBN 0-12-589320-5

Since the risk function of any such estimator is constant (independent of ξ) there will typically exist a minimum risk equivariant (MRE) estimator. If δ^* is any equivariant estimator the MRE estimator is given by

$$\delta(\underline{X}) = \delta^*(\underline{X}) - v(\underline{Y}) \tag{2}$$

where $\underline{Y} = (Y_1, \ldots, Y_{n-1})$ with $Y_i = X_i - X_n$ and where the function v is determined so as to minimize the conditional expectation

$$E_o\{\rho[\delta^*(\underline{X}) - v(\underline{y})]|\underline{y}\}, \tag{3}$$

the expectation being calculated for $\xi = 0$. (See for example Rao (1952) or Ferguson (1967 Sect. 4.7)). Let the risk of the MRE estimator be r(F) when F is the true distribution of the X's. Then for a given loss function ρ we shall call a distribution F_o that maximizes r(F) over a class $\mathcal{F}$ of distributions least informative within $\mathcal{F}$.

II. THE NORMAL DISTRIBUTION

Let the loss function be squared error, so that $\rho(t) = t^2$. In order to deal with comparable distributions, it is necessary to standardize their scale. As in the result for Fisher information, we shall for this purpose restrict attention to the class $\mathcal{F}_o$ of distributions with fixed variance σ_o^2.

Theorem 1

Among all distributions $F \in \mathcal{F}_o$, the normal distribution maximizes r(F), and for $n \geq 3$ it is the only distribution with this property.

Proof. It is well known that the MRE estimator in the normal case is $\bar{X}$. (This is easily seen from (2) and (3) by putting $\delta^*(\underline{X}) = \bar{X}$ and using the independence of $\bar{X}$ and $\underline{Y}$.) For the normal distribution N, we therefore have $r(N) = \sigma_0^2/n$ while for any other distribution $F \in \mathcal{F}_0$ the minimizing value r(F) is $\leq$ the risk achieved by $\bar{X}$, so that $r(F) \leq r(N)$ for all $F \in \mathcal{F}_0$. (The essential property is of course the fact that the risk of $\bar{X}$ is constant over $\mathcal{F}_0$.)

For n = 1,2, $\bar{X}$ is MRE for any symmetric F. However, for $n \geq 3$ we shall now show that $\bar{X}$ is MRE only when F is normal, so that $r(F) < r(N)$ for all $F \neq N$. Since the MRE estimator is unique and is given by $\bar{X} - E_0(\bar{X}|\underline{Y})$, $\bar{X}$ is MRE if and only if $E_0(\bar{X}|Y) = 0$. For $n \geq 3$ it was proved by Kagan, Linnik and Rao (1965) that this last equation holds if and only if F is normal, and this completes the proof.

Theorem 1 of course does not imply that the normal distribution is least informative (among all distributions with the same variance) in the sense of the comparison of experiments, as discussed for example in the review paper by Torgersen (1976). To see that, in fact, the normal distribution N does not possess this stronger property, it is enough to exhibit a different loss function, for which N does not maximize r(F). We shall in the next section consider this and some related questions for the case n = 2. It seems plausible that the conclusions will continue to hold for n > 2, but the proof is likely to be less simple.

III. THE CASE n = 2

The case of two observations is particularly simple because for large classes of loss functions and distributions F, $\bar{X}$ is MRE.

Theorem 2

If F is symmetric about 0 then $\bar{X}$ is MRE when

(i) ρ is convex and even;

or

(ii) $\rho(t) = \begin{cases} 0 \text{ if } |t| \leq a \\ 1 \text{ if } |t| > a \end{cases}$, and F is strongly unimodal.

Proof. The symmetry of F implies that the conditional distribution of $\bar{X}$ given $Y = X_2 - X_1$ is symmetric about 0. By (3), the problem then reduces to that of determining the constant v that minimizes $E[\rho(\bar{X} - v)]$ for the conditional distribution given y. From this, (i) follows immediately, while (ii) is a consequence of the fact that for strongly unimodal F, the conditional density of $\bar{X}$ given y is unimodal for each y. (This is easily seen from the fact that the conditional density of $Z = \bar{X}$ given y is proportional to $f(z - \frac{1}{2}y)f(z + \frac{1}{2}y)$ and that $-\log f(t)$ is convex.)

As an example, consider the double exponential distribution which, like the normal, is strongly unimodal. If without loss of generality we fix the value of the variance at 1, the double exponential (DE) density is $\frac{1}{\sqrt{2}}\exp(-\sqrt{2}|x|)$ and the density of the MRE estimator Z is

$$p_Z(z) = (2|z| + \frac{1}{\sqrt{2}})\exp(-2\sqrt{2}|z|). \tag{4}$$

Comparing this with the standard normal density $\phi(z)$, we see that for sufficiently large z

$$\phi(z) < p_Z(z) \tag{5}$$

and hence that for sufficiently large a

$$r(\text{DE}) = 2\int_a^\infty p_Z(z)dz > 2\int_a^\infty \phi(z)dz = r(N).$$

This example might owe its success at least in part to the fact that the standardization (by equating variances) is not well matched to the loss function and thus exerts too little control over F. A more natural standardization in the present case might be obtained by holding

$$F(a) - F(-a) \tag{6}$$

constant.

This modification does not however change the conclusion. A change in scale in the double exponential and/or normal distribution affects only the scale of these distributions and not their shape, and (5) therefore continues to hold for all sufficiently large z no matter what scales are adopted.

IV. EXPONENTIAL DISTRIBUTION

As a second example of a least informative property of the kind established for the normal distribution in Section II, consider location facmilies with lower threshold ξ, i.e. with cdf $F(x - \xi)$ where $F(x) = 0$ for $x < 0$ and > 0 for $x > 0$. We wish to estimate ξ with loss function $L(d,\xi) = 1$ if $|d - \xi| > a/2$ and $= 0$ otherwise. To standardize the scale, we shall restrict attention to the class $\mathcal{F}_2$ of distributions F with lower threshold 0 and fixed value F(a). (This is the one-sided version of the standardization (6) of Section III.)

Theorem 3

Among all distributions in $\mathcal{F}_2$ the exponential distribution maximes r(F).

Proof. Let $X_{(1)} <\ldots< X_{(n)}$ be the ordered observations and suppose that F is the exponential distribution in $\mathcal{F}_2$. By putting $\delta_0(\underline{X}) = X_{(1)}$ and using the independence of $X_{(1)}$ and the differences $X_{(i)} - X_{(1)}$ $(i = 2,\ldots,n)$, one sees from (2) and (3) that the MRE estimator is $X_{(1)} - v$ where the constant v is determined so as to maximize

$$P_\xi(|X_{(1)} - v - \xi| \leq a/2) = P_0(|X_{(1)} - v| \leq a/2).$$

Since $X_{(1)} - \xi$ has an exponential distribution with lower threshold 0, its density is strictly decreasing for $x > 0$ and the interval of length a with maximum probability is the interval $(0,a)$, with $v = a/2$. The MRE is therefore $X_{(1)} - \frac{a}{2}$ and its risk for any $F \in \mathcal{F}_2$ is

$$P_0(|X_{(1)} - \frac{a}{2}| > \frac{a}{2}) = (1 - F(a))^n.$$

The fact that this is constant over $\mathcal{F}_2$ completes the proof.

Whether the exponential distribution is the only distribution maximizing r(F) is an open question.

V. A MINIMAX FORMULATION

The property of the normal distribution in Section 2 can be given a somewhat different form by adopting a somewhat broader point of view. Quite generally, consider the problem of estimating a functional $\mu(F)$ in a family $\mathcal{F}$ with a given loss function. Then we shall say that any subfamily $\mathcal{F}_0$ of $\mathcal{F}$ is least informative for this problem if the maximum risk of the minimax estimator over $\mathcal{F}_0$ agrees with that over $\mathcal{F}$. (If $\mathcal{F}_0$ is independent of sample size, this means that the smallest sample size with which we can estimate $\mu(F)$ with risk not exceeding a given bound is the same whether or not we restrict attention to $\mathcal{F}_0$.)

As an example generalizing the location problem, consider the problem of estimating $\mu(F) = E_F(X_i)$ on the basis of a sample $X_1,\ldots,X_n$ from F with loss function $(d - \mu_F)^2/\mathrm{Var}_F(X_i)$, with F the class of all distributions on the real line with finite variance. The family F_0 of normal distributions is then least informative since $\bar{X}$ is minimax both over F_0 and F with the minimax risk for both families being 1/n.

The situation differs however from that in the location case, in that the normal family is no longer unique. The binomial and Poisson families provide two other examples in which $\bar{X}$ is minimax with minimax risk 1/n, and there are other exponential families with densities (with respect to a σ-finite μ)

$$p_\theta(x) = C(\theta)e^{\theta x} \tag{7}$$

for which this is the case. On the other hand, this property is not shared by all families (7) as is seen for example by letting $X_i = Y_i^2$ where Y is normal $N(0,\sigma^2)$, $\theta = -1/2\sigma^2$ and the minimax estimator of σ^2 is $\Sigma X_i/(n + 2)$. In fact the least informative families (7) are just those for which $\bar{X}$ is admissible, for which sufficient conditions were given by Karlin (1958).

VI. REFERENCES

Kagan, A.M., Linnik, Yu V. and Rao, C.R. (1965). "On a characterization of the normal law based on a property of the sample average." *Sankhyā Ser. A 27*, 405-406.

Kagan, A.M., Linnik, Yu V. and Rao, C.R. (1973). *Chacterization Problems in Mathematical Statistics*, Wiley, New York.

Karlin, S. (1958). "Admissibility for Estimation with Quadratic Loss." *Ann. Math. Statist. 29*, 406-436.

Rao, C.R. (1952). "Some Theorems on Minimum Variance Estimation." *Sankhyā Ser. A 12*, 27-42.

Torgersen, E.N. (1976). "Comparison of Statistical Experiments." *Scand. J. Statist. 3*, 186-208.

SIGNIFICANCE LEVELS, CONFIDENCE LEVELS, CLOSED REGIONS, CLOSED MODELS, AND DISCRETE DISTRIBUTIONS[1]

John W. Pratt

Graduate School of Business Administration
Harvard University
Boston, Massachusetts

I. INTRODUCTION

"All the data with which the statistician actually works comes from *discontinuous distributions*...When his theories are erected on a basis of a probability density function, or even a continuous cumulative, there is a definite extrapolation from theory to practice. It is, ultimately, a responsibility of the mathematical statistician to study discrete models and find out the dangerous large effects and the pleasant small effects which go with such extrapolation. *We all deal with discrete data, and must sooner or later face this fact.*" (Tukey, 1948, p. 30, including emphasis.)

One obvious question is how discreteness affects significance and confidence levels derived assuming continuity. Special cases have been handled in passing ("we easily obtain," Kolmogoroff, 1941, p. 462; "as usual," Tukey, 1949, p. 17) or in extenso (Tukey, 1948; Pratt, 1959), but published general results seem

[1] *Helpful comments by William S. Krasker, David Pollard, and Richard Savage and research support from the Associates of the Harvard Business School are gratefully acknowledged.*

ISBN 0-12-589320-5

surprisingly lacking except Noether's (1967) "projection" method and some exercises in Pratt and Gibbons (1981). Perhaps this is because special cases are easy to treat by a variety of methods, especially if tedious details are suppressed.

Chernoff's Advanced Probability course and his much-more-than-survey paper (1956) enlarged and codified with exemplary clarity an extremely useful body of asymptotic methods. Our purpose here is to point out that one can approach a discrete distribution via continuous ones without changing the sample size and still employ a tool from this kit to show easily that:

> The validity of significance and confidence levels based on closed acceptance regions is preserved by convergence in distribution and hence carries over quite generally from continuous models to their discrete counterparts.

This pleasant result applies, for example, to tests which accept the null hypothesis at or below critical values of continuous (or lower semicontinuous) test statistics, and to closed confidence intervals whose end points are continuous functions of the observations, under models assuming symmetry, identical distributions, shift, or permutation invariance. The usual nonparametric rank and other randomization procedures have these properties if ties are broken and signs attached to zeroes so as to favor rejection.

Though discrete distributions and nonparametric procedures provide motivating illustrations, we will pursue our ideas where they naturally lead and will present the results in a general form. The essential step applies to any hypothesis or model: take the set of distributions allowed and close it under convergence in distribution. "Our choice of the level of generality was to facilitate our writing and your reading." (Chernoff and Savage, 1958, p. 973.)

II. TESTS

Let P_n and P be probability distributions on a Euclidean space of finite or countably infinite dimension. By the usual definition, $P_n \to P$ in distribution (or law) if the corresponding convergence of cdf's holds at all points of continuity of the limit. We have

$P_n \to P$ in distribution iff

$$\text{for all closed } A,\ P(A) \geq \limsup P_n(A). \tag{1}$$

In more general spaces (of secondary concern here) this may be taken as the definition. (See, e.g., Chernoff, 1956).

Let $\mathcal{F}$ be a family of distributions and $\overline{\mathcal{F}}$ its closure under convergence in distribution, that is, the family of all limits in distribution of sequences in $\mathcal{F}$. Regard $\mathcal{F}$ and $\overline{\mathcal{F}}$ as statistical null hypotheses. By (1),

Theorem 1. A test whose acceptance region is closed has the same exact significance level for $\overline{\mathcal{F}}$ as for $\mathcal{F}$.

Specifically, given any $P \epsilon \overline{\mathcal{F}}$, the probability of acceptance $P(A)$ satisfies (1) for some $P_n \epsilon \mathcal{F}$. Hence any significance level for $\mathcal{F}$ (upper bound for the probability of rejection on $\mathcal{F}$) is also valid for $\overline{\mathcal{F}}$. The exact significance level (least upper bound) is therefore no greater for $\overline{\mathcal{F}}$ than for $\mathcal{F}$, and it cannot be smaller since $\overline{\mathcal{F}}$ contains $\mathcal{F}$.

Because closing the acceptance region can only enlarge it, Theorem 1 is equivalent to

Theorem 1'. If an acceptance region has significance level α for $\mathcal{F}$ then its closure has level α for $\overline{\mathcal{F}}$.

For tests expressed in terms of a test statistic, we have

Corollary 1. If T is a continuous (or lower semicontinuous) statistic, then the test rejecting when $T > k$ has the same exact significance level for $\overline{\mathcal{F}}$ as for $\mathcal{F}$.

Here k is any constant. T is lower semicontinuous means by definition that for every x' and every $\varepsilon > 0$ there exists a neighborhood of x' where $T(x) > T(x') - \varepsilon$. Equivalently, for every k, the set where $T > k$ is open. Thus the Corollary follows the Theorem. (The definition and equivalence are standard. For upper semicontinuity, reverse all three inequalities. We may also restate (1) to say that the probability of a closed region as a function of P is upper semicontinuous with respect to convergence in distribution.)

For an arbitrary test statistic T, closing the acceptance region $T \leq k$ gives

Corollary 1'. If the test rejecting when $T > k$ has level α for F then the test rejecting when $\overline{T} > k$ has level α for $\overline{F}$, where

$$\overline{T}(x) = \min\{T(x), \liminf_{y \to x} T(y)\}. \quad (2)$$

III. CONFIDENCE PROCEDURES

Direct translation of the foregoing results from tests to confidence regions is straightforward but leaves the closure conditions less transparent than they might be. In terms of the classic picture relating acceptance and confidence regions, the closure conditions relate to cross-sections with the parameter fixed (acceptance regions), but attention is now focussed on cross-sections with the observations fixed (confidence regions).

For each value of a parameter θ, let $F(\theta)$ be a family of distributions and $\overline{F}(\theta)$ its closure under convergence distribution as before. Let $C(x)$ be a confidence region for θ when x is observed. The corresponding test accepts $F(\theta)$ iff $\theta \in C(x)$. Thus, by Theorem 1,

Theorem 2. If $\{x: \theta \in C(x)\}$ is closed, then $C(x)$ has the same exact confidence level on $\overline{F}(\theta)$ as on $F(\theta)$.

Here θ is totally arbitrary and the Theorem applies to each value of θ individually. (A confidence level at θ is defined as any lower bound of $P\{x: \theta \in C(x)\}$ for $P \in \mathcal{F}(\theta)$ or $\bar{\mathcal{F}}(\theta)$ as appropriate, and the exact confidence level as the greatest lower bound.) For a real parameter θ we have

Corollary 2. A closed confidence interval whose lower and upper end points are continuous functions of x (possibly $\pm \infty$), or lower and upper semicontinuous respectively, has the same exact confidence level on $\bar{\mathcal{F}}(\theta)$ as on $\mathcal{F}(\theta)$.

Though useful, this corollary is misleading in that a topology on the θ-space is irrelevant. Given any $C(x)$, let $\bar{C}(x)$ be the confidence region resulting from closing the acceptance regions corresponding to $C(x)$. In terms of the classic picture, this means adding all limit points in the cross-sections with θ fixed and x varying. By Theorem 1',

Theorem 2'. If $C(x)$ has confidence level $1 - \alpha$ on $\mathcal{F}(\theta)$, then $\bar{C}(x)$ has level $1 - \alpha$ on $\bar{\mathcal{F}}(\theta)$.

Since the acceptance region for $\mathcal{F}(\theta)$ corresponding to $C(x)$ is those x for which $\theta \in C(x)$, its closure is all limits of x_n for which $\theta \in C(x_n)$, allowing $x_n = x$. Thus

$$\bar{C}(x) = \{\theta: \exists x_n \to x \text{ such that } \theta \in C(x_n)\}, \tag{3}$$

which can be reexpressed as

$$\bar{C}(x) = \lim_{\varepsilon \to 0} \bigcup_{|y-x| < \varepsilon} C(y) = C(x) \cup \limsup_{y \to x} C(y). \tag{4}$$

The lim sup in (4) is set-theoretic in θ-space, not topological. Even when θ has a topology, it is clear from the classic picture that $\bar{C}(x)$ need be neither closed (where the acceptance region is discontinuous) nor contained in the closure of $C(x)$ (where $C(x)$ is discontinuous). However,

Corollary 3. If $C(x)$ is an interval whose lower and upper end points are respectively lower and upper semicontinuous functions of x, then its closure contains $\overline{C}(x)$. If C(x) as a function of x, or equivalently, the corresponding acceptance region as a function of θ, is an upper semicontinuous correspondence, then C(x) is closed and $\overline{C}(x) = C(x)$. In either case, if also C(x) has confidence level 1 - α on $F(\theta)$, then its closure has level 1 - α on $\overline{F}(\theta)$.

C(x) is an upper semicontinuous correspondence means that $\theta \in C(x)$ if $x_n \to x$, $\theta_n \in C(x_n)$, and $\theta_n \to \theta$. This definition is used in other contexts in mathematical economics. It is equivalent to closure in (x, θ)-space in the classic picture and hence is stronger than necessary here. It is not equivalent to upper semicontinuity for functions. Incidentally, although C(x) as a function of x and the acceptance region as a function of θ are inverses in a sense, they are not inverse functions, since each is a function from points to sets (a correspondence).

In the following example, C(x) has continuous end points but is not always closed, and closing it would destroy one of its purposes, to exclude 0 as much as possible. (See Pratt, 1961, for other purposes and a similar, normal-theory example.) Nevertheless, $\overline{C}(x) = C(x)$ so Theorems 2 and 2' apply to show that C(x) has level 1 - α in the discrete case even though C(x) is not always closed. Of course neither C(x) nor the corresponding acceptance region is an upper semicontinuous correspondence.

Example: Let $W_1, \ldots, W_n$ be the order statistics of a sample of size n and let k and h be lower one- and two-tailed critical values for the one-sample sign test at level α $(1 \leq h \leq k < n/2)$. Let C(x) be the interval with end points $\min(W_{k+1}, 0)$ and $\max(W_{n-k}, 0)$, excluding the end point 0 when $0 < W_{h+1}$ or $0 > W_{n-h}$ but including the end points otherwise. (It is the confidence interval for the median θ derived from the most powerful sign test of each θ against 0, and an equal-tailed test for θ = 0, with closed acceptance regions.)

Similar examples could obviously be based on other procedures, such as the Wilcoxon signed-rank test and the Walsh averages, or the analogous two-sample procedures.

In Theorems 2 and 2', it is understood that the parameter value θ belongs to every $P \in \overline{F}(\theta)$. Formalizing this adds some insight. Given $F(\theta)$, define the set

$$\Theta(P) = \{\theta: P \in \overline{F}(\theta)\}. \tag{5}$$

$\Theta(P)$ is empty iff P is not a limit of distributions in some $F(\theta)$, i.e., $P \notin \cup_\theta \overline{F}(\theta)$. Otherwise $P \in \overline{F}(\theta)$ for some θ, in which case $\Theta(P)$ includes θ, but $\Theta(P)$ may include more than one point, even if P belongs to one of the original families $F(\theta)$.

Theorem 3. If C(x) has confidence level $1-\alpha$ on $F(\theta)$, then $\overline{C}(x)$ contains $\Theta(P)$ with probability at least $1 - \alpha$, that is, $P\{x: \overline{C}(x) \supset \Theta(P)\} \geq 1 - \alpha$ for all P.

Suppose, for example, $F(\theta)$ is the set of all continuous distributions P with unique or central median θ, that is, θ is the mid-point of the interval $\mu(P)$ consisting of all medians of P; then $\Theta(P) = \mu(P)$. Thus the closed interval between two order statistics contains all possible medians with probability no less than the usual confidence level whether P is discrete or continuous. (Of course this is not hard to show directly, once one thinks to. See, e.g., Pratt and Gibbons, 1981, p. 94.)

IV. REMARKS

We conclude with some remarks on what is required to check the closure conditions, on sequential and nonparametric procedures, on alternative approaches using tie-breaking and transformations, and on open instead of closed regions.

Verifying limits. The limits and limiting distributions referred to in Sections II and III occur in the space of observation vectors. To check that an acceptance region is closed or a confidence bound continuous, one must of course consider limits as all observations vary simultaneously. This is typically easy enough. And establishing the necessary limiting distributions is also easy if the observations are independent, since convergence of the joint distribution then follows from convergence of the distribution of each observation. For example, if $Y_1,\ldots,Y_m$, $Z_1-\mu,\ldots,Z_n-\mu$ are independently identically distributed with distribution G, and if $G_\nu \to G$ in distribution as $\nu \to \infty$, then the joint distribution of $Y_1,\ldots,Y_m$, $Z_1,\ldots,Z_n$ under G_ν approaches that under G; G might be discrete and G_ν continuous. (The classic picture relating tests and confidence regions might suggest using monotonicity, but since the observation vector is really multivariate, the monotonicity would have to be either multivariate or related to a real-valued statistic. Neither appears promising at first look.)

Sequential procedures. Sequential sampling and even sequential choice among a finite number of experiments are easily accommodated by including in X all observations that could ever be made. Then restrictions must be imposed to express the limitation of choices by what is known (e.g., by requiring measurability with respect to suitable σ-fields). One may also want the number of observations to be finite with probability 1. But such requirements complicate only the definition of the acceptance region, not checking whether it is closed. And it still suffices to consider convergence in distribution of the observations individually if they are independent, even if their possible number is unbounded. For the purpose at hand it appears inconvenient to describe sequential situations in other ways, for instance, by the number of observations actually made, the value of each, and index numbers identifying the experiments chosen.

Nonparametric procedures. We have already pointed out that it is easy to show by the approach of that paper rank and other randomization procedures based on closed acceptance regions and confidence intervals are conservative for discrete models. Test statistics T of the Kolomogorov-Smirnov type provide another example. Like rank-test statistics (which they are in the two-sample case), they are discontinuous but lower semicontinuous. Hence rejecting when $T > k$ is conservative in the discrete case, by Corollary 1, Corollary 1', or direct verification that the acceptance region is closed in the x-space. It is perhaps easiest to show that the corresponding confidence procedure is conservative by reference to the test, but it also follows from Theorem 2 and equation (4) for $\overline{C}(x)$. Here it is a blessing that no topology on the parameter space is needed.

All these results can be shown instead by either tie breaking or transformation, but both unfortunately seem to require checking a lot of special cases separately.

Tie breaking. In nonparametric tests, one theoretical approach to ties (and signed ranks of 0) is to break them at random. Though hardly recommendable for practice, this does preserve the significance level for the relevant continuous hypothesis under the corresponding discrete hypothesis. It follows, of course, that it is conservative to accept whenever a tie-breaking assignment exists for which acceptance would occur. In typical situations, it is easy to see that closing the acceptance region is equivalent. For a rank test, the values of the test statistic occurring without ties in every neighborhood of the observed point are just those obtainable by tie-breaking, and we are in the situation of Corollary 1 or Corollary 1' with a lower semicontinuous test statistic. For an observation-randomization test, such as one based on sample means, the test statistic one naturally thinks of is continuous, but it is not a test statistic in the strict (unconditional,

Corollary 1) sense because the randomization distribution to which it is referred depends on the observations. Nevertheless acceptance occurs in every neighborhood of the observed point if and only if it occurs for some tie-breaking assignment.

The effect of conservative tie-breaking on nonparametric confidence intervals for shift is to close them in all cases I am aware of. This suggests that under shift models the whole issue of ties is less important than it might otherwise appear. Even though conservative tie-breaking may greatly reduce power, it is also reducing Type I error. These two reductions should be nearly compensating when the significance level is chosen to balance the errors nearly optimally. Otherwise they should in principle be considered on their merits, not merely in connection with ties.

For the purpose at hand, compared to the approach of this paper, tie-breaking is piecemeal and less general, though more "elementary".

Transforming the observations. Noether (1967) treats certain nonparametric confidence intervals by "projection" of continuous distributions into discrete ones. Stripped of unnecessary restrictions, his method is to obtain a given discrete model by transforming continuous observations in such a way that if the parameter is strictly inside the confidence limits in the continuous case, then the transformed parameter is not strictly outside the transformed confidence limits. It follows that if the transformed confidence interval is closed, then its confidence level is at least that of the continuous case. This form of Noether's argument applies to his mixed model, too, and generalizes easily, but there remains the task of defining the transformations.

Open regions. The results of Sections II and III have counterparts for open acceptance and confidence regions, since taking complements in (1) gives

$$P_n \to P \text{ in distribution iff}$$
$$\text{for all open } A,\ P(A) \leq \liminf P_n(A). \qquad (6)$$

The situation is not symmetric, however, because the least upper bound of the Type I error probability is replaced by the greatest lower bound, which does not have the same significance. For confidence regions, a similar statement applies (with no pun). To state results in conventional terms for open sets seems to require assuming a constant error probability on $\mathcal{F}(\theta)$. Perhaps the most useful such result is

Theorem 4. If the lower and upper end points of an open confidence interval are continuous functions of x, or upper and lower semicontinuous respectively, and if the probability of coverage is constant on $\mathcal{F}(\theta)$, then it is no greater on $\overline{\mathcal{F}}(\theta)$.

REFERENCES

Chernoff, H. (1956). "Large-Sample Theory: Parametric Case." *Ann. Math. Statist. 27*, 1-22.

Chernoff, H. and Savage, I.R. (1958). "Asymptotic Normality and Efficiency of Certain Nonparametric Test Statistics." *Ann. Math. Statist. 29*, 972-994.

Kolmogorov, A.N. (1941). "Confidence Limits for an Unknown Distribution Function." *Ann. Math. Statist. 12*, 461-463.

Noether, G.E. (1967). "Wilcoxon Confidence Intervals for Location Parameters in the Discrete Case." *J. Amer. Statist. Ass. 62*, 184-188.

Pratt, J.W. (1959). "Remarks on Zeros and Ties in the Wilcoxon Signed Rank Procedures." *J. Amer. Statist. Ass. 54*, 655-667.

Pratt, J.W. (1961). "Length of Confidence Intervals." *J. Amer. Statist. Ass. 56*, 549-567.

Pratt, J.W. and Gibbons, J.D. (1981). *Concepts of Nonparametric Theory*. Springer-Verlag, New York.

Tukey, J.W. (1948). "Nonparametric Estimation, III. Statistically Equivalent Blocks and Multivariate Tolerance Regions--the Discontinuous Case." *Ann. Math. Statist. 19*, 30-39.

Tukey, J.W. (1949). The Simplest Signed-Rank Tests. Mimeographed Report No. 17, Statistical Research Group, Princeton University.